XXI International Conference on Spectral Line Shapes

(ICSLS 2012)

AA008876

Journal of Physics: Conference Series Volume 397

St. Petersburg, Russian Federation
3-9 June 2012

ISBN: 978-1-62276-863-9
ISSN: 1742-6588

Printed from e-media with permission by:

Curran Associates, Inc.
57 Morehouse Lane
Red Hook, NY 12571

Some format issues inherent in the e-media version may also appear in this print version.

Printed by Curran Associates, Inc. (2013)

For permission requests, please contact the Institute of Physics at the address below.

Institute of Physics
Dirac House, Temple Back
Bristol BS1 6BE UK

Phone: 44 1 17 929 7481
Fax: 44 1 17 920 0979

techtracking@iop.org

Additional copies of this publication are available from:

Curran Associates, Inc.
57 Morehouse Lane
Red Hook, NY 12571 USA
Phone: 845-758-0400
Fax: 845-758-2634
Email: curran@proceedings.com
Web: www.proceedings.com

General Aspects of Spectroscopy

Invited talks

Contributed talks

Poster presentations

High and Low Temperature Plasmas

Invited talks

Contributed talks

Poster presentations

Neutral Atoms and Molecules

Invited talks

Contributed talks

Poster presentations

Astrophysics and Atmosperes

Invited talks

Contributed talks

Poster presentations

Raman Spectroscopy

Poster presentations

Spectroscopy of Atoms/Ions at the Boundery of Two Phases

Invited talk

Contributed talks

Poster presentations

Technical Physics

Contributed talk

Poster presentation

XXI International Conference on Spectral Line Shapes (ICSLS 2012)

Alexander Z Devdariani

Abstract Metrics

PREFACE

The 21st International Conference on Spectral Line Shapes, ICSLS, was held in the historic main building of St Petersburg State University (St. Petersburg, Russia) on 3–9 June 2012. The event continued the tradition started in 1978 in Meudon Observatory in Paris. Representatives of line shape physics have since met every two years in different locations in Europe and North America. The most recent events were held in St John's, Newfoundland, Canada (2010), Valladolid, Spain (2008), and Auburn, AL (USA). Traditionally, the conferences consider experimental and theoretical issues of studying spectral line shapes, diagnostic utilization of spectral line profiles observed in absorption, emission or scattering of electromagnetic radiation by atoms, molecules, and clusters in different environments, including neutral environments, laboratory low and fusion plasmas, astrophysical conditions, and planetary atmospheres.

The Conference was attended by over 100 professionals from Europe, Asia, America, Africa and New Zealand. The conference program was put together in such a way so as to exclude any parallel sessions. Five afternoon sessions featured 19 invited talks and 20 oral contributions, and two evening sessions offered 61 poster presentations, including post-deadline posters. This setup allowed for a relaxed and unhurried discussion of results and facilitated productive networking.

The invited talks were selected by recommendation of members of the International Scientific Committee. The Organizers would like to thank all the members of the International Scientific Committee for their proposals on the agenda and their valuable advice. When considering candidates for oral contributions, the organizers took into account the suggestions and preferences of potential conference participants. When selecting the theses of poster presentations, the organizers focused on the topics in line with the theme of the conference and studies with well-formulated results. It must be noted that this year's conference included a noticeably larger number of reports on astrophysical applications and physics of planetary atmospheres. Another significant trend was related to a new area of inquiry: low temperature spectra, near-surface processes, and control and management of new technological processes. There were more presentations on elementary processes during the formation of spectral lines.

Most of the presented studies were included in the Conference Proceedings publication. For the first time ever the proceedings will be published as a volume of *Journal of Physics: Conference Series*, published by the Institute of Physics (UK), making it possible to access the materials of the conference online. All published studies underwent peer review. The organizers would like to thank all the reviewers who found time to review the submissions against a very tight deadline during the summer.

Professor N G Skvortsov, Vice-Rector for Research of St Petersburg State University, opened the conference. His speech was followed by a brief welcome message from Professor A Devdariani of St Petersburg State University, Deputy Chair of the Organizing Committee. An informal welcome party was held before the opening session. The cultural program of the event included various sightseeing tours, and a hydrofoil boat ride to the fountain city of Peterhof, where the delegates toured the park and attended a conference dinner. Organizers would like to express their appreciation to St Petersburg musicians, P Laul and D Kouzov for an excellent classical music concert.

The conference included a meeting of the International Scientific Committee the minutes of the event are included in this volume (article number 011004). Most importantly, the next meeting, 22nd ICSLS, has been scheduled for 2014 in The Center for Laser Applications, The University of Tennessee Space Institute, US. Professor Christian Parriger will coordinate the organization. Organizers of the St Petersburg Conference would like to wish him and his colleagues every success for the next conference in Tennessee.

The St Petersburg Conference was supported by the grants from St Petersburg State University, the Russian Foundation for Basic Research and Dynasty, Dmitry Zimin's Foundation for Non-Profit Programs. The Organizing Committee would like to thank them for their support. A word of thanks also goes to staff of the Research Department of St Petersburg State University for assistance with administrative issues, and to Mr S Slyusarev for creating the Conference logo and website. I would like to thank personally Professor A Kouzov, and Dr V Alexeev for providing much support in the process of preparing and holding the event in 2012.

Alexander Z Devdariani
St Petersburg University

Sponsors

St Petersburg University

Dynasty Foundation

Russian Foundation for Basic Research

Welcome to the 21st International Conference on Spectral Line Shapes

Abstract Metrics

Dear Participants and Guests
On behalf of the Rector's Office, welcome to St. Petersburg University, one of the oldest, largest and most prestigious universities in Russia. The 12 Colleges Building (named after the Collegia of Peter the Great's time, Ministries in the modern sense of the word) has been home to the University since 1804. St. Petersburg State University is the successor of the first Russian University founded by Peter the Great on 24 January 1724.

It's distinguished faculty members include mathematicians Chebyshev and Smirnov, physicists Lenz and Fock, chemists Mendeleev and Butlerov, embryologist Kovalevsky, and physiologist Sechenov. Several of our alumni won Nobel Prizes – Pavlov, Semenov, Kantorovich, Landau and Prokhorov. The University is the *alma mater* of two presidents of Russia, Vladimir Putin and Dmitry Medvedev. St. Petersburg State University today consists of 19 faculties on two campuses: in the historic downtown, on Vasilyevsky Island, and in Peterhof.

The Faculty of Physics carries out research projects through the efforts of 260 employees, of which 150 hold doctoral degrees, and 50 have been awarded Grand Doctorates. The work is done in a vast variety of areas, from the physics of elementary particles to the physics of the atmosphere.

Optics and spectroscopy has been a particularly significant area of research. This field of inquiry goes back to the work of Academician D S Rozhdestvensky more than 70 years ago, when he first suggested his famous hook method. After him the work was headed by S Frisch, corresponding member of the Academy of Sciences and editor-in-chief of Optics and Spectroscopy Journal. Today research projects in optics involve over 50 staff members, graduate and post-graduate students. Work involves projects on spectroscopy of cold and hot plasma, atomic and molecular spectroscopy, non-linear spectroscopy and spectroscopic analysis. Our scholars support active international ties with their colleagues worldwide.

The organizing committee of the conference has not forgotten about the cultural and tourism significance of the host city, with Hermitage and the Russian Museum, memorial museums of Pushkin and Dostoevsky, Mariinsky and Mikhailovsky Theaters being only a few of the many places to visit. Early June is the time of white nights, the best time to visit the environs of St. Petersburg with its many imperial palaces and parks, and attend multiple music and theater festivals. This is just the right time to take a break from physics overall and spectral line shapes

Minutes of the meeting of the international program committee

Abstract Metrics

The meeting of the International Program Committee occurred on 5 June 2012. The agenda consisted of the following items:
- Information on conference participants, contributions and grants
- Information on the financial support received by the conference
- Committee membership
- Organizers of the next two meeting of the ICSLS

Conference participants and contributions

There were about 100 registered participants for the ICSLS. They presented more than 100 contributions, namely, 19 invited talks, 20 oral contributions and more than 61 contributed papers. It was noted that only very few participants came from North America. Reasons quoted were finacial problems of laboratories and overlapping of several conferences.

Finacial support received
The conference received grants from St. Petersburg University, the Russian Foundation for Basic Research, and the non-profit Dynasty Foundation. About 40% of the budget was collected in the form of registration fees. Discounted fees and fee waives were provided for 40 participants.

Committee membership
The Committee instructed Professor A Devdariani to contact the absent members who had missed two successive conferences and ask them whether they intended to prolong their membership on the Committee, and inform other Committee members accordingly.

Organizers of the next meetings of the ICSLS
The 20th ICSLS held in St. John's, Newfoundland, Canada decided to hold the 22nd ICSLS at the University of Tennessee. The event will be organized by Christian Parriger. All issues regarding the next conference were discussed including budget, travel, conference site, accommodation, and proceedings. Torun, Poland was proposed for the 23d ICSLS by Roman Ciurylo. Roland Stamm proposed the Aix-Marseille University as a backup to the first proposal.

International Program and Local Organizing Committees

Abstract Metrics

International Program Committee

Dionisio Bermejo (Spain)
Roman Ciurylo (Poland)
Elisabeth Dalimier (France)
Alexander Devdariani (Russia)
Milan S Dimitrijevic (Serbia)
Robert Gamache (USA)
Marco A Gigosos (Spain)
Motoshi Goto (Japan)
Magnus Gustafsson (Sweden)
Jean-Michel Hartmann (France)
Carlos Iglesias (USA)
John Kielkopf (USA)
John C Lewis (Canada)
Valery Lisitsa (Russia)
Eugene Oks (USA)
Christian G Parigger (USA)
Gillian Peach (UK)
Adriana Predoi-Cross (Canada)
Roland Stamm (Germany)

Local Organizing Committee

Nikolay G Skvortsov (Chair, St Petersburg State University)
Evgenii B Aleksandrov (Ioffe Physico-Technical Institute, St Petersburg)
Vadim A Alekseev (Scientific Secretary, St Petersburg State University)
Sergey F Boureiko (St.Petersburg State University)
Yury N Gnedin (Pulkovo Observatory, St Petersburg)

Peer review statement

Abstract Metrics

All papers published in this volume of *Journal of Physics: Conference Series* have been peer reviewed through processes administered by the proceedings Editors. Reviews were conducted by expert referees to the professional and scientific standards expected of a proceedings journal published by IOP Publishing.

QED theory of the spectral line profile for few-electron atoms and ions

Oleg Yu. Andreev

Faculty of Physics, St. Petersburg State University, ul. Ulyanovskaya 1, RU-198504, St. Petersburg, Russia

E-mail: olyuan@gmail.com

Abstract. A review of the current status of the theoretical study of spectral line profile for systems with few electrons is presented. The line-profile approach (LPA) is introduced for the investigation of the energy levels within the framework of QED. In particular, the energy of the levels, transition probabilities and electron recombination processes are considered.

1. Introduction

The theory of the spectral line profile for the hydrogen atom has attracted renewed and broad interest due to the recent success in super-accurate resonance frequency measurements and the possibility for experimental tests of fundamental physics. Consequently, the study of the spectra of highly charged ions (HCI) has moved into the focus of experimentalists and theoreticians during the last few decades. Due to the small number of bound electrons involved, HCI appear as relatively simple systems allowing for rigorous theoretical investigations. The bound electrons move in the external electromagnetic field of the nucleus, which can become very strong for large nuclear charge number Z. If the total charge of electrons is small in comparison to the nuclear charge, the effect of screening is also small and binding effects dominate. Particularly, in heavy ions such as uranium the nuclear electric field exceeds in strength all other artificial electric fields accessible in laboratories. Accordingly, HCI provide a natural scenario for the study of Quantum Electrodynamics (QED) in very strong external fields.

2. Line-profile approach

The line-profile approach (LPA) is a method for description of the atomic systems (in particular, highly charged ions) within the framework of quantum electrodynamics (QED). It starts from the description of the atomic electrons as a set of noninteracting particles moving in the field of the nucleus (Furry picture) and described by the solutions of the Dirac equation. Within the framework of QED atomic electrons are interacting with each other through interaction with the quantum vacuum (the quantized electromagnetic, electron-positron fields) [1]. Accordingly, the set of electrons (together with the atomic nucleus) is not a conservative system, and the concept of the energy for this system needs to be carefully treated. Within the LPA [2] the energy levels are associated with resonances in the natural line profile for the process of resonant photon scattering. In order to keep the characteristics of the energy levels independent of the particular features of the process of scattering, the resonance approximation is employed. The resonance approximation consists of the description of the resonance area of the natural line profile by the Lorentz contour which is characterized by two parameters: position of the resonance and its width. The energy levels are connected with the corresponding resonances. The energy and width of an energy level are determined by position of the resonance and its width within the resonance approximation.

2.1. Energy levels

For the practical implementation of the LPA the process of the elastic photon scattering on atomic electron was employed. This procedure in the lowest QED perturbation theory order is depicted in Fig. 1.

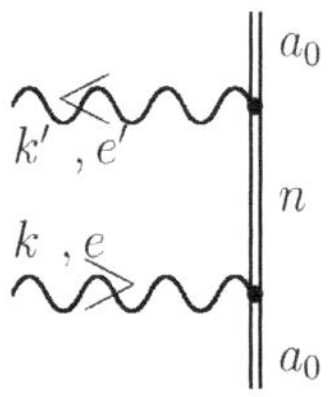

Figure 1. Feynman graph, describing the photon scattering on an atomic electron. The wavy lines with the arrows describe the absorption and emission of photons with momenta k, k' and polarizations e, e', respectively. The double solid line denotes the electron in the field of the nucleus, a corresponds to the ground electron state.

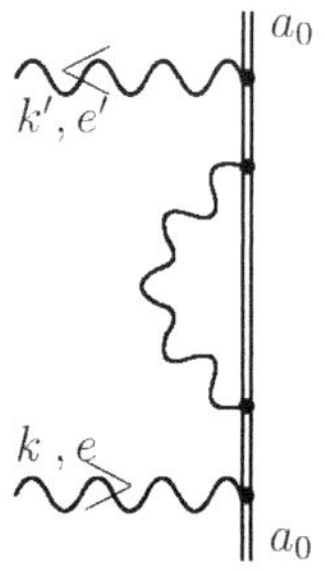

Figure 2. Feynman graph corresponding to the electron self-energy insertion into the electron propagator in Fig. 1. The wavy line denotes the virtual photon. The other notations are the same as in Fig. 1.

According to the standard Feynman rules (see, e.g., [3]), the S-matrix element for the graph depicted in Fig. 1 reads

$$S^{(2)} = (-ie)^2 \int d^4x_u d^4x_d \, \bar{\psi}_a(x_u)\gamma^{\mu_u} A^{(k',\lambda')*}_{\mu_u}(x_u) S(x_u, x_d)\gamma^{\mu_d} A^{(k,\lambda)}_{\mu_d}(x_d)\psi_a(x_d) \,, \qquad (1)$$

where $x^\mu = (t, \mathbf{r})$ denotes a space-time point, $\psi_a(x) = \psi_a(\mathbf{r})e^{-i\varepsilon t}$ is the one-electron wave function, $\bar{\psi} = \psi^+\gamma^0$ is the Dirac conjugated wave function and $A^{(k,\lambda)}_\mu(x) = A^{(k,\lambda)}_\mu(\mathbf{r})e^{-i\omega t}$ is the 4-vector of the electromagnetic field potential (photon wave function), $k^\mu = (\omega, \mathbf{k})$, λ are the wave vector and polarization. The frequency of the absorbed and emitted photons are $\omega = |\mathbf{k}|$ and $\omega' = |\mathbf{k}'|$, respectively.

The bound-electron propagator is represented in terms of an eigenmode decomposition with respect to one-electron eigenstates of the Dirac equation:

$$S(x_u, x_d) = \frac{i}{2\pi} \int d\omega_n e^{-i\omega(t_u - t_d)} \sum_n \frac{\psi_n(\mathbf{r}_u)\bar{\psi}_n(\mathbf{r}_d)}{\omega_n - \varepsilon_n(1 - i0)} \,. \qquad (2)$$

The sum over n runs over the entire Dirac spectrum, ε denotes the corresponding Dirac energy.

Inserting the expressions for the propagator and wave functions in Eq. (1) and integrating over time and frequency variables we arrive at

$$S^{(2)} = (-2\pi i)\delta(\omega' - \omega)e^2 \sum_n \frac{A^{(k',\lambda')*}_{an} A^{(k,\lambda)}_{na}}{\omega + \varepsilon_a - \varepsilon_n} \,. \qquad (3)$$

Here we employed the shorthand notation

$$A_{ab}^{(k,\lambda)} = \int d^3\mathbf{r}\,\bar{\psi}_a(\mathbf{r})\gamma^\mu A_\mu^{(k,\lambda)}(\mathbf{r})\psi_b(\mathbf{r})\,. \tag{4}$$

The amplitude (U) of the process of elastic photon scattering is related to the S-matrix element via

$$S = (-2\pi i)\delta(E_F - E_I)U\,, \tag{5}$$

where E_I and E_F represent the energies of the initial and final states, respectively.

The resonance scattering means that the photon frequency ω is close to the energy difference between two atomic levels $\omega = \varepsilon_n - \varepsilon_a + O(\alpha)$, where α is the fine-structure constant. Accordingly, we have to retain only one term in the sum over n in Eq. (3). Then the resonance amplitude looks like

$$U^{(0)} = e^2 A_{an}^{(k',\lambda')*}\frac{1}{\omega + \varepsilon_a - \varepsilon_n}A_{na}^{(k,\lambda)}\,. \tag{6}$$

This amplitude has a singularity at $\omega = -\varepsilon_a + \varepsilon_n$. To avoid this singularity and to obtain the Lorentz profile for the photon absorption and photon emission processes we have to consider the radiative insertions into the internal electron line in Fig. 1. The insertion of the electron self-energy is depicted in Fig. 2. In the resonance approximation the corresponding amplitude looks like

$$U^{(1)} = e^2 A_{an}^{(k',\lambda')*}\frac{1}{\omega + \varepsilon_a - \varepsilon_n}\hat{\Sigma}_{nn}(\omega + \varepsilon_a)\frac{1}{\omega + \varepsilon_a - \varepsilon_n}A_{na}^{(k,\lambda)}\,. \tag{7}$$

Here we introduced the energy-dependent matrix element of the electron self-energy

$$\Sigma_{ud}(\xi) = e^2\sum_n \frac{i}{2\pi}\int d\Omega\,\frac{I_{unnd}(|\Omega|)}{\xi - \Omega - \varepsilon_n(1 - i0)} \tag{8}$$

together with the shorthand notation

$$I_{u_1u_2d_1d_2}(\Omega) = \sum_{\mu_1\mu_2}\int d\mathbf{r}_1 d\mathbf{r}_2\,\bar{\psi}_{u_1}(\mathbf{r}_1)\bar{\psi}_{u_2}(\mathbf{r}_2)\gamma^{\mu_1}\gamma^{\mu_2}I_{\mu_1\mu_2}(\Omega, r_{12})\psi_{d_1}(\mathbf{r}_1)\psi_{d_2}(\mathbf{r}_2)\,. \tag{9}$$

It is assumed that the ultraviolet divergent matrix element (8) are renormalized in a standard way for the tightly bound electrons in atoms (see, *e.g.*, [3]).

We can consider insertion of two self-energy matrix elements into the internal electron line, see Fig. 3. Retaining in the electron propagators Eq. (2) only one term we get the following expression for the amplitude

$$U^{(2)} = e^2 A_{an}^{(k',\lambda')*}\frac{1}{\omega + \varepsilon_a - \varepsilon_n}\hat{\Sigma}_{nn}(\omega + \varepsilon_a)\frac{1}{\omega + \varepsilon_a - \varepsilon_n}\hat{\Sigma}_{nn}(\omega + \varepsilon_a)\frac{1}{\omega + \varepsilon_a - \varepsilon_n}A_{na}^{(k,\lambda)}. \tag{10}$$

Comparing Eqs. (6), (7), (10) we can see that the corrections to the amplitude corresponding to the insertion of the self-energy matrix elements compose a geometric progression. Applying the formula for convergent geometric progression we get

$$U = \sum_{l=0}^{\infty} U^{(l)} = e^2 A_{an}^{(k',\lambda')*}\frac{1}{\omega + \varepsilon_a - \varepsilon_n - \hat{\Sigma}_{nn}(\omega + \varepsilon_a)}A_{na}^{(k,\lambda)}. \tag{11}$$

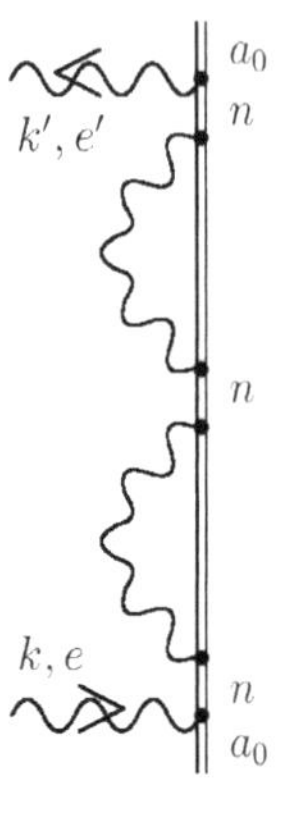

Figure 3. Feynman graph representing the higher-order electron self-energy correction within the line profile approach (SESE, loop-after-loop, irreducible).

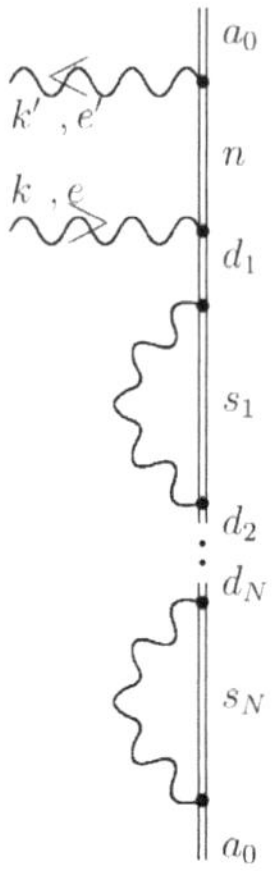

Figure 4. The Feynman graphs representing multiple insertions of the self-energy operator into the lower outer electron line.

We note that the insertion of the self-energy graphs into the outer lines in Fig. 2 leads to singularities. In the paper [4] it was shown that within the adiabatic formalism of Gell-Mann and Low [5] the contributions of the insertion of the self-energy graphs into the outer lines (see Fig. 4) compose a geometric progression. Accordingly, the expression for the adiabatic S-matrix corresponding to graphs in Fig. 4 yields

$$\sum_{l=0}^{\infty} S_{\lambda}^{(l)} = (-2\pi i)\delta(\omega-\omega')e^2\left[\sum_n \frac{A^*_{an}A_{na}}{\omega+\varepsilon_a+\hat{\Sigma}_{aa}(\varepsilon_a)-\varepsilon_n}+R_{\lambda}\right]\exp\left(\frac{\hat{\Sigma}_{aa}(\varepsilon_a)}{2i\lambda}\right). \quad (12)$$

Here, ε are the Dirac energies, λ is the adiabatic parameter ($\lambda\to 0$), R_λ represents the terms for which $\lim_{\lambda\to 0} R_\lambda = 0$. As the self-energy matrix element for the ground state has no imaginary part, then all the singularities compose an imaginary exponent, and the absolute value of the amplitude reads

$$|U| = e^2\left|\sum_n \frac{A^*_{an}A_{na}}{\omega+\varepsilon_a+\hat{\Sigma}_{aa}(\varepsilon_a)-\varepsilon_n}\right|. \quad (13)$$

Accordingly, the insertions to the outer lines lead to corrections to the energy of the ground state.

In the general case for one- or many-electron system the amplitude of the process

$$A \overset{\omega}{\to} N \overset{\omega'}{\to} A, \quad (14)$$

where A is the ground state of the system, N is the excited state, $\omega=\omega'$ are the photons, in the resonance approximation can be written as

$$U = T^{+}\frac{1}{\omega+E_a-V(\omega)}T. \quad (15)$$

Here, the matrices T describe complicated vertices, the matrix V is defined by the operator $\hat{V}$ which is build within the framework of the standard QED perturbation theory [2]. These operators are defined by their action on many-electron wave functions of noninteracting electrons.

$$V(\omega) \;=\; V^{(0)} + \Delta V(\omega)\,. \tag{16}$$

The matrix $V^{(0)}$ is a diagonal matrix, its eigenvalues are the sums of Dirac energies. The matrixΔ V describes various corrections: electron self-energy, vacuum polarization, one-photon exchange *etc.*

Taking squared absolute value of the amplitude Eq. (15) and integrating over the momentum and polarization of the absorbed photon we can get the absorption probability

$$dW_{a\to n}(\omega) \;=\; \frac{1}{2\pi}\frac{\Gamma_{A\to N}(\omega)\quad d\omega}{[\omega + E_A - V_N^{(0)} - \Re\Delta V_N(\omega)]^2 + [\Im\Delta V_N(\omega)]^2}\,. \tag{17}$$

Condition of the resonance reads

$$\omega^{\rm res} + E_A - V_N^{(0)} - \Re\Delta V_N(\omega^{\rm res}) \;=\; 0\,. \tag{18}$$

Interpolation of the line profile Eq. (17) by the Lorentz contour yields

$$dW_{A\to N}(\omega) \;=\; \frac{1}{2\pi}\frac{\Gamma_{A\to N}(\omega^{\rm res})\quad d\omega}{[\omega + E_A - E_N]^2 + \frac{1}{4}\Gamma_N^2}\,, \tag{19}$$

where we introduced the energy of the level

$$E_N \;=\; \omega^{\rm res} + E_A - V^{(0)} \;=\; V_N^{(0)} + \Re\Delta V_N(\omega^{\rm res}) \tag{20}$$

and the width of the level

$$\Gamma_N \;=\; -2\Im\Delta V_N(\omega^{\rm res})\,. \tag{21}$$

We note that the resonance approximation is plausible only either for the cases of one isolated energy level or a small group of energy levels. If we go beyond the resonance approximation, the energy level can not be describe by two parameters (the energy and the width). The corresponding corrections are called nonresonant corrections [6, 2]. Calculations of the energy levels (including quasidegenerate levels) for low-lying two-electron configurations within the framework of QED are presented in [7, 8, 9]. In these three works the covariant-evolution-operator method [10], the LPA [2] and the two-time Green's function method [11] were employed, respectively.

2.2. Transition probabilities

For consideration of transition from the initial state (I) to the final state (F) with emission of a photon (ω) within the framework of the LPA, we have to consider process of scattering

$$A \overset{\omega}{\to} I \overset{\omega_0}{\to} F \overset{\omega'}{\to} A\,. \tag{22}$$

In the resonance approximation the amplitude of this process can be written as

$$U \;=\; T^{+}\frac{1}{\omega + E_A - V(\omega')}\,\Xi(\omega_0)\,\frac{1}{\omega + E_A - V(\omega)}T\,. \tag{23}$$

The operator $\Xi(\omega_0)$ is derived within the framework of standard QED perturbation theory [4]. The resonance determined by the left denominator defines the initial state, the resonance caused by the right denominator defines the final state. The transition amplitude from the initial state (I) to the final state (F) is defined as

$$U_{I\to F} = \Xi(\omega_0)_{FI} . \tag{24}$$

The calculations of the transition probabilities for the two-electron states were performed by Drake [12] with the unified method (UM) and by Johnson [13] within the relativistic many-body perturbation theory (RMBPT). The QED calculation of the transition probabilities for the nondegenerate two-electron states is presented in [14], the calculations are performed within the two-time Green's function method [11]. The QED calculation of the transition probabilities for low-lying two-electron states (including quasidegenerate levels) within the framework of the QED is presented in [4].

2.3. One-electron capture

We consider process of electron recombination with a one-electron heavy ion. The process of electron recombination can be presented as

$$e(E_e) + U^{91+}(1s) \to \cdots \to U^{90+}(1s,1s) + n\gamma(\omega) . \tag{25}$$

If the process of electron recombination is registered by detection of a photon with frequency $\omega \approx E_e + E_{1s} - E(r)$, where $E(r)$ is the energy of a two-electron single-excited configuration $(r = (1s,2s), (1s,2p)\ )$, then the electron recombination is determined by two particular processes,

$$e(E_e) + U^{91+}(1s) \to U^{90+}(r) + \gamma(\omega) \to \cdots \tag{26}$$

and the dielectronic recombination (DR)

$$e(E_e) + U^{91+}(1s) \to U^{90+}(d) \to U^{90+}(r)\gamma(\omega) \to \cdots , \tag{27}$$

where d denotes the double-excited configurations $(2s,2s)$, $(2s,2p)$, $(2p2p)$. The cross section of electron recombination considered as a function of the incident electron energy (E_e) reveals resonances in the area where sum of the incident electron and 1s-electron energies $(E_I = E_e + E_{1s})$ is close to the energies of the double excited two-electron configurations. The DR process is the dominant process contributing to the cross-section in the resonance areas. The RR process is nonresonant. The frequency (ω) of emitted photon is defined by the energy conservation law. The RR and DR processes compose a single process of electron recombination, the RR and DR contributions to the cross-section can be separated only in the lowest order of the perturbation theory. Accordingly, in the general case the contributions of the RR and DR processes to the total cross-section become inseparable.

For calculation of the cross-section for the process of electron recombination within the LPA we introduce an auxiliary bound electron system which properties are explicitly related to the original system. We introduce a function

$$\psi^{\text{aux}}_{e_R}(\mathbf{r}) = \psi_e(\mathbf{r})\theta(R - |\mathbf{r}|) , \tag{28}$$

where $\psi_e(\mathbf{r})$ is the wave function of the incident electron, $\theta(R-|\mathbf{r}|)$ is the Heaviside step function, R is a radius of the large sphere. The function $\psi^{\text{aux}}_{e_R}(\mathbf{r})$ can be normalized to unity with the normalization factor N_R

$$\psi_{e_R}(\mathbf{r}) = N_R^{-1}\psi^{\text{aux}}_{e_R}(\mathbf{r}) , \qquad \|\psi_{e_R}\| = 1 . \tag{29}$$

We can introduce an artificial bound electron state e_R described by the wave function ψ_{e_R}. The energies and angular quantum numbers of the continuum electron state ψ_e and the bound electron state ψ_{e_R} are equal. Consider now two processes of transition from the initial state (I) to the final state (F) which differ by the initial states. The first initial state is $1s$ electron and the continuum electron e. The second initial state is $1s$ electron and the artificial bound electron e_R. The amplitudes of thirst process (U) and the second one (U_R) are related like

$$U = \lim_{R\to\infty} N_R U_R . \quad (30)$$

The LPA can be applied for the bound electron system $(1s, e_R)$ in the way it was performed for calculation of the transition probabilities [4]. The limit $(R \to \infty)$ can be evaluated numerically.

Calculations of the electron recombination with one-electron uranium are presented in [15, 16]. These calculation are performed within various approximations. The first *ab initio* QED calculation of the electron recombination with one-electron uranium and gadolinium is presented in [17, 18]. The measurement of the cross-section is presented in [19]. In the work [19] there is also presented QED calculation of the cross-section.

3. Conclusion

Summarizing, we have presented the complete and consequent formulation of QED theory of the line profile and applied it for describing dynamic processes in atomic systems. One of the important features of this approach is that it provides a most general way for the evaluation of all observable quantities of physical interest: energy levels, transition frequencies, all kinds of cross-sections, *etc.* Moreover, it allows for the exact description of the interelectronic interaction for both in- and out-states with the accuracy accepted in the Relativistic Many-Body Theory (RMBT); hence, QED and RMBT can be partly merged.

Our studies have been continued by going beyond the Lorentz (resonance) approximation. We have shown [2] that the nonresonant (NR) corrections to the Lorentz profile will set an absolute theoretical limit to the accuracy of the possible frequency measurements. This is closely related to the fact, that the concept of atomic levels, characterized by a level energy and width, becomes inadequate as soon as the distortions due to nonresonant corrections can no longer be neglected. Under such circumstances only the line profile remains as the quantum-mechanical observable that can be measured and compared with experimental data.

References

[1] Akhiezer A I and Berestetskii V B 1965 *Quantum Electrodynamics* (New York: Wiley Interscience)
[2] Yu Andreev O, Labzowsky L N, Plunien G and Solovyev D A 2008 *Phys. Rep.* **455** 135
[3] Labzowsky L, Klimchitskaya G and Yu Dmitriev 1993 *Relativistic Effects in the Spectra of Atomic Systems* (Bristol and Philadelphia: Institute of Physics Publishing)
[4] Yu Andreev O, Labzowsky L N and Plunien G 2009 *Phys. Rev. A* **79** 032515
[5] Gell-Mann M and Low F 1951 *Phys. Rev.* **84** 350
[6] Labzowsky L N, Solovyev D A, Plunien G and Soff G 2001 *Phys. Rev. Lett.* **87**(14) 143003
[7] Lindgren I, Åsén B, Salomonson S and A-M Mårtensson-Pendrill 2001 *Phys. Rev. A* **64** 062505
[8] Andreev O Y, Labzowsky L N, Plunien G and Soff G 2004 *Phys. Rev. A* **69** 062505
[9] Artemyev A N, Shabaev V M, Yerokhin V A, Plunien G and Soff G 2005 *Phys. Rev. A* **71** 062104
[10] Lindgren I, Salomonson S and Åsén B 2004 *Phys. Rep.* **389** 161
[11] Shabaev V M 2002 *Phys. Rep.* **356** 119
[12] G W F Drake 1979 *Phys. Rev. A* **19** 1387
[13] Johnson W R, Plante D R and Sapirstein J 1995 *Adv. At. Mol. Opt. Phys.* **35** 255
[14] Indelicato P, Shabaev V M and Volotka A V 2004 *Phys. Rev. A* **69** 062506
[15] Karasiov V V, Labzowsky L N, Nefiodov A V and Shabaev V M 1992 *Phys. Lett. A* **161** 453
[16] Zakowicz S, Scheid W and Grün N 2004 *J. Phys. B* **37** 131
[17] Yu Andreev O, Labzowsky L N and Prigorovsky A V 2009 *Phys. Rev. A* **80** 042514
[18] Andreev O Y, Labzowsky L N and Prigorovsky A V 2011 *Phys. Rev. A* **83**(6) 064501
[19] Bernhardt D, Brandau C, Harman Z, Kozhuharov C, Müller A, Scheid W, *et al.* 2011 *Phys. Rev. A* **83** 020701

Journal of Physics: Conference Series **397** (2012) 012002 doi:10.1088/1742-6596/397/1/012002

Peculiarities of matter excitation by ultra-short electromagnetic pulses

V Astapenko
Moscow Institute of Physics and Technology, Moscow, Russia
E-mail: astval@mail.ru

Abstract. The paper is devoted to theoretical analysis of matter excitation by ultra-short electromagnetic pulses (USP) with controllable shape in the frame of perturbation approach. We calculate the dependence of the total probability of photo-absorption of USP by an atom upon the pulse duration. It is shown that investigated probability has characteristic features which depend upon phase parameters of USP.

1. Introduction

Ultra-short electromagnetic pulses generation with controlled parameters is one of the fastest growing trends in modern experimental laser physics. By now it's possible to obtain pulses of duration 80–100 as in XUV in a line with pulses of a few fs with controlled parameters such as phase and chirp [1 – 7]. These pulses can be generated in teraherz, infrared, visible and far ultraviolet spectra and have duration from sub-cycle up to few cycles at carrier frequency.

This paper is devoted to theoretical study of atomic photo-absorption under the action of USPs with different durations and shapes in the frame of perturbation theory [8].

2. Ultra-short pulses with controllable shape

Consider the impact on the atom pulse of Gaussian shape with electric field varies in time according to the law:

$$E(t) = E_0 \exp\left(-t^2/\Delta t^2\right)\cos\left(\omega_0 t + \quad (t)\right), \tag{1}$$

where E_0 is amplitude of the electric field strength, Δt is parameter proportional to pulse duration (see below), ω_0 is carrier frequency, $\Phi(t)$ is carrier phase with respect to pulse envelope (CE-phase).

Let us consider two cases. In the first case we assume that CE phase is constant: $\Phi(t) = \varphi = \text{const}$, but from pulse to pulse its value can be changed in controllable way.

Parameter Δt can be expressed in terms of cycle number in the pulse as follows:

$$\Delta t = \frac{2\sqrt{2\pi}\, n_c}{\omega_0}, \tag{2}$$

where $n_c = \Delta t_p / T = \omega_0 \; t_p / 2\pi$ is number of cycles in pulse. Pulse duration Δt_p we define according to the following relation: $\Delta t_p \leftarrow W/w$ (long pulse limit) where W is total probability of photo-process (during all time action of USP) and w is the probability per unit time.

In the second case we have: $\Phi(t) = \kappa t^2$. Such pulse form is called chirped and parameter κ is temporal chirp. In chirped pulse considered here carrier frequency is linearly dependent on time: $\omega_c = \omega_0 + \kappa t$.

3. Photo-absorption in the local plasma frequency approximation

Photo-absorption cross-section is calculated here within purely classical approach. Atom is considered as inhomogeneous distribution of electron density $n(r)$, so that each spatial point has its own local plasma frequency $\omega_p(r)$. Photon absorption occurs at the distance from atomic nucleus where local plasma frequency matches the frequency of external radiation ω. Corresponding expression for cross-section has the form [9]:

$$\sigma_{ph}(\omega) = \frac{2\pi^2 e^2}{mc} \int n(r)\, \delta\left(\omega - \omega_p(r)\right) d^3 r \,. \tag{3}$$

Integration in the right-side of equation (3) using delta function gives the photo-absorption cross section within Brandt-Lundqvist approximation:

$$\sigma_{ph}^{B-L}(\omega) = \frac{4\pi^2 \omega}{c} r_\omega^2 \frac{n(r_\omega)}{\left| n_r'(r_\omega) \right|} \tag{4}$$

where r_ω is solution for the equation

$$\omega = \omega_p(r) \equiv \sqrt{\frac{4\pi e^2 n(r)}{m}} . \tag{5}$$

4. Atom (statistical model of Lenz-Jensen)

Electron density in the frame of statistical approach has the form:

$$n_{LJ}(r) = Z^2 f\left(x = r/r_{TF}\right) a_B^{-3}, \tag{6}$$

$$r_{TF} = \frac{b}{\sqrt[3]{Z}} a_B, \quad b = \sqrt[3]{\frac{9\pi^2}{128}} \quad 0.8853$$

where r_{TF} is Thomas-Fermi radius, $a_B = \hbar^2/me^2$ – Bohr radius, $f(x)$ is universal function of dimensionless distance. Expression of this function depends upon statistical model. Lenz and Jensen proposed the following function $f_{LJ}(x)$ [10]:

$$f_{LJ}(x) \cong 3.7 e^{-\sqrt{9.7x}} \frac{\left(1 + 0.26\sqrt{9.7x}\right)^3}{(9.7x)^{3/2}}. \tag{7}$$

This function we use in our calculations because it describes the electron density distribution more realistic in comparison with Thomas-Fermi function.

4. Total photo-absorption probability

Expression for total probability of photo-process under the action of ultra-short electromagnetic pulses can be obtained in the first order of perturbation theory as was shown in paper [8]. Corresponding expression has the form:

$$W_{tot}(n_c) = \frac{c}{(2\pi)^2} \int_0^\infty \sigma(\omega) \frac{\left|\mathbf{E}(\omega, n_c)\right|^2}{\hbar\omega} d\omega, \tag{8}$$

where $\sigma(\omega)$ is the spectral cross-section of the process, c is light velocity, $\mathbf{E}(\omega, n_c)$ is Fourier transform of electric field strength in the pulse.

Normalized total probability of photo-process at monochromatic limit ($n_c >> 1$) can be presented in the form:

$$\tilde{W}_{tot}^{(mon)} = \frac{W_{tot}^{(mon)}}{\left|\mathbf{E}_0(\omega_0)\right|^2} \to \frac{c\,\sigma(\omega_0)}{\hbar\omega_0^2} n_c. \tag{9}$$

Hence, probability of the photo-process in a monochromatic field (long pulse limit) increases linearly with the number of cycles in pulse n_c as follows from the traditional consideration. Moreover in monochromatic limit the total probability of the photo-process does not depend upon the CE phase φ as one can see from (9).

In the opposite case of zero-duration electro-magnetic pulse ($n_c << 1$) for $\Phi(t) = \varphi = \text{const}$ we have [8]:

$$W_{tot}(n_c) = \frac{4\pi}{3}\left(\frac{E_0}{h\omega}\right)^2 n_c^2 \langle 0|\hat{\mathbf{d}}^2|0\rangle \cos^2\varphi, \tag{10}$$

where $\langle 0|\hat{\mathbf{d}}^2|0\rangle$ is average value of squared dipole operator of atomic electrons, E_0 is amplitude of electric field strength in the pulse. Note that the sum rule was used in the derivation of right-side of equality (10).

It is of the interest to investigate the dependence of total probability of photo-process upon parameter n_c in the intermediate region (n_c : 1 and $n_c \leq 1$). For this purpose we used formulas (8) and (4) – (7) for the calculation of USP photo-absorption by Lenz- Jensen atom.

5. Results

The results of our calculations are presented in fig. 1–3 for Lenz-Jensen atom with nucleus charge number $Z = 30$ and for normalized total probability of atomic photo-absorption ($\tilde{W}_{tot}$). The latter quantity is defined by following relation

$$\tilde{W}_{tot} = \frac{W_{tot}}{|\mathbf{E}_0|^2}. \tag{11}$$

Photo-absorption probability under the action of USP with $\Phi(t) = \varphi = \text{const}$ is shown in fig. 1 for different values of CE phase and for given carrier frequency. One can see that for sub-cycle pulses there is the strong dependence on CE phase. Namely, photo-absorption probability dependence on cycle number in the pulse has maximum for cosine pulse for $n_c < 0.5$. The magnitude of this maximum decreases with the increase of CE phase. Straight lines in fig. 1 – 3 present the total probability of photo-absorption in monochromatic limit when the total probability of photo-absorption $\tilde{W}(n_c)$ is linear function of n_c.

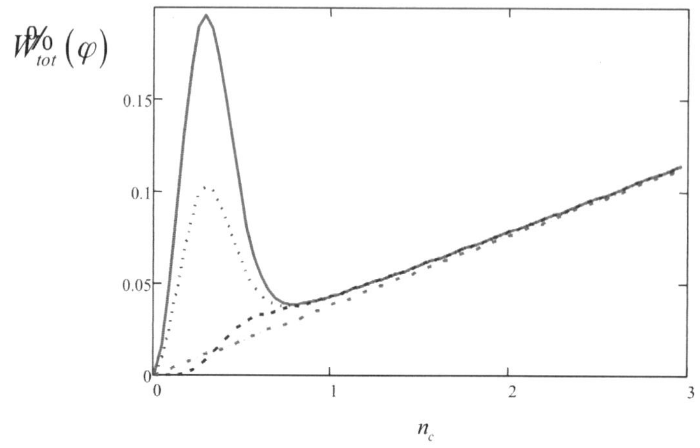

Figure 1. Total photo-absorption probability of USP with different values of CE phase calculated for Lenz-Jensen atom: solid line – φ=0, dotted line – φ=π/4, dashed line – φ=π/2, dashed-dotted line – monochromatic limit, $\hbar\omega_0 = 10$ a.u., Z=30.

It is shown in fig. 1 that total photo-absorption probability goes to the straight line for $n_c \geq 1$. Thus in considered case only sub-cycle pulses causes nonlinear dependence of $W(n_c)$ as a function of n_c.

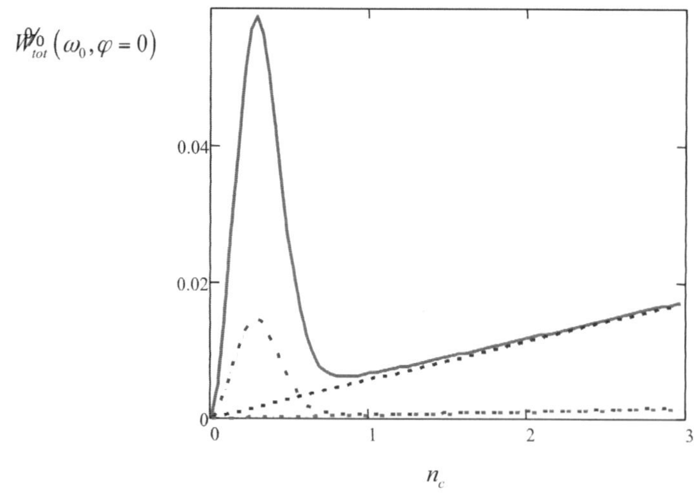

Figure 2. Total photo-absorption probability of cosine USP by Lenz-Jensen atom for two different carrier frequencies: solid line – $\hbar\omega_0 = 500$ eV, dashed-dotted line – $\hbar\omega_0 = 1000$ eV; straight lines show corresponding monochromatic limits.

Fig. 2 shows total photo-absorption probability dependence upon parameter n_c for cosine pulse and different values of carrier frequency. From this figure one can see that in case of Lenz-Jensen atom with charge Z=30 probability function ($\tilde{W}(n_c)$) in linear limit has smaller angle (with respect to abscissa axis) for greater photon energy.

Total probabilities of photo-absorption of chirped pulses by Lenz-Jensen atom are shown in fig. 3. In this case the dependence on cycle number n_c has oscillating character with maxima decreasing when parameter n_c increases (if dimensionless chirp $\alpha = \kappa \Delta t^2 \neq 0$).

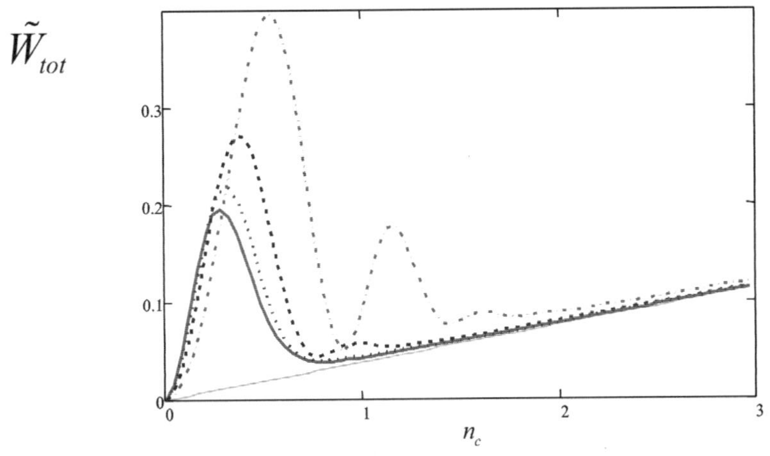

Figure 3. Total photo-absorption probability by Lenz-Jensen atom for chirped USP, $\hbar\omega_0 = 10$ a.u., Z=30: solid line: α=0, dotted – α=0.5, dashed line – α=1, dashed-dotted line – α=2, solid straight line – monochromatic limit.

Coincidence of photo-absorption probability for few-cycle chirped pulse and monochromatic one persists. However, the number of cycles at which this coincidence occurs is determined by the chirp parameter α. The larger chirp parameter the greater number of cycles should contain pulse for the same value of photo-absorption probability as in monochromatic case.

One can see from fig. 3 that the maxima of $\tilde{W}(n_c)$ function grow with increase of chirp value and shifts in the range of longer pulses.

6. Conclusion

In the frame of perturbation theory the analysis of atomic photo-absorption of electro-magnetic USP with controllable phase parameter is made in the intermediate region of pulse durations where number of cycles in the pulse (n_c) satisfies the following conditions $n_c : 1$ and $n_c \le 1$. It is shown that in this region the total probability of photo-absorption $\tilde{W}(n_c)$ displays peculiarities as a function of parameter n_c.

So in the case of USP with constant CE phase cosine pulse photo-absorption probability has a maximum for $n_c \le 0.5$ while in the case of sine USP such maximum is absent. In the case of chirped pulses photo-absorption probability has oscillating character as a function of cycle number per pulse in the duration range $n_c : 1$. Maxima of photo-absorption function $\tilde{W}(n_c)$ grow in magnitude with the increase of the chirp value and shifts in the range of longer pulse durations.

Acknowledgments
Work was supported by Russian Foundation for Basic Research (Grant No. 10-02-01095-a).

References

[1] Goulielmakis E., Schultze M., Hofstetter M. et al 2008 *Science* **320** 1614
[2] Morgner U. 2010 *Nature Photonics* **4** 14
[3] Krausz F., Ivanov M. 2009 *Rev. Mod. Phys.* **81** 163
[4] Bandulet H.-C., Comtois D., Bisson E. 2010 *Phys. Rev. A* **81**, 013803
[5] Kahra S., Leschhorn G., Kowalewski M. 2012 *Nature Physics* **8** 238
[6] Znakovskaya I., von den Hoff P., Zherebtsov S. 2009 *Phys. Rev. Lett.* 103002
[7] Jepsen P. U., Cooke D. G. and Koch M. 2011 *Laser Photon. Rev.* **5** 124
[8] Astapenko V.A. 2010 *Physics Letters* **A 374** 1585
[9] Brandt W., Lundqvist S. 1965 *Phys. Rev.* **139** A612
[10] Gombas P. 1949 *Die Statistische Theorie des Atoms und ihre Anwendungen* (Wien: Springer-Verlag)

Journal of Physics: Conference Series **397** (2012) 012003 doi:10.1088/1742-6596/397/1/012003

Optical lattice clocks: Hz-level spectral width with sub-Hz reproducibility

Tetsuya Ido
National Institute of Information and Communications Technology and JST-CREST
1-4-2 Nukui-kitamachi, Koganei, Tokyo 184-8795 Japan

E-mail: ido@nict.go.jp

Abstract. State-of-the-art optical clocks have surpassed microwave clocks in this century, causing discussions to redefine the SI second based on an optical transition. The superiority of optical standards was clearly revealed by two of optical-optical comparisons, one of which is a remote comparison of two strontium lattice clocks, and the other is a characterization of a single Ca^+ ion clock using a Sr lattice clock as a frequency reference. The former for the first time demonstrated the frequency reproducibility of physically separated clocks at the 10^{-16} level. The latter has confirmed the capability of single Ca^+ clocks to reach the $\sim 10^{-16}$ instability.

1. Introduction

Atomic clocks based on optical transitions have been proposed since the invention of the laser, where a laser as an optical oscillator is kept to be resonant to atomic transitions, providing stable ticks of clocks. To realize the frequency reproducibility of the optical clocks, it is critical to pursue "spectral line shapes" of Lorentzian with its width determined solely by the lifetime of the excited states. Electrically trapped single ions have been the most promising play ground to obtain such simple line shapes. The progress of the laser cooing technique has enabled single ions trapped in the vibrational ground state, which has a spatial distribution much smaller than the wavelength of the clock laser, realizing Doppler- as well as recoil- free spectrum. It is also possible to laser-cool neutral atoms. The invention of the magneto-optical trap has allowed us to trap large number of ultracold atoms, leading to quantum degenerate gas such as Bose-Einstein condensate. However, it has long been impossible to extract a "pure line shape" from ultracold neutral atoms. While the tiny spatial distribution of the atomic wave packet is realized by optical lattice potential made of the interference pattern of intensive far-detuned radiation, this trapping laser causes different amount of ac shark shift to ground and excited state, which causes considerable shift of the transition frequency. This drawback of using neutral atoms has been overcome by the invention of lattice clocks [1]. Choosing species and the wavelength of the far-detuned laser, identical shape of the trapping potential similar to ion traps is generated for ground and excited state. The recoil-free spectroscopy based on this idea is first performed using ${}^1S_0$-${}^3P_1$ transition of the ^{88}Sr, where the recoil-free nature is confirmed by the comparison of the line shapes with and without lattice potential [2]. The success of the recoil-free spectroscopy strongly pushed the quest for the lattice clock using the ultranarrow ${}^1S_0$-${}^3P_0$ transition of ^{87}Sr [3, 4].

^{87}Sr lattice clocks are currently operated in five laboratories. The inaccuracy of the state-of-the-art clocks is evaluated to be 1×10^{-16}, predominantly limited by black body radiation [5, 6]. This inaccuracy is smaller than that of state-of-the-art Cs fountain clocks. Therefore, as long as we employ frequencies based on the SI second, this situation causes trouble to investigate the agreement of the frequencies developed in physically separated laboratories. As shown in Fig. 1, the agreement of the absolute frequencies is currently in 10^{-15} level, limited not by lattice clocks but by the uncertainty of the realization of the SI second as well as of the frequency link between the Cs clocks and lattice clocks.

Fig.1 also indicates that Tokyo metropolitan area has two ^{87}Sr lattice clocks at University of Tokyo and National Institute of Information and Communications Technology (NICT). The distance of two laboratories is 24km in a crow line, which is short enough to implement an optical fibre link using a dark fibre without intermediate amplifiers. Dark fibre cables of 48km are fortunately available from NICT to the downtown Tokyo for research in telecommunication network, and we additionally rented commercially available 12km part from the downtown to U. Tokyo campus, giving the feasibility to compare two physically separated lattice clocks. This all-optical direct comparison no longer relies on the SI second. Therefore it is expected that the reproducibility of physically separated clocks is confirmed beyond the International Atomic Time (TAI).

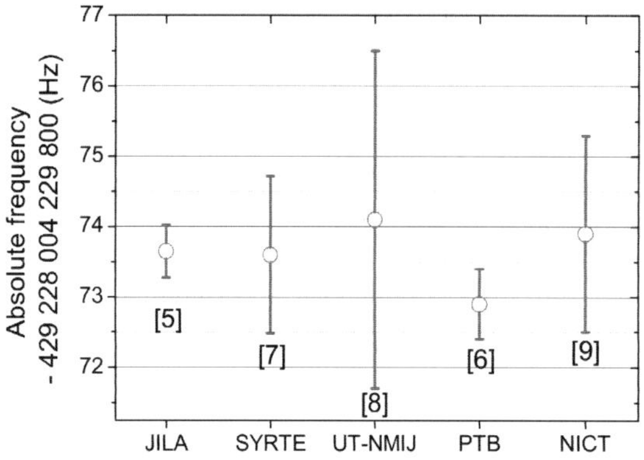

Fig.1 Absolute frequencies of the five ^{87}Sr lattice clocks in operation worldwide

The limitation of microwave standard also lies in the requirement of long signal integration time compared to optical standard. Optical frequency standards have a significant advantage in the speed of the comparison, requiring less than 1000 seconds instead of more than six hours to evaluate a fractional frequency difference with $\sim 10^{-16}$ uncertainty. Comparison also enables the rapid evaluation of systematic shifts by investigating the dependence of the frequency on various experimental parameters. We demonstrated an optical comparison of locally available Ca^+ clock against the lattice clock as a tool to characterize optical clocks.

2. Lattice clock at NICT

The detail of the ^{87}Sr lattice clock at NICT is found in [9]. The inaccuracy attributed to the Sr system was evaluated to be 5×10^{-16}, mainly determined by collision shift, 2nd Zeeman shift, and AC stark shift due to the black body radiation and lattice beam. When the fibre link experiment was demonstrated in 2011, the clock laser at 698nm was not as stable as other clock lasers. We effectively phase locked the clock laser to a Ca^+ clock laser via an optical frequency comb. The short term instability of the clock

Journal of Physics: Conference Series **397** (2012) 012003 doi:10.1088/1742-6596/397/1/012003

laser was 3×10^{-15}. Since then, the instability of the clock laser has been improved by implementing optical compensation against vibration-induced phase noise, details of which are described elsewhere. By this technique the short term instability is currently < 2×10^{-15} at 1s and no longer requires the stability transfer technique employed previously.

3. All optical link to another lattice clock in U. Tokyo

Figure 1 illustrates a schematic diagram of the frequency comparison via telecommunication fiber link. The clock signal at the National Institute of Information and Communications Technology (NICT) is optically transferred to The University of Tokyo (UT) using a telecommunication fibre link in Tokyo. At NICT, a Ti:sapphire-based optical frequency comb is phase-locked to the clock laser at 698 nm. The telecom laser with wavelength of 1538 nm is phase-locked to this optical frequency comb through its frequency doubled light at 769 nm and transferred to UT through the phase-noise-cancelled 60-km-long optical fibre. The Ti:sapphire-based optical frequency comb at UT is phase-locked to the frequency doubled light of the transferred light. Thus, the frequency difference between two Sr lattice clocks is calculated from the frequency of the beat signal between the clock laser and the nearest tooth of the optical frequency comb at UT.

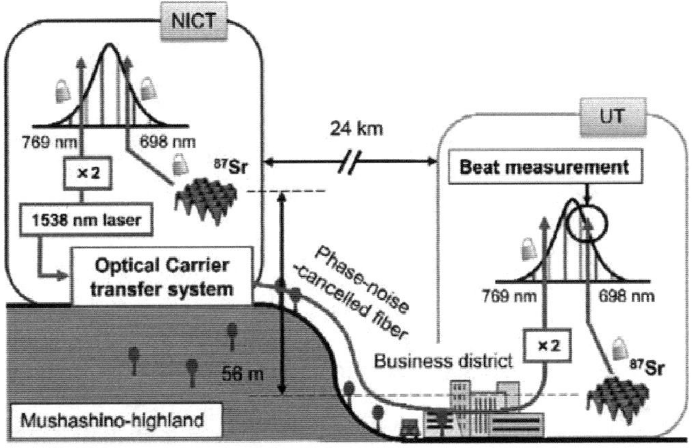

Fig.2 Schematic diagram of the remote frequency comparison

The frequency difference of the two clock frequencies f_{Sr} (NICT) – f_{Sr} (UT) is shown in Fig. 3. Each points obtained by a frequency counter have a gate time of 1 second. Ten seconds of signal integration reveals that the clock frequency in NICT is slightly higher than that of U. Tokyo. The average of the all data is 3.66 Hz. This difference is consistent with the difference of systematic corrections. The dominant differential correction is due to gravity shift. NICT has 56 m higher elevation than U. Tokyo, which corresponds to a 2.62 Hz larger blue shift. Subtracting the differential shift, the residual fractional difference has resulted in 1.0×10^{-16} with uncertainty of 7×10^{-16}. This uncertainty consists of the systematic uncertainties of two clocks, 5×10^{-16} for each. In terms of the uncertainty due to the fiber link, we confirmed that the instability of the frequency link reaches ~10^{-18} and that there is no frequency shift beyond the instability [10]. Frequency comparison implemented here is simply a frequency ratio measurement of the two clocks, which is sufficient to evaluate the frequency reproducibility of lattice clocks.

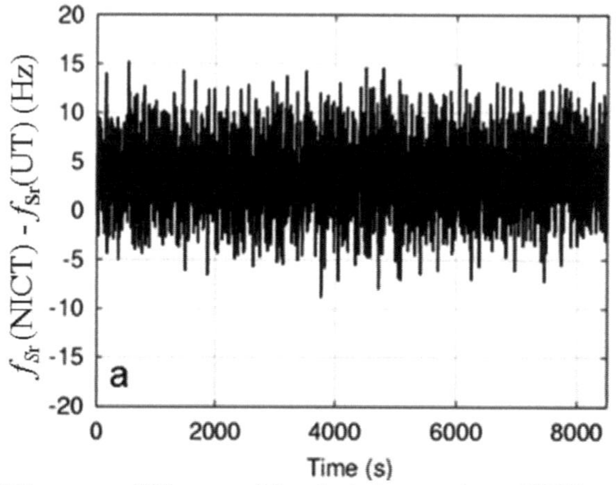

Fig.3 Frequency difference of the clock frequencies at NICT and U. Tokyo. The clock in NICT is slightly higher than that of U. Tokyo.

3. Lattice clock as a reference to characterize a Ca+ clock

Our institute has lately improved a single-ion clock based on the $^{40}Ca^{+}$ $^{2}S_{1/2}$-$^{2}D_{5/2}$ transition. By installing a magnetic shield onto the vacuum chamber, we have reduced stray ac electromagnetic field that previously limited the spectral width to 300Hz [11]. Furthermore, the optimization of the clock laser has reduced its spectral width less than 5Hz and noise cancelling technique have contributed to obtain narrower spectrum. The resultant spectral width and inaccuracy is 30Hz and 2×10^{-15}, respectively. The uncertainty is reduced for 14 times than our previous report [11]. To confirm this level of frequency reproducibility, we first measured absolute frequency by referring a hydrogen maser which is linked to the international Atomic Time (TAI) via locally available Universal Coordinated Time UTC(NICT). However, the resultant absolute frequency has a fluctuation larger than the systematic uncertainty. In this measurement, the frequency of the hydrogen maser that a frequency comb refers is calibrated by referring Circular T, where the time differences of every five days between UTC(NICT) and TAI are reported. On the other hand, it takes only several hours for our frequency measurement. Therefore, the fluctuation of hydrogen maser during five days, which affects the HM-based frequency measurement, is averaged out in the TAI calibration. Optical-optical comparison, in other words the measurement of frequency ratio instead of absolute frequency, does not suffer this problem. A number of this ratio measurement for certain duration is sufficient to confirm the reproducibility of both clocks. Fig. 4 shows the result of the ratio measurement between the Ca^{+} clock and the Sr lattice clock. Systematic shifts are corrected, whereas the error bars include only statistical uncertainty. Here, the standard deviation of the six data is calculated to be 6×10^{-16}. It is impossible to observe this level of the reproducibility by using the TAI link. The drift of the data at the last four points may indicate that some systematic shifts are slowly moving within the total systematic uncertainty of 2.3×10^{-15}.

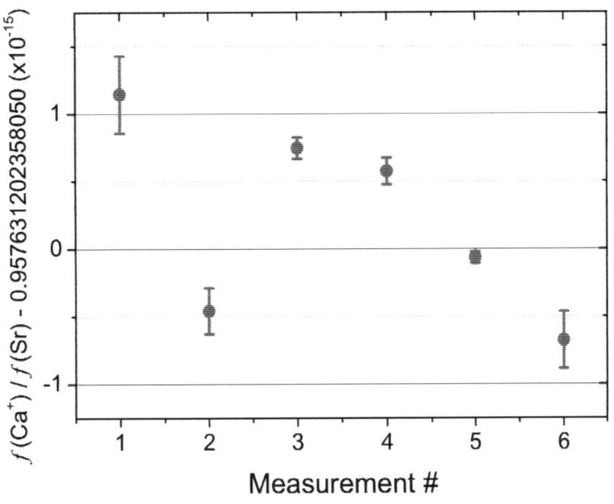

Fig. 4 The result of all optical comparison between Ca^+ clock and Sr lattice clock. The data includes the correction due to systematic shifts. The error bars only include statistical error.

4. Summary and Outlook

^{87}Sr lattice clock in NICT was compared to another lattice clock at U. Tokyo as well as a locally available single Ca^+ ion clock. All optical comparison did not only accelerate the comparison process but also evade the limitation of accuracy in locally available microwave reference. The comparison against the lattice clock at U. Tokyo confirmed the reproducibility of lattice clock for the first time at 10^{-16} level. The comparison against the Ca^+ clock demonstrated that the Ca+ clocks have an ability to reach 10^{-16} level of instability. These results imply that latest progress of optical clocks strongly requires frequency standards in optical region, making it realistic to redefine and share the SI second using optical transition instead of the microwave transition of Cs.

Acknowledgements

The author thanks H. Katori, M. Takamoto, and T. Takano in Univ. of Tokyo for the cooperation in the fiber-link experiment. The author is also grateful to many colleagues in NICT including M. Fujieda, M. Kumagai, H. Hachisu, Y. Li, S. Nagano, and A. Yamaguchi for the fiber-link experiment. The Ca^+ was operated by R. Kojima, C. Locke, K. Hayasaka, and K. Matsubara. This research was supported in part by the Japan Society for the Promotion of Science through its FIRST program.

References

[1] M. Takamoto, F.-L. Hong, R. Higashi, H. Katori, Nature 435, 321-324 (2005).
[2] T. Ido and H. Katori, Phys. Rev. Lett. **91**, 053001 (2003).
[3] M. Takamoto and H. Katori, Phys. Rev. Lett. **91**, 223001 (2003).
[4] A. D. Ludlow, M. M. Boyd, T. Zelevinsky, S. M. Foreman, S. Blatt, M. Notcutt, T. Ido, and J. Ye, Phys. Rev. Lett. **96**, 033003 (2006).

[5] G. K. Campbell, A. D. Ludlow, S. Blatt, J. W. Thomsen, M. J. Martin, M. H. G. de Miranda, T. Zelevinsky, M. M. Boyd, J. Ye, S. A. Diddams, T. P. Heavner, T. E. Parker, and S. R. Jefferts, Metrologia **45,** 539 (2008).

[6] St Falke, H Schnatz, J S R Vellore Winfred, Th Middelmann, St Vogt, SWeyers, B Lipphardt, G Grosche, F Riehle, U Sterr and Ch Lisdat, Metrologia **48**, 399-407 (2011).

[7] X. Baillard, M. Fouche, R. L. Targat, P. G. Westergaard, A. Lecallier, F. Chapelet, M. Abgrall, G. D. Rovera, P. Laurent, P. Rosenbusch, S. Bize, G. Santarelli, A. Clairon, P. Lemonde, G. Grosche, B. Lipphardt, and H. Schnatz, Eur. Phys. J. D **48**, 11 (2008).

[8] F. L. Hong, M. Musha, M. Takamoto, H. Inaba, S. Yanagimachi, A. Takamizawa, K. Watanabe, T. Ikegami, M. Imae, Y. Fujii, M. Amemiya, K. Nakagawa, K. Ueda, and H. Katori, Opt. Lett. **34** 692 (2009).

[9] A. Yamaguchi, N. Shiga, S. Nagano, Y. Li, H. Ishijima, H. Hachisu, M. Kumagai, and T. Ido, Appl. Phys. Express **5**, 022701 (2012).

[10] M. Fujieda, M. Kumagai, S. Nagano, A. Yamaguchi, H. Hachisu, and T. Ido, Opt. Express **19**, 16498 (2011).

[11] K. Matsubara, K. Hayasaka, Y. Li, H. Ito, S. Nagano, M. Kajita, and M. Hosokawa, Appl. Phys. Express, **1**, 067011 (2008).

Coulomb Broadening of Resonance Induced by Standing Wave

O V Belai[1] **and D A Shapiro**[1,2]

[1] Institute of Automation and Electrometry, Siberian Branch, Russian Academy of Sciences, 1 Koptjug Avenue, Novosibirsk, 630090 Russia

[2] Novosibirsk State University, 2 Pirogov Street, Novosibirsk 630090, Russia.

E-mail: shapiro@iae.nsk.su

Abstract.
Coherent preparation of quantum states of atoms and ions by laser light can lead to electromagnetically induced transparency and related effects. In particular, the standing wave at the adjacent transition induces a new type of nonlinear resonance in the probe-field spectrum of a three-level system. The resonance is due to the effect of high-order harmonics of atomic coherence. However, the resonance is broadened and the reason of the broadening was unclear. The paper describes how the velocity changing collisions broaden the resonance. The analytic theory taking into account up to 4-th order of the perturbation theory is presented. The numerical calculations are described. The analytical formulas are shown to describe the experiment qualitatively while the numerical computation demonstrates quantitative agreement. The Coulomb dephasing is discussed as the physical mechanism of the broadening.

1. Introduction

It is well-known that the light absorption in the atomic medium near a resonant frequency can be reduced substantially or even canceled with the help of strong driving field at the adjacent transition. This effect, so-called electromagnetically induced transparency (EIT) [1, 2], is a base of quite a number of atomic coherence effects, in particular, the resonant enhancement of electro-optical processes [3, 4], the lasing without inversion [5], suppression of two-photon absorption [6]. Under conditions of EIT the super-narrow resonances arise leading to the giant dispersion within the transparency window. This property is helpful for light deceleration [7–9] and makes the effect promising for application in the frequency standards [10], the precision magnetometery [11], and as a storage of quantum information [12, 13].

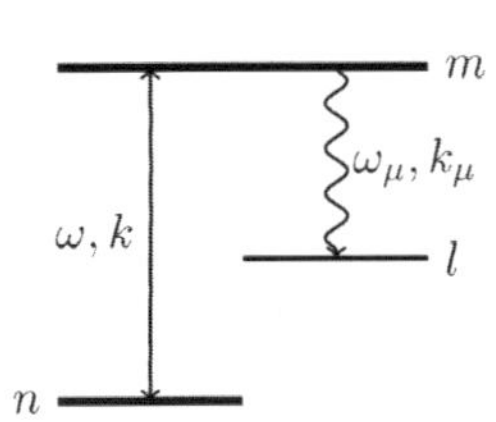

As a rule, in EIT experiments three-level Λ-systems interact with driving field presented by a running wave. Theoretical and experimental study of the standing wave at argon ionic lines in Λ-configuration is studied in [14, 15]. Strong field was generated on laser transition ArII $mn = 4p\,^2S_{1/2} - 4s\,^2P_{1/2}$ at wavelength $\lambda = 458$ nm. Probe field was a running wave at $ml = 4p\,^2S_{1/2} - 3d\,^2P_{3/2}$, $\lambda_\mu = 648$ nm. The absorption of probe field was measured as a function of detuning $\Delta = \omega - \omega_{mn}$. At the specific probe field detuning $\Delta_\mu \equiv \omega_\mu - \omega_{ml} = k_\mu \Delta / k$ the well known EIT resonance was observed. Along with it a new

Published under licence by IOP Publishing Ltd

structure at$\Delta_\mu = 0$ was discovered and its position in the spectrum was independent of the strong field detuning Δ.

At $\Delta = 0$ the positions of both resonances coincide but their widths and signs differ. The new structure looks like EIT peak with a narrow dip in the centre. At higher strong field detuning Δ the dip was transformed to a peak at the center$\Delta_\mu = 0$. The new structure was observed for non-typical situation with a wide lower level n, where the forbidden transition was wider than the allowed one$\Gamma_{nl} \gg \Gamma_{ml}$. In the conventional physics of a three-level system the opposite limiting$\Gamma_{nl} \ll \Gamma_{ml}$ case had been studied. For this reason the motionless structure was missed both in the perturbation theory and in numerical calculations [16]. The appearance of this structure was shown to be caused by the higher harmonics of the spatial coherence at transitions ml and nl induced by the standing wave.

At the same time the width of observed resonance occurs to be several times wider than the prediction of collisionless theory. It is naturally to assume that the broadening is a result of the velocity changing collisions. For ions in high current discharge the leading process is the Coulomb scattering [17]. In general theory of Coulomb scattering the higher harmonics of spatial coherence are neglected. The neglecting is valid for the Lamb dip in two-level system where the oscillation of populations are smoothed out after the averaging over velocities. The theory is also valid for running wave in three-level system. Calculation of the shape of known EIT resonance at$\Delta_\mu = k_\mu \Delta / k$ in the three-level system under the field of running wave demonstrates a good agreement with experiment [18]. It was shown that the resonance is caused by the Bennet dip in the population distribution over velocity. Its width becomes nearly 3 times greater than in collisionless case. For the new resonance$\Delta_\mu = 0$ the theory taking the scattering into account was absent.

The aim of the present paper is to explain how the collision broadening occurs. The key expressions of the broadening theory of the resonance of higher spatial harmonics are presented below, while the specific details of numerical and analytical calculation can be found in papers [19,20]. The account of collisions is important not only for the quantitative interpretation of experiments in argon discharge plasma. The velocity changing collisions could broaden the resonances and decrease the slope of dispersion curve in other experiments on EIT where the standing wave is exploited.

2. Perturbation theory

Let us consider the gas of three-level systems under the driving field of standing wave. An analytical expression for the probe field absorption at high intensity of the driving field cannot be derived even for collisionless gas. Higher spatial harmonics are taken into account by the Feldman — Feld continuous fractions [16]. This expression was obtained from the known continuous fraction for populations in two-level system in bichromatic field [21, 22]. In the case of equal relaxation constants and exact resonance the populations can be expressed in terms of Bessel function [23] (see also the survey by Stenholm [24]). In experiment the higher spatial harmonics were studied in cadmium spectrum. Atoms interacted with two counterpropagating waves of different frequency and amplitude [25, 26] (see also [27]). Under the frequency scanning the subradiative structure was observed in the absorption spectrum that included up to 5 peaks that became more frequent in the line center (so called "$1/n$-resonances"). Herewith the spectrum appeared to be very sensitive to the ratio of amplitudes of the waves, especially in the case of close amplitudes.

At weak saturation the solution can be obtained as a perturbation series. The second order of the perturbation series does not include the effects of higher spatial harmonics. The second harmonics appears only in the fourth order, then we restrict ourselves by the fourth order.

2.1. Equation for density matrix

The density matrix ρ_{ij} equation describes the interaction of three-level system with the parallel driving and probe waves

$$\mathrm{i}\,(\Gamma_{ij} + \partial_t + v\partial_x)\,\rho_{ij} = [V,\rho]_{ij}. \tag{1}$$

Here Γ_{ij} are the relaxation constants, t is time, x is the coordinate along the common direction of the wavevectors $\mathbf{k}\|\mathbf{k}_\mu$ of the driving and probe waves, v is the projection of the velocity vector to this common direction, $\hat{V} = -\mathbf{E}\cdot\hat{\mathbf{d}}/2\hbar$ is the operator of dipole interaction, $\hat{\mathbf{d}}$ is the dipole moment operator, $\hbar$ is the Plank constant, $\mathbf{E}$ is the electric field of the light wave, indices i, j possess the values l, m, n, square brackets denote the commutator.

We neglect the population ρ_{mm} of level m and polarization ρ_{nm} being small under experimental conditions. The equilibrium distribution of populations are Maxwellian. Let us specify the fields of running and standing waves in the form

$$\mathbf{E}_\mu(x,t) = \frac{1}{2}\mathbf{E}^0_\mu e^{\mathrm{i}k_\mu x - \mathrm{i}\omega_\mu t} + \text{c.c.}, \quad \mathbf{E}(x,t) = \mathbf{E}^0 \cos kx\, e^{-\mathrm{i}\omega t} + \text{c.c.}, \tag{2}$$

where $\mathbf{E}^0$ and $\mathbf{E}^0_\mu$ are the amplitudes, c.c. are complex conjugated terms. Denote Rabi frequencies of the driving and probe field as $\Omega_\mu = \mathbf{E}^0_\mu \cdot \mathbf{d}_{ml}/2\hbar, \Omega = \mathbf{E}^0 \cdot \mathbf{d}_{mn}/2\hbar$, where $\mathbf{d}_{ij}$ are matrix elements of the dipole moment.

Changing the variables:

$$\rho_{ml} = \rho_\mu(x,v)\exp(\mathrm{i}k_\mu x - \mathrm{i}\Delta_\mu t), \;\; \rho_{nl} = \rho_\nu(x,v)\exp(\mathrm{i}k_\mu x + \mathrm{i}\,\Delta t - \mathrm{i}\Delta_\mu t),$$

we get a closed steady state set for amplitudes ρ_μ and ρ_ν of polarizations:

$$\begin{aligned}(\Gamma_\mu - \mathrm{i}\Delta_\mu + \mathrm{i}k_\mu v + v\partial_x)\rho_\mu &= -2\mathrm{i}\Omega\cos(kx)\rho_\nu + \mathrm{i}\,\Omega_\mu N_l(v), \\ (\Gamma_\nu + \mathrm{i}\,\Delta - \mathrm{i}\Delta_\mu + \mathrm{i}k_\mu v + v\partial_x)\rho_\nu &= 2\mathrm{i}\,\Omega\cos(kx)\rho_\mu,\end{aligned} \tag{3}$$

where $\Gamma_\mu = \Gamma_{ml}$, $\Gamma_\nu = \Gamma_{nl}$, $\Delta = \omega - \omega_{mn}$, $\Delta_\mu = \omega_\mu - \omega_{ml}$.

These equations take into account only interaction of ions with light, however the Coulomb ion-ion scattering with velocity change affects the probe-field spectrum strongly. The scattering is small-angle then it can be described as a diffusion process in the velocity space. The scattering is independent of the ionic internal quantum state, then it can be taken into consideration by operator $\hat{S} = -D\partial_{vv}$ in the left sides of (1) or (3), where

$$D = \frac{\nu v_T^2}{2}, \quad \nu = \frac{16\sqrt{\pi}NZ^2e^4\Lambda}{3M^2v_T^3}. \tag{4}$$

Here D is the diffusion coefficient in the velocity space, ν is the effective transport frequency of ionic collisions, Ze is the ion charge, Λ is the Coulomb logarithm, N is the ionic density [17]. The full ionic density enters into expression (4), since the concentration of excited ions is small, and then the excited ions are scattered by the ions in the ground state.

2.2. Nonlinear absorption spectrum

Since the solutions to (1) are periodic functions of coordinate, the density matrix elements $\rho(x)$ are Fourier series over the spatial harmonics

$$\rho_{ij}(x,v) = \sum_{p=-\infty}^{\infty} \rho_{ij,p}(v)e^{\mathrm{i}pkx}. \tag{5}$$

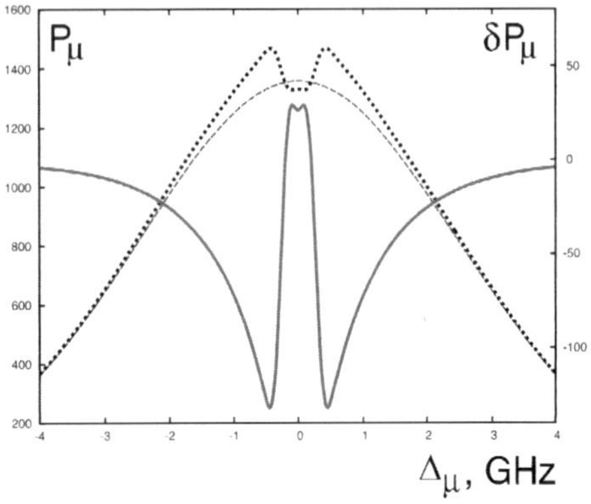

Figure 1. Absorption $\mathcal{P}_\mu(\Omega)$ (arb. units) as a function of the probe-field frequency detuning Δ_μ (dots), the same without the strong field $\mathcal{P}_\mu(0)$ (dashes); the ordinate axis is at the left. Their difference $\delta\mathcal{P}_\mu$, the spectrum of nonlinear absorption (solid curve,right axis).

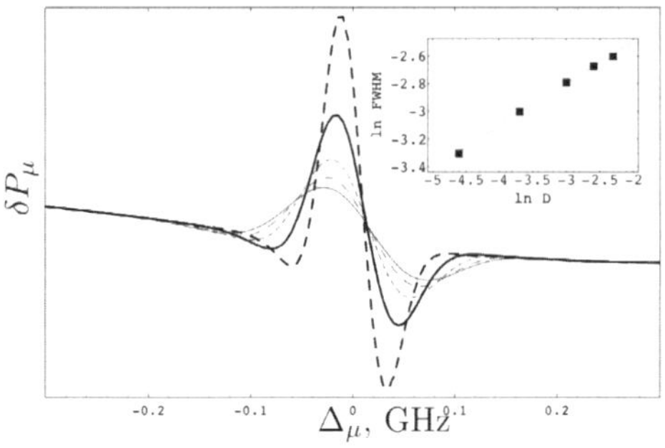

Figure 2. Central peak at$\Gamma_\nu = \Gamma_\mu =$ 1 GHz and different diffusion coefficient D: $D/D_0 = 0.1, 0.075, 0.05, 0.025, 0.001$, $D_0 = 10^{13}\,\mathrm{m}^2/\mathrm{s}^3$. Inset: the width of peak as a function of diffusion coefficient in logarithmic coordinates.

Nonlinear absorption spectrum is defined as the difference between probe-field absorption spectra when the driving field is turned off and turned on

$$\delta\mathcal{P}_\mu = \mathcal{P}_\mu(0) - \mathcal{P}_\mu(\Omega), \quad \mathcal{P}_\mu \propto \mathrm{Im}\int\limits_{-\infty}^{\infty} r_{ml,0}(v)\,dv. \tag{6}$$

Its advantage is absence of the broad Doppler contour being subtracted, as shown in Fig. 1. Scales of axis y for linear and nonlinear absorption differs by an order of magnitude, then the nonlinear spectrum is better to study small nonlinear contributions of the strong driving field. Expression for $\delta\mathcal{P}_\mu$ includes only zeroth Fourier component of the polarization $r_{ml,0}(v)$, since one must average the absorption over coordinate x and other components give no contribution.

The dip or peak broaden compared to the collisionless case. The broadening due to velocity changing collisions could depend only on the diffusion coefficient and the wavenumber $k_\mu \sim k$. There is the only their combination with frequency dimension $\delta\Delta_\mu \sim D^{1/3}k^{2/3} \sim \nu^{1/3}(kv_T)^{2/3}$. For checking the hypothesis we plotted the series of curves $\delta\mathcal{P}_\mu(\Delta_\mu)$, Fig. 2, at different diffusion coefficients. The curves are calculated by the formulas of perturbation theory at small Rabi frequency Ω. The full width at half maximum (FWHM) as a function of D is shown in the inset. The slope 0.31 is obtained by the least square fitting in agreement with estimated exponent 1/3. This dependence has a simple physical explanation.

The broadening takes place because of the dephasing of wave in the reference frame of an ion due to the Coulomb scattering. At time t the active excited ion is scattered by ions in ground state by random angle ϑ. Its average value is zero $\langle\vartheta\rangle = 0$, while the mean square value is nonzero according to the diffusion law $\langle\vartheta^2\rangle = \nu t$. The ion started in the antinode of standing wave and having zero projection of the velocity onto the wavevector acquires velocity $\delta v \sim (\nu t)^{1/2}v_T$, then passes the distance $\delta x \sim \nu^{1/2}t^{3/2}v_T$. When the phase incursion of n-th harmonics reaches π the ion leaves the interaciton region, and then the effect of n-th harmonics vanishes. The characteristic duration of this process is of the order of $t \sim (nkv_T)^{-2/3}\nu^{-1/3}$. For estimation we take $n = 2$, then the broadening is $\delta\Delta_\mu \simeq D^{1/3}k^{2/3}$. At $\delta\Delta_\mu \sim \Gamma_\mu$ the dip (in

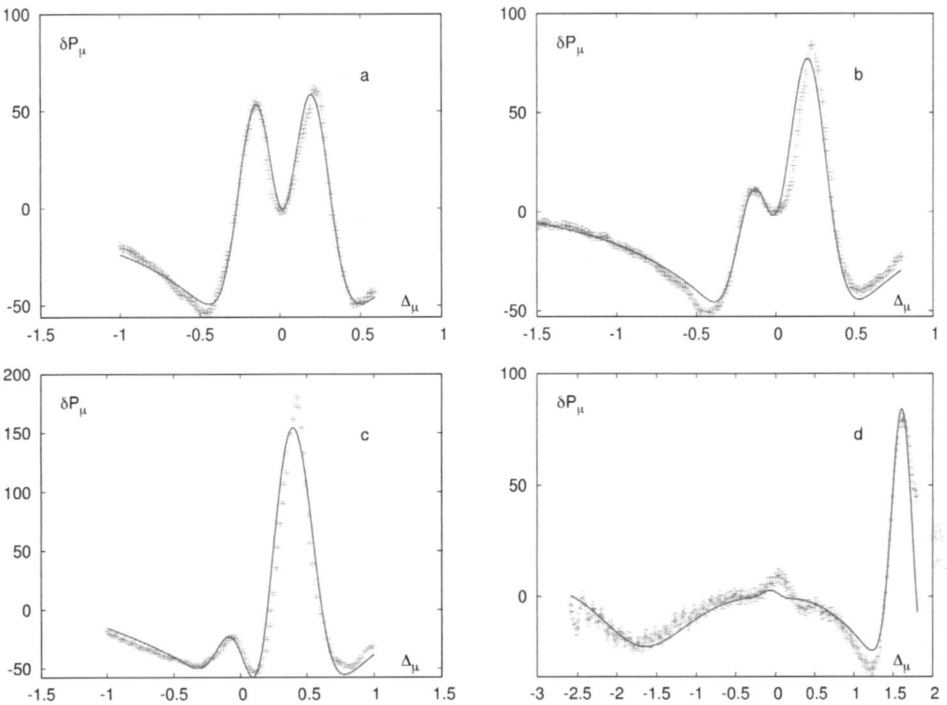

Figure 3. Nonlinear absorption $\delta\mathcal{P}_\mu$ (arb. units) as a function of probe-field detuningΔ_μ (GHz): experimental data (dots) and least-square fitting (curve).

resonance case) is washed off and the peak (in nonresonance case) becomes wider. Notice that the considered effect of Coulomb dephasing corresponds to the density dependence $N^{1/3}$, while the known effect of Coulomb broadening of the Lamb dip gives $N^{1/2}$ [28].

3. Numerical calculation

Three level scheme is studied numerically. In the previous section we neglect the population of intermediate level m, and then take into account only the population of the final level l. This approximation answer the experimental conditions in ArII spectrum, where the initial state n is relatively wide, when the intermediate and final states m, l are narrow:$\Gamma_n \approx 3 \times 10^9$ s^{-1}, $\Gamma_m \approx 1.5 \times 10^8$ s^{-1}, $\Gamma_l \approx 8 \times 10^7$ s^{-1}. The populations of these states are nearly $N_n \approx 10^9$ cm^{-3}, $N_m \approx 5 \times 10^9$ cm^{-3}, $N_l \approx 10^{11}$ cm^{-3}, then $N_n, N_m \ll N_l$. In calculations all the populations are taken into account. For this purpose the full set of 9 coupled Fokker — Plank equations are solved numerically instead of simplified set (3).

To compare the computations with experiment the nonlinear spectra are fitted with the least squares. The full set of linear equations includes 13 parameters: N_j, Γ_j, the unperturbed population and the width of level $j = n, m$, Γ_{ij}, the homogeneous width of transition ij, $i, j = n, m, l$, Δ, Δ_μ, the detunings of strong and probe fields, k, k_μ, the wavenumbers of strong and probe fields, Ω, the Rabi frequency of strong field, ν, the effective frequency of scattering. The majority of parameters are known from independent measurements. The least-square fitting is carried out by 3 physical parameters: the effective frequency of ionic scattering ν, the Rabi frequency of strong field Ω, and the driving field detuningΔ . Nonlinear resonances at different

Δ shown in Fig. 3 along with the results of fitting demonstrate a good agreement. Subfigures a–d are plotted at different detuning Δ: from resonance to nonresonance case.

4. Conclusions

Thus, the shape of the standing wave resonance is calculated both numerically and analytically. The absorption is found in the 4th order of the perturbation theory in qualitative agreement with measurement. The quantitative calculations taking into consideration small population of the upper level and saturation of mn transition allows fitting the experimental data with accuracy $5 \div 10\%$. The agreement occurs at experimental plasma and laser parameters.

Acknowledgments

Authors are grateful to S. A. Babin and D. V. Churkin for kindly presented experimental data, for E. V. Podivilov and M. G. Stepanov for helpful discussions. The work is partially supported by the Program "Optical spectroscopy and frequency standards" from the Physical Science Department of Russian Academy of Sciences and the Government program NSh–2979.2012.2.

References

[1] Harris S E 1997 *Physics Today* **50** 36–42
[2] Fleischhauer M, Imamoglu A and Marangos J P 2005 *Rev. Mod. Phys.* **77**(2) 633–673
[3] Jain M, Xia H, Yin G Y, Merriam A J and Harris S E 1996 *Phys. Rev. Lett.* **77** 4326–4329
[4] Zibrov A S, Lukin M D and Scully M O 1999 *Phys. Rev. Lett.* **83** 4049–4052
[5] Zibrov A S, Lukin M D, Nikonov D E, Hollberg L, Scully M O, Velichansky V L and Robinson H G 1995 *Phys. Rev. Lett.* **75** 1499–1502
[6] Gao J Y, Yang S H, Wang D, Guo X Z, Chen K X, Jiang Y and Zhao B 1999 *Phys. Rev. A* **61** 023401
[7] Kash M M, Sautenkov V A, Zibrov A S, Hollberg L, Welch G R, Lukin M D, Rostovtsev Y, Fry E S and Scully M O 1999 *Phys. Rev. Lett.* **82** 5229–5232
[8] Budker D, Kimball D F, Rochester S M and Yashchuk V V 1999 *Phys. Rev. Lett.* **83** 1767–1770
[9] Mikhailov E E, Rostovtsev Y V and Welch G R 2003 *J. Mod. Opt.* **50** 2645 – 2654
[10] Knappe S, Wynands R, Kitching J, Robinson H G and Hollberg L 2001 *J. Opt. Soc. Am. B* **18** 1545–1553
[11] Novikova I, Matsko A B and Welch G R 2001 *Opt. Lett.* **26** 1016–1018
[12] Phillips D F, Fleischhauer A, Mair A, Walsworth R L and Lukin M D 2001 *Phys. Rev. Lett.* **86** 783–786
[13] Chaneliere T, Matsukevich D N, Jenkins S D, Lan S Y, Kennedy T A B and Kuzmich A 2005 *Nature* **438** 833–836
[14] Babin S A, Podivilov E V, Potapov V V, Churkin D V and Shapiro D A 2002 *Zh. Eksp. Teor. Fiz.* **121** 807–818 [JETP, **94** (4) 694–703 (2002)]
[15] Babin S A, Churkin D V, Podivilov E V, Potapov V V and Shapiro D A 2003 *Phys. Rev. A* **67** 043808
[16] Feldman B J and Feld M S 1972 *Phys. Rev. A* **5** 899–918
[17] Babin S A and Shapiro D A 1994 *Phys. Rep.* **241** 119–216
[18] Babin S A, Stepanov M G, Churkin D V and Shapiro D A 2004 *Zh. Eksp. Teor. Fiz.* **125** 1092–1099 [JETP, **98** (5) 953–959 (2004)]
[19] Belai O V, Shapiro D A and Yakovenko S N 2005 *Kvant. Elektr.* **25** 1027–1032
[20] Belai O V and Shapiro D A 2006 *Opt. Spektr.* **100** 971–978
[21] Stenholm S and Lamb W E 1969 *Phys. Rep.* **181** 618–635
[22] Feldman B J and Feld M S 1970 *Phys. Rev. A* **1** 1375–1396
[23] Rautian S G and Sobel'man I I 1963 *Zh. Eksp. Teor. Fiz.* **44** 934–945 [Sov. Phys. JETP **17** (3) 635-642 (1963)]
[24] Stenholm S 1978 *Phys. Rep.* **43** 151–221
[25] Bonch-Bruevich A M, Vartanyan T A and Chigir' N A 1979 *Zh. Eksp. Teor. Fiz.* **77** 1899–1909 [Sov. Phys. JETP **50** (5) 901–906 (1979)]
[26] Mak A A, Przhibelskii S G and Chigir N A 1983 *Izv. Acad. Nauk. SSSR, Ser. Fiz.* **47** 1976–1983
[27] Papademetriou S, Chakmakjian S and C R Stroud J 1992 *J. Opt. Soc. Am. B* **9** 1182–1188
[28] Smirnov G I and Shapiro D A 1979 *Zh. Eksp. Teor. Fiz.* **76** 2084–2093 [Sov. Phys. JETP, **49** (6) 1054–1058 (1979)]
[29] Belai O V and Shapiro D A 2004 *Pis'ma v Zh. Eksp. Teor. Fiz.* **79** 257–261 [JETP Lett., **79** (5) 203–207 (2004)]

Inverse scattering for inelastic atomic processes at thresholds

P A Macri[1] and R O Barrachina[2]

[1] Instituto de Investigaciones Físicas de Mar del Plata (IFIMAR)-Departamento de Física, FCEyN, Universidad Nacional de Mar del Plata-CONICET, Funes 3350, 7600 Mar del Plata, Argentina.

[2] Centro Atómico Bariloche and Instituto Balseiro (Comisión Nacional de Energía Atómica, CNEA, and Universidad Nacional de Cuyo, UNC), R8402AGP S. C. de Bariloche, Río Negro, Argentina

E-mail: macri@mdp.edu.ar, barra@cab.cnea.gov.ar

Abstract. We have observed resonance structures in detached electron spectra arising from fast collisions of Li^- with gas targets and from photo-detachment of Li^-. Using a generalization of the Wigner threshold law, we were able to fit the experimental data to a high degree of accuracy. Moreover, we were able to extract numerical values of different electron-atom parameters as the scattering length, the effective range of the potential and the number of supported bound states. Results for the scattering length obtained from this two differentiated experiments agree with each other within a maximum error of 2%.

1. Introduction

One of the ultimate purposes of studying scattering processes in Physics is the determination of the effective interactions between the components of a system. Usually these effective interactions are assumed known and cross sections and other observables are computed by means of scattering theory. Then, comparison with experiments let us know to what extend we succeeded in approximating all fundamental but otherwise intractable interactions or, at least, if we could manage to include all the necessary contributions to describe a given process. A more direct –although less used– method to determine the interactions would instead start from the experiments themselves. In elastic scattering, for example, phase-shift information at fixed angular momentum as function of energy (or alternatively, fixed-energy observations as a function of the angular momentum) can be used to feed a mathematical algorithm which would produce a potential that reproduces such observations [1, 2]. Potential ambiguities can be eliminated if additional data is known, such as the bound state energies and wave functions normalization factors. For inelastic processes, on the other hand, the multichannel inverse scattering theory is still under development [3, 4, 5]. Of particular interest are the energy regions where new (e.g., excitation, ionization or dissociation) channels open. At these thresholds, the interactions act over space scales greater than those usually found in Atomic or Molecular Physics [6]; and only a few low-energy parameters are expected to govern the cross sections. The energy dependence of cross sections near the threshold of opening reaction channels has been of great interest for years. A complete description of the two-body threshold behavior for the opening of an inelastic channel is given by $\sigma_{inel} \propto k^{2\ell+1}$ where k and ℓ stand for the relative linear and angular momentum, respectively. This formula was proposed by Wigner more than 50 years ago [7], showing that the longest-range forces determine the energy dependencies near threshold. To go further than the leading energy dependence of the Wigner threshold law, one may proceed through the well-known multi-channel effective-range expansions [8, 9]. In the simpler case of the elastic scattering for an $s-$wave, it simply reads [10]

$$k \cot \delta_0 = -1/a_0 + r_0 k^2/2 + ...,$$

where a_0 is the scattering length and r_0 is the effective range. When the interaction behaves as $V(r) \sim r^{-s}$ for large r, $k \cot \delta_0$ is not longer analytic around $k = 0$ and the effective range expansions have to be modified [11, 12, 13].

On the other hand, in practice Wigner's law usually describes the observations only over a very small range of energies above threshold. Rapid deviations have been observed for energies as low as a few meV when an electron and a neutral atom interact above threshold, and even for $s-$wave final states [14]. This shortcoming is of great theoretical and practical interest. In photodetachment of negative ions, for example, the knowledge of the threshold behavior is critical for the determination of electron affinities. At a first sight, it might seem that only the data that are closer to the threshold are useful. However, since the experimental observations in the close vicinity of the threshold usually have the worst signal-to-noise ratio, the proper procedure is to extrapolate the data down to the threshold, thereby making use of all the experimental information. It is therefore important to know the correct functional form that has to be employed in the extrapolation process. One can expect that an extended threshold law will depend on the interaction of the two-body fragments. In this sense, a threshold law in a broader energy range can be regarded as a useful tool to extract different interaction properties as the scattering length, the number of supported bound states, the long range behavior, etc.

In this work we propose an inverse scattering fitting technique based on the elastic $\ell-$wave Jost function of the final continuum electronic state. In section 2 we will analyze how the Wigner law has to be modified at threshold to take into account resonances effects and the presence of long range potentials. With this generalized threshold law, in section 3 we are able to represent observations on the full energy range for different kind of experiments and, moreover, to extract numerical values of interaction properties as the scattering length. Finally, in section 4 we present our conclusions and outlook for future works.

2. Theoretical description

During the last years [15, 16, 17, 18], we showed that the ℓ-wave Jost functions establish a clear link between a resonant structure in a multichannel collision and the low-energy elastic dispersion by the same final-state interaction. In this context, for a system composed of an arbitrary number of particles, two of them moving with small relative velocities with respect to the threshold of an opening channel, we propose to generalize the Wigner law as

$$\sigma_{inel} \propto k^{2\ell+1} \frac{g_{inel}(k)}{|f_\ell(k)|^2}, \tag{1}$$

In (1), $f_\ell(k)$ is the elastic ℓ-wave Jost function for the two-body system in the final state continuum, as given by the $r \to 0$ limit of the corresponding normalized radial wave function

$$\psi_{\ell,k}(r) \underset{r\to 0}{\approx} \frac{(kr)^{\ell+1}}{(2\ell+1)!!} \frac{1}{f_\ell(k)} .$$

Here ℓ represents the lowest final angular momentum allowed by selection rules, or the quantum number of a resonant wave at threshold. Meanwhile, the well behaved function $g_{inel}(k)$ accounts for the transition from the initial to the final state. If the potential is short-ranged and the system is far from a resonance, the Jost function is nearly constant and the prefactor of Eq. (1) provides the usual Wigner law for the threshold of an opening reaction channel [7]. However, in the near presence of resonances, the Jost function strongly depends on k and the denominator produces a sharp a deviation of σ_{inel} from the usual $k^{2\ell+1}$ dependence. At the same time, the small-k behavior of the $\ell-$wave Jost function is dominated by such parameters as the scattering length a_ℓ in the corresponding effective range expansion of the Jost function about $k = 0$ [19]. Precisely, this dependence of σ_{inel} on the parameters that are characteristic of the elastic two-body scattering at low energies is the key element of the proposed method for obtaining information about them from an inelastic cross section. The underlying idea is that the small-k behavior of the cross section in a inelastic collision reproduces many aspects of the elastic cross section for the same two-body system of relative momentum k. In principle, these parameters can be obtained by fitting the cross section at threshold with a suitable parametrization. For this reason Eq. (1) provides an alternative approach for the study of a two-particle continuum at extremely low-energies that might not be easily accessible in direct elastic scattering experiments. For example, for an $s-$wave scattering on a short-range potential, it is usually assumed that [13]

$$\frac{1}{|f_0(k)|^2} \propto \frac{1}{1+(a_0 k)^2}. \tag{2}$$

However, when the active interactions are long ranged, the Jost functions are not analytical at $k = 0$ (as occurs with the phase shifts δ_ℓ) and the above expression is not longer valid. In these cases a specifically constructed Jost function for the long-range potential is needed. Furthermore, even in the case of a short range interaction, the parameter accompanying k^2 in Eq. (2) is not necessarily equal to a_o^2, and it can even become negative [19].

As one of many systems for which Eq. (1) can be applied, in this work we will study threshold reactions for an electron-atom system, more specifically the electron-Lithium system. The four electrons of the Li^- ion make of it an interesting system to study, because of the significant role played by electron correlation in the binding of the outermost electron in this weakly bound system. In this case, the electron-atom interaction is dominated at large distances by the polarization potential $V(r) \sim -\beta/(2mr^4)$, where β is the dipole polarizability of the system. One decade ago we demonstrated that the validity of effective range expansions for potentials with large polarizabilities is limited to a very small energy range, and does not provide a reliable starting point for the parametrization of resonant structures in multichannel collisions [19]. It is important to point out that switching to a modified effective range theory [20] or to a quantum defect multichannel approach [21] does not guarantee a solution to this shortcoming, since they also include some approximations in powers of k. However, it can be shown that for the analysis of the Jost function at threshold, the intricate polarization effects of the electron-atom interaction can be cast in terms of a very simple model potential

$$V(r) = -\left[1 - \Theta(R-r)\left(1 - \frac{r^4}{R^4}\right)\right]\frac{\beta^2}{2mr^4},$$

where$\Theta(\ r)$ is the Heaviside's step function. For this model potential, the Jost functions have been analytically obtained [19]. Thus, instead of dealing with a very complicated generalized effective range expansion with many parameters, this analytical expression, with at most two free parameter, the range R and the polarizability β, can be used to fit the experimental data. Once these parameters are adjusted, the other coefficients in the expansion can be easily obtained [19]. For instance,

$$a_o = \frac{\beta}{2\sin(\beta/R)}\left(\frac{(2\beta/R)\cos(2\beta/R) - \sin(2\beta/R)}{(2\beta/R)\cos(\beta/R) - \sin(\beta/R)}\right).$$

3. Determination of electron - atom low energy interaction parameters

Negative alkali-metal ions as Li^- have valence electronic configurations ns^2. Thus the first excited state in Li^- photodetachment occurs as $\hbar\nu$+Li^- $(2s^2\,{}^1S)$ $\longrightarrow$$Li^-$ $(2p\,{}^2P)$ + ϵs. In this case, above threshold we consider

$$\sigma_{inel} = \sigma_{bg} + k\frac{g\left(k^2\right)}{\left|f_0\left(k;R\right)\right|^2}, \tag{3}$$

where σ_{bg} is an experimental background, and k is the relative momentum $k = \sqrt{E_\gamma - E_{th}}$ with E_γ and E_{th} the photon and threshold energy, respectively. Here, $g\left(k\right)$ is a function which does not depend sensibly on k. In this work we keep the polarizability fixed at the tabulated value [22] and vary only the polarization potential range R whose explicit dependence is showed in the Jost function $f_0\left(k;R\right)$. A quadratic expression on k for the function $g(k)$ was also considered. In figure 1, we show the fitting with our theoretical model of Li^- photodetachment data in the vicinity of the $Li(2^2P)$ threshold measured by Sandström *et al* [23]. As can be seen, our method provides a perfect match over the whole energy range, well beyond the scope of the Wigner law. We obtained a threshold energy $E_{th} = 2.465856(65)$ eV and a scattering length $a_0 = -26.0 \pm 2.7$ au. We note that this result does not agree with the value $a_0 = 61 \pm 6$ au obtained in [23] using a modified effective range theory [20] or a single channel version of a many channel quantum defect theory [21]. A number of reasons might explain this difference. First, in our fit we use all the available data points. In [23], on the other hand, the $s-$wave cross section was fitted to only two experimental data points near threshold. This $s-$wave cross section falls below the experimental data after an energy around $E = 2.47$ eV. It was assumed that this difference might be due to the neglect of the $d-$wave cross section. However, this assumption seems to be at odds with the fact that a continuum electron with only ~ 4 meV above threshold should be very-well described by an $s-$wave alone. Secondly, the presence of a virtual state close to the threshold advises to be extremely cautious on the fitting model employed in a region where, as it is the case here, the experimental cross section exhibits a sharp variation. In our case, we do not approximate the denominator of Eq.(3). Meanwhile in [23] a zero order approximation of $K_{20}^{P0}(\ell)$ was employed in the denominator of Eq.(A3).

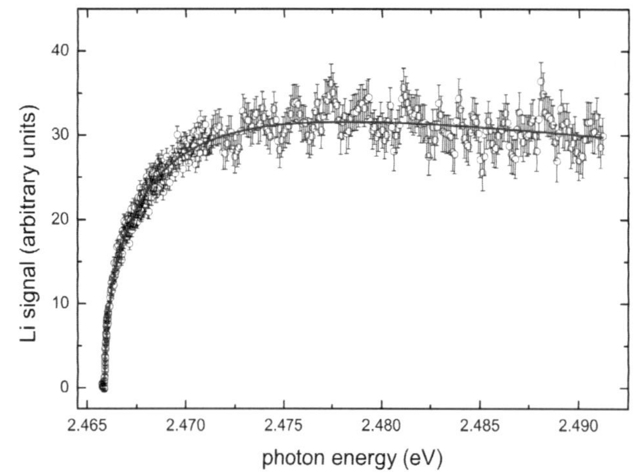

Figure 1. Cross section for the photodetachment of Li^- above the $Li(2p_1/2)$ threshold. Symbols: experiments of Ref.[23]. Solid line: Least-squared fit of experimental data using Eq.(3)

In this work, we also consider the electron loss experiment of 100 KeV Li^- ions colliding with He atoms measured by Lee et al [24]. Although very few bound states of negative ions exist, many unstable excited states have been found embedded in the continuum of the neutral atom. Hence, several channels might contribute to the measured double differential cross section (DDCS) in the forward direction. First, an electron is ionized leaving the Lithium atom in the Li($2p\ ^2P$) state as in the previous photodetachment experiment. But here the atom can also reach the Li($2s\ ^2S$) state. This could be achieved in a direct way or through the shape resonance given by the $Li^-(2s2p\ ^3P)$ state. Therefore, these two last contributions should be added coherently. We do this by using the standard method of Balashov [25] modified to include Wigner's law at threshold. We fit the electron loss to the continuum using the following model for the projectile frame DDCS

$$\frac{d\sigma^{Li}}{dE'd\Omega'} = k'\frac{d(\mathbf{k}')}{f_o\,(k';R)} + k'c(\mathbf{k}') + k^2\frac{a(\mathbf{k}')\epsilon + b(\ \mathbf{k}')}{1+\epsilon^2}, \tag{4}$$

where $\mathbf{k}'$ is the electron momentum in the projectile frame with modulus k' and angle θ'. The reduced energy $\epsilon = 2(E' - E_R)/\Gamma$ measures the departure from the resonant energy E_R and Γ is the half lifetime of the metastable resonant state. The functions $a(\mathbf{k}')$, $b(\mathbf{k}')$, $c(\mathbf{k}')$ and $d(\mathbf{k}')$ are supposed to be well-behaved. We expand them up to first order in k' and in terms of Legendre polynomials in $\cos\theta'$. In figure 2 we show the fitting of the experimental data of Ref. [24] employing Eq.(4) after transforming it to the laboratory frame and convoluting it with the reported detector resolution. As can be seen in the figure, the fitted model shows a very good agreement over the whole energy range. Furthermore, by varying the polarization potential range R in the $s-$wave Jost function $f_o\,(k';R)$, we managed to determine a scattering length $a_0 = -26.6 \pm 5$ au for the electron in the continuum of the atom in the Li($2p\ ^2P$) state. This result is in complete agreement with the value previously obtained from the photodetachment experiment. We also find that the electron in the continuum of the Li($2p\ ^2S$) atom has a resonance at $E_R = 0.0058(9)$ eV with half life $\Gamma = 0.0059(7)$ eV, corresponding to the configuration $Li^-(2s2p\ ^3P)$ which is in agreement with the experimental determinations [24].

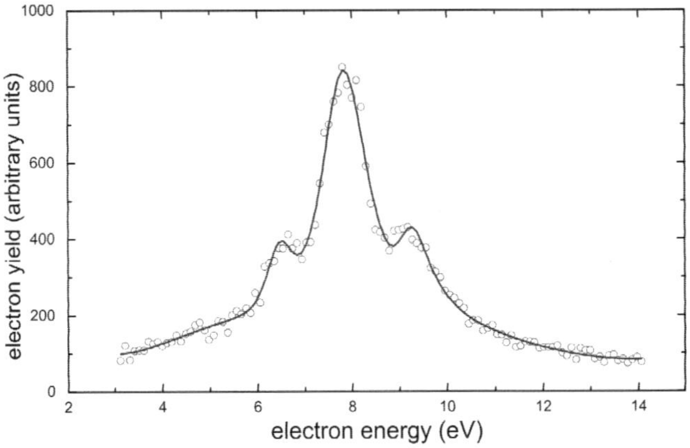

Figure 2. Electron-loss energy spectrum for 100 keV Li^- ions incident on He targets. Symbols: experiments of Ref.[24]. Solid line: Least-squared fit of experimental data using Eq.(1)

4. Conclusions

In this work, we presented a physical motivated model to fit inelastic atomic processes at threshold. This model is based on the determination of the polarization potential range R

which parameterizes the elastic $\ell-$wave Jost function for an electron ending in the continuum of a neutral atom. To test our model we considered experimental data corresponding to two different inelastic atomic processes leading to the same final electron-atom interaction, namely, the Li^- photodetachment in the vicinity of the $Li(2^2P)$ threshold [23] and the electron loss of 100 KeV Li^- ions colliding with He atoms [24]. Our fitting technique provides a kind of inverse scattering method giving values for the scattering lengths which impressively agree within 2 percent, even though they were extracted from two different kinds of experiment. These results seem to validate our technique and encourage us to pursue its future application to other systems, as for instance, Li^- photodetachment experiments with $Li(2s\ ^2S)$ in the final state close to the opening of the $Li(2p\ ^2P)$ threshold [14], and benchmark computations for electron Li(2s)→Li(2p) excitation close to threshold [26]. Finally, we want to emphasize that starting from the fitted value of the range R it would be possible, in principle, to determine the number of bound states of the negative ion. Furthermore, it would be also feasible to acquire the electron-atom dipole polarizability by treating β as another adjustable parameter.

Acknowledgments
We gratefully thank D. H. Lee, D. Hanstorp and D. J. Pegg for enlightened discussions and for providing us with the experimental data in tabulated form. This work was supported by the Consejo Nacional de Investigaciones Científicas y Técnicas (Grant no PIP 112-200801011269) and Universidad Nacional de Cuyo (grant 06/C348). P.A.M and R.O.B are also members of the Consejo Nacional de Investigaciones Cientficas y Tcnicas (CONICET), Argentina.

References

[1] Levitan B M 1984 *Inverse Sturm-Liouville Problems* (Moscow: Nauka)
[2] Chadan K and Sabatier P C 1989 *Inverse Problems in Quantum Scattering Theory* (New York: Springer).
[3] Baye D and Sparenberg J-M 2004 *J. Phys. A* **37** 10223
[4] Sparenberg J-M, Samsonov B F, Foucart F and Baye D 2006 *J. Phys. A* **39** L639
[5] Samsonov B F, Sparenberg J-M and Baye D 2007 *J. Phys. A* **40** 4225
[6] Sadeghpour H R, Bohn J L, Cavagnero M J, Esryk B D, Fabrikant I I, Macek J H and Rau A R P 2000 *J. Phys. B* **33** R93
[7] Wigner E P 1948 *Phys. Rev.* **73** 1002
[8] Ross M H and Shaw G L 1960 *Ann. Phys.* **9** 391
[9] Kermode M W 1967 *Nucl. Phys. A* **99** 605
[10] Bethe H A 1949 *Phys. Rev.* **76** 38
[11] O' Malley T F, Spruch L and Rosenberg L 1961 *J. Math. Phys.* **2** 491
[12] Levy B and Keller J 1963 *J. Math. Phys.* **4** 54
[13] Taylor J R 1972, *Scattering Theory* (New York: Wiley).
[14] Bae Y K and Peterson J R 1985 *Phys. Rev. A* **32** 1917
[15] Barrachina R O 1997 *Nucl. Instrum. Methods Phys. Res. B* **124** 198
[16] Macri P A and Barrachina R O 1998 *J. Phys. B* **31** 1303
[17] Macri P and Barrachina R O 2003 *Nucl. Instrum. Methods Phys. Res. B* **205** 543
[18] Barrachina R O and Macri P A 2004 *Few-Body Systems* **34** 175
[19] Macri P A and Barrachina R O 2002 *Phys. Rev. A* **65** 062718
[20] Fabrikant I I 1982 *Opt. Spectrosc. (USSR)* **53** 131
[21] Watanabe S and Greene C H 1980 *Phys. Rev. A* **22** 158
[22] Christophorou L.G. and Illenberger E 1993 *Phys. Lett. A* **173** 78
[23] Sandström J, Haeffler G, Kiyan I, Berzinsh U, Hanstorp D, Pegg D J, Hunnell J C, and Ward S J 2004 *Phys. Rev. A* **70** 052707
[24] Lee D H, Pegg D J and Hanstorp D 2000 *APS DAMOP meeting, Storrs.* Lee D H, Brandon W D, Hanstorp D and Pegg D J 1996 *Phys. Rev. A* **53** R635
[25] Balashov V 1972 *Sov. Phys. JETP* **36** 858
[26] Griffin D C, Mitnik D M, Colgan J and Pindzola M S 2001 *Phys. Rev. A* **64** 032718

A Stark broadening simulation using a renewal process for the electric microfield

R Hammami[a], D Boland[a], H Capes[a], M Christova[b], M Koubiti[a], Y Marandet[a], J Rosato[a], R Stamm[a]
[a]PIIM, Aix-Marseille Université and CNRS, centre Saint Jérôme, Marseille, 13397, France.
[b]Technical university Sofia, Sofia, Bulgaria

E-mail: roland.stamm@univ-amu.fr

Abstract. Stark broadening of atomic lines in plasmas is investigated by generating the electric microfield with a renewal process. In this model, the microfield is stepwise constant, with jumping times sampled from a waiting time distribution (WTD). Our model is a true simulation of the renewal process, using random number generators for generating different probability density functions (PDF). The use of an equilibrium static microfield PDF and an exponential waiting time distribution reproduces the so called model microfield method (MMM) results. The work presented is an application to the hydrogen Lyman-α line, for which we compare our renewal process model and ab initio simulations calculations, with the aim of studying the role of non exponential waiting time distributions.

1.Introduction

Different approaches are used today for computing line profiles broadened by Stark effect. Particle based approaches have been used in early developments [1,2,3], and have recently been revisited for an inclusion of multiple simultaneous emitter-perturbers interactions [4]. Computer simulations are commonly used today with a combination of particle simulations and a numerical integration of the Schrödinger equation [5,6]. Such ab initio simulations provide useful references for developing other less computer intensive models. Using an efficient procedure for mixing Stark components, the Frequency Fluctuation Model is an example of such an approach, available as a line shape code able to compute an arbitrary atomic or ionic line [7]. This approach may be viewed as an example of a model microfield method, since it uses statistical properties of the microfield, together with a procedure for its time evolution. Model Microfield Methods (MMM) have first been proposed for Stark broadening by Brissaud and Frisch in 1971 [8], using a stochastic process. In the last decades, the MMM has been developed and applied to a large range of astrophysical conditions [9]. We report here about an alternative approach which consists of a numerical simulation for the stochastic process for the plasma microfield following a proposal made by Frerichs [10]. In section 2, we recall that the calculation is performed in the framework of the renewal process model, using stepwise constant values of the microfield. The simulation technique is described in section 3, and our first results are presented and discussed in section 4.

Published under licence by IOP Publishing Ltd

2. Renewal process for the electric microfield

We consider a microfield which is stepwise constant, and jumps to a new value after a waiting time following a microfield dependent Waiting Time Distribution (WTD) v(t|E) for the first step. Also for the first step, the microfield modulus E is distributed according to a Probability Density Function (PDF) P(E). We assume that the plasma is isotropic, with each new value of the microfield vector characterized by its modulus and a random direction. In the following, the average over all directions of the field will be performed analytically on the evolution operator of the atom. For a stationary renewal process, the microfield PDF for all but the first step is changed to Q(E), and the WTD is changed to w(t|E). The relations between the first step and next steps PDF and WTD is obtained by writing the stationarity conditions, i.e. imposing that a measure starting at time t=0, is identical to a measure at an arbitrary time. The following relations then hold [11]:

$$Q(E) = \frac{\nu_E P(E)}{\langle \nu_E \rangle_S}, \tag{1}$$

and

$$\mathrm{w}(t|E) = -\frac{\dot{\mathrm{v}}(t|E)}{\nu_E}. \tag{2}$$

Symbol $<..>_S$ denotes a static average taken over P(E), and ν_E is a jumping frequency equal to the inverse of the time on a step averaged over the initial WTD. It should be noted that the relations (1) and (2) are different from the corresponding relations in reference [11], since they are not expressed with the WTD at time t=0, but with the microfield dependent jumping frequency. This more general formulation is found in the work of Haus and Kerr on stochastic processes [12]. From these relations, it is clear that the statistical properties of a stationary renewal process are determined by the microfield PDF P(E) and WTD v(t|E). In the case of a Markovian process, one recovers the Kangaroo process, with a WTD:

$$\mathrm{v}(t|E) = \mathrm{w}(t|E) = \mathrm{v}(E)\exp(-\mathrm{v}(E)t). \tag{3}$$

The choice of an exponential WTD has been made for most applications of the so called model microfield method, since it allows one to write a rather simple solution for the line shape. This choice implies that the stochastic process has a jumping rate which is constant over time, and it may be interesting to investigate if this has any impact on the line shape. Renewal processes are commonly used in survival analysis (for instance in the study of lifetime of manufactured products), and the first choice for the WTD is an exponential function. For processes which are supposed to be affected by a time dependent jumping rate, an alternative WTD is often given by a Weibull function:

$$\mathrm{u}(t|E) = \nu_E \alpha \{\nu_E t\}^{\alpha-1} \exp\left[-\{\nu_E t\}^{\alpha}\right], \tag{4}$$

where the shape parameter α is a real positive number, and ν_E is a microfield dependent jumping frequency equal to the inverse of the average time on a step $<t>$ obtained with the Weibull WTD. It is now necessary to calculate the new Q(E) and w(t|E) with this WTD. One should note here by considering equation (2), that since u(t|E) and w(t|E) are non-negative probability densities, u(t|E) must be a monotonically decreasing function [11]. This restricts the use of a Weibull distribution to values of $\alpha<1$, cases for which the jumping rate decreases over time. Using equations (1) and (2), it is possible to express Q(E) and w(t|E), but we first need to write the average time on a step as:

$$\langle t \rangle = \int_0^\infty t u(t|E)\,dt = \frac{\Gamma(1+\frac{1}{\alpha})}{\nu_E}, \tag{5}$$

where Γ is the gamma function. Using equation 4, the WTD w(t|E) for all but the first step may be written as:

$$w(t|E) = \nu_E \alpha \{\nu_E t\}^{\alpha-2} \left[1-\alpha+\alpha\{\nu_E t\}^{\alpha}\right] \exp-\{\nu_E t\}^{\alpha} \tag{6}$$

The survival probability of a microfield during a time interval t is obtained by the time integral of the WTD from t to infinity. Using this probability, it is then possible to write the microfield autocorrelation function for the renewal process Γ_{RP} as:

$$\Gamma_{RP}(t) = \int_0^\infty dE\,P(E)E^2 \exp-\{\nu_E t\}^{\alpha}. \tag{7}$$

This autocorrelation function for the renewal process should be chosen so as to reproduce the true function obtained from plasma kinetic theory. In the following , we will use the theoretical expression known from the work of Rosenbluth and Rostoker [13]:

$$C_p(\tau) = \langle \vec{\beta}(\tau).\vec{\beta}(0) \rangle = \frac{3}{\sqrt{\pi}}\left[\frac{r_0}{\lambda_D}\right]\frac{1}{\tau}\left[1+\tau^2-\sqrt{\pi}\,\tau\left(\tau^2+\frac{3}{2}\right)\exp(\tau^2)erfc(\tau)\right], \tag{8}$$

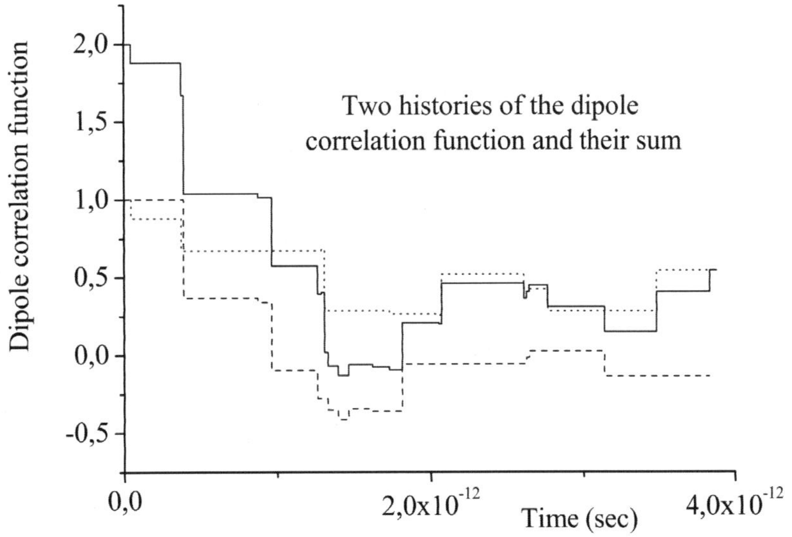

Figure 1. Sum of two histories for the dipole correlation

where $\tau=t\omega_p/2^{1/2}$ and $\omega_p=(4\pi Ne^2/m)^{1/2}$ is the plasma frequency for particles of mass m, density N and temperature T. We denoted the average distance between charged particles by r_0, and the Debye length by $\lambda_D=(kT/4\pi Ne^2)^{1/2}$. The dimensionless microfield $\beta=E/E_0$ is reduced in units of the Holtsmark field that we define as $E_0=e/r_0^2$. Equalizing equation (7) and (8) allows one to obtain the frequency ν_E as a function of the microfield.

3. Line shape calculation with a stochastic process

The line profile is obtained with a Fourier-Laplace transform of the dipole autocorrelation function $C_{dd}(t)$:

$$C_{dd}(t) = Tr\left\langle \vec{d}U^+(t).\vec{d}U(t)\rho \right\rangle . \tag{9}$$

In this expression, the trace is over atomic states, $\vec{d}$ is the atomic dipole operator, ρ the density matrix, and the brackets denote an average over the charged particles. The atomic evolution operator obeys to a Schrödinger equation:

$$i\hbar \frac{dU(t)}{dt} = [H_0 + V(t)]U(t), \tag{10}$$

where H_0 is the atomic Hamiltonian and $V(t) = -\vec{d}.\vec{E}(t)$ the dipolar interaction potential of the atom with the plasma microfield. As described in section 2, the microfield is assumed to follow a stepwise constant stochastic process with prescribed PDF and WTD, and we use the computer to generate microfield histories with pseudorandom numbers algorithms, associated to transformation or rejection methods. At each step, the microfield is constant, and so is the evolution operator. The evolution operator for n steps may be written as a product of constant operators in time intervals t_i-t_{i-1}:

$$U(t_n,0)=U(t_n,\, t_{n-1})\; U(t_{n-1},\, t_{n-2})\ldots\ldots U(t_1,0). \tag{11}$$

The evolution operator is thus obtained by an iteration from one step to the following:

$$U(t_{j+1},t_j) = \exp\left(\frac{i}{\hbar}\vec{d}.\vec{E}_j\,\Delta t\right)U(t_j), \tag{12}$$

where $\vec{E}_j$ is the constant microfield applied on the atom during $\Delta t = t_{j+1} - t_j$. With this approach the solution of the Schrödinger equation is obtained by multiplying matrices representing $U(t_j - t_{j-1})$, a procedure which is much more efficient on the computer than those used for ab initio simulation. Summing finally over a large number of such time histories of the microfield, it is possible to obtain a smooth averaged evolution operator, since every history has a different time grid. This calculation may be applied to both the ionic and electronic components, using for instance the joint jump approximation [10]. Our first calculations using this simulation of the stochastic process have been applied to the ionic component only. The effect of electron broadening may be retained in our calculation with an impact electronic operator.

4. Results

Calculations performed for this work have been applied to the Lyman α line of hydrogen, neglecting the fine structure. We first reproduce the calculations of the MMM, with a PDF P(E) calculated by Hooper [14], and an exponential WTD. An example of the sum of two histories is shown on figure 1. Repeating this summation procedure over several thousands microfield histories, it is possible to obtain a smooth dipole correlation function. We have first studied the angular average over the microfield angles. It is possible to generate for each field, two microfield angles, and retain the angular dependence for each $U(t_i - t_{i-1})$. After an average over a large number of field histories, we have checked that such a calculation leads to the same averaged evolution operator as a calculation performed using an analytical angular average on each $U(t_j - t_{j-1})$. Calculating then the dipole correlation function for a density N=10^{17} cm^{-3}, and a temperature of 10^4 K, and performing a Laplace-Fourier transform, our calculation reproduces the profiles of the MMM for the Lyman α line in the case of ion broadening and impact electrons [15]. We have plotted on figure 2 the dipole autocorrelation function for Lyman α broadened by ions only and calculated with our technique (dashed line), and have compared it to an ab initio simulation (solid line). It can be seen that a standard MMM calculation applied to ions alone is significantly different from the ab initio simulation. This is a strong indication that the stochastic process may be improved for the ionic component. There is however no clear indication of how this should be done, and our proposal is to change the WTD as explained in section 2. The work underway consists in using the four functions P(E), Q(E), u(t|E) and w(t|E) in the case of a Weibull WTD, in an attempt to improve the agreement between our present technique and the ab initio simulation.

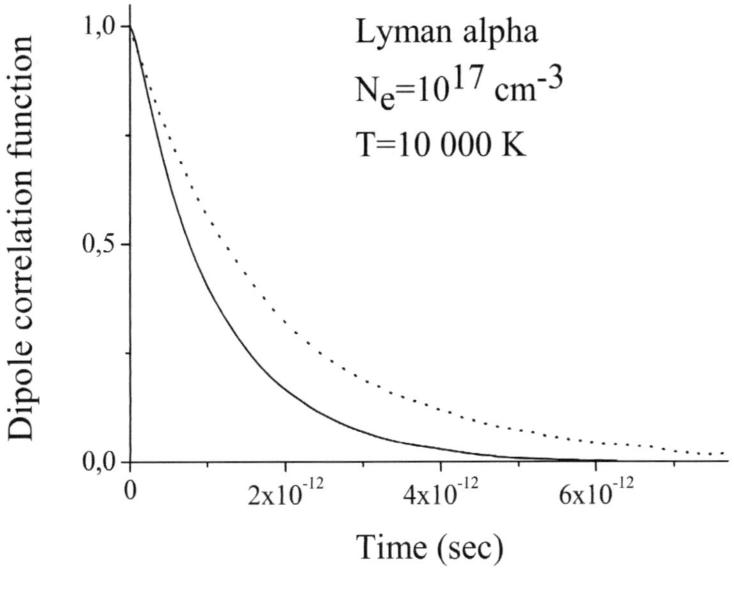

Figure 2. Lyman α dipole correlation function calculated with our simulation of the stochastic model (dotted line), and an ab initio simulation (solid line)

5.Conclusion

We have developed a simulation calculation with a renewal process for the plasma microfield, coupled to an efficient algorithm for calculating the dipole autocorrelation function of the emitter. Revisiting the basic assumptions for the stochastic process applied to Stark broadening, we have proposed the use of a Weibull WTD for the test of a time dependent jumping process. We have derived expressions for the four probability density functions required for the line shape calculation, as well as a procedure for obtaining the microfield dependence of the jumping frequency. The first calculations have shown the applicability of our model for reproducing the standard MMM calculations for the Lyman α line. Applied to ion broadening only, and compared to ab initio simulations, our calculations are a clear indication that the stochastic process could be improved. Work is in progress for using the Weibull or other non exponential WTD

Acknowledgments
This work was carried out within the framework of the European Fusion Development Agreement and the French Research Federation for Fusion Studies. It is supported by the European Communities under the contract of Association between Euratom and CEA. The views and opinions expressed herein do not necessarily reflect those of the European Commission. We also acknowledge the support of the French National Research Agency (contract ANR-11- BS09-023, SEDIBA).

References

[1] Griem H R 1964 *Plasma Spectroscopy* (New York: McGraw-Hill)
[2] Voslamber D 1969 *Z. Naturforsch.* **24a** 1458–72
[3] Smith E W, Cooper J and Vidal C R 1969 *Phys. Rev.* **185** 140–51
[4] Rosato J, Capes H and Stamm R, this volume
[5] Stamm R and Voslamber D 1979 *J. Quant. Spectr. Rad. Transfer* **22** 599-609
[6] Stambulchik E and Maron Y 2010 *High Energy Density Physics* **6** 9-14
[7] Talin B, Calisti A, Godbert L, Stamm R, Lee R W and Klein L 1995 *Phys. Rev. A* **51** 1918-28
[8] Brissaud A and Frisch U 1971 *J. Quant. Spectrosc. Radiat. Transfer* **11** 1767-83
[9] Stehlé C 1994 *Astron. Astrophys. Suppl. Ser.* **104** 509-27
[10] Frerichs M 1989 *Z. Phys. D - Atoms, Molecules and Clusters* **11** 315-21
[11] Seidel J 1980 *Z. Naturforsch.* **35a** 679-89
[12] Haus J W and Kerr K W 1987 *Physics Reports* **150** 263-406
[13] Rosenbluth M and Rostoker N 1962 *Phys. Fluids* **5** 776-86
[14] Hooper C F 1968 *Phys. Rev.* **165** 215-22
[15] Stehlé C and Hutcheon R 1999 *Astron. Astrophys. Suppl. Ser.* **140** 93-7

Modeling of hydrogen Stark line shapes with kinetic theory methods

J Rosato, H Capes and R Stamm

Laboratoire PIIM, UMR 7345 Université d'Aix-Marseille / CNRS, Centre de Saint-Jérôme, Case 232, F-13397 Marseille Cedex 20, France

E-mail: joel.rosato@univ-amu.fr

Abstract. The unified formalism for Stark line shapes is revisited and extended to non-binary interactions between an emitter and the surrounding perturbers. The accuracy of this theory is examined through comparisons with ab initio numerical simulations.

1. Introduction

Since a few years ago, modeling efforts have been carried out in order to provide analytical formulas for the Stark broadening of the first Lyman lines of hydrogen, for opacity simulation purposes in magnetic fusion research [1]. One of the most challenging issues concerns the description of the ion dynamics. At densities and temperatures relevant to tokamak edge and divertors, the impact approximation is not far from being valid for ions, with typically $b_W/r_0 \leq 0.5$ for Ly-α and Ly-β with b_W, r_0 being the Weisskopf radius and the mean interparticle distance. At such regimes, an approach based on first principles like the unified theory [2,3] is particularly suitable for the description of the ions. In this work, we present an extension of the unified theory applicable beyond the binary limit and suitable for numerical calculations. The theoretical framework is presented in Sec. 2, the model is presented in Sec. 3, and a practical formula for calculations is derived in Sec. 4. Comparisons with numerical simulations are done in Sec. 5.

2. The unified theory

The theoretical development done hereafter follows a previous work reported in [4]. We focus on hydrogen line broadening due to ions. A spectral line shape is determined from the Laplace transform of the average evolution operator $\Phi_0(t)$:

$$I(\omega) = \frac{1}{\pi}\mathrm{Re}\sum_{\vec{\varepsilon}}\left\langle\left\langle \vec{d}\cdot\vec{\varepsilon}\left|\tilde{\Phi}_0(-i\omega)\right|\rho\vec{d}\cdot\vec{\varepsilon}\right\rangle\right\rangle. \quad (1)$$

Here the double ket notation for Liouville space $\left|...\right\rangle\rangle$ has been used, ρ is the atomic density operator, $\vec{d}\cdot\vec{\varepsilon}$ is the dipole projected onto the polarization vector, and $\tilde{\Phi}_0$ is the Laplace transform of the evolution operator averaged over the perturber trajectories (classical path assumption). The latter obeys a Schödinger-like equation (we set $\hbar = 1$)

$$\left(\frac{\partial}{\partial t} + iH_0\right)\Phi_0 = -i\mathcal{N}\int d1V(1)\Gamma_1(1), \quad (2)$$

with the initial condition $\Phi_0(t = 0) = 1$. H_0 is the Liouvillian accounting for the atomic energy level structure and the right-hand side denotes a source term accounting for interactions with the perturbers. The integral is performed over the phase space coordinates $\vec{r}_1, \vec{v}_1$ of one perturber ($d1$ is a shortcut notation for $d^3r_1d^3v_1$), $V(1) = -\vec{d}\cdot\vec{E}(\vec{r}_1)$ denotes the Stark term resulting from the electric field due to the perturber under consideration, and $\Gamma_1(1,t)$ is the atom - perturber 1 correlation function. The correlation function is defined from a cluster representation technique similar to that used in kinetic theory (e.g. [5]). In the most simplified treatment of the plasma, the perturbers can be described as independent particles following straight-line trajectories (weak coupling approximation) and the medium can be assumed uniform and stationary. In this framework, the evolution of $\Gamma_1(1,t)$ is governed by the following equation

$$\left\{\frac{\partial}{\partial t} + i\left[H_0 + V(1)\right] + \vec{v}_1\cdot\frac{\partial}{\partial \vec{r}_1}\right\}\Gamma_1(1) = -if_1(1)V(1)\Phi_0 - i\mathcal{N}\int d2V(2)\Gamma_2(1,2)\,, \tag{3}$$

with the initial condition $\Gamma_1(1,t = 0) = 0$. $f_1(1)$ is the one-particle phase space distribution. The integral in the right-hand side involves $\Gamma_2(1,2,t)$, the atom - particle 1 - particle 2 correlation function. Originally, the unified theory was developed for applications in cases where correlations between the collisions are negligible. Γ_1 is calculated assuming $\Gamma_2 = 0$. In this framework, the line shape can be written as a sum of generalized Lorentzian functions, whose widths and shifts are frequency-dependent and given by matrix elements of the collision operator

$$K(s) = \mathcal{N}\int_0^\infty dte^{-st}\int d1V(\vec{r}_1 + \vec{v}_1t)Q(1,t)V(\vec{r}_1)f_1(1)\,, \tag{4}$$

with $s = -i\omega$. Here $Q(1,t)$ is the propagator of the atom under the influence of one perturber. It obeys the time-dependent Schrödinger equation

$$\left\{\frac{\partial}{\partial t} + i\left[H_0 + V(\vec{r}_1 + \vec{v}_1t)\right]\right\}Q(1,t) = 0\,, \tag{5}$$

with the initial condition $Q(1,t = 0) = 1$, and it is proportional to a time-ordered exponential (Dyson series)

$$Q(1,t) = e^{-iH_0t}\mathcal{T}\exp\left[-i\int_0^t d\tau e^{iH_0\tau}V(\vec{r}_1 + \vec{v}_1\tau)e^{-iH_0\tau}\right]. \tag{6}$$

3. Multiple collisions

We go beyond the binary approximation by assuming Γ_2 finite. We consider the following truncature relation, similar to that proposed by Kirkwood in kinetic theory [6]

$$\tilde{\hat{\Gamma}}_2(\vec{k}_1,\vec{v}_1,\vec{k}_2,\vec{v}_2;s) = \tilde{\hat{\Gamma}}_1(\vec{k}_2,\vec{v}_2,s + i\vec{k}_1\cdot\vec{v}_1)\tilde{\Phi}_0^{-1}(s + i\vec{k}_1\cdot\vec{v}_1)\tilde{\hat{\Gamma}}_1(\vec{k}_1,\vec{v}_1;s) + 1 \leftrightarrow 2\,. \tag{7}$$

This expression can be obtained formally from solving the evolution equation for Γ_2 perturbatively, assuming V small [4]. Here, the hat denotes Fourier transform with respect to the $\vec{r}$ -variable and $1 \leftrightarrow 2$ denotes permutation between the variables. This closure relation provides a closed nonlinear evolution equation for Γ_1. A practical simplification for numerical calculations is provided by neglecting the second term in Eq. (7). We obtain a collision operator similar to that obtained within the binary approximation [Eq. (4)] in terms of an effective propagator $Q_{eff}(1,t)$ accounting for multiple collisions

$$K(s) = \mathcal{N}\int_0^\infty dte^{-st}\int d1V(\vec{r}_1 + \vec{v}_1t)Q_{eff}(1,t)V(1)f_1(1)\,, \tag{8}$$

$$\left\{\frac{\partial}{\partial t} + i\left[H_0 + V(\vec{r}_1 + \vec{v}_1t)\right]\right\}Q_{eff}(1,t) + \int_0^t d\tau M(\tau)Q_{eff}(1,t-\tau) = 0\,. \tag{9}$$

$Q_{eff}(1,t)$ can be interpreted as describing the evolution of the atom under the influence of one collision represented by the interaction term V, given a set of collisions occurring in its past history. These collisions are characterized by the kernel $M(t)$. This term is identical to the inverse Laplace transform of the collision operator. In contrast with the binary case, Eq. (8) is not a closed expression for the collision operator because the evolution equation of Q_{eff} involves the collision operator, through the M term. In practice, a calculation should be done by iterations. A further simplification is provided by assuming $M(t) \equiv K_0\delta(t)$ with $K_0 = K(-i\omega_0)$ in Eq. (9). This leads to a simple relation between Q_{eff} and Q

$$Q_{eff}(1,t) = e^{-K_0 t} Q(1,t) \tag{10}$$

As in the binary case, the multiple integral in Eq. (8) can be evaluated with the use of cutoffs accounting for the oscillatory behavior of Q_{eff} at small impact parameters (strong collisions). The special case of hydrogen can be treated analytically following the same method as that used for obtaining the exact solution of the binary problem (e.g. [7] and Refs. therein). In the following we consider the first approach and derive an analytical formula for the collision operator suitable for numerical calculations.

4. An analytical formula for the collision operator

We assume that collisions with a small impact parameter give no significant contribution to the integrals in Eq. (8). On the other hand, we consider that the atomic wavefunction is not sensitive to perturbers located at large distance. In this framework, an approximation for the evolution operator is provided by

$$Q(1,t) \approx \begin{cases} 0 & \text{if } r_1 < r_m \\ 1 & \text{otherwise} \end{cases}, \tag{11}$$

where r_m is a characteristic radius that remains to be determined. A possible approach consists in estimating the matrix elements of the integral $\int_0^t d\tau V(\vec{r}_1 + \vec{v}_1\tau)$ that appears in the Dyson series and comparing them to unity [8]. Such a procedure yields the following relation

$$r_m = \begin{cases} b_W & \text{if } b_W \le v_1 t \\ \sqrt{b_W v_1 t} & \text{otherwise} \end{cases}, \tag{12}$$

where $b_W = n^2/m_e v_1$ is the Weisskopf radius. This relation exhibits the separation between the short and long time regimes, which correspond to incomplete and complete collisions, respectively. The interaction term has been estimated as $V \sim H(r_m/v_1 - \tau)b_W v_1/r_m^2$, H being the Heaviside function. From the approximation Eq. (11), the space integral in Eq. (8) can be calculated analytically, in a fashion similar to what is done for the electric field autocorrelation function (e.g. [9]). We assume a Coulomb field and retain Debye screening through a cut-off at λ_D. Hence,

$$\int d^3r_1 V(\vec{r}_1 + \vec{v}_1 t) Q_{eff}(1,t) V(1) = \frac{4\pi (b_W v)^2}{3(n^2 e a_0)^2} \vec{d} \cdot e^{-K_0 t} \vec{d} \times \begin{cases} \left(\frac{1}{r_m} - \frac{1}{\lambda_D}\right) & \text{if } v_1 t \le r_m \le \lambda_D \\ \left(\frac{1}{v_1 t} - \frac{1}{\lambda_D}\right) & \text{if } r_m \le v_1 t \le \lambda_D \\ 0 & \text{otherwise} \end{cases}. \tag{13}$$

The time integral in Eq. (8) can be evaluated analytically. We focus on Ly-α and we choose a system of units such that $\lambda_D = 1 = v_T$ (thermal velocity). The matrix element $\gamma(s) = \langle 210|K(s)|210\rangle$, which provides the line width and shift, is given by

$$\gamma(s) = \frac{3\pi N b_{WT}^2}{4}\left\{\begin{aligned}&\left[\operatorname{erf}(b_{WT}) - \frac{2b_{WT}}{\sqrt{\pi}}e^{-b_{WT}^2}\right]\left[\sqrt{\frac{\pi}{s' b_{WT}}}\operatorname{erf}\left(\sqrt{\frac{s'}{b_{WT}}}\right) - \frac{1-e^{-s'/b_{WT}}}{s'}\right]\\ &+\int_{b_{WT}}^{\infty} dv F(v)\left[\begin{aligned}&\sqrt{\frac{\pi}{s' b_{WT}}}\operatorname{erf}\left(\frac{\sqrt{s' b_{WT}}}{v}\right) + \frac{1}{v}E_1\left(\frac{s' b_{WT}}{v^2}\right)\\ &-\frac{1}{v}E_1\left(\frac{s'}{v}\right) - \frac{1-e^{-s'/v}}{s'}\end{aligned}\right]\end{aligned}\right\}, \quad (14)$$

where b_{WT} is the Weisskopf radius evaluated at the thermal velocity, $s' = -i\Delta\omega + \gamma(0)$ with $\Delta\omega$ being the frequency detuning, $F(v) = 4v^2\exp(-v^2)/\sqrt{\pi}$ is the Maxwellian velocity distribution function, and erf, E_1 are the error and first exponential integral functions, respectively. Equation (14) is suitable for numerical calculations. The binary approximation ("usual" unified theory) corresponds to setting $s' \equiv -i\Delta\omega$, and the complete collision approximation ("impact" theory) corresponds to setting $s' \to 0$. An explicit calculation for $b_{WT} << \lambda_D$ yields

$$\gamma(s) \approx \frac{3\sqrt{\pi} N b_{WT}^2}{4}\left[2 + E_1(b_{WT}{}^2)\right], \quad (15)$$

which corresponds to the standard result for hydrogen [10].

5. Application

We have applied the collision operator formula Eq. (14) to calculations of hydrogen line shapes in ideal cases. Figure 1 and figure 2 present a plot of Ly-α broadened due to ions at $b_{WT}/r_0 = 0.01$ and 0.5, respectively, assuming $r_0/\lambda_D = 0.5$ in both cases. The calculations have been carried out using the unified theory (binary approximation) and the non-binary model (referred to as "Q_{eff}-model"). A numerical result from an ab initio simulation code is also shown in the figure. This code uses the "collision-time technique", which has been developed and designed in the past so as to provide correct statistics at near-impact regimes where $b_{WT}/r_0 << 1$ [11]. Both the binary and non-binary models are in a good agreement with the simulation in the first case (the small deviation visible at the line center falls within the typical uncertainty range of the simulation). This confirms the applicability of the unified theory at conditions where the atom - perturber interactions are supposed binary, typically when $b_{WT}/r_0 << 1$. It also illustrates that the non-binary model includes the binary limit. In contrast, the second case shows a strong discrepancy between the unified theory and the simulation. The binary assumption leads to an overestimate of the line width and predicts a different structure, with a dip at the line center. This deviation stems from an improper use of the unified theory. The non-binary model yields a better result, with no dip and an overestimate of the width no larger than 25%. A plot of $|I(\Delta\omega = 0) - I_{sim}(\Delta\omega = 0)|/I_{sim}(\Delta\omega = 0)$ in terms of b_{WT}/r_0 is presented in Fig. 3 and shows the trend of the deviation in terms of the non-binarity.

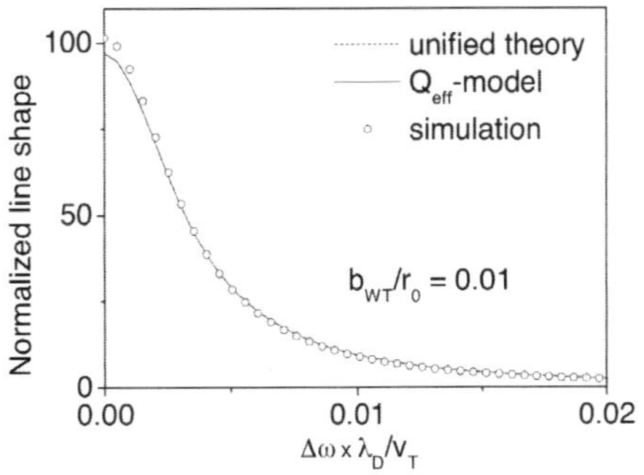

Figure 1. Plot of the Ly-α profile obtained within the unified theory (dashed line), the non-binary Q_{eff}-model (solid line) and a simulation method (circles), at conditions where the atom - perturber interactions are binary. The analytical models are both in a good agreement with the simulation.

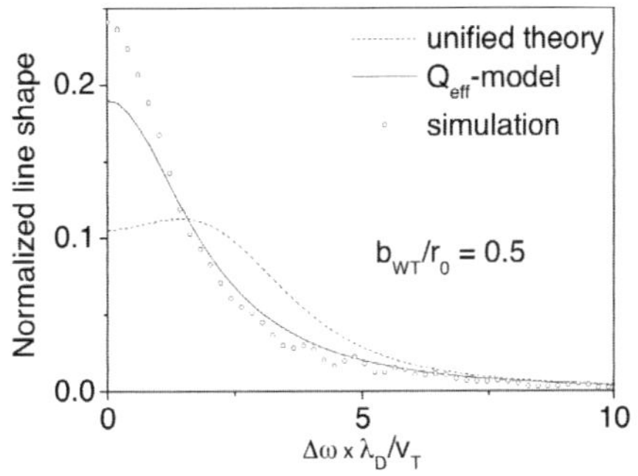

Figure 2. Plot of the Ly-α profile at conditions where the interactions are no longer binary. The Q_{eff}-model gives a correct estimate of the line shape whereas using the unified theory yields an overestimate of the width and predicts an incorrect structure.

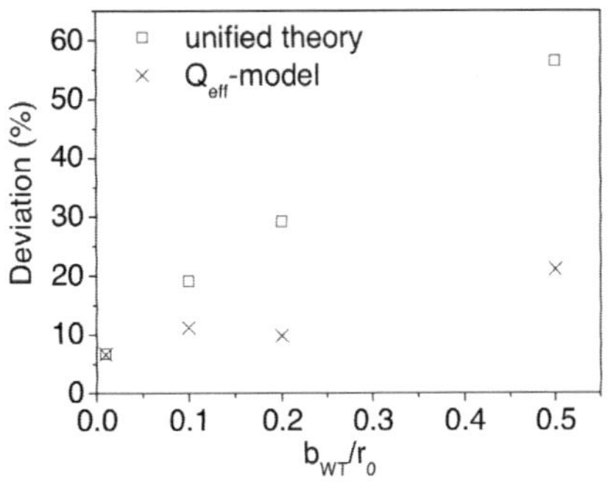

Figure 3. Plot of the deviation to the simulation results. The Q_{eff}-model provides a good estimate of the line shape, even for a value of b_{WT}/r_0 as large as 0.5, whereas the unified theory gives a significant overestimate of the width as soon as b_{WT}/r_0 exceeds 0.1.

6. Conclusion

We have demonstrated the possibility of retaining non-binary effects in Stark broadening models through a simple approach inspired from kinetic theory. The average atomic evolution operator is calculated from a hierarchy of equations similar to that of Bogoliubov-Born-Green-Kirkwood-Yvon (BBGKY), able to account for many-particle correlations with arbitrary accuracy. When one single collision is considered, the correlations result in an alteration of the atom evolution due to cumulative collisions [12]. The model developed in this work provides results in a very good agreement with numerical simulations. The method can be applied either to ions or electrons, and it can be generalized to isolated lines. Also, the kinetic theory treatment is transposable to quantum plasma [13]. Further work would address the study of multiple correlations in the presence of a strong magnetic field like in tokamak plasmas. In this case the Zeeman effect can be important, so that the detailed structure of the levels should be retained.

Acknowledgements

This work was carried out within the framework of the European Fusion Development Agreement and the French Research Federation for Fusion Studies. It is supported by the European Communities under the contract of Association between Euratom and CEA. The views and opinions expressed herein do not necessarily reflect those of the European Commission.

References

[1] Rosato J, Reiter D, Kotov V, Marandet Y, Capes H, Godbert-Mouret L, Koubiti M and Stamm R 2010 *Contrib. Plasma Phys.* **50** 398–403
[2] Voslamber D 1969 *Z. Naturforsch.* **24a** 1458–72
[3] Smith E W, Cooper J and Vidal C R 1969 *Phys. Rev.* **185** 140–51
[4] Rosato J, Capes H, Marandet Y, Mekkaoui A and Stamm R (in press) *Transport Theor. Stat. Phys.*
[5] Balescu R 1997 *Statistical Dynamics – Matter Out of Equilibrium* (London: Imperial College Press)
[6] Kirkwood J G 1935 *J. Chem. Phys.* **3** 300–13
[7] Greene R L, Cooper J and Smith E W 1975 *J. Quant. Spectrosc. Radiat. Transfer* **15** 1025–36
[8] Stamm R and Voslamber D 1975 *C. R. Acad. Sc. Paris Série B* **t. 280** 223–6
[9] Brissaud A, Goldbach C, Léorat J, Mazure A and Nollez G 1975 *J. Phys. B: At. Mol. Phys.* **9** 1129–46
[10] Griem H R, Kolb A C and Shen K Y 1959 *Phys. Rev.* **116** 4–16
[11] Hegerfeldt G and Kesting V 1988 *Phys. Rev. A* **37** 1488–96
[12] Capes H and Voslamber D 1972 *Phys. Rev. A* **5** 2528–36
[13] Voslamber D 1972 *Phys. Lett.* **40A** 266–8

Dicke Narrowing Effect for $r^{-\nu}$-type Collisional Potential

Piotr Wcisło, Roman Ciuryło

Institute of Physics, Nicholas Copernicus University, Grudziadzka 5/7, 87-100 Toruń, Poland

E-mail: `piotr.wcislo@fizyka.umk.pl`

Abstract. In this paper we present results of spectral profile shape calculations for the $r^{-\nu}$-type collisional potential. The attention is drown to the Dicke narrowing, however we also refer to the speed-dependent effects. Our considerations were performed within the impact approximation. It was also assumed that dephasing and velocity-changing contributions to the collisions are statistically independent. We found that changing the potential type the collisional effects influence on the line shapes can vary by more then 10%. Therefore it is relevant not only for the ultraaccurate spectroscopy, where model accuracy up to 10^{-6} is required, but also in more common applications.

1. Introduction

There is a need for application of *ab initio* collision operators in the spectral line shapes modeling [1] including Dicke narrowing effect [2, 3] and speed-dependent effects [4, 5]. The class of $r^{-\nu}$-type collisional potentials applied to the line shape problem was introduced by Blackmore [1]. The billiard ball model (hard spheres collisions) based on results obtained by Lindenfeld [6] was applied to the case including speed-dependent collisional broadening and shifting [7]. We extended in Refs.[8, 9] the approach described in Refs. [7, 10] to the class of $r^{-\nu}$-type collisional potentials in the way similar to Blackmore [1]. Expressions for the matrix elements of the velocity-changing collision operators given by Lindenfeld and Shizgal [12] have been used. Contrary to conclusions in [1] we have found significant differences between profiles calculated for various potential powers ν, especially when perturbers are much heavier then absorbers and collisional broadening is negligible. It was shown that, independently from potential power ν, our results converge to the Galatry profile [13] in the soft-collision limit, where perturber to absorber mass ratio α goes to zero. The pure Dicke narrowing case [1] was compared with the case, where speed-dependent collisional broadening as well as Dicke narrowing were taken into account [7]. It was carefully analysed in which part of collisional operator the variation of the potential power ν has the biggest influence on the spectral line shape.

2. Model

In our consideration the attention is drown to the system of absorbers highly diluted in a perturber bath. By absorbers we mean particles that resonantly or nearly resonantly interact with the light (here the interaction is approximated with two-level structure), while by perturbers we mean some other particles which are not influenced by the light. It is well known that to find the stationary state of such system taking into account both absorber-light interaction and absorber-perturber collisions one has to solve transport/relaxation equation [1, 10, 11], which can be transformed to the following form [8, 9, 10]

$$1 = -i(\omega - \omega_0 - \vec{k}\vec{v})h(\omega, \vec{v}) - (\hat{S}^f_{VC} + \hat{S}^f_D)h(\omega, \vec{v}), \tag{1}$$

where ω and $\vec{k}$ are frequency and wave vector of absorbed light, respectively. Unperturbed transition frequency is denoted by ω_0 and $\vec{v}$ is an absorber velocity. The $h(\omega, \vec{v})$ function, which related to optical coherence, allows us to calculate the normalized line shape profile

$$I(\omega) = \frac{1}{\pi}\mathrm{Re}\int f_m(\vec{v})h(\omega, \vec{v})d^3\vec{v}, \tag{2}$$

where $f_m(\vec{v}) = (\sqrt{\pi}v_m)^{-3}e^{(-v/v_m)^2}$ is a Maxwellian velocity distribution, $v_m = \sqrt{2k_BT/m_1}$ is the most probable absorber speed, T is a temperature, k_B is the Boltzmann constant and m_1 is an absorber mass.

The important task is to provide a description of absorber-perturber collisions for the repulsive power low potential $V(r) = V_0(d/r)^\nu$. Since it is assumed that dephasing and velocity-changing contributions to collisions are statistically independent, see Refs. [14, 15, 16, 17, 18], the collisional operator in Eq. (1) can be written as a simple sum of dephasing operator $\hat{S}^f_D$ and velocity-changing operator $\hat{S}^f_{VC}$. Following Bermann as well as Ward *et al.* [4, 5] the dephasing operator $\hat{S}^f_D$ for the $r^{-\nu}$-type collisional potential can be written as a confluent hypergeometric function of the first kind [19]

$$\hat{S}^f_D = (\Gamma + i\Delta)(1+\alpha)^{-\frac{\nu-3}{2\nu-2}}{}_1F_1\left(-\frac{\nu-3}{2\nu-2}, \frac{3}{2}, -\alpha x^2\right), \tag{3}$$

where Γ and Δ are average collisional width and shift, respectively. The normalized absorber velocity is defined as $x = v/v_m$. It should be noted that the formula (3) has been derived within the straight line trajectories approximation.

To handle the velocity-changing contribution to collisions one has to take the Boltzmann collisional operator [1, 12, 20]

$$\hat{S}^f_{VC}h(\omega, \vec{v}) = n_2\int d^3\vec{v}_2\int d\tilde{\Omega}\left(\frac{d\sigma}{d\tilde{\Omega}}\right)|\vec{v} - \vec{v}_2|f_{m_2}(\vec{v}_2)\left(h(\omega, \vec{v}') - h(\omega, \vec{v})\right), \tag{4}$$

where $d\sigma/d\tilde{\Omega}$ is a differential cross section, $\vec{v}_2$ is a perturber velocity, $f_{m_2}(\vec{v}_2)$ is a Maxwellian velocity distribution of perturbers, n_2 is a concentration of perturbers and $d\tilde{\Omega}$ is a scattering solid angle. The absorber velocity after collision is denoted by $\vec{v}'$.

Equation (1) together with Eqs. (3) and (4) constitute an integral equation for the $h(\omega, \vec{v})$ function. According to our knowledge the most efficient method of solving this problem is to expand the operators and unknown function $h(\omega, \vec{v})$ in the Burnett functions basis [7, 12]. This technique has been already applied to analyse experimental spectral line shapes [21, 22, 23, 24]. That allows to convert the integral equation into an infinite set of self-coupled linear equations, which, however, may be truncated. Matrix elements of $\hat{S}^f_D$ can be easily evaluated by performing one dimensional numerical integration [7]. More complex task is to evaluate S^f_{VC} matrix elements where more sophisticated numerical techniques have to be applied [12] (in the case of $\nu = \infty$ there is an analytical expression[12]).

3. Results

The model introduced in the previous section is characterized by a six physical parameters: the Doppler width $\omega_D = kv_m$ (half width at $1/e$ of maximum), the collisional width Γ (half width at half maximum - HWHM), the frequency of the velocity changing collisions ν_{diff}, the perturber to absorber mass ratio $\alpha = m_2/m_1$, the exponent from the power-low potential ν and the spectral detuning $\Delta\omega = \omega - \omega_0$ or, alternatively, normalized spectral detuning defined as $u = (\omega - \omega_0)/\omega_D$. There is also collisional shift Δ from Eq. (3), but hereafter, it is assumed to be zero. Moreover, the convergence of numerical methods is determined by the Burnett basis dimensions L_{max} and N_{max}, see notation in [7]. Within the following calculations it was assumed that $L_{max} = N_{max} = 10$. Blackmore [1] was the first who has provided calculations of line profiles for the $r^{-\nu}$-type collisional potential describing velocity-changing contribution to the collisions. Therefore we call such profiles the Blackmore profiles and denote by B_νP, where the subscript ν is the potential power. Consequently, when the model contains also speed-dependent dephasing part then we call the profiles speed-dependent Blackmore profiles and designate them by SD_hB_νP, where the subscript h indicates that the hypergeometric speed-dependence, given by Eq. (3), is applied.

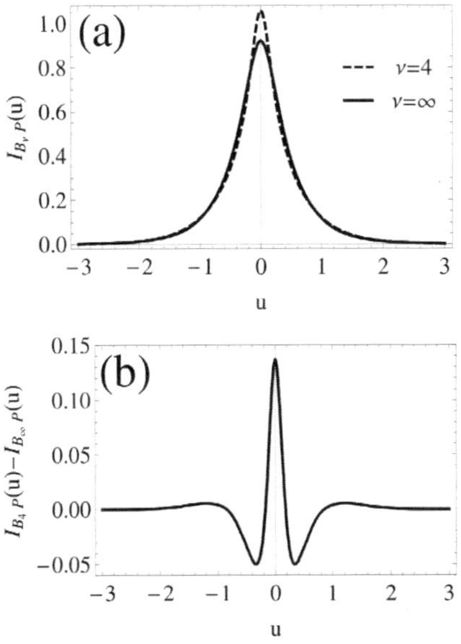

Figure 1. (a) The dashed line represents the Blackmore profile for the potential power $\nu = 4$, while the solid line represents the Blackmore profile for $\nu = \infty$ (e.i. billiard balls profile). (b) Difference between these two profiles ($I_{B_\nu P}$-$I_{B_\infty P}$). Above results were obtained for $\nu_{diff}/\omega_D = 0.8$, $\Gamma/\omega_D = 0.0$, $\alpha = 20$ and for basis dimensions $N_{max} = L_{max} = 10$.

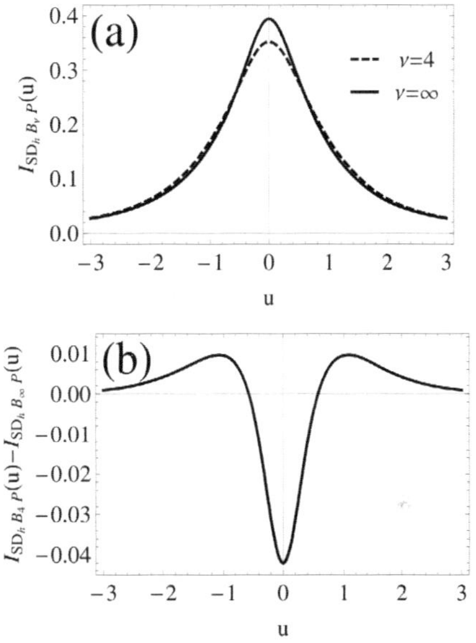

Figure 2. (a) The dashed line represents the speed-dependent Blackmore profile for the potential power $\nu = 4$, while the solid line represents the speed-dependent Blackmore profile for $\nu = \infty$. (b) Difference between these two profiles ($I_{SD_hB_\nu P}$-$I_{SD_hB_\infty P}$). Above results were obtained for $\nu_{diff}/\omega_D = 0.8$, $\Gamma/\omega_D = 0.7$, $\alpha = 20$ and for basis dimensions $N_{max} = L_{max} = 10$.

In the Fig. 1 the results for a zero dephasing part ($\Gamma = 0$) are presented. In this case the Dicke narrowing contributes to the spectral profile significantly , because for $\Gamma = 0$ there is only a pure competition between Doppler broadening and velocity-changing collisions. If the profiles for $\nu = 4$ (dashed line) and for $\nu = \infty$ (solid line) are compared then it can be concluded that in this case the type of collisional potential have an influence on the line shape even higher then 10%. The difference between these profiles can be seen in more detail in the Fig. 1 (b). It should be noted that such high differences can be observed when ν_{diff}/ω_D ratio is close to one and this difference vanishes in two limit cases when the ratio ν_{diff}/ω_D goes to zero or infinity. This conclusion is not strictly true for extremely high α [8].

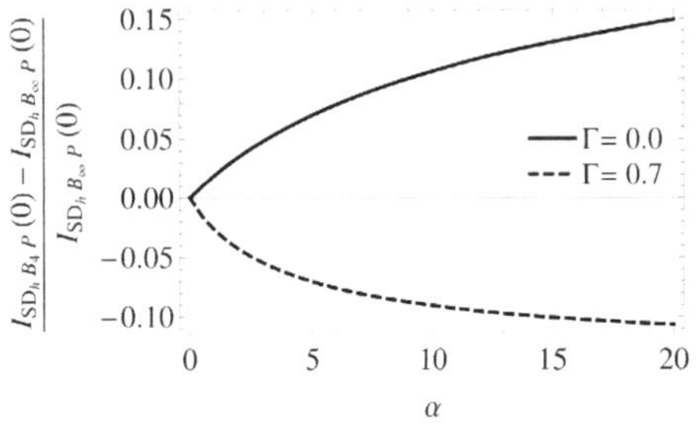

Figure 3. Differences between $I_{SD_h B_\nu P}(0)$ and $I_{SD_h B_\infty P}(0)$ normalized to the $I_{SD_h B_\infty P}(0)$ as a function of α. The solid line corresponds to the case of $\Gamma/\omega_D = 0$ (in this case $SD_hB_\nu P$ is reduced to the $B_\nu P$), while the dashed line corresponds to the $\Gamma/\omega_D = 0.7$. Above results were obtained for $\nu_{diff}/\omega_D = 0.8$ and for basis dimensions $N_{max} = L_{max} = 10$.

The spectral profiles for the speed-dependent cases ($\Gamma = 0.7$) are given in the Fig. 2 (a). The nonzero dephasing collisional contribution suppresses the Dicke narrowing effect substantially (even in the cases when the dephasing is independent from the speed). Moreover, the presence of speed-dependency leads to the contrary result then in pure Dicke narrowing case. Finally the speed-dependent Blackmore profile for $\nu = 4$ is rather broadened then narrowed (as it was in the case from Fig. 1) with respect to the speed-dependent Blackmore profile for $\nu = \infty$, see the Fig. 2 (a). Again, the (b) part of the picture illustrates the difference between these two profiles. Note that the bigger narrowing for $\nu = \infty$ is related to more dramatic speed-dependence of collisional widths for this case comparing to $\nu = 4$ case [9].

To illustrate how the mass ratio α determines the potential power ν influence on the line shape the differences between the speed-dependent Blackmore profiles for $\nu = 4$ and $\nu = \infty$ normalized to the speed-dependent Blackmore profiles for $\nu = \infty$ were plotted, see Fig. 3. The solid line corresponds to the case without dephasing contribution to the collisions (compare with results from the Fig. 1), while the dashed line corresponds to the case with speed-dependent dephasing, where $\Gamma = 0.7$ (compare with results from the Fig. 2). As it is expected, in the limit of soft collisions ($\alpha \to 0$) both lines from the Fig. 3 converge to zero, because in this limit all profiles go to Galatry profile [13] independently from collisional potential related to the velocity-changing contribution. Note that here is no need to distinguish between Galatry [13] profile and speed-dependent Galatry [25] profile, since the hypergeometric speed-dependency given by Eq. (3) disappears for $\alpha = 0$.

4. Conclusions

In this paper we have shown that the type of collisional potential posses significant impact on the spectral profile shape and in particular cases it could be even higher then 10%. Moreover, it was presented that this impact, in the case of $\Gamma = 0$, is caused by the velocity-changing collisions, while for $\Gamma \neq 0$ is rather dominated by the speed dependency.

5. Acknowledgments

The research is a part of the program of the National Laboratory FAMO in Toruń, Poland. The research is partially supported by the Foundation for Polish Science TEAM Project co-financed by the EU European Regional Development Fund, and is partially supported by the National Science Centre, Project No. DEC-2011/01/B/ST2/00491.

References

[1] Blackmore R 1987 *J. Chem. Phys.* **87** 791

[2] Dicke R H 1953 *Phys. Rev.* **89** 472
[3] Wittke J P and Dicke R H 1956 *Phys. Rev.* **103** 620
[4] Berman P R 1972 *J. Quant. Spectrosc. Radiat. Transf.* **12** 1331
[5] Ward J, Cooper J, and Smith E W 1974 *J. Quant. Spectrosc. Radiat. Transf.* **14** 555
[6] Lindenfeld M J 1980 *J. Chem. Phys.* **73** 5817
[7] Ciuryło R, Shapiro D A, Drummond J R, May A D 2001 *Phys. Rev.* A **65** 012502
[8] Wcisło P, Ciuryło R, *Phys. Rev.* A (part I, submitted for publications)
[9] Wcisło P, Ciuryło R, *Phys. Rev.* A (part II, submitted for publications)
[10] Shapiro D A, Ciuryło R, Drummond J R, May A D 2001 *Phys. Rev.* A **65** 012501
[11] May A D 1999 *Phys. Rev.* A **59** 3495
[12] Lindenfeld M J, Shizgal B 1979 *Chem. Phys.* **41** 81
[13] Galatry L 1961 *Phys. Rev.* **122** 1218
[14] Rautian S G and Sobelman I I 1966 *Usp. Fiz. Nauk* **90** 209 [1967 *Sov. Phys. Usp.* **9** 701]
[15] Pine A S 1994 *J. Chem. Phys.* **101** 3444
[16] Demeio L, Green S and Monchik L 1995 *J. Chem. Phys.* **102** 9160
[17] Berman P R, Mossberg T W, Hartmann S R 1982 *Phys. Rev. A* **25** 2550
[18] Ngo N H, Tran H, Gamache R R 2012 *J. Chem. Phys.* **136** 154310
[19] Olver F W J, Lozier D L, Boisvert R F, Clark C W (editors) 2010 *NIST Handbook of Mathematical Functions* Cambridge University Press
[20] Ferziger J H and Kaper H G 1972 *Mathematical theory of transport processes in gases* North-Holland Publishing Company - Amsterdam
[21] Ciuryo R, Lisak D, and Szudy J 2002 *Phys. Rev.* A **66** 032701
[22] Wehr R, Vitcu A, Ciurylo R, Thibault F, Drummond J R, and May A D 2002 *Phys. Rev.* A **66** 062502
[23] Lisak D, Hodges J T, Ciuryło R 2006 *Phys. Rev.* A **73** 012507
[24] Wehr R, Ciurylo R, Vitcu A, Thibault F, Drummond J R, May A D 2006 *J. Mol. Spectrosc.* **235** 54
[25] Ciuryło R, Jaworski R, Jurkowski J, Pine A S, and Szudy J 2001 *Phys. Rev. A* **63** 032507

Quantum mechanical and uniform quasi-classical approaches for total cross-sections: the radiative collisions $He(1^1S) + Ne + \hbar\omega \to He(2^1S) + Ne$ at low energies

A Devdariani[1,3], E Chesnokov[1] and F Rebentrost[2]

[1]Department of Optics and Spectroscopy, St. Petersburg University, St. Petersburg 198904, Russia
[2] Max-Planck-Institut für Quantenoptik, D 85748 Garching, Germany

E-mail: snbrn2@yandex.ru

Abstract. The total cross sections of the radiative collisional process $He(1^1S) + Ne + \hbar\omega \to He(2^1S) + Ne$ have been calculated in the framework of the exact quantum mechanical and uniform quasi-classical approaches for the row of energies $1 - 220\ cm^{-1}$. The dependence of the total cross sections on the frequency and polarization of light has been investigated. Although the process involves only atoms in the symmetric S-states the results demonstrate the azimuthal scattering asymmetry. The disagreement between the quantum and quasi-classical approaches does not exceed 10 per cent for collision energies $50 - 220\ cm^{-1}$, but increases up to 25-40 per cent with collision energy decreasing to $1 - 10\ cm^{-1}$ on retention of a functional dependence on angle. The reason of the disagreement is mainly due to the loss of accuracy of the WKB approximation for the ground state wave function at low energies.

1. Introduction

The present work has been motivated by two reasons. Firstly, in recent years, considerable interest has been shown in analysing the cross sections of the radiative-collisional processes that lead to the collisionally induced radiative excitation of metastable states for one of the colliding atoms. The first crossed beam experiment devoted to the investigation of the angular and spectral distributions produced in collisionally induced radiative excitation of metastable Ca(4s3d ^{1}D) state have been reported in [1].The total and differential cross sections for the process

$$He(1^1S) + Ne + \hbar\omega \to He(2^1S) + Ne \tag{1}$$

have been calculated in [2-4] starting from the uniform quasiclassical approximation [2,5].

Secondly, it has been shown recently [6] that different radiative characteristics of quasimolecules formed by colliding atoms are determined by different regions of interatomic distances. But the best

[3] To whom any correspondence should be addressed.

way to check the input quantum chemical data set that is potential energy surfaces and optical transition moments seems to be to compare the results of experiments with calculations of the scattering cross sections for the reactions like (1). As for the theoretical part of the work it seems that the analytical approaches would be favourable for two reasons at least. The influence of input data variations on radiative characteristics is immediately evident from analytical results and such approaches are best suited for fitting the results of calculations.

This work is devoted to the calculation of the total cross sections of the collision process (1) as the functions of the frequency and polarization of light at low collision energies 1-200 cm^{-1}. The present results have been obtained in the frame of a uniform quasi-classical approach. As a rule quasi-classical approaches produce analytical results but although the general criteria of the approaches are quite clear they do not lead to precise boundaries and limitations in specific cases in the domain low collision energies which we are especially interested in. Thus, the present work includes the exact quantum mechanical calculations too, and the comparison of the exact quantum mechanical, and the uniform quasi-classical approaches is aimed at a conclusion about the applicability limit of the uniform quasi-classical approximation utilized in the work. We should stress that the reaction (1) gives a unique example of the reaction that involves atoms in the symmetric S-states both for the initial and final states. Until recently the polarization effects in scattering have been studied for aligned and oriented atoms in non-symmetrical states, e.g. [7].

2. Quantum mechanical and uniform quasi-classical approaches

The mechanism of the collisional process (1) was described in detail in [8,9]. At low radiation intensities ($I \leq 10\ GW/cm^2$) the total and the differential cross sections of process (1) can be determined using the distorted-wave method [10]. In this case the dependence of the cross section on the intensity is linear. In our calculations we used a scale factor corresponding to an intensity $I = MW/cm^2$. The total cross sections for the polarizations parallel and perpendicular to the initial relative interatomic momentum are expressed via the S - matrix elements by the formulas [4]

$$\sigma_{\parallel} = (\pi/k_i^2)\cdot\sum_{l=0}^{\infty} 1/(2l+1)\cdot\left|l\cdot S_{l-1\to l} - (l+1)\cdot S_{l+1\to l}\right|^2, \tag{2}$$

$$\sigma_{\perp} = (\pi/2k_i^2)\cdot\sum_{l=1}^{\infty} l(l+1)/(2l+1)\cdot\left|S_{l-1\to l} + S_{l+1\to l}\right|^2 .$$

For the case of an in-collision plane linear polarization of the light the differential cross sections have the form

$$\frac{d\sigma(\theta_p)}{d\Omega} = \frac{k_f}{k_i}\left(\cos^2\theta_p\cdot\left|f_0(\theta)\right|^2 + \sin^2\theta_p\left|f_{(1)}(\theta)\right|^2 \pm \sin(2\theta_p)\,\mathrm{Re}\left[f_0(\theta)f_{(1)}^*(\theta)\right]\right), \tag{3}$$

and for the right-hand and the left-hand circular polarizations we obtain

$$d\sigma_{r,l}/d\Omega = (k_f/k_i)\left(1/2\cdot\left|f_0(\theta)\right|^2 + 1/2\cdot\left|f_{(1)}(\theta)\right|^2 \pm \mathrm{Im}\left[f_0(\theta)f_{(1)}^*(\theta)\right]\right). \tag{4}$$

Here

$$f_0(\theta) = (4k_i k_f)^{-1/2} \sum_{l=0}^{\infty} \left[l \cdot S_{l-1\to l} - (l+1) \cdot S_{l+1\to l} \right] \cdot P_l(\cos\theta), \tag{5}$$

$$f_{(1)}(\theta) = (4k_i k_f)^{-1/2} \sum_{l=1}^{\infty} \left[S_{l-1\to l} + S_{l+1\to l} \right] \cdot P_l^1(\cos\theta)$$

are the spherical components of the scattering amplitude [4], and θ_p is the polarization angle (angle between the initial relative interatomic momentum and the light polarization vector). The upper and lower signs in Eq. (3) correspond to the leftward (the final relative interatomic momentum and the polarization vector are in the same half-plane) and rightward scatterings, respectively. The upper and lower signs in Eq. (4) correspond to the leftward and the rightward scatterings for a right-hand circular polarization and vice versa for a left-hand case.

In the framework of the distorted wave method the S - matrix elements have the form

$$S_{l_i \to l_f} = -i2\pi \exp[i(\delta_i^{l_i} + \delta_f^{l_f})] \int_0^{\infty} \Psi_{E_f}^{l_f}(R) \cdot V(R) \cdot \Psi_{E_i}^{l_i}(R) dR, \tag{6}$$

where $\Psi_E^l(R)$ are the radial, regular (at zero) real wave functions normalized to the δ - function of energy, $V(R) = E_0 d(R)/2$ is the interaction matrix element, and δ^l are the elastic scattering phase shifts.

The exact quantum mechanical approach consists in a numeral evaluation of the S - matrix elements starting from the wave functions $\Psi_E^l(R)$ for the initial and the excited quasimolecular states obtained by numerical integration of the radial Schroedinger equations

$$d^2\Psi_E^l / dR^2 + k^2(R)\Psi_E^l = 0, \tag{7}$$

where $k(R) = \sqrt{2\mu(E - U(R)) - J^2 / R^2}$ is the momentum function, $\mu = 6089$ is the reduced mass of the colliding atoms, and $J = \sqrt{l(l+1)}$.

On the other hand, the uniform quasi-classical approach deals with the uniform Langer's (Airy's) approximations [10] for the radial wave functions

$$\Psi_E^l(R) = (2\mu)^{1/2} \left(\xi(R)/k^2(R)\right)^{1/4} Ai(-\xi(R)), \tag{8}$$

$$\xi(R) = \left(3/2 \cdot S(R)\right)^{2/3}, \quad S(R) = \int_{R_{tp}}^{R} k(R) dR, \quad J = l + 1/2,$$

and the overlap integral (6) is calculated analytically using the uniform saddle points method [5,11]. Moreover, as has been emphasized in [9], it is necessary for asymptotically forbidden processes like (1) to take into account the steep (exponential) dependence of the interaction matrix element on the interatomic distance $V(R) = V_0(R)\exp(-\gamma R)$ (V_0 is a smooth function of R, in accordance with [9] $V_0 = 2.78 \cdot 10^{-3}$, $\gamma = 0.92$) that leads to the following correction of the well-known Franck-Condon approximation. Instead the condition

$$\Delta U = \Delta\omega,$$

where $\Delta U = U_f - U_i$ is the difference potential, $\Delta\omega$ is the detuning of the radiation frequency ω from the frequency of the forbidden atomic line $\omega_0 = 166278\ cm^{-1}$, the saddle points are determined by the equations

$$(k_f - k_i)(R_\pm) = \pm i\gamma \ . \tag{9}$$

The final formula for the S - matrix element then has the form [5]

$$S_{l_i \to l_f} = (F^+[+] + F^-[-])\exp(iA), \tag{10}$$

where

$$[\pm] = B^{1/4} Ai(-B) \pm iB^{-1/4} Ai'(-B),$$

$$F^\pm = -i2^{1/2}\pi\mu V_0(R_\pm)[k_f k_i (k'_f - k'_i)(R_\pm)]^{-1/2},$$

$$2A = i\gamma(R_+ + R_-) + S_i(R_+) - S_i(R_-) - S_f(R_+) + S_f(R_-) + 2(\delta_i^{J_i} + \delta_f^{J_f}),$$

$$(4/3)B^{3/2} = i\gamma(R_+ - R_-) + S_i(R_+) + S_i(R_-) - S_f(R_+) - S_f(R_-),$$

and

$\delta^J = \lim_{R\to\infty}[S(R) - kR] + J\pi/2$ is the quasi-classical limit for the phase shift.

3. Total cross-sections

Starting from the S - matrix elements we have calculated the total and differential cross sections of the process (1) by exact numeral summation over the partial waves (see Eqs. (2-5)). The data have been obtained for the range of the collision energies E=1-200 cm^{-1} and for the different detunings (including both wings and the center of the forbidden atomic line) and polarizations of light. The number of partial waves which must be included in summation never exceeds 40 for the energies considered. To specify the application limit of the uniform quasi-classical approximation both the exact quantum mechanical and the uniform quasi-classical calculations have been performed in all cases. Figure 1 shows the total cross section averaged over the polarizations of light

$$\sigma = (\sigma_\parallel + 2\sigma_\perp)/3$$

together with the polarization coefficient

$$P = \frac{\sigma_\perp - \sigma_\parallel}{\sigma_\perp + \sigma_\parallel}$$

as the functions of the detuning $\Delta\omega$.

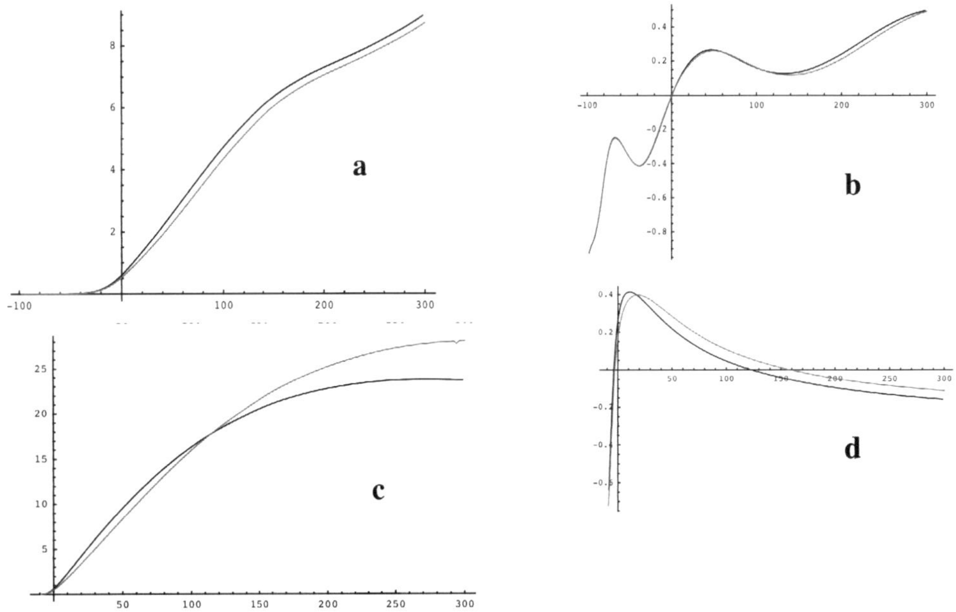

Figure 1. Total cross sections $\sigma, 10^{-22}\,cm^2$ (a,c) of the process (1) and the polarization coefficients P (b,d) versus detuning $\Delta\omega = \omega - \omega_0, cm^{-1}$; (a,b) $E = 100\,cm^{-1}$ and (c,d) $E = 10\,cm^{-1}$; (black line) exact quantum mechanical calculation, (red line) uniform quasi-classical result. The scale factor corresponds to $I = MW/cm^2$.

Let us note that disagreement between the exact quantum mechanical and the uniform quasi-classical approximations never exceeds 10 per cent in the case of the collision energy $E = 100\,cm^{-1}$ but increase up to 25 per cent when the collision energy decreases to $E = 10\,cm^{-1}$. In the first case disagreement is mainly due to the use of the uniform saddle points method in the quasi-classical calculations instead of the exact numeral evaluation of the overlapping integrals (6). With decreasing energy the difference in the ground state wave functions obtained in the frame of the two different approaches becomes more essential and for an energy of $E = 10\,cm^{-1}$ makes the main contribution to the observed disagreement. Note also that for specific detunings ($\Delta\omega = 0$ for $E = 100\,cm^{-1}$ and $\Delta\omega = -4\,cm^{-1}$, $123\,cm^{-1}$ for $E = 10\,cm^{-1}$) the total cross sections does not depend on the polarization of light ($P = 0$).

4. Conclusion

The radiative-collisional process (1) with atoms in the symmetric S-states have been analyzed in large detail for collision energies $1 - 220\,cm^{-1}$. For all collision energies considered total cross sections depend on the frequency and the polarization of light. It has been shown that there are some specific frequencies at which the total cross section does not depend on the polarization of light. The dependencies of cross sections on collision energy and absorbed photon frequency open up fresh opportunities for tests of the potential energies and dipole moments.

The comparison of the exact quantum mechanical calculations with the uniform quasi-classical calculations results in the conclusion that the exactness of the uniform quasi-classical approach depends mainly on the exactness of the WKB approximation for the elastic radial wave functions corresponding to the different collision channels. For the collisions considered the low limit of the collision energies for the validity of the uniform quasi-classical approximation is $E \sim 10\ cm^{-1}$. It should be stressed that this conclusion has been obtained starting from the exponential approximations for the ground state and the excited state quasi-molecular potentials [9].

References

[1] Figl C, Grosser J, Hoffmann O and Rebentrost F 2003 *Chem. Phys. Lett.* **380** 196
[2] Devdariani A Z and Chesnokov E A 2005 *Optics and Spectr.* **99** 858
[3] Devdariani A Z and Chesnokov E A 2005 *Optics and Spectr.* **99** 866
[4] Chesnokov E A 2005 *JETP* **101** 761
[5] Devdariani A Z 1989 *Sov. Phys. JETP* **69** 266
[6] Devdariani A, Chesnokov E, Zagrebin A, Lednev M, Petsalakis I D, Theodorakopoulos G, Lieberman H-P and Buenker R J 2006 **330** 101
[7] Hertel I, Schmidt H, Bährings A, Meyer E 1985 *Rep.Prog.Phys.* **48** 375
[8] Zagrebin A L and Tserkovnyi S I 1995 *Optics and Spectr.* **79** 511
[9] Devdariani A Z, Zagrebin A L, Rebentrost F, Tserkovnyi S I and Tchesnokov E A 2002 *JETP* **95** 413
[10] Mott N F and Massey H S W 1865 *The Theory of Atomic Collisions* (Oxford: Claredon)
[11] Connor J N L 1981 *J. Chem. Phys.* **74** 1047

XXI International Conference on Spectral Line Shapes (ICSLS 2012) IOP Publishing
Journal of Physics: Conference Series **397** (2012) 012010 doi:10.1088/1742-6596/397/1/012010

Elementary statistical models for collision-sequence interference effects with arbitrary persistence of velocity

H T Wheeler[1] **and J C Lewis**[2]

[1]605 - 160 Third Street West, North Vancouver BC Canada V7M 0A9
[2]Department of Physics and Physical Oceanography, Memorial University of Newfoundland, St. John's NL Canada A1B 3X7

E-mail: htwsci@shaw.ca, john.lewis@mun.ca

Abstract. Elementary statistical models for collision-sequence interference effects usually assume Gaussian-distributed velocities with zero persistence of velocity. Here we extend the treatment to arbitrary persistences of velocity in the range zero to just less than unity. For vector collision-sequence interference the results show the interference dip narrowing with increasing persistence of velocity as expected. Only partial agreement is obtained with some observations in a molecular dynamics simulation in Lennard-Jonesium.

1. Introduction

1.1. Background

The elementary statistical models developed by Lewis et al. [1-3] to describe collision-sequence interference effects have mostly used Gaussian-distributed velocities with zero persistence of velocity. Persistences of velocity which run from zero to just less than unity can be modelled by an AR(1) (Box-Jenkins) process [4] allowing the effect of varying this parameter on vector and scalar interference in simple models to be determined.

Such processes lead to persistence-of-velocity terms $\langle \boldsymbol{v}_i \cdot \boldsymbol{v}_{i+1} \rangle$, $\langle \boldsymbol{v}_i \cdot \boldsymbol{v}_{i+2} \rangle$ ... $\langle \boldsymbol{v}_i \cdot \boldsymbol{v}_{i+k} \rangle$, which form a geometrical progression where $\boldsymbol{v}_i$ is the velocity before the *ith* collision in a collision sequence. More complicated behaviour, however, can be modeled.

1.2. The General Model

We assume an infinite set of statistically independent basis vectors $\boldsymbol{u}_i$ with identical normal distributions and variance σ^2

$$\langle \boldsymbol{u}_i \cdot \boldsymbol{u}_j \rangle = \delta_{i,j} \left\langle \sigma^2 \right\rangle \tag{1}$$

and a series a_i where the squared series

$$\sum_{i=0}^{\infty} a_i^2 = S \tag{2}$$

is convergent. Correlated velocities $\boldsymbol{v}_k$ for $k = 0, 1, 2, 3$ ... can be defined through

$$\boldsymbol{v}_k = \frac{1}{\sqrt{S}} \sum_{i=0}^{\infty} a_i u_{k+i} \tag{3}$$

The velocities in eq.(3) will have a normal distribution with variance σ^2

$$\langle v_k^2 \rangle = \sigma^2 \tag{4}$$

The value of $\langle \boldsymbol{v}_k \cdot \boldsymbol{v}_{k+j} \rangle$ for $j > 1$ will depend on the choice of the series a_i. The vectors $\boldsymbol{u}_i$ could have other distributions but normal is commonly used.

2. The AR Model

A case of particular interest is to take the series a_i as a geometric series. Define velocities $\boldsymbol{v}_k$ in terms of the $\boldsymbol{u}_i$ and a ratio γ with $|\gamma| < 1$ as

$$\boldsymbol{v}_k = \sqrt{(1-\gamma^2)} \sum_{i=0}^{\infty} \gamma^i \boldsymbol{u}_{k+i} \tag{5}$$

since eq.(3) becomes

$$S = \sum_{i=0}^{\infty} \gamma^{2i} = \frac{1}{1-\gamma^2} \tag{6}$$

Then the variance of the $\boldsymbol{v}_k$ will be the same as that of the $\boldsymbol{u}_k$

$$\langle \boldsymbol{v}_i \cdot \boldsymbol{v}_i \rangle = \langle v^2 \rangle = \sigma^2 \tag{7}$$

We have upon assuming a stationary and reversible process

$$\langle \boldsymbol{v}_i \cdot \boldsymbol{v}_j \rangle = \gamma^{|i-j|} \langle v^2 \rangle \tag{8}$$

where the ratio γ is seen to be the persistence of velocity. The definition used here is

$$\gamma = \left\langle \left[\frac{\langle \boldsymbol{v}_i \cdot \boldsymbol{v}_{i+j} \rangle}{\langle v^2 \rangle} \right]^{(1/j)} \right\rangle \tag{9}$$

3. Vector Collision Sequences

3.1. Equal Collision Intervals

As in [1] define impulses $\boldsymbol{f}_k = \boldsymbol{v}_{k+1} - \boldsymbol{v}_k$. From eq. (8)

$$\langle \boldsymbol{f}_k^2 \rangle = \langle (\boldsymbol{v}_{k+1} - \boldsymbol{v}_k)^2 \rangle = 2 \langle v^2 \rangle (1-\gamma) \tag{10}$$

$$\langle \boldsymbol{f}_k \cdot \boldsymbol{f}_j \rangle = \langle (\boldsymbol{v}_{k+1} - \boldsymbol{v}_k) \cdot (\boldsymbol{v}_{j+1} - \boldsymbol{v}_j) \rangle = -\langle v^2 \rangle \gamma^{|k-j|-1} (1-\gamma)^2 \tag{11}$$

Upon assuming equal time intervals between collisions, which can be scaled so that $t_k = k$ for $k \in [0, N-1]$ for N collisions, we have

$$\boldsymbol{f}(t) = \sum_k \boldsymbol{f}_k \delta(t-k) \tag{12}$$

and the corresponding Fourier transform, with $\omega = 2\pi/N$

$$\tilde{\boldsymbol{f}}(t) = \sum_{k=0}^{N-1} \boldsymbol{f}_k e^{\iota\omega k} \tag{13}$$

With the use of eqs.(10) and (11) we have

$$\frac{1}{N}\left\langle \tilde{\boldsymbol{f}}\cdot\tilde{\boldsymbol{f}}^*\right\rangle = 2\left\langle v^2\right\rangle(1-\gamma)\left[1-(1-\gamma)\sum_{j=1}\gamma^{j-1}\cos\omega j\right] \tag{14}$$

Following [1], after some manipulation, the power spectrum is derived as

$$S(\omega) = 2\left\langle v^2\right\rangle(1-\gamma^2)\left[\frac{1-\cos\omega}{1-2\gamma\cos\omega+\gamma^2}\right] \tag{15}$$

showing the line shape is dependent on γ.

3.2. Poisson Distributed Intercollisional Intervals

Following [2] assume the collisions times t_k follow a Poisson distribution with frequency $\bar{\nu}$. The intervals Δ_i will be exponentially distributed. The N collisions lie in the interval $[0, T]$. For large N the random time T can be taken as $N/\bar{\nu}$ and then

$$\frac{1}{N} = \frac{\bar{\nu}}{T} \tag{16}$$

so the average over N can be replaced by an average over T. For the current model, with $\boldsymbol{f}_k = \boldsymbol{\mu}_k$ and $\tilde{\omega} = \omega - \omega_0$, as in the development from [2, eq. 6], we have

$$\frac{S(\tilde{\omega})}{\bar{\nu}} = \left\langle \tilde{\boldsymbol{f}}\cdot\tilde{\boldsymbol{f}}^*\right\rangle \tag{17}$$

$$= \left\langle \boldsymbol{f}^2\right\rangle + 2\Re\mathfrak{e}\left\{\sum_{j=1}^{\infty}\langle \boldsymbol{f}_k\cdot\boldsymbol{f}_{k+j}\rangle\left[\frac{\bar{\nu}}{\bar{\nu}+\iota\tilde{\omega}}\right]^j\right\} \tag{18}$$

Writing $\kappa = (1-\gamma)\bar{\nu}$, the power spectrum is found to be

$$S(\tilde{\omega}) = 2\left\langle v^2\right\rangle\kappa\left[\frac{\tilde{\omega}^2}{\kappa^2+\tilde{\omega}^2}\right] \tag{19}$$

This shows the shape of the interference dip depends on the product κ and not on $\bar{\nu}$ and γ in a separable manner. The line shape of the dip is an inverted Lorentzian for all values of γ.

The results of this and the preceding section 3.1 show the interference dip narrowing with increasing persistence of velocity, as predicted earlier [5-8].

4. Scalar Interference

For $\gamma > 0$, only $\langle f_k f_{k+1}\rangle$ and $\langle f_k f_{k-1}\rangle$ are different from $\langle f\rangle^2$ (see [1]). We have from numerical calculations, using the Box-Muller method to construct random normal vectors [10], that $\langle f_k f_{k+1}\rangle$ decreases with increasing γ and

$$\langle f_k f_{k+1}\rangle \lesssim (1-\gamma)\langle f_k f_{k+1}\rangle_{\gamma=0} \tag{20}$$

$$\langle f_k f_{k+1}\rangle \to \langle f\rangle^2 \ as\ \gamma \to 1 \tag{21}$$

so that $\mathrm{cov}(f_k, f_{k+1}) \to 0$ as $\gamma \to 1$. This is to be expected from the argument in [1] on the origin of the covariance: the larger the persistence of velocity, the smaller is the variation in speed from one collision to the next.

4.1. An Analytic Solution for Two Dimensions

Given Gaussian velocities $\boldsymbol{v}_k$ and $\boldsymbol{v}_{k+1}$ in two dimensions with unit variance, these will not be independent for $\gamma > 0$. The method of [1] can be used by a change of variable since

$$|\boldsymbol{v}_{k+1} - \boldsymbol{v}_k| = |(1-\gamma)\boldsymbol{v}_k - \sqrt{1-\gamma^2}\boldsymbol{u}_{k+1}| \tag{22}$$

where $\boldsymbol{v}_k$ and $\boldsymbol{u}_{k+1}$ are independent. Since $\boldsymbol{u}_{k+1}$ and $\boldsymbol{v}_k$ have Gaussian distributions with unit variance, we have

$$\langle f^n \rangle = \frac{1}{4\pi^2} \int d^2v_k d^2u_{k+1} \, \exp[-(1/2)(v_k^2 + u_{k+1}^2)] \, |(1-\gamma)\boldsymbol{v}_k - \sqrt{1-\gamma^2}\boldsymbol{u}_{k+1}|^n \tag{23}$$

Set

$$\boldsymbol{a} = \frac{(1-\gamma)}{\sqrt{2}}\boldsymbol{v}_k - \sqrt{\frac{1-\gamma^2}{2}}\boldsymbol{u}_{k+1} \tag{24a}$$

$$\boldsymbol{b} = \frac{\sqrt{1+\gamma}}{2}\boldsymbol{v}_k + \frac{\sqrt{1-\gamma}}{2}\boldsymbol{u}_{k+1} \tag{24b}$$

Then

$$\frac{a^2}{2(1-\gamma)} + b^2 = \frac{v_k^2 + u_{k+1}^2}{2} \tag{25}$$

and

$$|(1-\gamma)\boldsymbol{v}_k - \sqrt{1-\gamma^2}\boldsymbol{u}_{k+1}|^n = |a|^n(\sqrt{2})^n \tag{26}$$

The transformation is

$$\begin{pmatrix} \boldsymbol{v}_k \\ \boldsymbol{u}_{k+1} \end{pmatrix} = \begin{pmatrix} \boldsymbol{a}/\sqrt{2} & \sqrt{1+\gamma}\boldsymbol{b} \\ -\sqrt{(1+\gamma)}\boldsymbol{a}/\sqrt{2(1-\gamma)} & \sqrt{1-\gamma}\boldsymbol{b} \end{pmatrix} \tag{27}$$

giving the Jacobian $J(\gamma)$ as

$$J(\gamma) = \begin{vmatrix} \partial \boldsymbol{v}_k/\partial \boldsymbol{a} & \partial \boldsymbol{v}_k/\partial \boldsymbol{b} \\ \partial \boldsymbol{u}_{k+1}/\partial \boldsymbol{a} & \partial \boldsymbol{u}_{k+1}/\partial \boldsymbol{b} \end{vmatrix} = \begin{vmatrix} 1/\sqrt{2} & \sqrt{1+\gamma} \\ -\sqrt{1+\gamma}/\sqrt{2(1-\gamma)} & \sqrt{1-\gamma} \end{vmatrix} = \frac{\sqrt{2}}{\sqrt{1-\gamma}} \tag{28}$$

Then eq. (23) is evaluated

$$\langle f^n \rangle = \frac{J(\gamma)^2(\sqrt{2})^n}{4\pi^2} \int d^2a d^2b \, \exp\{-a^2/[2(1-\gamma)] - b^2\}|a|^n \tag{29a}$$

$$= \frac{2}{1-\gamma}(\sqrt{2})^n \int_0^\infty db \, b \exp(-b^2) \int_0^\infty da \, \exp\{-a^2/[2(1-\gamma)]\}|a|^{n+1} \tag{29b}$$

$$= 2^n(1-\gamma)^{(n/2)} \, \Gamma\left(\frac{n}{2}+1\right) \tag{29c}$$

This result is in agreement with previous results for $\gamma = 0$ and with numerical calculations for $\gamma > 0$.

4.2. Comparison to the $\gamma = 0$ Value

From numerical calculations in three dimensions and from eq. (29c), we have for a Gaussian velocity distribution with unit variance

$$\langle f^n \rangle = (1-\gamma)^{n/2} \langle f_0^n \rangle \tag{30}$$

where $\langle f_0^n \rangle$ is the quantity evaluated for $\gamma = 0$.

5. Comparison to Lennard-Jones Molecular Dynamics Results

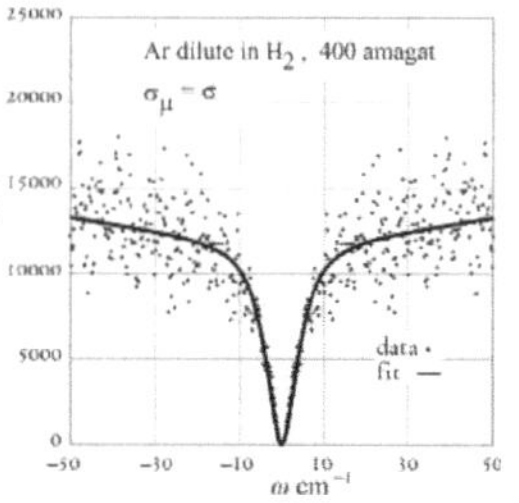

Fig. 1 Dip for $\sigma_\mu = \sigma$.

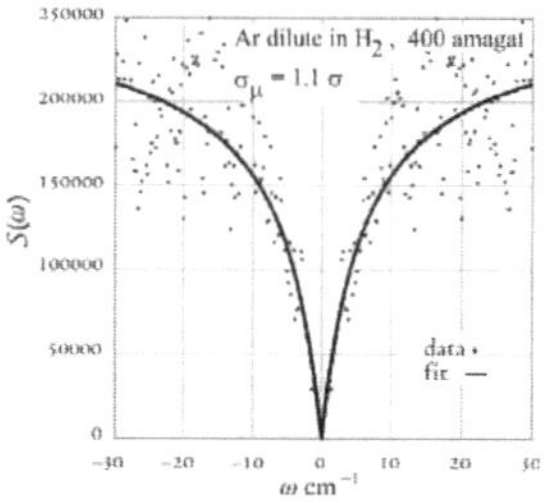

Fig. 2 Dip for $\sigma_\mu = 1.1\sigma$.

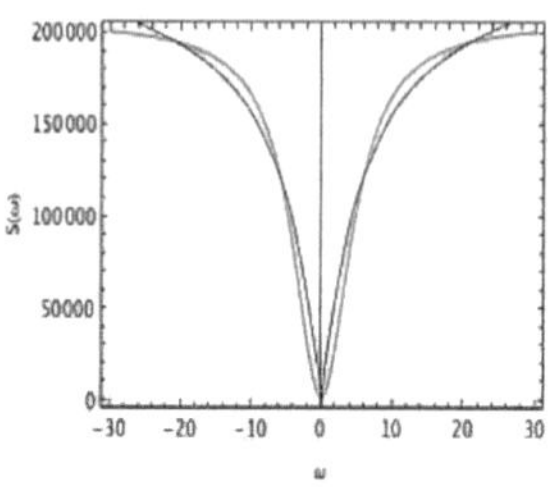

Fig. 3 AR-model vs Fig. 2

Figs. 1 and 2 show interference dips from molecular dynamics simulations of Lennard-Jonesium mixtures where the parameters were chosen for Ar and H_2, with the model induced dipole moment also of Lennard-Jones type. These data were first published in [9]. Taking the induced dipole moment proportional to the force, $\boldsymbol{\mu}_k \propto \boldsymbol{f}_k = \boldsymbol{v}_{k+1} - \boldsymbol{v}_k$, the data of Fig. 1 were well fit using the AR-model eq.(19). The value of κ so obtained was then used in a stochastic simulation to fit Fig. 2 with $\boldsymbol{\mu}_k = c_\mu \boldsymbol{f}_k |\boldsymbol{f}_k|^\beta$. Note that the physical parameters must be kept the same so the two fits are not independent. In this second fit both γ and β were varied by generating random velocities with a Gaussian distribution and the desired persistence of velocity. Fig. 3 shows the best fit overlaid with the cusp from Fig. 2. This best fit is for $c_\mu \approx 0.34$, $\beta \approx 0.47$ and $\gamma \approx 0$. It was not possible to reproduce the cusp with an expected large value of γ as the larger γ is the broader the resulting interference dip: normally increasing γ would narrow the dip but here the fit to the Fig. 2 case is constrained by the value of κ derived from fitting the Fig. 1 case.

6. Conclusions

The theory developed in [1, 2] for the zero persistence of velocity case has been extended to the arbitrary persistence of velocity case using the AR-model although certain quantities could only be evaluated numerically. It is fairly straightforward to generate velocities $\boldsymbol{v_k}$ using eqs. (3) or (5), and then express the induced dipole moment $\boldsymbol{\mu}_k$ as some function of the force (impulse) $\boldsymbol{f}_k = \boldsymbol{v}_{k+1} - \boldsymbol{v}_k$ to do stochastic simulations for cases that are too complex to obtain an analytical solution.

The AR-model was successful in reproducing the Lennard-Jones results of [9] for the case $\sigma_\mu = \sigma$ but, using consistent physical parameters, here κ, a satisfactory fit to the case $\sigma_\mu = 1.1\sigma$ could not be obtained. The reason for this is not clear, but may reflect differing roles for the soft and hard parts of the interaction in the simulation results so that the simple fitting formula used, $\boldsymbol{\mu}_k = c_\mu \boldsymbol{f}_k |\boldsymbol{f}_k|^\beta$, is not adequate.

7. Acknowledgement

We thank Terrence Tricco for his assistance in carrying out and analysing the molecular dynamics simulations discussed above and reported in [9].

References

[1] Lewis J C 2008 *Phys. Rev.* A **77** 062702
[2] Lewis J C 2009 *International Journal of Spectroscopy* 561697
[3] Lewis J C and Herman R M 2011 *International Review of Atomic and Molecular Physics* **2** 25
[4] Box G, Jenkins G M and Reinsel G C 1994 *Time Series Analysis: Forecasting and Control* (Prentice-Hall) 3rd ed.
[5] Lewis J C and van Kranendonk J 1972 *Can. J. Phys.* **50** 352–67

[6] Lewis J C and van Kranendonk J 1972 *Can. J. Phys.* **50** 2902–13
[7] Lewis J C 1985 *Phenomena Induced by Intermolecular Interactions* ed Birnbaum G (New York, Plenum Press) 215–57.
[8] Lewis J C and R M Herman 2003 *Phys. Rev. A* **68** 032703
[9] Lewis J C 2004 *Proc. of the 17th Int. Conf. on Spectral Line Shapes* ed Dalimier E (Frontier Group) 358–60.
[10] Box G E P and Muller M E 1958 A note on the generation of random normal deviates *The Annals of Mathematical Statistics* **29** No. 2 610–1

Journal of Physics: Conference Series **397** (2012) 012011 doi:10.1088/1742-6596/397/1/012011

Spectroscopy of cooperative muon-gamma-nuclear processes: Energy and spectral parameters

A V Glushkov
Odessa State University - OSENU, P.O. Box 24a, Odessa-9, SE-65009, Ukraine

E-mail: glushkov@paco.net , dirac13@mail.ru

Abstract. We present a consistent relativistic theory of discharge of a metastable nucleus with emission of γ quantum and further μ^- conversion. The numerical calculation is carried out for the scandium nucleus (A=49, N=21) with using the Dirac-Woods-Saxon model.

1. Introduction

Muonic atoms have always been useful tools for nuclear spectroscopy employing atomic-physics techniques [1-12]. Muonic atoms also play an important role as catalysts for nuclear fusion [1,2]. It should be also mentioned the growing importance of muon spectroscopy in molecular physics. Electrons, muons (other particles such as kaons, pions etc) originally in the ground state of the target atom can be excited reversibly either to the bound or continuum states. With appearance of the intensive neutron pencils and laser sources studying the gamma-muon-nuclear interactions is of a great importance [4-16]. The rapid progress in laser technology even opens prospects for nuclear quantum optics via direct laser-nucleus coupling [7-12]. It is known that a negative muon captured by a metastable nucleus may accelerate a discharge of the latter by many orders of magnitude [3,4,6].

A principal possibility of storage of the significant quantities of the metastable nuclei in the nuclear technology processes and their concentrating by chemical and laser methods leads to problem of governing their decay velocity [3,4]. The μ-atom system differs advantageously of the usual atom; the relation r_n/r_a (r_n is a radius of a nucleus and r_a is a radius of an atom) can vary in the wide limits in dependence upon the nuclear charge. Because of the large muon mass and the correspondingly small Bohr radius, the muonic wave function has a large overlap with the nucleus and thus effectively probes its structural features, such as finite size, deformation, polarization etc. For a certain relation between the energy range of the nuclear and muonic levels a discharge of the metastable nucleus may be followed by the ejection of a muon, which may then participate in the discharge of other nuclei. The estimates of probabilities for discharge of a nucleus with emission of gamma quantum and further muon or electron conversion are presented in ref. [4,6,11,12]. Despite the relatively long history, studying processes of the muon-atom and muon-nucleus interactions hitherto remains very actual and complicated problem. Here we present a consistent relativistic theory of discharge of a metastable nucleus with emission of γ quantum and further μ^- conversion. The accurate numerical calculation is carried out for scandium nucleus (A=49, N=21) with using the Dirac-Woods-Saxon model.

2. Relativistic theory of metastable nucleus discharge during negative muon capture

For simplicity, we consider the model of a nucleus as the 1-QP system [6]. Further we suppose that a nucleus consists off a twice-magic core and a single proton and single muon, which move in the nuclear core field. The proton and muon interact through the Coulomb potential. This interaction will be accounted for in the first order of the atomic PT or in the second order of the QED PT [13-16]. Surely a majority of the known excited nuclear states have the multi-body character and it is hardly possible to describe their structure within the one-QP model [17]. Nevertheless, the studied effects of the muon-proton interaction are not covered by the one-particle model. It is possible to consider a dynamical interaction of two particles through the core too. It accounts for finite core mass. However, this interaction may decrease a multipolarity of the nuclear transition only by unity. Strongly forbidden transitions of high multipolarity are of a great interest. We calculate the decay probabilities to different channels of the system, which consists of the proton (in an excited state $\Phi_{N_1 J_1}$) and muon (in the ground state Ψ^{μ}_{1s}). Three channels should be taken into account [4,6,12]: i). a radiative purely nuclear 2^j -poled transition (probability P_1); ii). a non-radiative decay, when proton transits to the ground state and the muon leaves a nucleus with energy: $E = \Delta E^{p}_{N_1 J_1} - E^{i}_{\mu}$; $\Delta E^{p}_{N_1 J_1}$ is the energy of nuclear transition; E^{i}_{μ} is a bond energy of muon in the *1s* state (probability P_2); iii). transition of a proton to the ground state with excitation of muon and emission of γ-quantum with energy $h\omega = \Delta E^{p}_{N_1 J_1} - \Delta E^{\mu}_{nl}$ (probability P_3). Feynman diagrams, corresponding to different muon-nuclear decay channels, are shown in figure 1.Diagram A (fig.1) corresponds to the first channel (i), diagram B - to the second channel (ii) and diagrams $C_1$and C_2 – to the third channel (iii).

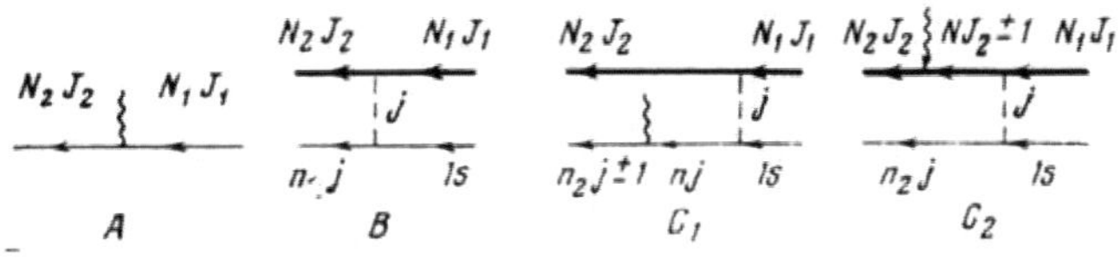

Figure 1. Feynman diagrams corresponding to different channels of a decay (see text).

The thin line on the diagrams (figure 5) corresponds to the muon state, the bold line – to the proton state. The initial and final states of proton and muon are designated by indices on the lines. The dashed line with the index j means the Coulomb interaction between muon and proton with an exchange of the 2^j-pole quanta. The wavy line corresponds to operator of the radiative dipole transition. This effect is due to the muon-proton interaction. The diagram A (fig.1) has the zeroth order on the muon-proton interaction; other diagrams (fig.1) are first order. The probability of purely radiative nuclear 2^j - pole transition is defined by convention (r_n=5·10^{-13} cm) [17]:

$$P_1 = 2 \cdot 10^{20} \cdot \frac{j+1}{j[(2j+1)!!]^2} (\frac{3}{j+3})^2 (\frac{\Delta E[MeV]}{40})^{2j+1} . \quad (1)$$

(standard notation). Diagrams C_1 and C_2 (figure 1) account for an effect of the QP interaction on the initial state. Surely there are other versions of these diagrams, but their contribution to probabilities of the decay processes is significantly less than contributions of the diagrams $C_1$and C_2 .

Within the relativistic energy approach [6,12-16] the total probability is divided into the sum of the partial contributions, connected with decay to the definite final states of a system. These contributions are equal to the corresponding transition probabilities (P_i). If $\Delta E^{p}_{N_1 J_1} > E^{i}_{\mu}$ the probability determination reduces to relativistic calculation of probability for two-QP system autoionization decay. An imaginary excited state energy in the lowest QED PT order is written as follows:

$$\mathrm{Im}\Delta E = e^2\,\mathrm{Im}\,i\lim\iint d^4x_1 d^4x_2 \exp[\gamma(t_1+t_2)]\{D(r_{c_1t_1}, r_{c_2t_2})\cdot$$
$$<\Phi_I\,|(j_{cv}(x_1)j_{cv}(x_2))|\,\Phi_I> + D(r_{p_1t_1}, r_{p_2t_2}) <\Phi_I\,|(j_{pv}(x_1)j_{pv}(x_2))|\,\Phi_I> +$$
$$+ D(r_{\mu_1t_1}, r_{\mu_2t_2}) <\Phi_I\,|(j_{\mu v}(x_1)j_{\mu v}(x_2))|\,\Phi_I>\} \quad (2)$$

Here $D(r_1t_1,r_2t_2)$ is the photon propagator; j_{cv}, j_{pv}, $j_{\mu v}$ are the 4-dimensional components of a current operator for the particles: core, proton, muon; $x=(r_c,r_p,r_\mu,t)$ is the four-dimensional space-time coordinate of the particles, respectively; γ is an adiabatic parameter. Further one should use the exact electrodynamic expression for the photon propagator. Below we are limited by the lowest order of the QED PT, i.e. the next QED corrections to ImΔE will not be considered. Finally, the imaginary part of energy of the excited state can be represented as a sum of the corresponding QP contributions [12]:

$$\mathrm{Im}\ \Delta E = \mathrm{Im}\ E_c + \mathrm{Im}\ E_p + \mathrm{Im}\ E_\mu,$$
$$\mathrm{Im}\,E_a = -Z_a^2/4\pi\sum_F \iint dr_{c1}dr_{c2}\iint dr_{p1}dr_{p2}\iint dr_{\mu 1}dr_{\mu 2}\cdot \quad (3)$$
$$\cdot\,\Phi_I^*(1)\Phi_F^*(2)\cdot T_a(1,2)\Phi_F(1)\Phi_I(2),$$

$$T_a(1,2) = \frac{\exp(w_{IF}\,r_{a12})}{r_{a12}}(1-\alpha_1\alpha_2),$$

Here $r_{a12} = |r_{a1} - r_{a2}|$; Φ_c, Φ_p, Φ_e are the second quantization operators of field of the core particles, proton and muon. Sum on F means the summation on the final states of a system. The technique of calculating the matrix elements in Eq. (3) is very well carried out and in details presented in Refs. [12-16,18,19].

3. Results and conclusions

The nucleus of ${}^{49}_{21}Sc_{28}$ contains a single proton above the twice magic core ${}^{48}_{20}Ca_{28}$. The scheme of the corresponding energy levels for this nucleus is presented in figure 2. The transitions of proton and muon on the first and second stages are noted by the solid and dotted lines. The level $p_{1/2}$ is connected with the ground level $f_{7/2}$ by the E4 transition and with the low-lying level $p_{3/2}$ by the E2 transition. The levels $p_{3/2}$ and $f_{7/2}$ are connected by the E2 transition. One could consider also magnetic transitions between these levels. The life-time for the isolated nucleus in the excited state is of order 10^{-11} s.

There are possible the different approaches to modelling the proton motion in a field of a nuclear core [17-20]. The first model corresponds to the well-known Dirac-Wood-Saxon model [12]. Another approach uses the Bloumkvist-Wahlborn potential [11] (see also Refs. [4,6,20,21]). The electric core potential is given by the Gauss model potential [22]. In the relativistic Dirac-Woods-Saxon model the wave functions are determined by solution of the Dirac equations with the Woods-Saxon potential, where the parameters V_0, a_0, R_0 are fitting using the levels energies. Namely, the parameters are calculated from the fitting condition for the experimental and theoretical energies of the ground and first excited states. For the values V_0=-47.6, R_1==2, R_2=7.65, R=4.75 it has been obtained for the proton states: $E(f7/2)$=-9.62, $E(p3/2)$=-6.53, $E(p1/2)$=-5.24 and for the muon states: $E(1s)$=-1.05, $E(2s)$=-0.272 $E(2p)$ =-0.281 (the units of energy 1 MeV and length units 10^{-13} cm are used).It should be noted that in the last years the relativistic mean field model with using the Dirac-Woods-Saxon orbital set is developed too.

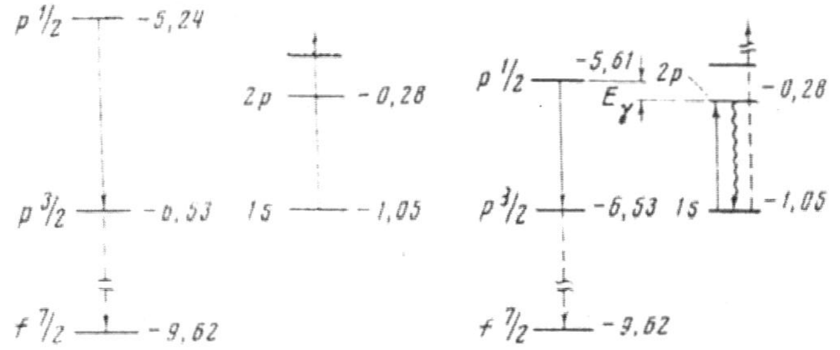

Figure 2 Schemes of energy levels of proton (the left part of the figure) and muon (the right part of the figure) in ${}^{49}_{21}Sc_{28}$. Transitions of proton and muon on the first and second stages are noted by the solid and dotted lines.

We present numerical data (calculation with two nuclear potentials) for the scandium nucleus. The probabilities of the muon-atomic decay (in s^{-1}) for a most interesting nuclear transitions are:
i). for the Bloumkvist-Wahlborn potential [11]:

$$P_2(p_{1/2}\text{-}p_{3/2})=3{,}93\cdot 10^{15},\ P_2(p_{1/2}\text{-}f_{7/2})=3{,}15\cdot 10^{12},\ P_2(p_{3/2}\text{-}f_{7/2})=8{,}83\cdot 10^{14},$$

ii). for the Woods-Saxon potential:

$$P_2(p_{1/2}\text{-}p_{3/2})=3{,}87\cdot 10^{15},\ P_2(p_{1/2}\text{-}f_{7/2})=3{,}09\cdot 10^{12},\ P_2(p_{3/2}\text{-}f_{7/2})=8{,}75\cdot 10^{14}.$$

For both potentials the presented values are significantly higher that the corresponding non-relativistic estimates of Refs. [4,6]. For example, according to [4]: $P_2(p1/2\text{-}p3/2)=3.3\cdot 10^{15}$. If a muon-atomic system is in the initial state $p_{1/2}$, than the cascade discharge occurs with an ejection of the muon on the first stage and the gamma quantum emission on the second stage. To consider a case when the second channel is closed and the third channel is opened, let us suppose that $E^p(p_{1/2})\text{-}E^p(p_{3/2})=0.92$ MeV (figure 2). The nuclear transition energy is not sufficient to provide transition of the muon to the continuum state. However, it is sufficient for the excitation to the $2p$ state. It is important to note here that this energy is not lying in the resonant range. The diagram C_1 (see figure 1) describes the proton transition $p_{1/2}$-$p_{3/2}$ with the virtual excitation of muon to states of the series nd and with emitting gamma quantum of the following energy:

$$hw=E^p(p_{1/2})+E^{\mu}(1s)\text{-}E^p(p_{3/2})\text{-}\ E^{\mu}(2p).$$

Further the dipole transition $2p$-$1s$ can occur. The calculated value for the probability of this transition is $P_3=1.9\cdot 10^{13}$. It should be noted that the value P_3 is more than the probability value for the radiation transition $p_{1/2}$-$p_{3/2}$ and the probability value for the transition $p_{1/2}$- $f_{7/2}$. The transition $p_{3/2}$-$f_{7/2}$ occurs during $\sim 10^{-15}$ s without emission, but with the ejection of a muon.

The experimental possibilities of search of the metastable nucleus discharge effect have been discussed in Refs. [4,6,12] and require a choice of the special type nuclei (a target). Probability of the μ^- capture by the excited nucleus must be comparable or being more than a probability of the capture by other (non-excited) nuclei of a target. As result, the target must be prepared as the excited nuclei concentrate with the minimal size of order or more than the free μ^- running length l in relation to a capture by a nucleus. The condition for fewest excited nuclei in a target is $N_{\min}>l^3 n_0$, where n_0 is a density of the target atoms. For initial μ slowing to energies of 0.1-0.3MeV, the free running length is ~0,1 cm. The required number of metastable nuclei is $N_{\rm min}>10^{19}$ for the density $n_0=10^{22}$cm^{-3}. The radioactivity of such a target is $R=N_{\rm min}/T$, where T is the decay time. For example, for $T=10^2$ days one can get the estimate $R\sim10^3$Ci (1Ci=$3.7\cdot10^{10}$ decays per s). In conclusion, let us note that a development of electron-μ-nuclear spectroscopy of atoms (nuclei) is of a great theoretical and practical interest. The development of adequate approaches to studying the cooperative e-,μ-γ-nuclear processes promises the rapid progress in our understanding of a nuclear decay. Such an approach is useful, providing perspective for developing advanced nuclear models, search for new cooperative effects on the boundary of atomic and nuclear physics, carrying out new methods for treating (muonic chemistry tools) the spatial structure of molecular orbitals, studying the chemical bond nature and checking different models in quantum chemistry, atomic and nuclear-laser spectroscopy [4-12].

References

[1] Gerstain S S, Petrov Yu V and Ponomarev L I 1990 *Phys.-Uspekhi* **160** 3
[2] Kravtsov A V and Mikhailov A I 1994 *Phys. Rev.* A **49** 3566
[3] Letokhov V S and Goldanskii V I 1974 *JETP.* **67** 513
[4] Ivanov L N and Letokhov V S 1976 *JETP.* **70** 19
[5] Ivanov L N and Letokhov V S 1985 *Com. Mod. Phys.* D **4** 169
[6] Glushkov A V and Ivanov L N 1992 Energy approach to discharge of metastable nuclei during negative muon capture 1992 *Preprint of Institute for Spectroscopy of Russian Academy of Sciences* Moscow-Troitsk N92-4-AS
[7] Shahbaz A, Müller C, Bürvenich T J and Keitel C H 2008 *arXiv:phys.* 0812.2565v1 [nucl-th]
[8] Shahbaz A, Müller C, Bürvenich T J and Keitel C H 2009 *Nucl.Phys.* A **821** 106
[9] Shahbaz A, Müller C, Staudt A, Bürvenich T J and Keitel C H 2007 *Phys. Rev. Lett.* **98** 263901
[10] Bürvenich T J, Evers J and Keitel C H 2006 *Phys. Rev.* C **74** 044601
[11] Glushkov A V, Malinovskaya S V, Vitavetskaya L A and Dubrovskaya Yu V 2006 *Recent Advances in Theory of Chemical and Physical Systems, Series: Progress in Theoretical Chemistry and Physics* vol 15, ed J-P Julien, J Maruani, D Mayou, S Wilson and G Delgado-Barrio (Amsterdam: Springer) p 301
[12] Glushkov A V, Khetselius O Yu and Svinarenko A A 2011 *Advances in the Theory of Quantum Systems in Chemistry and Physics, Series: Progress in Theoretical Chemistry and Physics* vol 22, ed P Hoggan, E Brandas, G Delgado-Barrio and P Piecuch (Berlin: Springer) p 51
[13] Glushkov A V, Ivanov L N and Ivanova E P 1986 *Autoionization Phenomena in Atoms* (Moscow: Moscow State University Press) p 58
[14] Ivanov L N, Ivanova E P and Aglitsky E V 1988 *Phys. Rep.* **166** 315
[15] Glushkov A V and Ivanov L N 1992 *Phys. Lett.* A **170** 33
[16] Glushkov A V 2012 *Advances in the Theory of Quantum Systems in Chemistry and Physics, Series: Progress in Theoretical Chemistry and Physics* vol 26, Ch 12, ed K Nishikawa, J Maruani, E Brandas, G Delgado-Barrio and P Piecuch (Berlin: Springer) p 131
[17] Bohr O and Motelsson B 1971 *Structure of Atomic Nucleus* (New York: Plenum)
[18] Glushkov A V, Khetselius O Yu and Malinovskaya S V 2008 *Europ. Phys. Journ* ST **160** 195
[19] Glushkov A V, Khetselius O Yu and Lovett L 2010 *Advances in the Theory of Atomic and Molecular Systems: Dynamics, Spectroscopy, Clusters and Nanostructures, Series: Progress in Theoretical Chemistry and Physics* vol 20, ed P Piecuch, J Maruani, G Delgado-Barrio, S Wilson (Berlin: Springer) p 125
[20] Glushkov A, Lovett L, Khetselius O, Loboda A, Dubrovskaya Yu, Gurnitskaya E 2009 *Int. J. Mod. Phys.* A **24** 611.
[21] Glushkov A V, Rusov V D, Ambrosov S V and Loboda A V 2003 *New Projects and New Lines of Research in Nuclear Physics*, ed G Fazio and F Hanappe (Singapore: World Scientific) p 142
[22] Khetselius O Yu 2009 *Int. J. Quant. Chem.* **109** 3330

Spectroscopy of cooperative electron-gamma-nuclear processes in heavy atoms: NEET effect

O Yu Khetselius

Odessa State University – OSENU, P.O. Box 24a, Odessa-9, SE-65009, Ukraine

E-mail: nuckhet@mail.ru

Abstract. A consistent relativistic energy approach, based on the S-matrix Gell-Mann and Low formalism (energy approach) combined with the relativistic many-body perturbation (PT) theory, to studying NEET (nuclear excitation by electron transition) effect is presented. The calculation results of the NEET probabilities for the $^{189}_{76}Os$, $^{193}_{77}Ir$, $^{197}_{79}Au$ atoms are presented and compared with available experimental and other theoretical data.

1. Introduction

This paper goes on our work on studying the cooperative electron-gamma-nuclear processes [1-4]. The important example of the cooperative electron-gamma-nuclear process is so called NEET (nuclear excitation by electron transition) effect [5-7]. Naturally, the similar NEEC (nuclear excitation by electron capture) process should be reminded too. In both NEEC and NEET, which are at the borderline between atomic and nuclear physics, electronic orbital energy is converted directly into nuclear energy. These effects offer therefore the possibility to explore the spectral properties of heavy nuclei through the typical atomic physics experiments. In fact, the NEET is a fundamental but rare mode of decay of an excited atomic state in which the energy of atomic excitation is transferred to the nucleus via a virtual photon. This process is naturally possible if within the electron shell there exists an electronic transition close in energy and coinciding in type with nuclear one. In fact the resonance condition between the energy of nuclear transition w_N and the energy of the atomic transition w_A should be fulfilled. Obviously, the NEET process corresponds to time-reversed bound-state internal conversion. Let us remind that firstly the NEET and NEEC effects have been postulated in 1973 by Morita and Goldanskii–Letokhov-Namiot [5-7] (see also review [1] and Refs.[8-10]). Unlike the NEEC effect, the NEET process has been observed experimentally in $^{197}_{79}Au$ by Kishimoto et al (Institute of Material Structure Science, KEK, and Japan Synchrotron Radiation Research Centre, Japan) and in $^{189}_{76}Os$ by Ahmad et al (Argonne National Laboratory, USA) [11-13]. Below in table 1 we present a summary of the experimental works on NEET in $^{189}_{76}Os$. It should be noted that each of the experimental techniques has certain inherent difficulties. Analysis of this problem has been presented in Refs. [1,13]. It explains quite large difference between the results of different experiments. Saying briefly, the cited difficulties are reduced to the problem of revealing a NEET signal among the surrounding other effects. Really, use of an electron beam can cause direct Coulomb excitation of a nucleus. In this case it is hardly possible to distinguish this component from that due to the NEET process. Use of a broad continuous spectral distribution of synchrotron or bremsstrahlung X-rays results in contribution from a direct nuclear photoabsorption into the nuclear state or into a range of nuclear levels that can feed that state or the lower-lying metastable state.

Table 1. Experimental data on the NEET probabilities P_{NEET} for the isotope of ${}^{189}_{76}Os$.

	year	Experimental techniques	P_{NEET}
Otozai et al	1973	e^- bombardment 75-85 keV	$1\cdot10^{-6}$
Otozai et al	1978	e^- bombardment 72-100 keV	$(1.7\pm0.2)\cdot10^{-7}$
Saito et al	1981	200 keV bremsstrahlung	$(4.3\pm0.2)\cdot10^{-8}$
Shinohara etal	1987	„white" synchrotron radiation	$(5.7\pm1.7)\cdot10^{-9}$
Lakosi et al	1995	300 keV bremsstrahlung	$(2.0\pm1.4)\cdot10^{-8}$
Ahmad et al	2000	monochromatic 100-keV X rays	$<9\cdot10^{-10}$

The theoretical models for the NEET effect were developed in Refs. [5-17] (see details in [1]). Many of the early estimates involved the using simplifying approximations that led to results at considerable variance. More lately Tkalya [16,17] has proposed a model for description of the NEET process near the K-shell ionization threshold of an atom. The QED PT with empirical estimates of the nuclear and electron matrix elements and the Dirac-Fock code by Band and Fomichev (taking into account the finite nuclear size) were used. New theoretical approach by Ahmad et al [13] is based on using the time-dependent amplitude coupled equations. These authors calculated electron wave functions using the GRASP code and tabulated values of the nuclear transition matrix elements. Therefore, the theoretical models involved the use of different consistency level approximations led to results at quite considerable variance. It is obvious that more sophisticated relativistic many-body methods should be used for correct treating the NEET effect. Really, the nuclear wave functions have the many-body character. The correct treating the electron subsystem processes requires an account of the relativistic, exchange-correlation and nuclear effects. Really, the nuclear excitation occurs by electron transition from the M shell to the K shell. So, there is the electron-hole interaction and it is of a great importance a correct account for the many-body correlation effects, including inter-shell correlations, the post-act interaction of the removing electron and hole, possibly the continuum pressure etc [1,17,18]. In any case the theoretical calculations for NEET occurring in scattering measurements are particularly useful, especially in finding candidate isotopes and transitions suitable for experimental observation. In our paper a new, consistent relativistic energy approach to calculation of probabilities of the NEET is presented. In our approach the NEET process probability and cross-section are determined within the S-matrix Gell-Mann and Low formalism (energy approach) combined with the relativistic many-body PT [18-23]. The calculation results of the NEET probabilities for the ${}^{189}_{76}Os$, ${}^{193}_{77}Ir$, ${}^{197}_{79}Au$ atoms within different theoretical models are presented and compared with available experimental data.

2. Relativistic energy approach to process of nuclear excitation by electron transition

The relativistic energy approach is based on the S-matrix Gell-Mann and Low formalism combined with the relativistic many-body PT [1,19-23]. Let us remind that in atomic theory, a convenient field procedure is known for calculating the energy shifts ΔE of the degenerate states [20]. Secular matrix M diagonalization is used. In constructing M, the Gell-Mann and Low adiabatic formula for ΔE is used. A similar approach, using this formula with the electrodynamical scattering matrix, is applicable in the relativistic theory [19-21]. In contrast to the non-relativistic case, the secular matrix elements are already complex in the PT second order (1storder of the inter-electron interaction). Their imaginary parts relate to radiation decay (transition) probability. The total energy shift of the state is as:

$$\Delta E = \mathrm{Re}\Delta E + \mathrm{i}\ \mathrm{Im}\Delta E, \tag{1}$$

$$\mathrm{Im}\ \Delta E = -\Gamma/2, \tag{2}$$

where Γ is interpreted as the level width, and the decay possibility P=Γ. The whole calculation of energies and decay probabilities of a non-degenerate excited state is reduced to calculation and diagonalization of the complex matrix *M*. To start with the Gell-Mann and Low formula it is necessary to choose the PT zero-order approximation. Usually, the one-electron Hamiltonian is used, with a central potential that can be treated as a bare potential in the formally exact PT [21]. The total probability of radiative decay (excitation, de-excitation) is connected with imaginary part of ΔE of the system "atom plus field" [18-21]. The corresponding corrections of the PT for ImΔE can be represented as a sum on the virtual states. In the lowest PT the separated terms of these sums correspond to the additive contributions of different physical channels into the total decay probability. Naturally the channels interference effects will appear in the next PT orders. The fundamental parameter of the cooperative NEET process is a probability P_{NEET} (cross-section) of the nuclear excitation by electron transition. In fact it can be defined as the probability that the decay of the initial excited atomic state will result to the excitation of and subsequent decay from the nuclear state. Within the energy approach a decay probability is connected with an imaginary part of energy shift for the system (nuclear subsystem plus electron subsystem) excited state. An imaginary part of the excited state energy shift in the lowest PT order can be written as [1]:

$$\begin{aligned} \operatorname{Im}\Delta E = e^2 \operatorname{Im} i \cdot \lim_{\gamma\to 0} \iint d^4x_1 d^4x_2 e^{\gamma(t_1+t_2)} \cdot \\ \cdot\{D(r_{N1t1}, r_{N2t2}) < \Psi_I \,|\, (\hat{J}_N(x_1)\hat{J}_N(x_2)) \,|\, \Psi_I > + \\ + D(r_{e1t1}, r_{e2t2}) < \Psi_I \,|\, (\hat{j}_e(x_1)\hat{j}_e(x_2)) \,|\, \Psi_I >\} \end{aligned} \quad (3)$$

Here $D(r_1t_1,r_2t_2)$ is the photon propagator (for example in the Lorenz gauge); $\hat{j}_N$, $\hat{j}_e$— are the 4-dimensional components of a current operator for the nuclear and electron (hole) subsystems; $x=(r_n,r_e, t)$ is the four-dimensional space-time coordinate of the particles, respectively; γ is an adiabatic parameter. The nuclear current can be written as follows:

$$J^P(R,t) = \psi^+_{N*}\hat{J}^P\psi_N, \quad (4)$$

where $\hat{J}^P$ is operator of an nuclear electromagnetic transition, ψ_N is a nuclear wave function. The current operator for electron is

$$\bar{j}^\mu_e = \bar{\hat{\psi}}_e \gamma^\mu \hat{\psi}_e, \quad (5)$$

where γ^μ are the Dirac matrices. The Hamiltonian of the interaction of the electronic hole current j^μ_{fi} and the nuclear current $J^\nu_{fi}(R)$ is written as :

$$H_{int} = \int d^3r\, d^3R\; j^\mu_{fi}\, D_{\mu\nu}(x_N, r-R)\; J^\nu_{fi}(R) \quad (6)$$

The energy shift can be further represented as the PT set. After integration transformations the final expression for the imaginary part of energy shift can be represented as a sum of the corresponding nuclear-electron (hole) contributions:

$$\operatorname{Im}\Delta E = \operatorname{Im} E_e + \operatorname{Im} E_N,$$

$$\operatorname{Im} E_a = -\frac{z_a^2}{4\pi}\sum_F \iint dr_{e1}dr_{e2} \iint dr_{N1}dr_{N2} \cdot \Psi^*_I(1)\Psi^*_F(2)\hat{T}_a(1,2)\Psi_F(1)\Psi_I, \quad (10)$$

$$\hat{T}_a(1,2) = \frac{\sin(\omega_{IF} r_{a12})}{r_{a12}}$$

Here, as usually, $r_{a12} = |r_{a1} - r_{a2}|$, ω_{IF} is the energy of transition between the initial I and final F states; the sum on F means the summation on the final states of a system. Naturally, the form of operator in (10) is determined by a gauge of the photon propagator (look discussion in Ref. [21]). In the zeroth approximation the dependence Ψ_F, Ψ_I on the nuclear and electron coordinates (R_N , $R_{e(h)}$) is factorized ($\sim \Phi_e \Phi_N$). Therefore, the combined electron (hole)- nuclear one-photon transitions occur as each of the operators T_N, T_e in (10) contains the combination of the nuclear and electron variables. After factorization and some transformations the expression (10) can be presented in the following form:

$$\begin{aligned} \operatorname{Im} E_a = -\frac{z_a^2}{4\pi} \sum_{F_e F_N} \iint dR_{N1} dR_{N2} \iint dR_{e1} dR_{e2}\, \Phi^*_{Ie}(R_{e1}) \Phi^*_{IN}(R_{N1}) \Phi^*_{Fe}(R_{e2}) \cdot \\ \cdot \Phi^*_{FN}(R_{N2}) \frac{\sin \omega_{IF} R_{a12}}{R_{a12}} \Phi_{Fe}(R_{e1}) \Phi_{FN}(R_{N1}) \Phi_{Ie}(R_{e2}) \Phi_{IN}(R_{N2}). \end{aligned} \quad (11)$$

The expansion of the operator $\frac{\sin(\omega_{IF} R_{a12})}{R_{a12}}$ on the spherical harmonics generates the decay probability multipole expansion. It can be written in the following known form:

$$\frac{\sin|\omega| R_{12}}{R_{a12}} = \frac{\pi}{2\sqrt{R_1 R_2}} \sum_{\lambda=0}^{\infty} (\lambda) J_{\lambda+1/2}(|\omega| R_{a1}) J_{\lambda+1/2}(|\omega| R_{a2}) P_\lambda(\cos \widehat{R_{a1} R_{a2}}), \quad (12)$$

where J is the Bessel function of the first kind and $(\lambda)=2\lambda+1$. In fact this expansion coincides with the known power expansion; naturally the strict decreasing contribution on multipolarity corresponds to them. In our problem the power expansion parameters are the combinations $\omega^a_{IF} R_e$, $\omega^N_{IF} R_N$. Further the effects of purely nuclear transition, purely electron-(hole) transition and combined electron – nuclear transition in (11) can be distinguished. The corresponding technique of work with these expansions is well developed [8,19-21] and often used in our precious papers (look [1-4]). Finally the NEET probability P_{NEET} is connected with the imaginary part of energy of the excited nuclear-electron state. It can be shown that P_{NEET} can be presented in the following form [17]:

$$P_{NEET} = \left(1 + \frac{\Gamma_i}{\Gamma_f}\right) \frac{M^2_{int}}{(\omega_N - \omega_A)^2 + (\Gamma_i + \Gamma_f)^2 / 4} \,. \quad (13)$$

Here, as usually, $\Gamma_{i,f}$ are the widths of the initial and final electron states; M^2 is averaged over initial states and summed over the final states the square modulus of the Hamiltonian of the electron hole current- nuclear current interaction. It can be written (M_1 –K transition) as follows (see Ref. [17]):

$$M^2_{int} = 4\pi e^2 \omega_N^{2(\lambda+1)} \frac{(j_i \frac{1}{2} \lambda O | j_f \frac{1}{2})^2}{[(2\lambda+1)!!]^2} \left| R_\lambda^{E/M}(\omega_N) \right|^2 B(E/M\lambda; J_i \to J_f). \quad (14)$$

Here $B[E/(M)\lambda;\ J_i\text{-}J_f\,]$ is the reduced nuclear probability, $|R_\lambda^{E/M}(\omega_N)|$ are the atomic radial matrix elements of electric (magnetic) [E/M] multipolarity λ; $j_{i,f}$ and $J_{i,f}$ are the angular momenta of the electronic and nuclear states correspondingly. The atomic radial matrix elements $|R_\lambda^M(\omega_N)|$ of electric (magnetic) [E/M] multipolarity λ are expressed by means the integral:

$$\int_0^\infty drr^2 Z_\lambda^{(1)}(\omega r)[g_i(r)f_f(r)+f_i(r)g_f(r)], \qquad (15)$$

where *f(r)* and *g(r)* are the large and small components of the Dirac electronic wave functions and

$$Z_\lambda^{(1)} = \left[\frac{2}{|\omega|\alpha Z}\right]^{\lambda+1/2} \frac{J_{\lambda+1/2}(\alpha|\omega|r)}{r^\lambda \Gamma(\lambda+3/2)} \qquad (16)$$

Other details can be found in Refs. [1-4,18-21,23].

3. Results and conclusions

In concrete calculation of the NEET probabilities for different atomic/nuclear systems one should calculate the corresponding matrix elements. As we will consider below M1 (E2) transition from the ground state to the first excited state in the nuclei $^{189}_{76}Os$, $^{193}_{77}Ir$, $^{197}_{79}Au$, it should be noted that the values of $B[E/(M)\lambda;\ J_i\text{-}J_f\,]$ are usually taken from the Nuclear Data tables [33] or can be estimated according the known formula (look [24,25]). In order to calculate the electronic wave functions and matrix elements we have used the relativistic many-body PT formalism [1-4,18,23]. It allows take into account accurately the relativistic, exchange-correlation, nuclear, radiative corrections (the PC code "Superatom-ISAN"). The corresponding code contains the atomic and nuclear blocks. The zeroth approximation electronic wave functions are found from the Dirac (or Dirac-Kohn-Sham) equation with potential, which includes the SCF potential, the electric and polarization potentials of a nucleus. As an account of the finite nuclear size has a sensitive effect on the energy levels of the bound electron, we usually use the smooth Gaussian (or Fermi) function of the charge distribution in a nucleus. The correlation corrections of the second and high orders are taken into account within the Green functions method (with the use of the Feynman diagram's technique). There have taken into account all correlation corrections of the second order and dominated classes of the higher orders diagrams [18]. The magnetic inter-electron interaction is accounted in the lowest (on α^2 parameter; α is the hyperfine structure constant). The radiative corrections are taken into account effectively, namely, the Lamb shift self-energy part is accounted within the generalized Ivanov-Ivanova non-perturbative procedure and the polarization part - in the generalized Uehling-Serber approximation [18]. The important feature of the whole method is using the optimized one-quasiparticle representation in the zeroth approximation, which is constructed within the method [21]. The nuclear part of the general method includes a set of the nuclear shells models, including the relativistic mean-field approach and the Dirac-Bloumkvist-Wahlborn and Dirac-Woods-Saxon models [2,26-28]. The calculation results on the NEET probability for the $^{189}_{76}Os$, $^{193}_{77}Ir$, $^{197}_{79}Au$ atoms together with the alternative theoretical (by Tkalya and Ahmed et al) [13,16,17] and experimental data (see [11-13 and Refs. therein] are listed in the table 2. Let us note that in $^{189}_{76}Os$ during the NEET process the initial K-vacancy state decays via an electronic transition from the M shell. The KM_I (70.822 keV, M1), KM_{IV} (71.840 keV, E2) and KM_V (71.911keV, E2) atomic transitions can give the contribution.

Table 2. Theoretical and experimental probabilities P_{NEET} (M1) for $^{189}_{76}Os$, $^{193}_{77}Ir$, $^{197}_{79}Au$.

Nucleus	Energy of nuclear excitation (keV)	Experimental values	Theory: Tkalya and Ahmed etal	Present work
$^{189}_{76}Os$	69.535	$<9.5\cdot10^{-10}$	$1.2\cdot10^{-10}$ $1.3\cdot10^{-10}$	$1.9\cdot10^{-10}$
$^{193}_{77}Ir$	73.04	$(2.8\pm0.4)\cdot10^{-9}$	$2.0\cdot10^{-9}$	$2.7\cdot10^{-9}$
$^{197}_{79}Au$	77.351	$(5.7\pm1.2)\cdot10^{-8}$ $(4.5\pm0.6)\cdot10^{-8}$	$3.4\cdot10^{-8}$ $4.5\cdot10^{-8}$	$4.6\cdot10^{-8}$

The corresponding nuclear state at 69.535 keV can be excited via M1 or E2 transitions from the $3/2^-$ nuclear ground state. The following energy parameters ω_N=69.535 keV, $\omega_A=E_{M1}-E_K$=70.822 keV, Γ_K=42.6 eV, Γ_M=12.8 eV are used for the $^{189}_{76}Os$ atom. Correspondingly, the energy parameters for $^{197}_{79}Au$ are as follows: ω_N=77.351 keV, ω_A=77.325 keV, Γ_K=52 eV, Γ_M =14.3 eV and for $^{193}_{77}Ir$: ω_N=73.04 keV, ω_A=72.937 keV, Γ_K =45 eV, Γ_M =12.8 eV. Analysis of all presented theoretical data shows that these results are consistent with each other and are in the physically reasonable agreement with the experimental results.

References

[1] Khetselius O Yu 2012 *Advances in the Theory of Quantum Systems in Chemistry and Physics, Series: Progress in Theoretical Chemistry and Physics* vol 28, ed. K Nishikawa, J Maruani, E Brandas, G Delgado-Barrio and P Piecuch (Berlin: Springer) p 57
[2] Glushkov A V, Khetselius O Yu and Lovett L 2010 *Advances in the Theory of Atomic and Molecular Systems: Dynamics, Spectroscopy, Clusters and Nanostructures, Series: Progress in Theoretical Chemistry and Physics* vol 20, ed P Piecuch, J Maruani, G Delgado-Barrio, S Wilson (Berlin: Springer) p 125
[3] Glushkov A V, Khetselius O Yu and Malinovskaya S V 2008 *Europ. Phys. Journ* ST **160** 195
[4] Glushkov A V, Khetselius O Yu and Malinovskaya S V 2008 *Molec. Phys.* **106** 1257
[5] Morita M 1973 *Progr. Theor. Phys.* **49** 1574
[6] Letokhov V S and Goldanskii V I 1974 *JETP.* **67** 513
[7] Goldanskii V I and Namiot V A 1976 *Phys. Lett.* B **62** 393
[8] Ivanov L N and Letokhov V S 1985 *Com. Mod. Phys.* D **4** 169
[9] Glushkov A V, Ivanov L N and Letokhov V S 1991 Nuclear quantum optics *Preprint of ISAN* Troitsk N AS-1
[10] Okamoto K 1980 *Nucl. Phys.* A **341** 75
[11] Kishimoto S, Yoda Y, Seto M, Kobayashi Y, Kitao S, Haruki R, Kawauchi T, Fukutani K and Okano T 2000 *Phys. Rev. Lett.* **85** 1831
[12] Kishimoto S, Yoda Y, Kobayashi Y, Kitao S, Haruki R, Masuda R and Seto M 2006 *Phys. Rev.* C **74** 031301
[13] Ahmad I, Dunfird R, Esbensen H, Gemmell D S, Kanter E P, Run U and Siuthwirth S H 2000 *Phys. Rev.* C **61** 051304
[14] Shinohara A, Saito T, Shoji M, Yokoyama A, Baba H, Ando M and Taniguchi K 1987 *Nucl. Phys.* A **472** 151
[15] Morel P, Daugas J M, Gosselin G, Méot V and Gogny D 2004 *Nucl. Phys.* A **746** 608
[16] Tkalya E V 1992 *Nucl. Phys.* A **539** 209
[17] Tkalya E V 2007 *Phys. Rev.* A **75** 022509
[18] Khetselius O Yu 2009 *Int. Journ. Quant. Chem.* **109** 3330
[19] Glushkov A V, Ivanov L N and Ivanova E P 1986 *Autoionization Phenomena in Atoms* (Moscow: Moscow State University Press) p 58
[20] Ivanov L N, Ivanova E P and Aglitsky E V 1988 *Phys. Rep.* **166** 315
[21] Glushkov A V and Ivanov L N 1992 *Phys. Lett.* A **170** 33
[22] Glushkov A V, Loboda A V, Gurnitskaya E P and Svinarenko A A 2009 *Phys. Scripta* T **135** 014022
[23] Khetselius O Yu 2009 *Phys. Scripta* T **135** 014023
[24] Wu S C and Niu H 2003 *Nucl. Data Sheets* **100** 1
[25] Xiaolong H and Chunmei Z 2006 *Nucl. Data Sheets* **104** 283
[26] Serot B and Walecka J 1986 *Adv. Nucl. Phys.* **16** 1
[27] Bender M, Heenen P and Reinhard P 2003 *Rev. Mod. Phys.* **75** 121
[28] Glushkov A V, Lovett L, Khetselius O Yu, Loboda A V, Dubrovskaya Yu V, Gurnitskaya E P 2009 *Int. J. Mod. Phys.* A **24** 611

Spectroscopy of hadronic atoms: Energy shifts

I N Serga, Yu V Dubrovskaya, A S Kvasikova, A N Shakhman and D E Sukharev

Odessa State University – OSENU, P.O. Box 24a, Odessa-9, 65009, Ukraine

E-mail: nucserga@mail.ru

Abstract. An analysis of the electromagnetic and strong interactions contributions to the transitions energies in the X-ray spectra of some pionic atomic systems is presented within the relativistic theory.

1. Introduction

In this paper we give an analysis of the electromagnetic and strong interactions contributions to the transitions energies in the X-ray spectra of pionic atomic systems. As it is known, one of the fundamental questions in the modern hadron's physics is that the hadron masses being much higher than the mass of their quark content. The current mass of the up (u) and down (d) quarks is two orders of magnitude smaller than a typical hadron's mass of about 1 GeV. This extraordinary phenomenon is proposed to originate from spontaneous breaking of chiral symmetry of massless quarks in strong interaction physics [1-5]. At present time one of the most sensitive tests for the chiral symmetry breaking scenario in the modern hadron's physics is provided by studying the exotic hadron-atomic systems. One could turn the attention on the some differences between the pionic and kaonic systems. In the kaonic case one deals with the strangeness sector, and, for example, the strong interaction effect amounts to a very small shift and width in pionic hydrogen of 7 and 1 eV respectively, Nowadays the transition energies in pionic (kaonic, muonic etc.) atoms are measured with an unprecedented precision and from studying spectra of the hadronic atoms it is possible to investigate the strong interaction at low energies measuring the energy and natural width of the ground level with a precision of few meV [1-10]. The mechanism of creation of the hadronic atoms is well known now (e.g. [1,2]).. Really, such an atom is formed when a negative pion enters a medium, looses its kinetic energy through ionization and excitation of the atoms and molecules and eventually is captured, replacing the electron, in an excited atomic orbit. The further de-excitation scenario includes the different cascade processes such as the Auger transitions, Coulomb de-excitation, scattering etc. When a pion reaches a low-n state with the little angular momentum, strong interaction with the nucleus causes its absorption. The strong interaction is the reason for a shift in the energies of the low-lying levels from the purely electromagnetic values and the finite lifetime of the state corresponds to an increase in the observed level width. The possible energy shifts caused by the pion-induced fluorescence X-rays were checked in the measurement of the pion beams at PSI in Switzerland. The systematic error of the fit functions is found to be 2 eV. The most known theoretical models to treating the hadronic atomic systems are presented in refs. [1-5,7,8]. The most difficult aspects of the theoretical modeling are reduced to the correct description of the kaon (pion)-strong interaction [1-3] as the electromagnetic part of the problem is reasonably accounted for. Besides, quite new aspect is linked with the possible, obviously, very tiny electroweak and hyperfine interactions.

2. Relativistic approach to pionic atoms spectra

All available theoretical models to treating the hadronic (kaonic, pionic) atoms are naturally based on the using the Klein-Gordon equation [10-12]:

$$m^2c^2\Psi(x) = \{\frac{1}{c^2}[i\hbar\partial_t + eV_0(r)]^2 + \hbar^2\nabla^2\}\Psi(x), \tag{1}$$

where c is a speed of the light, h is the Planck constant, and $\Psi_0(x)$ is the scalar wave function of the space-temporal coordinates. Usually one considers the central potential $[V_0(r), 0]$ approximation with the stationary solution:

$$\Psi(x) = \exp(-iEt/\hbar)\varphi(x), \tag{2}$$

where $\varphi(x)$ is the solution of the stationary equation:

$$\{\frac{1}{c^2}[E + eV_0(r)]^2 + \hbar^2\nabla^2 - m^2c^2\}\varphi(x) = 0 \tag{3}$$

Here E is the total energy of the system (sum of the mass energy mc^2 and binding energy ε_0). In principle, the central potential V_0 naturally includes the central Coulomb potential, the vacuum-polarization potential, the strong interaction potential. Standard approach to treating the last interaction is provided by the well known optical potential model (c.g. [2]). Practically in all works the central potential V_0 is the sum of the following potentials. The nuclear potential for the spherically symmetric density $\rho(r|R)$ is [6,13]:

$$V_{nucl}(r|R) = -((1/r)\int_0^r dr' r'^2 \rho(r'|R) + \int_r^\infty dr' r' \rho(r'|R)) \tag{4}$$

The most popular Fermi-model approximation the charge distribution in the nucleus $\rho(r)$ (c.f.[1,2]) is as follows:

$$\rho(r) = \rho_0 / \{1 + \exp[(r-c)/a)]\}, \tag{5}$$

where the parameter $a=0.523$ fm, the parameter c is chosen by such a way that it is true the following condition for average-squared radius: $\langle r^2\rangle^{1/2} = (0.836 \cdot A^{1/3} + 0.5700)$fm. The effective algorithm for its definition is used in refs. [7,18-20] and reduced to solution of the following system of the differential equations:

$$V'_{nucl}(r,R) = (1/r^2)\int_0^r dr' r'^2 \rho(r',R) \equiv (1/r^2)y(r,R), \tag{6}$$

$$y'(r,R) = r^2\rho(r,R), \tag{7}$$

$$\rho'(r) = (\rho_0/a)\exp[(r-c)/a]\{1 + \exp[(r-c)/a)]\}^2 \tag{8}$$

with the corresponding boundary conditions. Another, probably, more consistent approach is in using the relativistic mean-field (RMF) model, which been designed as a renormalizable meson-field theory for nuclear matter and finite nuclei [14].To take into account the radiation corrections, namely, the effect of the vacuum polarization there are traditionally used the Ueling potential and its different modifications such as [4,13].

The most difficult aspect is an adequate account for the strong interaction. In the pion-nucleon state interaction one should use the following pulse approximation expression for scattering amplitude of a pion on the "i" nucleon [1,2]:

$$f_i(r)=\{b_0'+b_1'(t\tau)+[c_0'+c_1'(t\tau)]kk'\}\delta(r-r_i); \tag{9}$$

where t and τ are the isospines of pion and nucleon. The nucleon spin proportional terms of the kind $\sigma[kk']$ are omitted. The constants in (9) can be expressed through usual s-wave (α_{2T}) and p-wave $(\alpha_{2T,2J})$ scattering length (T and J -isospin and spin of he system πN. The corresponding parameters in the Compton wave length λ_π terms are as follows:

$$b_0'=(\alpha_1+2\alpha_3)/3=-0.0017\,\lambda_\pi.$$
$$b_0'=(\alpha_3-\alpha_1)/3=-0.086\,\lambda_\pi.$$
$$c_0'=(4\alpha_{33}+2\alpha_{13}+2\alpha_{31}+\alpha_{11})/3=-0.208(\lambda_\pi)^3.$$
$$c_1'=(2\alpha_{33}-2\alpha_{13}+\alpha_{31}-\alpha_{11})/3=-0.184(\lambda_\pi)^3. \tag{10}$$

The scattering amplitude for pin on a nucleus is further received as a coherent sum of the πN-scattering lengths. рассеяния. In approximation of the only s-wave interaction the corresponding potential can be written in the Dezer form [1,2]:

$$V_N(r)=-2\pi\hbar^2\mu_\pi^{-1}[ZA^{-1}a_p+(A-Z)A^{-1}a_n]\rho(r). \tag{11}$$

The s-wave lengths of the $\pi^{-1}p$-scattering $a_p=(2\alpha_1+\alpha_3)/3$ и $\pi^{-1}n$- scattering $a_n=\alpha_3$; scattering are introduced to Eq. (11). Because of the equality between $a_n=b_0'+b_1'$ and $a_p=b_0'-b_1'$ (with an opposite sign) the theoretical shift of the s-level with $T=0$ $(A=2Z)$ from Eq. (12) is much less than the observed shift. So, the more correct approximation must take into account the effects of the higher orders. In whole the energy of the hadronic atom is represented as the sum:

$$E\approx E_{KG}+E_{FS}+E_{VP}+E_N; \tag{12}$$

Here E_{KG}-is the energy of a pion in a nucleus (Z,A) with the point-like charge (dominative contribution in (12)), E_{FS} is the contribution due to the nucleus finite size effect, E_{VP} is the radiation correction due to the vacuum-polarization effect, E_N is the energy shift due to the strong interaction V_N.The last contribution can be defined from the experimental energy values as:

$$E_N=E-(E_{KG}+E_{FS}+E_{VP}) \tag{13}$$

From the other side the strong pion-nucleus interaction contribution can be found from the solution of the Klein-Gordon equation with the corresponding meson-nucleon potential. In this case, this contribution E_N is the function of the potentials (8)-(11) parameters.

3. Results and conclusions

In table 1 the data on the transition energies in some pionic atoms (from. Refs. [1-3]): the measured values form the Berkley, CERN and Virginia laboratories, the theoretical values for the $2p-1s$, $3d-2p$, $4f-3d$, $5g-4f$ pionic transitions (E^N_{th1} - values from the Klein-Gordon-Fock equation with the pion-nucleus potential [2]; E_{KGF}- values from the Klein-Gordon-Fock equation with the finite size nucleus potential (our data), E^N_{th2} - values from the Klein-Gordon-Fock equation with the generalized pion-nucleus potential [12]: our data).

Table 1. Transition energies (keV) in the spectra of some pionic atoms

Element	E_{EXP}				E_{th}	
	Berkley	CERN	Virginia	E_{KGF}	E^N_{th1}	E^N_{th2}
		Transition *4f-3d*				
Cs^{133}	560.5±1.1	562.0±1.5		553.330	561.47	562.12
		Transition *5g-4f*				
Tl^{204}		561.67±0.25		556.562	560.93	561.63
Pb^{207}		575.56±0.25		570.614	575.21	575.78
U^{238}	731.4±1.1	732.0±0.4	730.88±0.75	724.317	729.80	730.52

It is easily to understand that when there is the close agreement between theoretical and experimental shifts, the corresponding energy levels are not significantly sensitive to strong nuclear interaction, i.e the electromagnetic contribution is dominative. In the opposite situation the strong-interaction effect is very significant. The analysis of the presented data indicate on the necessity of the further more exact experimental investigations and further improvement of the pion-nuclear potential modelling. Really, under availability of the "exact" values of the transitions energies one can perform the comparison of the theoretically and experimentally defined transition energies in the X-ray spectra in order to make a redefinition of the pion-nucleon model potential parameters using Eqs. (9)-(11). Taking into account the increasing accuracy of the X-ray pionic atom spectroscopy experiments, one can conclude that the such a way will make more clear the true values for parameters of the pion-nuclear potentials and correct the disadvantage of widely used parameterization of the potentials (9)-(11).

References

[1] Deslattes R, Kessler E, Indelicato P, de Billy L, Lindroth E and Anton J 2003 *Rev. Mod. Phys.* **75** 35
[2] Menshikov L I and Evseev M K 2001 *Phys. Uspekhi* **171** 150
[3] Hayano R S, Hori M, Horvath D and Widman E 2007 *Rep. Prog. Phys.* **70** 1995
[4] Mohr P J 1993 *Phys. Scripta* **46** 44
[5] Pavlovich V N, Zelentsova T N and Serga I N 2010 *Photoelectronics* **19** 65
[6] Glushkov A V, Rusov V D, Ambrosov S V and Loboda A V 2003 *New Projects and New Lines of Research in Nuclear Physics*, ed G Fazio and F Hanappe (Singapore: World Scientific) p 142
[7] Anagnostopoulos D F, Biri S, Boisbourdain V et al 2003 *Nucl. Instr. Methods* B **205** 9
[8] Anagnostopoulos D F, Gotta D, Indelicato P and Simons L M 2003 *arXiv: phys.* 0312090v1
[9] Batty C J, Eckhause M, Gall K P et al 1989 *Phys. Rev.* C **40** 2154
[10] Santos J P, Parente F, Boucard S, Indelicato P and Desclaux J P 2005 *Phys. Rev.* A **71** 032501
[11] Indelicato P and Trassinelli M 2004 *arXiv:phys.* 0510126v1
[12] Serga I N 2011 *Photoelectronics* **20** 43
[13] Khetselius O Yu 2009 *Int. Journ. Quant. Chem.* **109** 3330
[14] Serot B and Walecka J 1986 *Adv. Nucl. Phys.* **16** 1

Secondary peaks in the atomic ionization by a resonant laser pulse

M G Bustamante[1], V D Rodríguez[1] and R O Barrachina[2]

[1] Departamento de Física and Instituto de Física de Buenos Aires, FCEyN, Universidad de Buenos Aires C.C. 67, Suc. 28, 1428 Buenos Aires, Argentina
[2] Centro Atómico Bariloche and Instituto Balseiro (Comisión Nacional de Energía Atómica, CNEA, and Universidad Nacional de Cuyo, UNC), R8402AGP S. C. de Bariloche, Río Negro, Argentina. Also member of CONICET, Argentina.

E-mail: vladimir@df.uba.ar, barra@cab.cnea.gov.ar

Abstract. The above-threshold ionization of atoms by XUV short laser pulses with frequencies resonant with a ground-excited state transition is investigated. A theory based on a variational expression with trial wave functions for the final and the initial states is presented. For the former we use a Coulomb-Volkov wave function, and for the latter a close-coupling solution of the time-dependent Schrödinger equation considering a few bound states and their depletion towards the continuum. We find that this discrete-continuum Coulomb-Volkov theory provides a very good description of the photoelectron spectrum. In particular it accounts for the splitting of the above-threshold ionization peaks by Rabi oscillations and the appearance of secondary-peaks observed in full numerical simulations.

1. Introduction

Resonances in above-threshold ionization (ATI) have been observed in different contexts. For instance, experiments with rare gases exhibit a rescattering plateau enhancement for sharply defined laser intensities [1, 2]. In a previous article [3], we reported the appearance of doublets in the photoelectron spectra of positronium irradiated by short laser pulses and showed that they are related to a bound–bound resonance known as Autler–Townes doublets [4]. Furthermore, the ATI peaks splitting, a consequence of Rabi oscillations, was described by means of a two-state Coulomb-Volkov approximation [3]. The ATI photoelectron spectrum also shows resonance enhancement effects as recently investigated [5]. Previous non-perturbative semi-analytical theories [6, 7], besides employing the Coulomb-Volkov approximation for describing the time-dependent final continuum wave function, have also relied on a field-unperturbed bound initial state. This simple description of the initial state is not reliable for processes where an intermediate field-dressed state is populated by a resonant tuned laser. For this reason, we have recently improved the description of the initial state by employing an exact two-state close-coupling calculation, and applied it to the analysis of the positronium 1s-2p$_0$ transition [3]. The same calculation was also employed for hydrogen atoms [8]. These results might be interpreted by considering the Rabi flopping dynamics as a zeroth-order approximation of the problem of an infinite number of bound states coupled to the continuum under a rotating wave approximation.

In this article, we propose a further improvement over these previous approximations [3, 5]. Our new model accounts for the depletion of the initial state towards the continuum increased by ionization from the resonant excited state. The paper is organized as follows. We introduce our theoretical formulation in section 2 and present the corresponding results in section 3. Finally, section 4 contains conclusions and perspectives. Atomic units are used throughout this article.

2. Theory

Under non-relativistic conditions and within the electric dipole approximation, the electron wave function $\Psi(\mathbf{r},t)$ for a hydrogenic atom satisfies the time-dependent Schrödinger equation (TDSE) which, written in the length gauge, reads

$$i\frac{d\Psi(\mathbf{r},t)}{dt} = [-\frac{1}{2}\nabla^2 + V(r) + \mathbf{r}\cdot\mathbf{F(t)}]\Psi(\mathbf{r},t)\,. \quad (1)$$

Here $\mathbf{r}$ is the position of the electron with respect to the positive ion and $V(r)$ is the interaction between the electron and the rest of the target. The external laser radiation is given by the electric field $\mathbf{F}(t) = \mathbf{F}_0 \sin(\omega t + \varphi)\sin^2(\pi t/\tau)$. The transition amplitude from the state i at $t = 0$ to the state f at $t = \tau$ can be approximated by the following variational expression [9]:

$$a_{fi}^{-} = \lim_{t\to 0}\langle\chi_f^{-}(t)|\chi_i^{+}(t)\rangle - i\int_0^{\tau} dt'\ \langle\chi_f^{-}(t')|H - id/dt'|\chi_i^{+}(t')\rangle\,, \quad (2)$$

where the non-hermitian operator $H - id/dt'$ applies onto $\chi_f^{-}(t')$. The trial wave functions $\chi_f^{-}(t)$ and $\chi_i^{+}(t)$ approximate the exact solutions to the TDSE (1), subject to the following initial and final conditions:

$$\lim_{t\to 0}\chi_i^{+}(t) = \phi_i^{+}(\mathbf{r},t) = \varphi_i^{+}(\mathbf{r})\exp(-i\epsilon_i t)\,,$$
$$\lim_{t\to \tau}\chi_f^{-}(t) = \phi_f^{-}(\mathbf{r},t) = \varphi_f^{-}(\mathbf{r})\exp(-i\epsilon_f t)\,.$$

Here, $\varphi_f^{-}(\mathbf{r})$ and $\varphi_i^{+}(\mathbf{r})$ are eigenfunctions of the atomic Hamiltonian associated with the eigenenergies ϵ_f and ϵ_i, respectively. In references [5, 6, 7], the Coulomb-Volkov approximation was used to represent the final trial wavefunction

$$\chi_f^{-}(\mathbf{r},t) = \phi_f^{-}(\mathbf{r},t)\exp[i\mathbf{A}(t)\cdot\mathbf{r} - i\ \mathbf{k}\cdot\int_{\tau}^{t} dt'\mathbf{A}(t')]\,,$$

where $\mathbf{A}(t) = \int_t^{\tau} dt'\ \mathbf{F}(t')$ is the vector potential. Replacing in Eq. (2), we obtain

$$a_{fi}^{-} = \int_0^{\tau} dt\ \exp\left(i\frac{k^2}{2}t + i\ \mathbf{k}\cdot\int_{\tau}^{t} dt'\mathbf{A}(t') + \frac{i}{2}\int_{\tau}^{t} dt' A^2(t')\right)$$
$$\int d\mathbf{r}\ \chi_i^{+}(\mathbf{r},t)e^{i\mathbf{A}(t)\cdot\mathbf{r}}\mathbf{A}(t)\cdot[i\mathbf{k} + \nabla]\varphi_f^{-*}(r)\,. \quad (3)$$

The choice of the appropriate trial wave function $\chi_i^{+}(\mathbf{r},t)$ depends on the process to be described. For photon energies above the ionization threshold and for small enough intensities so as to ensure that the total ionization probability is far from the saturation regime, the unperturbed initial wave function $\varphi_f^{-}(\mathbf{r})\exp(-i\epsilon_f t)$ can be safely used. This has been done in the framework of the Coulomb-Volkov ($CV2^{-}$) theory where a favorable comparison with exact TDSE computations has been found within the aforementioned conditions [6, 7]. However, for photon energies below the ionization threshold a better description of the initial trial wave function is required. In this case, the electron will absorb several photons during the ionization process, passing through many virtual intermediate states. Even when the photon energy is not in tune, the wide spectrum of the short laser pulse enables the population of the excited states. This effect is observed in the appearance of secondary peaks on the ATI photoelectron spectrum and can be interpreted in terms of different multi-photon ionization channels launched from every relevant field-dressed excited state. A simple way to improve the CV2$^-$ taking these channels into account is to consider a different choice for the trial wave function [10], namely $\chi_i^+(\mathbf{r},t) = \sum_j b_{ji}(t)\phi_j(\mathbf{r},t)$, where $b_{ji}(t)$ is the transition amplitude at time t from the initial state ϕ_i to the intermediate state ϕ_j. In reference [10], $b_{ji}(t)$ was evaluated within a first Born approximation giving place to the so called modified Coulomb-Volkov (MCV2$^-$) approximation. Although MCV2$^-$ correctly describes the position of the emergent secondary peaks, some sizable quantitative differences with TDSE remains. In [5] some improvements were achieved by evaluating the amplitudes $b_{ji}(t)$ with a close-coupling calculation involving some few bound states including the initial and the near resonant ones among others. This close-coupling CoulombVolkov (CC-CV2$^-$) theory provides a correct treatment of atomic ionization processes induced by a laser resonant with a given ground-bound state transition. The laser intensities for these theoretical treatment should be relatively small since any population depletion towards the continuum is not accounted for. In this work, we improve the CC-CV2$^-$ theory by allowing the coupling between discrete states and the continuum, and between discrete states. In particular, we take into account the coupling of the initial state i with the resonant state r as well as with some other bound states. On the other hand, we also account for the coupling between the resonant state r and the continuum. Using the set of close coupling equations we obtain two integro-differential equations system (IDEs) of Volterra type:

$$\dot{a}_i(t) = -iF(t)z_{ir}e^{-i(\varepsilon_r-\varepsilon_i)t}a_r(t) - \int_0^t dt'F(t)F(t')e^{i\varepsilon_i(t-t')}a_i(t')h_i^D(t-t') \quad (4)$$

$$\dot{a}_r(t) = -iF(t)z_{ri}e^{-i(\varepsilon_i-\varepsilon_r)t}a_i(t) - \int_0^t dt'F(t)F(t')e^{i\varepsilon_i(t-t')}a_r(t')h_r^C(t-t') . \quad (5)$$

In this case the laser pulse is assumed to be linearly polarized in the z direction. $z_{ir} = \langle i|z|r\rangle$ is the corresponding matrix element for the transition between the initial and the resonant states. The kernel of the first equation is obtained by adding over the discrete spectrum, while the kernel of the second equation can be computed by means of the Nordsieck method [11], namely

$$h_i^D(t-t') = \sum_{n\neq i,r} |z_{in}|^2 e^{-i\varepsilon_n(t-t')} \quad , \quad h_r^C(t-t') = \int d\mathbf{k}|z_{r\mathbf{k}}|^2 e^{-i\varepsilon_k(t-t')} .$$

In these equations, the superscript D stands for *Discrete* and the C for *Continuum*. We solve these equations using the Goldfine method [12]; namely, we use a Taylor expansion to propagate the initial and resonant transition amplitudes from time t to time $t+\delta t$. The required time derivatives are obtained using the original equations. Once $a_i(t)$ and $a_r(t)$ were obtained, we can calculate $a_n(t)$ by direct integration,

$$a_n(t) = -i\int_0^t dt' z_{ni}F(t')e^{-i(\varepsilon_i-\varepsilon_n)t'}a_i(t') .$$

In terms of $a_i(t)$, $a_r(t)$ and $a_n(t)$, the trial wave function reads

$$\chi_i^+(\mathbf{r},t) = a_i(t)\phi_i(\mathbf{r},t) + a_r(t)\phi_r(\mathbf{r},t) + \sum_n a_n(t)\phi_n(\mathbf{r},t) , \quad (6)$$

Finally, the transition amplitude for the new discrete-continuum Coulomb-Volkov theory (DC-CV2$^-$) accounting for bound-state continuum coupling is obtained by replacing Eq. (6) in Eq. (3). The CD-CV2$^-$ transition amplitude represents a series of CV2$^-$-like amplitudes for transitions starting from intermediate states n. The interpretation is rather straightforward: every bound state is populated with an amplitude given by $b_{ji}(t)$, where a multi-photon promotion to the continuum is considered by using a CV2$^-$-like amplitude.

3. Results

We apply our model to the study of the ionization of Hydrogen by a laser pulse with a frequency $\omega = 0.375$ a.u., resonant with the $1s - 2p_0$ transition ($i = 1s$ and $r = 2p_0$). The pulse length is $\tau = 670.2$ a.u., *i.e.* corresponding to 40 cycles. The laser polarization is linear, and chosen in the z-direction. Therefore, the magnetic quantum number is a constant of motion in the present application. We examine two laser field amplitudes $E_0 = 0.04$ and 0.05 a.u., large enough to produce an appreciable depopulation towards the continuum. We included $3p_0$, $4p_0$ and $5p_0$ bound non-resonant states and checked that the convergence is achieved for principal quantum numbers $n \leq 5$. We compare our results with those obtained by numerically solving the TDSE. For this latter purpose we use the QPROP code [13]. In this code the initial wave function for the system is expanded in spherical harmonics and propagated with the Crank-Nicholson approximation for the short-time propagator. The spectral analysis of the real-time propagation computed with the Qprop package is performed through the window-operator technique [13] allowing the computation of the total and the angular resolved ionization spectra without the knowledge of the eigenstates.

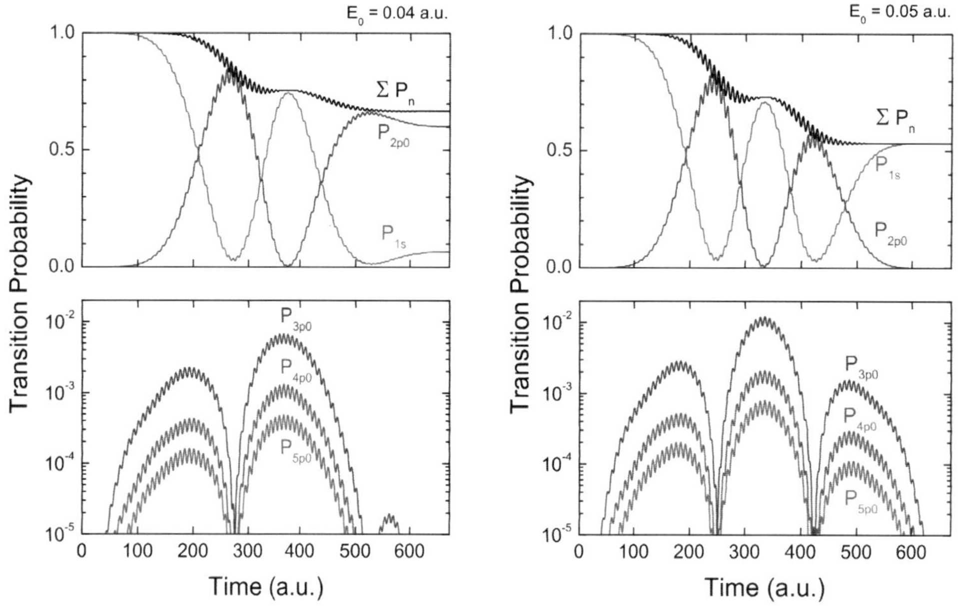

Figure 1. Transition probabilities as a function of time for a 40 cycles length laser pulse ($\tau = 670$ a.u. $= 16.25$ fs.). The laser electric field intensities are $E_0 = 0.04$ a.u. (left panel) and 0.05 a.u. (right panel).

3.1. Ionization probabilities and ground state population decay

We first analyze the quality of the trial initial wave function obtained from the numerical solution of the IDEs of Eq. (5). The left panel of Figure 1 shows the elastic and transition probabilities as a function of time for a laser electric field intensity of $E_0 = 0.04$ a.u. Both, the $1s$ and the $2p_0$ populations display a Rabi oscillations pattern. We can distinguish one and a half

Rabi oscillations. Since ionization is important, a considerable depletion of the bound states population $\sum P_n$ is exhibited as a function of time. A small amplitude and high frequency oscillation behavior of the probabilities are due to counter-rotating terms. Also the transitions probabilities to $3p_0$, $4p_0$ and $5p_0$ (botton figure) exhibit Rabi like oscillations presenting two maxima before going to zero near the end of the pulse. Since the population of these states occurs via the $1s$ state, it is reasonable that the corresponding maxima are almost coincident in time. For times closer to the end of the pulse, the $1s$ state population does not increase and therefore all the np probabilities remain small. In the right panel we show similar results but for a larger laser field amplitude of $E_0 = 0.05$ a.u.. The features are similar but the Rabi frequency is slightly increased ($\Omega_R \propto E_0$). In this case a full two cycles of Rabi oscillations is observed. The $n = 3 - 5$ excitation probabilities display three maxima. The total bound state population decreases to 0.53, that is considerably smaller than the value 0.67 corresponding to the smaller perturbation in the left panel. Since the $1s$ population shows three clear maxima, also the np probabilities do.

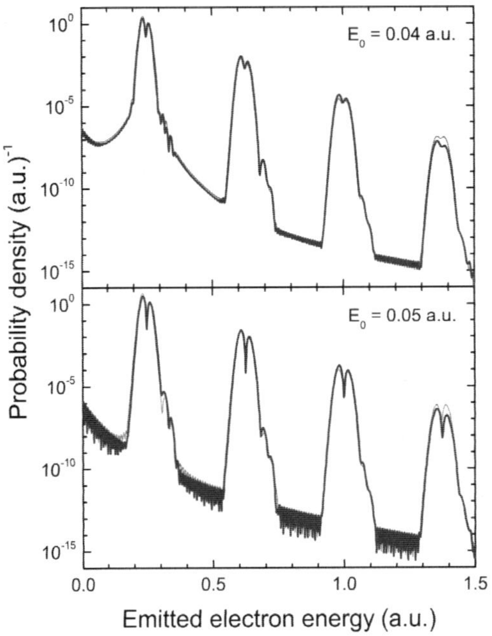

Figure 2. Hydrogen ionization spectra for a 40 cycles laser pulse (t=670 u.a=16.25 fs.) and two different laser electric field intensities. The present DC-CV2$^-$ theory (thick (blue) line) is compared with the solution of the TDSE obtained with the Qprop code (thin (red) line).

3.2. Ionization spectra

The ionization spectra were computed with the new modified Coulomb-Volkov theory. The trial wave function for the initial state was approximated by an expansion up to the fifth excited bound eigenstate of the atomic Hamiltonian. The corresponding probability amplitudes were determined from the solution of the set of coupled IDEs in Eq. (5). In figure 2 the results obtained for photons with a frequency $\omega = 0.375$ a.u. and laser intensities $E_0 = 0.04$ and 0.05 a.u. are displayed. The first four principal ATI peaks are plotted in each spectrum. We clearly notice the peak splitting due to Rabi oscillations [3]. The separation between the subpeaks increases linearly with the maximum electric field amplitude. The presence of a secondary peak structure to the right of the main secondary peaks can be clearly distinguished. The theory is clearly able to show this feature that can be related to excitation followed by subsequent ionization. To probe this hypothesis, we have checked that the structure disappears when the excited states are removed from the trial initial wave function.

In general the agreement with the quantum propagation provided by the Qprop package [13] is very good. This is a clear indication that the use of an improved initial wave function accounting for the bound state depletion towards the continuum is correct. Some differences between QPROP and DC-CV2 may be observed in the background where the former results display a smaller amplitude oscillation. Regarding this, we should remember that the QPROP results are convoluted with a window in energy that naturally tends to average faster oscillations. Furthermore, some numerical error might arise due to the quite small values of the spectral probabilities in the background.

4. Conclusions

By comparing our DC-CV2$^-$ approximation with the numerical solution of the TDSE, we have been able to establish the importance of including the pathways through the bound states when the laser frequency has been tuned accordingly. An excellent agreement is achieved in the description of the photoelectron spectrum. In particular the secondary peak spectrum is explained and related to the transient population of intermediate states followed by absorbtion of one or more photons. On the contrary, previous CC-CV2 models tend to underestimates both the ionization spectrum and the total ionization probabilities, and only achieve a quantitative agreement when the laser field amplitude is in a perturbative regime. This is a strong indication that the coupling with the continuum states should be included in the theory. This is particularly important when a higher bound state is populated under resonance condition.

Acknowledgments

This work has been supported by the Consejo Nacional de Investigaciones Científicas y Técnicas (Grant no PIP 112-200801-011269) and Universidad Nacional de Cuyo (Grant no 06/C348). V.D.R. and R.O.B are also members of the Consejo Nacional de Investigaciones Científicas y Técnicas (CONICET), Argentina.

References

[1] Hansch P, Walker M A, and Van Woerkom L D 1997 *Phys. Rev. A* **55** R2535
[2] Paulus G G, Grasbon F, Walther H, Kopold R and Becker W 2001 *Phys. Rev. A* **64** 021401(R)
[3] Rodríguez V D 2006 *Nucl. Instrum. and Methods Phys. Res. B* **247** 107
[4] Autler S H and Townes C H 1955 *Phys. Rev.* **100** 703
[5] Rodríguez V D, Arbó D and Macri P 2011 *J. Phys. B: At. Mol. Opt. Phys.* **44** 125603
[6] Duchateau G, Cormier E and Gayet R 2002 *Phys. Rev. A* **66** 023412
[7] Macri P A, Miraglia J E and Gravielle M S 2003 *J. Opt. Soc. Am. B* **20** 1801
[8] Girju M G, Hristov K, Kidun O and Bauer D 2007 *J. Phys. B: At. Mol. Opt. Phys.* **40** 4165
[9] Demkov Yu N 1963 *Variational Principles in the Theory of Collisions* (Oxford: Pergamon Press)
[10] Rodríguez V D, Cormier E and Gayet R 2004 *Phys. Rev. A* **69** 053402
[11] Nordsieck A 1954 *Phys. Rev* **93** 785
[12] Goldfine A 1977 *Mathematics of Computation* **31** 691
[13] Bauer D and Koval P 2006 *Comput. Phys. Comm.* **174** 396

Journal of Physics: Conference Series **397** (2012) 012015 doi:10.1088/1742-6596/397/1/012015

Spectroscopic problems in ITER diagnostics

V S Lisitsa, L A Bureyeva, A B Kukushkin, M B Kadomtsev,V A Krupin, M G Levashova, A A Medvedev, E E Mukhin, V A Shurygin, S N Tugarinov and K Yu Vukolov
TPhI NRC "Kurchatov Institute", Moscow, Russia
E-mail: lisitsa@nfi.kiae.ru

Abstract. Problems of spectroscopic diagnostics of ITER plasma are under consideration. Three types of diagnostics are presented: 1) Balmer lines spectroscopy in the edge and divertor plasmas; 2) Thomson scattering, 3) charge exchange recombination spectroscopy. The Zeeman-Stark structure of line shapes is discussed. The overlapping of isotopes H-D-T spectral line shapes are presented for the SOL and divertor conditions. The polarization measurements of H-alpha spectral lines for H-D mixture on T-10 tokamak are shown in order to separate Zeeman splitting in more details. The problem of plasma background radiation emission for Thomson scattering in ITER is discussed in details. The line shape of P-7 hydrogen spectral line having a wave length close to laser one is presented together with continuum radiation. The charge exchange recombination spectroscopy (CXRS) is discussed in details. The data on D_α, HeII and CVI measurements in CXRS experiments on T-10 tokamak are presented.

1. Introduction

The spectroscopic diagnostics of ITER plasma includes three important types of diagnostics: 1) Balmer lines spectroscopy in the edge and divertor plasmas; 2) Thomson scattering, 3) charge exchange recombination spectroscopy.

Diagnostics of ITER plasma based on Balmer spectral line shapes measurements is one of the main spectroscopic diagnostics. The problem of spectral line shapes calculations faces with the development of the spectral line broadening theory for the broad domain of plasma densities and temperatures which change along lines of sight. The number of effects making the influence on observed line shapes is to be taking into account. One of the main effect is the presence of strong magnetic field resulting in the strong Zeeman splitting being of the same order of magnitude as the atom interaction with electric fields of surrounding plasma particles (Stark effect). This results in hard enough calculations of Zeeman-Stark structures of atomic energy levels at arbitrary orientation of magnetic and electric fields. Moreover, the electric fields are not static due to intense thermal motion of charged particles. The new version of FFM method provides fast codes for line shapes calculations [1-3]. The main goal of the calculations is to resolve between H, D, T isotopes spectra in order to determine the isotope content in ITER. The isotope spectra have to be calculated for different lines of sights. Some experimental results for isotope composition observations in T-10 tokamak are demonstrated below. General problem of edge plasma radiation blended by strong divertor radiation emission is discussed as well.

Thomson scattering in ITER is one of the one of the most important method for electron temperature measurements in thermonuclear plasmas. The signal of Thomson scattering is relatively

Journal of Physics: Conference Series **397** (2012) 012015 doi:10.1088/1742-6596/397/1/012015

small and sensitive to the of plasma background radiation emission. The wave length of hydrogen P-7 spectral line is close to one of laser radiation. So the line shape of P-7 spectral line is of importance for Thomson scattering diagnostics. The intensities of nitrogen impurity spectral lines relative to the Thomson diagnostics spectral range are also of interest.

The charge exchange recombination spectroscopy (CXRS) is also the important method of active plasma diagnostics in ITER. The detail information on cross sections for charge exchanges between fast beam atoms and carbon ions are needed. The problem of halo atomic structure together with their charge exchange on carbon ions is also of importance. Some experimental data on Doppler line shapes of D_α, HeII and CVI charge exchange recombination spectra on T-10 tokamak are presented below.

2. Line broadening of hydrogen spectral lines in strongly magnetized plasmas. Balmer lines spectroscopy in the edge and divertor plasmas

An analysis of H-alpha spectrum for various observation chords in ITER SOL, carried out with the help of fast FFM numerical codes BALMER-SZDYN [4] for line shapes in thermonuclear plasmas with strong magnetic fields combined with the SOLPS4.3 code [5] for plasma parameters distribution along lines of sights is presented below. The results are given for the case of the vertical chord observation from the upper port UPP#2 in ITER for the conditions of the ITER SOL/divertor regime 1514. The position of observation chords are shown on figure 1. The distributions of plasma parameters (electron/ion temperatures and electron density) along horizontal chord are shown on figure 2 [5].

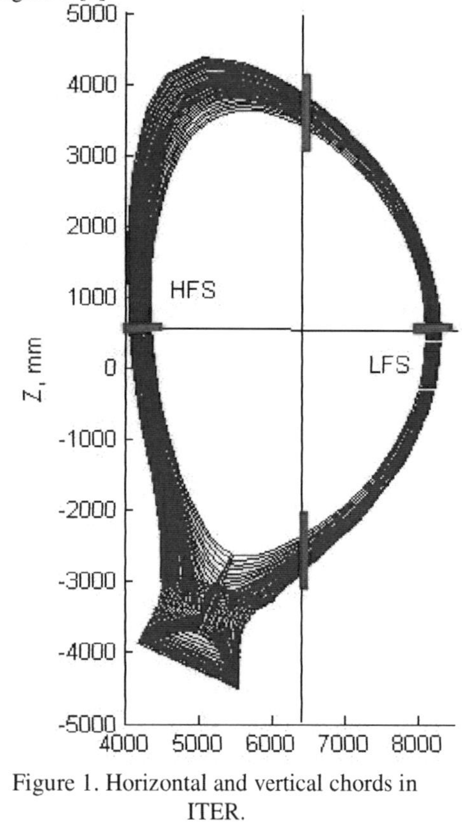

Figure 1. Horizontal and vertical chords in ITER.

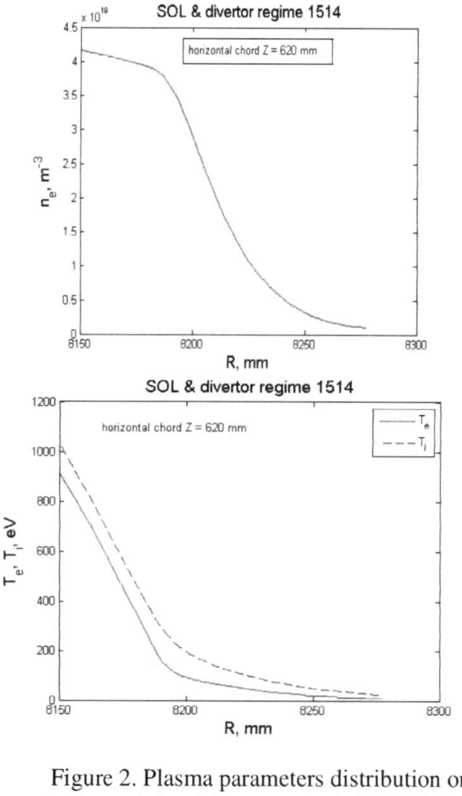

Figure 2. Plasma parameters distribution on horizontal chord (outer region). SOLPS4.3 – code [5].

D-alpha, D-beta emissivity on horizontal chord is presented on figure 3. Line shapes of hydrogen isotopes in SOL for direction of observation perpendicular to magnetic field are shown on figure 4 [4]. The same line shapes of hydrogen isotopes (H, D, T) for divertor stray light and the composite spectrum for a [5%, 65%, 35%] mixture of isotopes are presented on figure 6 [4]. One can see a strong overlapping of isotopes line shapes especially for dense plasmas in divertor.

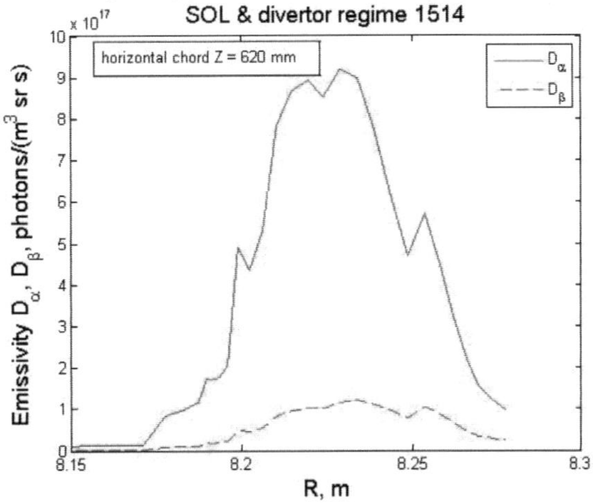

Figure 3. D-alpha, D-beta emissivity on horizontal chord SOLPS4.3 – code [5].

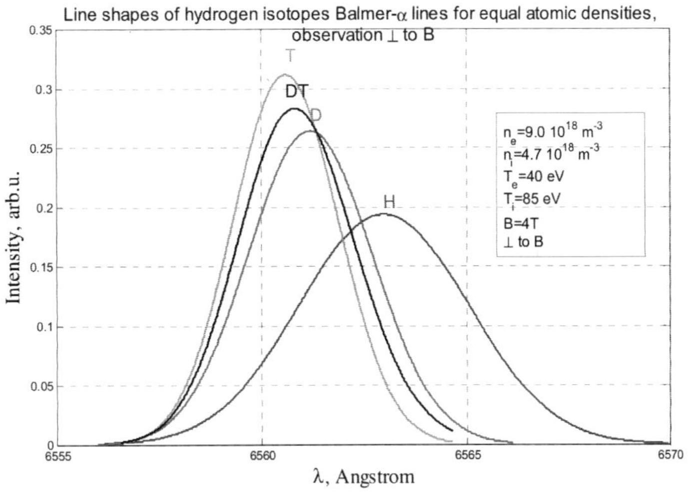

Figure 4. Total line shape of hydrogen isotopes in SOL [4].

The problem of SOL radiation observation is very difficult because of strong blending of the SOL radiation by intense divertor radiation emission. The reflected divertor emissivity dominates over SOL near 2-3 orders of magnitude. One need special tricks (light traps) to observe the SOL radiation on the background of divertor radiation.

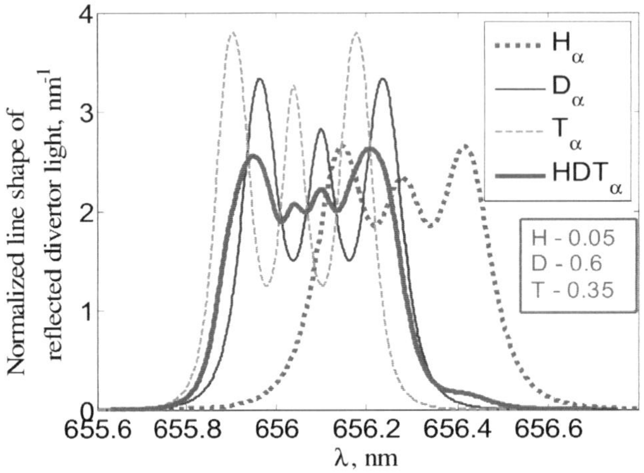

Figure 5. Normalized line shapes of hydrogen isotopes (H, D, T) for divertor stray light and the composite spectrum for a [5%, 65%, 35%] mixture of isotopes [4].

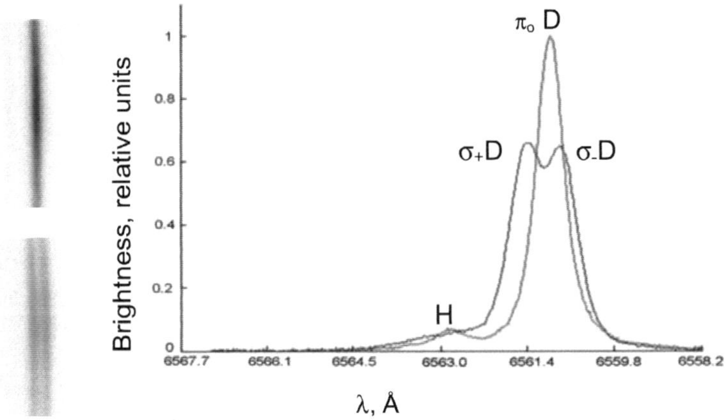

Figure 6. Observation Balmer-alpha line shapes of isotopes H, D in T-10 tokamak (Medvedev A A, private communication): left hand side is observed positions of Stark components, right hand side is the fitting by Gaussian shapes.

There are some possibilities for separation of isotopes line shapes. One of them is connected with observation of atomic line shapes in polarized light. Figures 6 and 7 demonstrate such a possibility from experimental spectra observations on T-10 tokamak. One can see that the polarization measurements make it possible to separate well between π and σ components of the isotopes D, H mixture.

3. Radiation Background in Thomson scattering

The radiation plasma background is of importance for Thomson diagnostics because of a small value of reflected laser radiation signal. In particular the hydrogen P-7 line wavelength (10049,3 Å) is very close to laser signal. That is why the detail shape of the line is needed for foundation of Thomson diagnostic spectroscopy. Figure 7 presents lines of sight for Thomson diagnostics in ITER divertor.

The spectral intensity distributions along two lines of sight proposed for Thomson scattering observations in divertor [6] are shown on figures 8, 9.

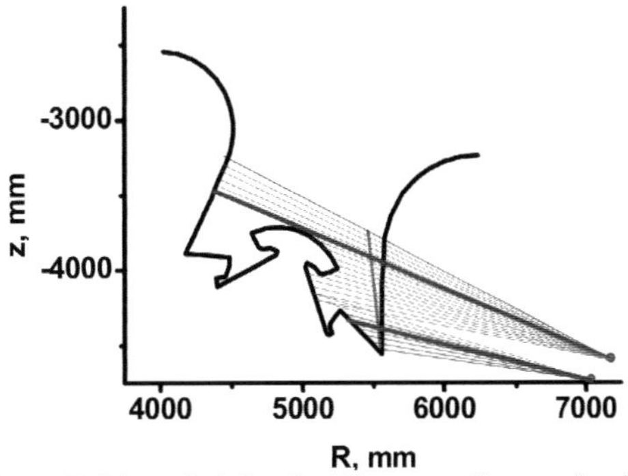

Figure 7. Line of sights for Thomson diagnostics in ITER divertor. The blue lines corrsesponds to the chords N5 and N21.

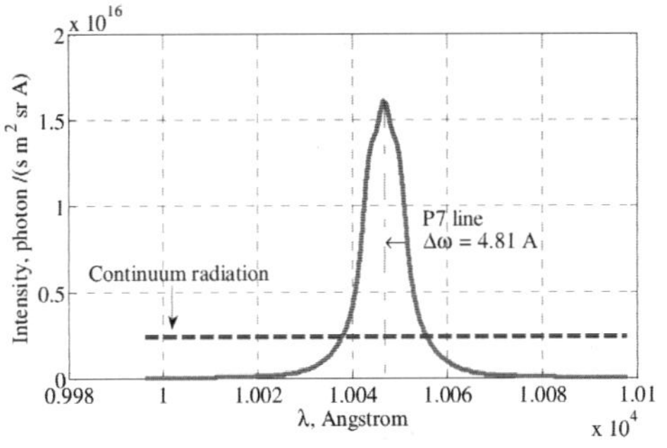

Figure 8. The intensity distribution of P7 spectral line along the chord N21 in divertor (red line) The contribution of plasma continuum is shown by the blue line.

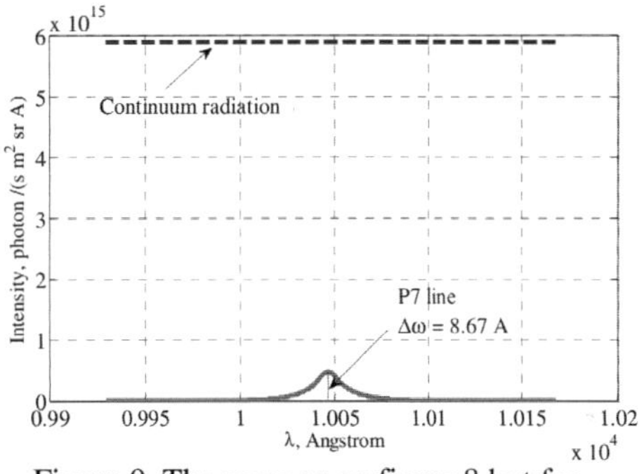

Figure 9. The same as on figure 8 but for the chord N5.

One can see a large difference between line and continuum radiation intensities for both chords. Note that the background for Thomson scattering can be due to the presence of different impurities in plasmas. Such for example, the radiation of neutral nitrogen contains strong radiative transitions at wavelengths 10105,13 – 10114,64 Å (line10112,48 Å from NIST data) in the spectral range close to laser signal. The problem is in detail line shapes calculations of the corresponding radiative transitions

4. Charge Exchange Recombination Spectroscopy (CXRS)

The Charge Exchange Recombination Spectroscopy (CXRS) is one of the effective methods of active plasma diagnostics in ITER. It is based ob charge exchange recombination of impurity ions in plasma on fast (near 100 keV) hydrogen neutral beam atoms injected into the hot plasma [7]. Selective charge exchange cross sections are needed for interpretation of experimental data. The example of such cross section for reaction $C^{6+} + H(n=2) \ = \ C^{5+}(n_Z, l_Z) + H^+$ is presented in figures 10, 11.

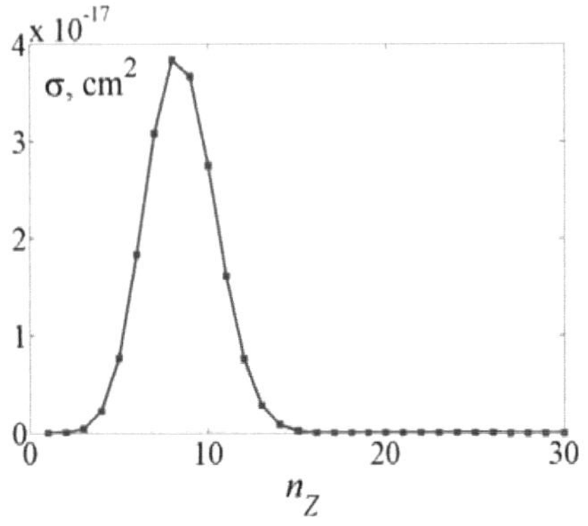

Figure 10. Selective in principal quantum number n_Z charge exchange cross sections of H beam with energy 100 keV on C^{6+} ions [8].

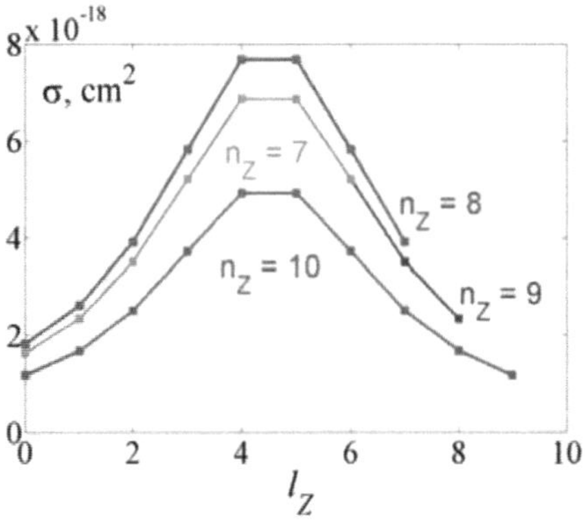

Figure 11. Selective in orbital momentum quantum number l_Z charge exchange cross sections of H beam with energy 100 keV on C^{6+} ions [8].

The most convenient spectral line for observations for carbon (Z=6) impurity corresponds to the radiative transitions from the ion energy level with principle quantum number n_Z=8 to the level n_Z=7 with the wave length 5290 Å. Observations of line intensities provide the information on local impurity contents in the plasma. The Doppler line shapes provide the information on ion temperature.

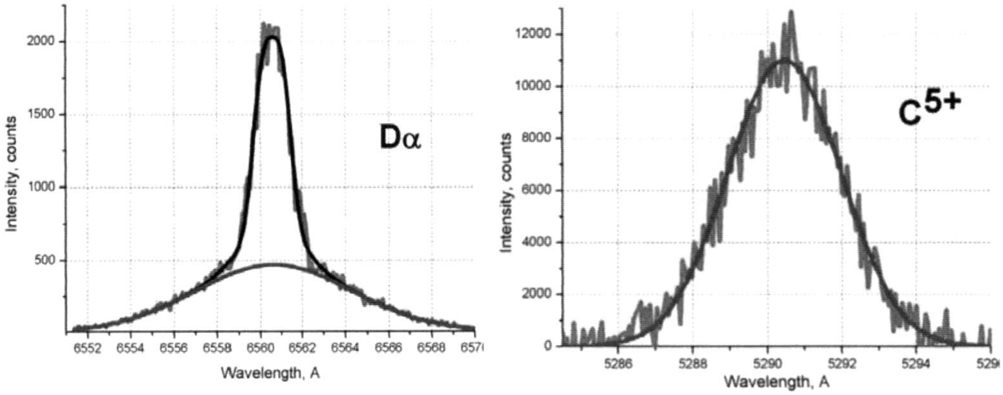

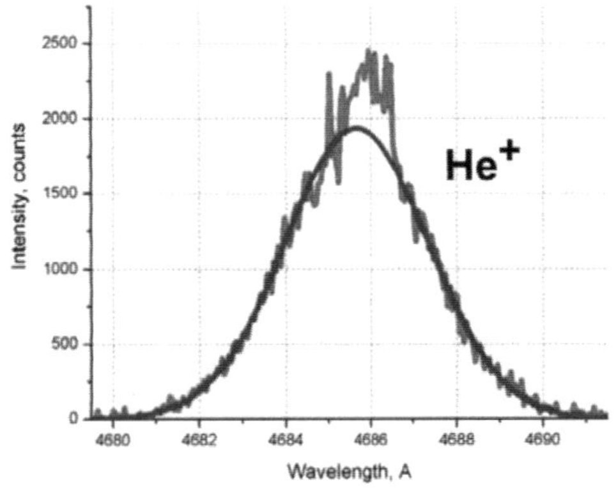

Figure 12. Doppler line shapes of D-alpha, HeII and CVI ions observed in CXRS experiments on T-10 tokamak [9]

Problems with interpretation of the CXRS signals are connected with cascade processes such as:

$$H_F + D_T^+ \rightarrow H_F^+ + D_T(n);$$

$$D_T(n) + C^{+6} \rightarrow D^+ + C^{+5}(n_Z \approx 4n) \rightarrow C^{+5}(n_Z \rightarrow 8) + \hbar\omega \rightarrow C^{+5}(n_Z = 7) + \hbar\omega_{8-7}$$

Here index F corresponds to fast beam atoms and index T corresponds to thermal plasma ions.

The cascade processes are especially difficult for interpretation because of halo effect. The halo arises around the neutral beam due to the charge exchange of fast atoms on thermal ions. The charge exchange to the ground atomic state is followed by the electron excitation of halo atoms into excited atomic states. The charge exchange from these excited states grows very fast with principle quantum numbers of the atoms (proportional to n^4). So the atomic energy states of impurities are effectively populated by the charge exchange from excited states of halo atoms resulting in radiative-collisional cascades to the observed impurity carbon energy level n=8. Such cascade processes result in a complicated density dependence of the CXRS signals.

5. Conclusions

The spectroscopic diagnostic methods are widely used in ITER plasma diagnostics. The passive method relates to Balmer spectroscopy; the main problems are connected with extracting of isotope effects from H-alpha observations; the problem of SOL intensity observation at the background of strong divertor radiation being of importance. The radiation trapping in divertor plasma plays an important role [10, 11]. The plasma background coming both from continuum radiation as well as line radiation from hydrogen isotopes and impurities is essential problem for Thomson scattering observation. The line shapes of line radiation are of importance in the case. Account for radiative-collisional cascade transitions in active CXRS with account of halo effect is of importance for interpretation of experimental data.

References

[1] Calisti A, Mossé C, Ferri S, Talin B, Rosmej F, Bureyeva L A and Lisitsa V S 2010 *Phys. Rev.* E **81** 016406

[2] Ferri S, Calisti A, Mossé C, Talin B, Gigosos M A, Gonzalez M A and Lisitsa V S 2011 *Phys. Rev.* E **84** 026407

[3] Stambulchik E and Maron Y 2008 *J. Phys.* B: *At. Mol. Opt. Phys.* **41** 095703

[4] Lisitsa V S, Vukolov K Yu, Kadomtsev M B, Levashova M G and Medvedev A A 2010 Numerical code BALMER-SZDYN for spectroscopy of hydrogen isotopes Balmer lines in a strong magnetic field *Abstracts Zvenigorod conference on plasma physics, Russia,* http://www.fpl.gpi.ru/Zvenigorod/XXXVIII/E/en/PI_Lisitsa_e.doc

[5] Kukushkin A S, Pacher H D, Loarte A *et al* 2009 *Nucl. Fusion* **49** 075008

[6] Lisitsa V S, Mukhin E E, Kadomtsev M B *et al* 2012 *Plasma Phys. Reports* **38** 157

[7] Tugarinov S N, Beigman I L and Vaistein L A 2004 *Plasma Phys. Reports* **30** 1

[8] K.R. Cornelius, K. Wojtkowski and R.E. Olson. J. Phys. B, 33 (2000) 2017-2035

[9] Krupin V A et al The CXRS measurements of ion temperature at T-10 2011 *38th Zvenigorod international conference on plasma physics and controlled fusion, Russia, Zvenigorod, February* 14-18

[10] Rosato J, Kotov V and Reiter D 2010 *J Phys. B* **43** 144024

[11] Rosato J, Reiter D, Kotov V, Marandet Y, Capes H, Godbert-Mouret L, Koubiti M and Stamm R 2010 Progress on Radiative Transfer Modelling in Optically Thick Divertor Plasmas *Contrib. Plasma Phys.* **50** 398

Wideband High-Resolution Spectroscopy on Al-pellet Ablation Plasmas in Large Helical Device

M. Hasuo[1], K. Fujii[1], T. Shikama[1], S. Morita[2], M. Goto[2] and H. Tanaka[1]

[1] Graduate School of Engineering, Kyoto University, Kyoto 606-8501, Japan
[2] National Institute for Fusion Science, Toki 509-5292, Japan

hasuo@kues.kyoto-u.ac.jp

Abstract. An aluminum pellet has been injected into a high-temperature plasma produced in the Large Helical Device (LHD). Emission from an ablation cloud of the pellet in a whole visible range is simultaneously measured with an echelle spectrometer having an instrumental resolution from 0.055 nm at 400 nm to 0.10 nm at 750 nm. More than 50 of Al I, II, III and IV lines are resolved. We estimate the electron temperature and density of the ablation cloud from the intensity distribution and Stark broadening of the emission lines, respectively. The electron temperature and density are found to show increase and decrease, respectively, as an increase in the charge state of aluminum. We also determine the Stark broadening coefficients of several Al II lines from the observed Stark widths.

1. Introduction

Pellets containing various elements have been injected into a magnetically confined fusion plasma for the purposes of impurity transport study and disruption control of the plasma [1,2]. In the Large Helical Device (LHD), for example, pellets of carbon, aluminum, titanium and so on were injected to examine the dependence of transport coefficients on the impurity nuclear charge [2].

A pellet plunged into the plasma is immediately ablated due to heat flux from the plasma. A high density plasma called ablation cloud forms in the vicinity of the pellet, which emits many atomic and ionic emissions. From spectroscopic studies of the emissions, not only impurity transport in the main plasma, but also ablation mechanism, atomic processes in the ablation cloud, and acquisition of atomic data like line-broadening coefficients can be done. Wideband spectra containing many emission lines have been measured for hydrogen, carbon and aluminum pellets [3,4,5]. From such measurements, the population distribution of excited levels of atoms or ions in the ablation cloud has been investigated and plasma parameters such as electron temperature, neutral atom density and ion density have been estimated. On the other hand, high resolution spectra of the emission lines have been measured for carbon pellets, and the electron density of the cloud has been estimated from the Stark broadening [4].

In this work, we use an echelle spectrometer to perform wideband and high-resolution spectroscopic measurements at one time for a single ablation cloud.

2. Experimental setup

The LHD has a torus-shape vacuum vessel with a major radius of about 3.9 m and a minor radius of about 1 m, in which a plasma of about 30 m^3 is confined. The plasma is heated by electron cyclotron heating, ion cyclotron resonance heating and neutral beam injection. Major radius position of the magnetic axis is 3.90 m and the field strength at the magnetic axis is 2.54 T (shot number: 104159).

We use a pure aluminum pellet having a cylindrical shape whose diameter and length are 0.8 mm. The pellet is accelerated in a 6 m long acceleration tube with a pneumatic pipe-gun system, which uses helium gas of 18 atm pressure, and then injected horizontally into the plasma from an outer port of the LHD (1-O port). Figure 1 shows a schematic illustration for a part of the equatorial plane of the LHD, in which the pellet trajectory is indicated by the thick arrow.

We place a lens having a focal length of 10 mm and an optical fiber having a core diameter of 1 mm at an observation port close to the pellet acceleration tube. The line of sight is parallel to the injection direction of the pellet as shown in figure 1. Emission from the pellet ablation cloud generated in the LHD plasma is recorded by a CCD camera (Andor, DV435-BV, 1024×1024 pixels) through an echelle spectrometer we developed. Since the exposure time of CCD is set to be 95 ms, which is much longer than the period of luminous radiation from the ablation cloud of several ms, the entire emission during the pellet ablation is accumulated in a single measurement. The CCD is cooled to be -25°C to reduce thermal noise.

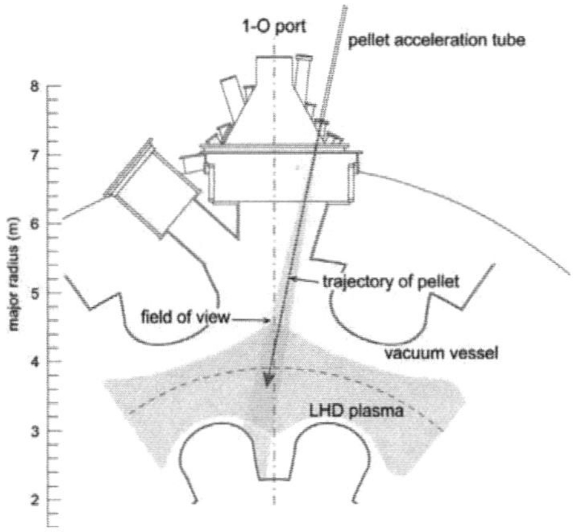

Figure 1. Top view of the equatorial plane of the LHD around the 1-O port.

For the purpose of the wavelength calibration and resolution estimation of the echelle spectrometer, we measure emission lines of a Th-Ar hollow cathode lamp (Heraeus, P858A) because it has many emission lines in the visible region and the line widths are small enough. It is found that the instrumental function of the spectrometer can be approximated by a Gaussian profile. We evaluate the instrumental width (FWHM) by fitting with a Gaussian function at the entrance slit width of 25 μm as a function of the wavelength. Typical instrumental widths range from 0.055 nm at 400 nm to 0.10 nm at 750 nm. We also calibrate the relative sensitivity of the spectrometer for the whole wavelength region using a tungsten standard lamp with an integrating sphere (Labsphere, USS-600C).

3. Results and discussion

The observed spectrum is shown in figure 2. It is noted that the intensity is shown in a logarithmic scale. More than 100 emission lines are resolved, in which more than 50 of Al I, Al II, Al III and Al IV lines are identified with available databases [6,7]. The identified lines are indicated in figure 2 by arrows. The lines denoted by 'a'-'d' are used to estimate electron density of the ablation cloud from their Stark widths as explained below. Since the LHD plasma contains hydrogen, helium, argon, etc, emission lines of these elements are also seen in figure 2.

Many of observed lines are found to show broadenings, which cannot be reproduced by a Gaussian profile. The broadenings may be mainly due to Stark broadening having a Lorentzian profile because electron density in the ablation cloud is expected to be high. Furthermore, influence of the Zeeman splitting on the line profile may not be negligible. We fit the observed line profiles as follows; we calculate the splitting and relative emission intensities of the Zeeman components with the method based on the perturbation theory for B=1.5 T, which is the approximate magnetic field strength at the location where the radiation intensity of the ablation cloud takes its maximum, and then fit the entire line profile by a sum of several shifted Voigt functions according to the Zeeman splitting, which have a common Lorentz width, W_L, and a fixed Gaussian width, W_G, determined previously as the instrumental width. The adjustable parameters for the fit are W_L and the area of the spectrum. A typical example of the observed spectrum and the result of the fit is shown in figure 3.

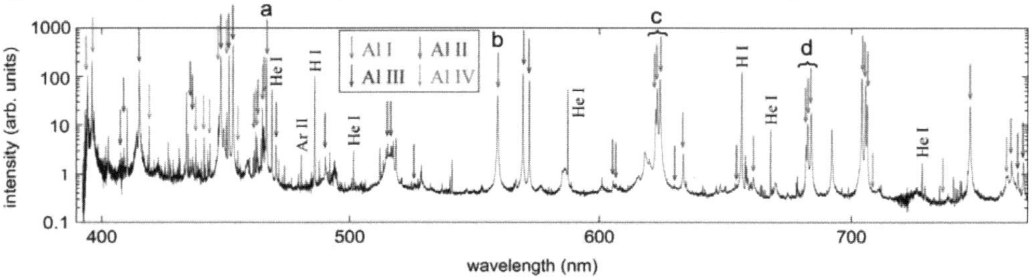

Figure 2. Observed spectrum of the aluminum pellet ablation cloud.

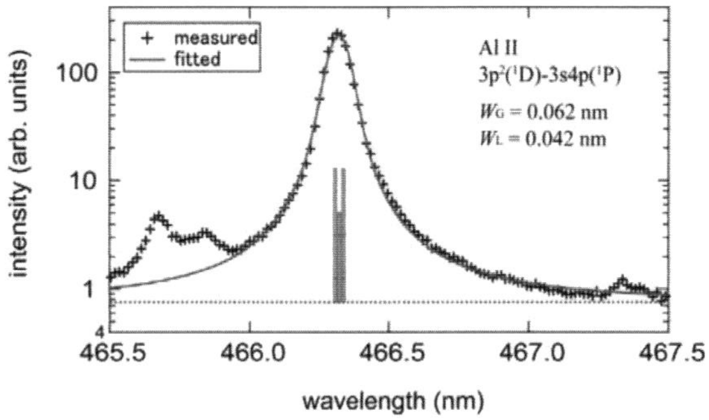

Figure 3. Enlarged spectrum of the line denoted by 'a' in figure 2. The fitted result is shown by the red curve. The vertical bars show the Zeeman splitting and the intensities of the Zeeman components.

Emission line intensity is given as the area of the spectrum. The line intensity, $I(p,q)$, from a level p to a level q is proportional to the upper level population, $n(p)$, as

$$I(p,q) = h\nu(p,q)\,A(p,q)\,n(p) \qquad (1)$$

where h is Planck's constant. $\nu(p,q)$ and $A(p,q)$ are the frequency of light and the spontaneous transition probability, respectively.

Populations of excited levels of atom and ions in the ablation cloud can be estimated from the observed line intensities with equation (1) for well-resolved emission lines enough to evaluate area of the spectra with available transition probabilities [6,7]. The results for neutral aluminum atoms (Al), singly-ionized aluminum ions (Al^+) and doubly-ionized aluminum ions (Al^{2+}) are shown in figure 4 by the Boltzmann plot against the upper level energy. The population distribution seems to be on a straight line for each charge state.

It is known that in a case where atoms or ions of an excited level p in a plasma is in thermo dynamic equilibrium (LTE), population in the level p follows the Saha-Boltzmann distribution;

$$\frac{n_z(p)}{g_z(p)} = \frac{1}{2}\left(\frac{h^2}{2\pi mkT_e}\right)^{3/2} \exp\left(\frac{\chi_z(p)}{kT_e}\right) n_e \frac{n_{z+1}(1)}{g_{z+1}(1)} \qquad (2)$$

where m is the electron mass, k is the Boltzmann constant. $n_Z(p)$, $g_Z(p)$ and $\chi_Z(p)$ are the population, statistical weight and ionization potential of the level p, respectively, of a charge state Al^{Z+} [8]. p=1 stands for the ground state level. n_e and T_e are the electron density and temperature, respectively. In this case, the Boltzmann plot of the upper level populations is on a straight line, the slop of which is determined by the electron temperature. From the straight lines fitted to the population distributions as shown in figure 4, we estimate T_e of the ablation cloud for Al, Al^+ and Al^{2+} plasmas (more strictly, in the region of ablation cloud where most of Al atoms, Al^+ ions or Al^{2+} ions are emitting), the values of which are shown in figure 4.

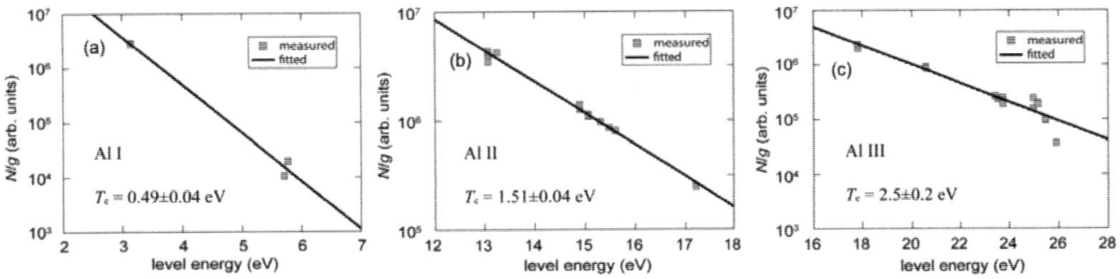

Figure 4. Boltzmann plot of upper level populations for (a) Al, (b) Al^+ and (c) Al^{2+} plasmas.

Regarding the Lorentz width, the natural broadening determined by the spontaneous transition is negligibly small in comparison with the observed width. Therefore, we treat the observed Lorentz width as the Stark width; $W_S = W_L$. The Stark width is proportional to the electron density as

$$W_s = C_s(T_e)n_e \quad (3)$$

where C_S is the Stark broadening coefficient.

W_S of the Al II lines denoted by 'a', 'b', 'c' and 'd' in figure 2 have been determined in previous calculations and experiments at several T_e as shown in figure 5 [9,10,11]. The curves in the figure are the results of the fit with exponential functions. From the curves and n_e in previous calculations and experiments, we determine C_S of the lines at T_e = 1.51 eV, which is the temperature of the Al^+ plasma. Since the experimental data only at T_e = 0.9 eV is available for the Al II line denoted by 'b', we adopt the value of C_S at T_e = 0.9 eV as that at T_e = 1.51 eV assuming T_e dependence of C_S is small in this temperature region. The estimate n_e from these lines are listed in table 1. It is found that scatter of the values are within 15 % of the average. We adopt the average value as the electron density of the Al^+ plasma.

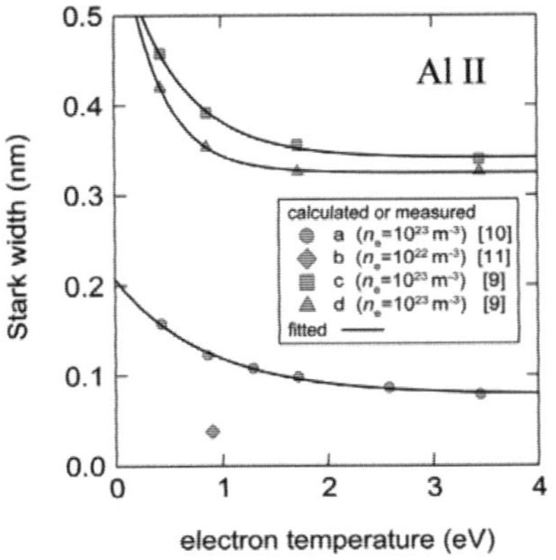

Figure 5. Stark widths of the Al II lines [9,10,11], which are denoted by 'a', 'b', 'c' and 'd' in figure 2. The curves are the fitted result by exponential functions.

Table 1. Electron densities estimated from the Stark widths of the Al II lines denoted by 'a', 'b', 'c' and 'd' in figure 2.

	λ (nm)	Transition (lower - upper)	C_S (nm m^3)	W_S (nm)	n_e (m^{-3})
a	466	$3p^2(^1D)$ - $3s4p(^1P)$	1.01×10^{-24}	0.042	4.1×10^{22}
b	559	$3s4p(^1P)$ - $3s4d(^1D)$	3.8×10^{-24}	0.182	4.8×10^{22}
c	624	$3s4p(^3P)$ - $3s4d(^3D)$	3.58×10^{-24}	0.172	4.80×10^{22}
d	684	$3s4p(^3P)$ - $3s5s(^3S)$	3.30×10^{-24}	0.165	5.01×10^{22}
					avg 4.7×10^{22}

Similar analyses can be done for the Al I and Al III lines. In table 2, we summarize T_e and n_e of the ablation cloud for Al, Al^+ and Al^{2+} plasmas. It is found that T_e increases while n_e decreases as an increase in the charge state. This tendency suggests that aluminum in different charge states emit radiation in different regions of the ablation cloud and/or at different time during the ablation. It has been reported, however, that intensity ratios of emission lines of various carbon ions change little during the carbon pellet ablation [4]. Supposing that the ablation mechanism due to heat flux from the LHD main plasma is the same for carbon and aluminum pellets, we attribute the observed tendency to an inhomogeneous structure of the ablation cloud; the plasma region for neutral atoms is close to ablating pellet where n_e is high and T_e is low and the region for a higher charge state is farther from the pellet where the cloud is expanding in the LHD main plasma with increasing T_e and decreasing n_e.

Table 2. Electron temperatures and densities estimated from the observed emission lines of aluminum atoms and ions.

	T_e (eV)	n_e (m^{-3})
Al (Al I)	0.49±0.04	3.0×10^{23}
Al^+ (Al II)	1.51±0.04	4.7×10^{22}
Al^{2+} (Al III)	2.5±0.2	3.6×10^{22}

From the observed Stark widths with n_e estimated above, we determine Stark broadening coefficients of several Al II lines, the values of which have not been reported. The determined values are listed in tables 3 together with the data used for the n_e estimation. It is demonstrated that the ablation plasma spectroscopy with a simultaneous wideband and high-resolution measurement can be a powerful tool to determine Stark broadening coefficients.

Table 3. Stark broadening coefficients of Al II lines (at T_e=1.5 eV).

λ (nm)	Transition (lower - upper)	W_S (nm)	C_S (nm m^3)	
			This work	Previous studies
463	$3s5s(^1S)$ - $3s8p(^1P)$	0.051	1.1×10^{-24}	
465	$3s4d(^1D)$ -$3s9f(^1F)$	0.031	6.6×10^{-25}	
466	$3p^2(^1D)$ - $3s4p(^1P)$	0.042	8.9×10^{-25}	1.01×10^{-24} [10]
559	$3s4p(^1P)$ - $3s4d(^1D)$	0.182	3.9×10^{-24}	3.8×10^{-24} [11]
624	$3s4p(^3P)$ - $3s4d(^3D)$	0.172	3.7×10^{-24}	3.58×10^{-24} [9]
634	$3s3d(^1D)$ - $3s5p(^1P)$	0.265	5.7×10^{-24}	
683	$3s4p(^3P)$ - $3s5s(^3S)$	0.165	3.5×10^{-24}	3.30×10^{-24} [9]
705	$3s4s(^3S)$ - $3s4p(^3P)$	0.077	1.7×10^{-24}	
747	$3s3d(^1D)$ - $3s4f(^1F)$	0.188	4.0×10^{-24}	

4. Conclusion

We simultaneously observed emission spectrum of an ablation cloud of an aluminum pellet injected into an LHD plasma in the entire visible region with high-resolution. More than 50 of Al I, II, III and IV emission lines were resolved in the spectrum. Population distributions of excited levels of atoms and ions in the Al, Al^{+} and Al^{2+} plasmas converted from the observed line intensities were analyzed. LTE states were thought to be generated in the Al, Al^{+} and Al^{2+} plasmas and the electron temperatures of the plasmas were estimated. The electron densities of the plasmas were also estimated from the observed Stark widths by comparison with the previously reported Stark widths. It was found that the electron temperature and density increases and decreases, respectively, as an increase in the charge state of aluminum. The phenomenon implies a laminar structure of the ablation cloud; higher density, lower temperature and lower ionized plasma inside and lower density, higher temperature and higher ionized plasma outside. On the other hand, Stark broadening coefficients of several Al II lines were newly determined from the observed Stark width with the estimated electron density.

Acknowledgments

The authors are grateful to technical staffs of the work shop of the department of mechanical engineering and science in Kyoto University for their support in manufacturing of the spectrometer. The support of LHD experiment group is also greatly acknowledged.

References

[1] Tamura N, Sudo S, Khlopenkov K V, Kato S, Sergeev V Yu, Muto S, Sato K, Funaba H, Tanaka K, Tokuzawa T, Yamada I, Narihara K, Nakamura Y, Kawahata K, Ohyabu N, Motojima O and LHD experimental groups 2002 *Plasma Phys. Control. Fusion* **45** 27

[2] Nozato H, Morita S, Goto M, Takase Y, Ejiri A, Amano T, Tanaka K, Inagaki S and LHD Experimental Group 2004 *Phys. Plasmas* **11** 1920

[3] Goto M, Sakamoto R and Morita S 2007 *Plasma Phys.Control.Fusion* **49** 1163

[4] Goto M, Morita S and Koubiti M 2010 *J.Phys.B:At.Mol.Opt.Phys.* **43** 144023

[5] Koubiti M, Goto M, Morita S and Stamm R 2009 *J.Plasma Fusion Res. SERIES* **8** 991

[6] NIST Atomic Spectra Database ver.4 (http://physics.nist.gov/PhysRefData/ASD/)

[7] Kelleher D E and Podobedova L I 2008 *J.Phys.Chem.Ref.Data* **37** 709

[8] Fujimoto T 2004 *Plasma Spectroscopy* (Oxford: Clarendon Press)

[9] Griem H 1974 *Spectral Line Broadening by Plasmas* (Academic Press)

[10] Allen A, Blaha M, Jones W, Sanchez A and Griem R 1975 *Phys.Rev.A* **11** 477

[11] Konjevic N, Lesage A, Fuhr J R and Wiese W L 2002 *J.Phys.Chem.Ref.Data* **31** 819

Journal of Physics: Conference Series **397** (2012) 012017 doi:10.1088/1742-6596/397/1/012017

Charge Exchange Signatures in X-Ray Line Emission Accompanying Plasma-Wall Interaction

O Renner[1], E Dalimier[2], R Liska[3], E Oks[4] and M Šmíd[1,3]

[1] Institute of Physics, v.v.i., Academy of Sciences CR, 18221 Prague, Czech Republic
[2] Sorbonne Universités, Pierre et Marie Curie, F-75252 Paris Cedex 05, and École Polytechnique, LULI, F-91128 Palaiseau Cedex, France
[3] Czech Technical University in Prague, FNSPE, 11519 Prague, Czech Republic
[4] Physics Department, Auburn University, Auburn, AL 36849, USA

E-mail: renner@fzu.cz

Abstract. Directional flows of well-collimated energetic ions produced at laser-burnt-through foils were used to investigate transient phenomena accompanying the plasma interaction with surfaces of solid targets (generally known as plasma-wall interaction, PWI). The plasma jet launched from the rear surface of the 0.8-μm-thick Al foil irradiated at oblique laser incidence with an intensity of 3×10^{14} W/cm^2 was incident on the quasi-massive C target. The plasma x-ray self-emission was analyzed by focusing survey and high-dispersion spectrometers. The time-integrated, spatially resolved narrow-band spectra recorded close to the C surface exhibit a dip structure in the red-wing profiles of the hydrogenic Al Lyγ line which was attributed to the charge exchange between two stationary Coulomb centers represented by the Al XIII and fully stripped C ions. This identification of the charge exchange signatures in x-ray line emission is supported by hydrodynamic simulations of environmental conditions in the near-wall plasma and by predictions of the dips positions following from complementary theoretical models. The agreement between the experiment and theory validates the first high precision x-ray spectroscopic identification of charge exchange phenomena accompanying the PWI.

1. Introduction

Being subjected to extremely high heat loads and high particle and neutron fluxes, the plasma-facing materials of present-day experimental fusion reactors experience particularly hostile environmental conditions. Many of currently used inner-wall components cannot withstand the plasma scenarios foreseen for the future commercial power plants [1]. Further optimization of plasma-facing materials needed for development of new technological concepts requires a detailed understanding of the PWI processes that occur when solid surfaces are bombarded by intense particle fluxes.

Various mechanisms of the energy transfer, material erosion and migration in the near-wall region include ion deceleration and stopping, shock waves generation, formation of highly excited Rydberg states or hollow atoms with multiple inner vacancies, charge transfer processes, and ion neutralization [for references, see e.g. 2]. The purpose of this paper is to contribute to x-ray spectroscopic diagnostics of charge exchange (hereafter CE) phenomena accompanying the PWI. We describe the experimental configuration used, present high-resolution spectroscopic data, and interpret the observed modulation of Al Lyγ profiles in terms of CE between multicharged atoms. The identification of CE phenomena is supported by hydrodynamic simulations and by predictions of two complementary theoretical models.

2. PALS experiment

The experiment has been performed on the iodine laser system of the PALS Research Centre [3]. The scheme of the experimental setup is shown in figure 1. The laser beam delivering 50-90 J of frequency-tripled radiation (438 nm) in a pulse length of 0.24-0.3 ns (full width at half maximum) was incident onto the primary, jet-producing target (0.8-μm-thick Al foil) at an angle inclined by α = 30° from the target normal.

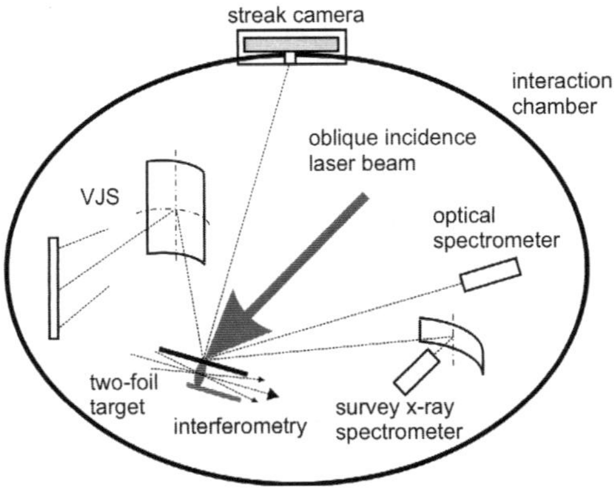

Figure 1. Schematic drawing of the diagnostic complex used for x-ray characterization of plasma-wall interactions.

The reason for the oblique incidence of the laser beam onto the target is explained in figure 2, where the jets formation in transmission geometry is simulated by the Prague Arbitrary Lagrangian Eulerian hydrocode PALE [4]. The 2D distribution of plasma parameters indicates that the foil burns through well before the laser pulse maximum, i.e., at normal incidence (α = 0°) the plasma jet strikes the secondary target pre-ionized by the action of the transmitted radiation. In the oblique incidence case (α = 30°), the laser beam does not hit the secondary foil and the expanding plasma jet launched in a direction of normal to the foil surface interacts with the unperturbed target, thus creating a better-characterized environment for PWI studies. The optimum conditions for formation of sufficiently high aspect-ratio jets were found when defocusing the beam to the focal spot radius of 150 μm [5].

The standard diagnostic complex used in the PALS PWI experiments consists of the time-resolved x-ray imaging of the plasma expansion, optical spectroscopy, interferometry and several x-ray spectrometers. The primary diagnostics applied in the present study was a high-resolution vertical dispersion Johann spectrometer (VJS) fitted with a cylindrically bent crystal of quartz (100) [6]. Hereafter we shall concentrate on evaluation of high-resolution spectra obtained by the VJS. The interpretation of experimental data provided by supporting diagnostics can be found in papers [2, 5].

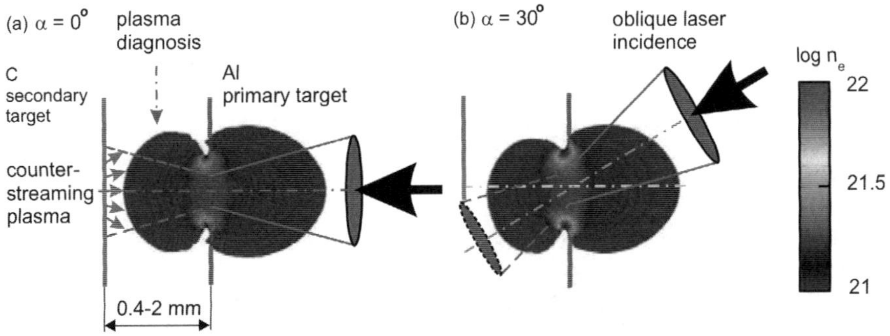

Figure 2. Alternate schemes of the plasma jets production and their interaction with secondary targets. The simulated distribution of the plasma electron density n_e corresponds to the laser pulse maximum.

3. Corelation of observed spectra and simulated environmental conditions

One of characteristic properties of the VJS is a simultaneous production of two sets of mirror symmetric spectra with one-dimensional (1D) spatial resolution perpendicular to a direction of plasma observation. The existence of the axis of symmetry facilitates the computational reconstruction of the spectra and considerably increases the reliability of identification of tiny spectral features assuming that they are observed in both parts of corresponding spectral records [6].

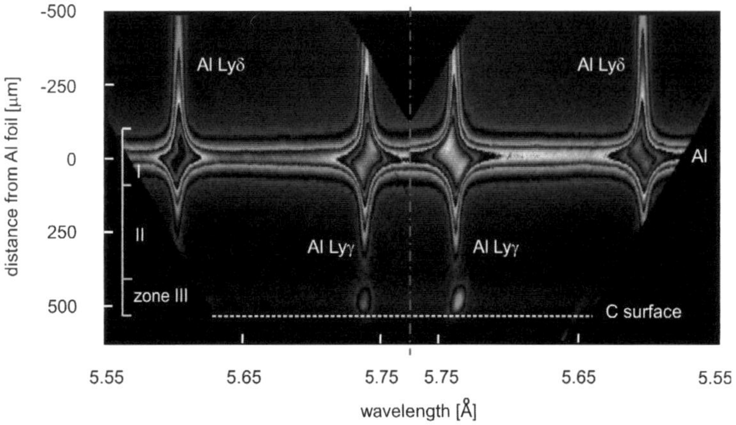

Figure 3. Spatially resolved spectra of Al Lyγ, δ self-emission from double-foil Al/C target recorded by the VJS. The laser strikes the primary Al-foil target at oblique incidence from above.

The spatially resolved x-ray spectrum presented in figure 3 was recorded at the oblique incidence of the laser beam (57 J, 0.27 ns, 2.5×10^{14} W/cm^2) striking the primary Al target from above. The Al foil and the secondary quasi-massive C target (250-μm-thick pyrolytic graphite foil) were separated by a distance of 590 μm. An expansion of the plasma jets produced at both front and rear surfaces of the Al foil (spatial coordinate 0) is visualized via the emission of the Lyman series members γ and δ of the H-like Al. The recorded line profiles are characteristic for three zones marked in figure 3. In zone I extending within approximately ±100 μm from the Al target, the emitted lines display distinct Doppler shifts due to directional reversal of the plasma jets outflow at opposite sides of the burnt-through foil. Close to the Al target, the optical thickness of the Al Lyγ emission increases, thus the observed shifts result from a combination of the lateral and normal plasma expansion with respect to the target surface. The detailed evaluation of the line shifts at the distances of ±25 μm from the Al foil provided the effective lateral velocity component $\sim1.5\times10^7$ cm/s which agrees with the near-surface semispherical plasma expansion observed in interferometric measurements [5].

In zone II extending up to the distance of about 450 μm from the Al foil, the plasma propagates in a laminar flow and no large-scale plasma turbulences are visible [5]. Consequently, the distribution of the measured effective plasma parameters follows predictions of a simple 1D hydrodynamic modeling. With the increasing distance from the primary target, the intensity of the observed Al Lyγ emission gradually decreases but then it again intensifies in the zone III. The enhanced emission close to the C surface indicates deceleration, trapping and subsequent dissipation of Al ions at the secondary target.

The interpretation of the spectra emitted from the near-wall zone requires the knowledge of the environmental plasma conditions. The analysis of the spectra recorded by the spherically bent mica crystal spectrometer [7] in the zone III provided the electron temperature $T_e \approx$ 450–550 eV and electron density n_e within $0.5–5\times10^{21}$ cm^{-3}. The limited spatial resolution of the spectrometer (~20 μm) does not permit the more detailed determination of the plasma parameters distribution in the near-wall zone; to remedy this, a series of simulations has been performed to evaluate the time-dependent distribution of the macroscopic plasma parameters.

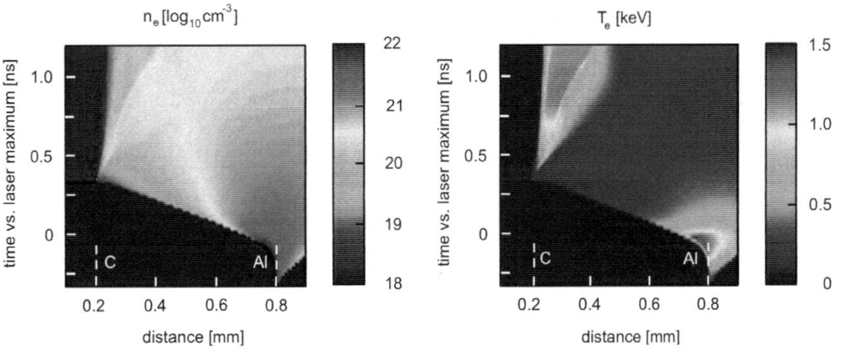

Figure 4. Temporal-spatial distribution of plasma parameters simulated by codes PALE and MULTIF.

Interpenetrating plasmas were modeled by a combination of the 2D hydrodynamic code PALE with the 1.5 D multi-fluid code MULTIF [8]. The PALE code provided the temporally resolved 2D maps of hydrodynamic quantities. For time t = 340 ps after the laser pulse maximum, the simulated plasma outflow approaches closely the secondary target. From relevant 2D distribution of plasma parameters, the 1D profile corresponding to the central cylindrical axis of symmetry was extracted and used as input for the MULTIF description of the Al plasma interaction with the cold C surface, production of the counter-streaming C plasma and subsequent collision and interpenetration of both plasmas [9].

Immediately after the impact of Al jet onto C surface, the simulations indicate a fast creation of a relatively thin (~20 μm) near-wall layer of highly ionized, strongly interpenetrating Al and C ions. The resulting temporal-spatial distribution of the electron density and temperature is shown in figure 4; as discussed in more detail in section 4, the found environmental conditions are favorable for observation or charge exchange phenomena in the near-wall layer. The validity of these simulations is limited by the 1D simplification used when running the MULTIF and by neglecting the C-surface preheat due to combined effects of the scattered laser light, fast particles and shock waves. On the other hand, the simulations provide a good qualitative insight into processes occurring within the inter-target space.

4. Charge exchange signatures in spatially resolved spectra of Al Lyγ

The Al Lyγ profiles of the near-wall-emitted spectra from figure 3 are presented in figure 5a. The mirror-symmetric line profiles display a reproducible fine structure, the development of which is more distinct in the 10-points FFT-smoothed right-hand side spectra shown in figure 5b. As the distance from the C target surface decreases to less than 10 μm, a depression surrounded by two peaks appears in the very center of the line. This is a manifestation of a significant self-absorption in agreement with the strong density gradient predicted by the plasma simulations; this depression helps to identify the center of the line. Second, with the decreasing distance the line centers shift distinctly to the red. Assuming that these shifts can entirely be ascribed to the so called plasma polarization shift [10], its evaluation in terms of quantum-mechanical impact theory would indicate steep density gradients (starting from $n_e \approx 4\times10^{22}$ cm^{-3} observed for the distance of 5 μm below the target up to $n_e \approx 4\times10^{21}$ cm^{-3} in the distance of 37 μm above the C surface). On the other hand, the mechanisms responsible for the observed line shifts may also include the effects of the ions deceleration, radiative transfer in the turbulent plasma close to the target surface, and an enhanced satellites emission, thus the above given estimates should be taken like the upper limit for the density estimates. Both these features however indicate that the plasma-wall experiments exhibit *high-density effects*.

The third most exotic feature consists in the appearance of prominent dips in the near red wing, whose separations from the line center show a weak dependence on the distance from the target surface, and thus on n_e. This feature is a *quasi-molecular effect*. A weak density dependence of their positions is a characteristic of the charge-exchange-caused dips (X-dips, for brevity), in distinction to the Langmuir-waves-caused dips (observed, e.g., in [11]) whose positions scale as $n_e^{1/2}$.

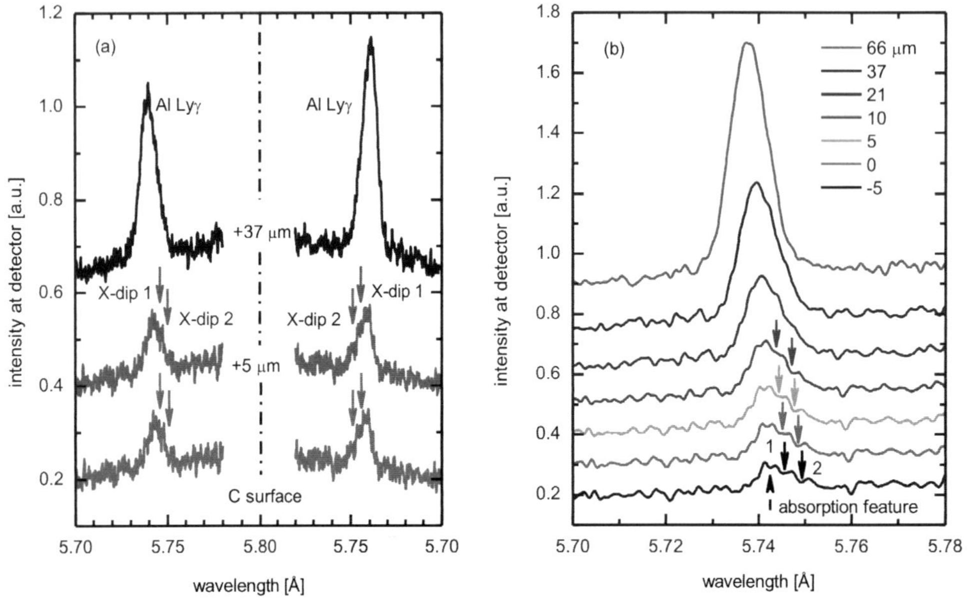

Figure 5. Reproducible fine structure in mirror-symmetric Al Lyγ profiles taken at given distances from the C surface (a) and dips evolution in the FFT-smoothed right-hand side spectra (b).

Physically, X-dips correspond to anti-crossings of energy terms of the one-electron quasi-molecule of nuclear charges Z_1 and Z_2: anti-crossings enhance charge-exchange. For the quasi-molecule $Al^{13+}eC^{6+}$ ($Z_1 = 13$ and $Z_2 = 6$) and the energy terms of Al XIII perturbed by C^{6+} and the terms of C VI perturbed by Al^{13+}, the theory predicts three anti-crossings at different inter-nuclear distances (depending only weakly on n_e) and thus up to three X-dips to observe. For a clear observation of a particular X-dip, the inter-ionic distance, where the corresponding anti-crossing of terms occurs, should be close to the most probable inter-ionic distance (controlled by n_e) [12]. Therefore it is unlikely to observe all three x-dips in the same line profile. The present experiment is favorable for observing dips1 and 2 (dip 3, the farthest from the line center is unlikely to be observed) while in previous experiments at LULI facility [13,14], only dips 2 and 3 could be observed.

The positions of the X-dips 1 and 2 were carefully determined with respect to the shifted centers of the spectral profiles. At distances ±5 μm from the target surface, the found positions are 3.1±0.2 mÅ for dip 1 and 6.6±0.2 mÅ for dip 2; at distances ≥10 μm, dip 2 slightly shifted to 6.2±0.2 mÅ. With respect to simulated density profiles close to the C surface, this means that only a weak dependence of the dips position on the electron density has been detected. This dependence could be supported by the theoretical analysis of charge-exchange phenomena and two complementary theoretical models.

In the first model [12,15], the dip positions were calculated analytically in the parabolic coordinates using the multipole expansion for the interaction of the hydrogenic ion Al XIII with the fully-stripped ion C VI, and without taking into account the screening by plasma electrons (n_e=0). We retained a large number of terms in the multipole expansion to achieve the accuracy adequate for the comparison with experiments. Using this model the predicted positions of the dips 1 and 2 are 3.1 mÅ and 6.7 mÅ.

In the second model [14,16], we calculated the dip positions for the quasi-molecule numerically by diagonalizing the Hamiltonian in the elliptical coordinates, in which the problem of two Coulomb centers allows the separation of variables (simulations by the quasi-molecular code IDEFIX). These simulations yield exact terms of the quasi-molecule $Al^{13+}eC^{6+}$ with the subsequent allowance for the plasma electron screening (depending on n_e). As explained below, the usage of the more realistic *quasi-molecular model* IDEFIX *shifted the x-dips* positions slightly further *to the red*, while the allowance for the *electron screening shifted them to the blue*.

Within the quasi-molecular model and the quasi-static approximation (valid in the line wings), the Al Lyγ transition emitted from a dicenter plasma Al-C is controlled by the energy difference $E_u(R) - E_l(R)$ between the upper u and lower l levels concerned (R being the inter-ionic distance). For spectral line shapes the average over R using the nearest neighbor probability distribution is required.

Let us define the center of the line when the electron screening is neglected,

$$\Delta E_0 = E_u(R \to \infty) - E_l(R \to \infty).$$

This reference is the same for the analytical model "*analytic*" and the quasi-molecular model without taking into consideration the electron screening "*mol*" (*mol* stands for molecular). For the quasi-molecular model with the allowance of the electron screening "*mol + PPS*" (*PPS* stands for plasma polarization shift), the line center of the line is

$$\Delta E_0^{PPS} = \Delta E_u^{PPS}(R \to \infty) - E_l^{PPS}(R \to \infty).$$

It is well known that the PPS is red, thus $\Delta E_0^{PPS} - \Delta E_0 < 0$. Now we focus on charge exchange corresponding to the anti-crossing at the position R_{CE}. This anti-crossing is responsible of an X-dip. Let us introduce the position $\Delta E_{CE}^{analytic}$ of the anti-crossing level with respect to the level l at the anti-crossing distance $R_{CE}^{analytic}$ for the analytical model. In the same way ΔE_{CE}^{mol} and $\Delta E_{CE}^{mol+PPS}$ stand for the positions of the anti-crossing levels with respect to the level l at the distances R_{CE}^{mol} and $R_{CE}^{mol+PPS}$, respectively, for the two versions "*mol*" and "*mol+PPS*" of the quasi-molecular code IDEFIX [14,16].

The positions of the X-dips with respect to the line center are $\Delta E_{CE}^{analytic} - \Delta E_0$, $\Delta E_{CE}^{mol} - \Delta E_0$, and $\Delta E_{CE}^{mol+PPS} - E_0^{PPS}$ within the three models "*analytic*", "*mol*", and "*mol+PPS*", respectively.

When compared to the "*analytic*" model, the model "*mol*", *valid for any electron density, yields a red correction* C_1 for the distance of the X-dip with respect to the line center

$$C_1 = \Delta E_{CE}^{mol} - \Delta E_{CE}^{analytic} < 0.$$

The model "*mol+PPS*" depending on n_e gives an added correction C_2 with respect to the "*mol*" model

$$C_2 = (\Delta E_{CE}^{mol+PPS} - \Delta E_{CE}^{mol}) + (\Delta E_0 - \Delta E_0^{PPS}).$$

The second term in this sum is positive and increases with n_e. The first term is negative and its absolute value is negligible with respect to the second term because it relates to small inter-ionic distance not very sensitive to the electron screening. Thus the *correction* C_2 *is always blue* (C_2>0).

For the dip 2, the positions 7 mÅ and 6 mÅ, calculated for $n_e = 1\times10^{22}$ cm^{-3} using the models "*mol*", and "*mol+PPS*", respectively [14,16], agree with the previous discussion (C_1<0 and C_2>0). The fact that the experimental X-dip 2 position 6.2±0.2 mÅ measured at distances ≥10 μm is closer to the prediction of the "*mol+PPS*" model (6 mÅ) than to the prediction of the "*analytic*" model (6.7 mÅ), is the first experimental evidence of some (weak) dependence of the X-dip positions on n_e.

In the close near-wall area, i.e., at distances ±5 μm from the C-foil surface, the measured positions of both dips, 3.1±0.2 mÅ for dip 1 and 6.6±0.2 mÅ for dip 2, respectively, are close to the "analytic" values 3.1 mÅ and 6.7 mÅ, confirming that C_2 increasing with n_e can cancel C_1.

To conclude, the experiment performed at PALS is the first to yield with a high precision the X-dips positions in Al XIII Lyγ red wing. Their appearance in both mirror-symmetric parts of the spectra and their small density evolution, confirmed by the theoretical models, support the spectroscopic manifestations of charge-exchange phenomena in plasma-wall interaction.

Acknowledgments
This research was supported by the Czech Science Foundation, grant No. P205/10/0814. The authors gratefully acknowledge an assistance provided by the scientific and technical staff of the PALS laser.

References

[1] Jacob W, Linsmeier C and Rubel M 2011 *Phys. Scr.* T **145** 011001
[2] Renner O, Liska R and Rosmej FB 2009 *Laser Part. Beams* **27** 725
[3] Jungwirth K, Cejnarova A, Juha J, Králiková B, Krása J, Krouský E, Krupičková P, Láska L, Mašek K, Mocek T, Pfeifer M, Präg A, Renner O, Rohlena K, Rus B, Skála J, Straka P and Ullschmied J 2001 *Phys. Plasmas* **8** 2495
[4] Liska R, Limpouch J, Kucharik M and Renner O 2008 *J. Phys.: Conf. Series* **112** *022009*
[5] Renner O, Pisarczyk T, Chodukowski T, Kalinowska Z, Krouský E, Pisarczyk T, Šmíd M, Ullschmied J and Dalimier E 2011 *Phys. Plasmas* **18** 093503
[6] Renner O, Mißalla T, Sondhauß P, Krouský E, Förster E, Chenais-Popovics C and Rancu O 1997 *Rev. Sci. Instrum.* **68** 2393
[7] Renner O, Krouský E, Šmíd M, Liska R, Váchal P, Khattak FY, Dalimier E and Oks E 2011 *Europ. Phys. J.: Web of Conferences*, submitted
[8] Chenais-Popovics C, Renaudin P, Rancu O, Gilleron F, Gauthier JC, Larroche O, Peyrusse O, Dirksmöller M, Sondhauß P, Mißalla T, Uschmann I, Förster E, Renner O and Krouský E 1997 *Phys. Plasmas* **4** 190
[9] Smid M, Renner O and Liska R 2012 submitted to 32th ECLIM, Warsaw Sept 10-14,
[10] Nguyen H, Koenig M, Benredjem D, Caby M and Coulaud G 1986 *Phys. Rev.* A **33** 1279
[11] Renner O, Dalimier E, Oks E, Krasniqi F, Dufour E, Schott R and Förster E 2006 *J. Quant. Spectr. Rad. Transfer* **99** 439
[12] Oks E and Leboucher-Dalimier E 2000 *J. Phys. B: At. Mol. Opt. Phys.* **33** 3795
[13] Leboucher-Dalimier E, Oks V, Dufour E, Sauvan P, Angelo P, Schott R and Poquerusse A 2001 *Phys. Rev. E, Rapid Communications* **64** 065401
[14] Dalimier E, Oks E, Renner O and Schott R 2007 *J. Phys. B: At. Mol. Opt. Phys.* **40** 909
[15] Oks E and Leboucher-Dalimier E 2000 *Phys. Rev E, Rapid Communications* **62** R3067
[16] Leboucher-Dalimier E, Oks E, Dufour E, Angelo P, Sauvan P, Schott R and Poquerusse A 2002 *Eur. Phys. J. D* **20** 269

Spectral line formation with wave effects in optically thick plasmas

J Rosato

Laboratoire PIIM, UMR 7345 Université d'Aix-Marseille / CNRS, Centre de Saint-Jérôme, Case 232, F-13397 Marseille Cedex 20, France

E-mail: joel.rosato@univ-amu.fr

Abstract. An extension of the radiative transfer theory is presented and applied to the formation of spectral lines in optically thick media. In the model, the radiation field is described with the quantum phase space formalism adapted to second quantization. Comparisons with the standard radiative transfer theory in ideal cases indicate that the latter may be inaccurate if the light's coherence length is of the same order as or larger than the photon mean free path.

1. Introduction

Recent theoretical works have reported on an extension of the radiative transfer theory suitable for problems involving the wave nature of light. A derivation of the radiative transfer equation (RTE) from the first principles has been presented in [1]. The treatment involves the quantum phase space formalism adapted to second quantization and the result is an integro-differential equation for the one-photon Wigner function, a quantity interpretable as a phase space probability density with quantum features (e.g., it can be negative in domains violating the Heisenberg uncertainty relation). This "quantum" radiative transfer equation (hereafter referred to as QRTE) reduces to the usual RTE in the limit of small wavelength and coherence length. The applicability of this limit has been examined in ideal cases [2–4]. It has been shown that the spatial coherence of light leads to a distortion of the Wigner function if the coherence length is of the same order as or larger than the other spatial scales of interest. Hence, in such conditions, the use of the RTE for describing the formation of line shapes would be inappropriate and could lead to significant misestimates. In this work, we report on new calculations. Section 2 gives a brief overview of the standard RTE and the terminology used in the theory, the QRTE is presented in Sec. 3 and results are shown in Sec. 4.

2. Standard radiative transfer formalism

The fundamental quantity of interest in radiative transfer is the spectral energy flux $I(\omega,\vec{n},\vec{r},t)$ [J m^{-2} s^{-1} sr^{-1} (rad/s)$^{-1}$], referred to as "specific intensity" (e.g. [5]). The RTE is a transport equation for this quantity, and it is obtained from balance considerations. In its usual form, this equation reads

$$\left(\frac{1}{c}\frac{\partial}{\partial t} + \vec{n}\cdot\vec{\nabla} + \chi\right)I = \eta\,, \tag{1}$$

where $\eta(\omega,\vec{n},\vec{r},t)$ and $\chi(\omega,\vec{n},\vec{r},t)$ are source and loss terms, respectively, and express the interaction between the radiation field and the constituents of the matter present in the medium under consideration, for example atoms, molecules, or ions. For spectral line radiation with an atomic transition $u \rightarrow d$, these terms are given by

$$\eta(\omega,\vec{n},\vec{r},t) = \frac{\hbar\omega}{4\pi} N_u(\vec{r},t) A_{ud} \phi(\omega,\vec{n},\vec{r},t), \tag{2}$$

$$\chi(\omega,\vec{n},\vec{r},t) = \frac{\hbar\omega}{4\pi}\left[N_d(\vec{r},t)B_{du} - N_u(\vec{r},t)B_{ud}\right]\phi(\omega,\vec{n},\vec{r},t), \tag{3}$$

where A_{ud}, B_{du}, and B_{ud} are the Einstein coefficients for spontaneous emission, absorption, and stimulated emission, respectively; N_u, N_d are the densities of atoms in the upper and lower states; and ϕ is the local spectral line shape normalized such that $\int d\omega \int d\Omega \phi(\omega,\vec{n},\vec{r},t) = 4\pi$. It denotes the probability density function of emitting or absorbing a photon at the frequency ω. The fact that the stimulated emission is treated as a negative part in the absorption is only a matter of convention, which allows one to write the RTE in the compact form given in Eq. (1). The Einstein A and B coefficients are related by $g_u B_{ud} = g_d B_{du} = g_u A_{ud} 4\pi^3 c^2/\hbar\omega_0^3$, where g_u and g_d denote the statistical weights of the upper and lower level, respectively, and ω_0 is the Bohr frequency of the transition under consideration.

The RTE looks like a Boltzmann equation. An examination of the loss term χ shows that it has the dimensions of an inverse length, which suggests that it can be interpreted as an inverse monochromatic mean free path (when it is positive) for the radiation energy. Because the energy exchange processes occur through the creation and destruction of photons, it suggests that we can interpret χ as an inverse mean free path for photons also. A similar argument for η leads to the conclusion that the radiative transfer equation is equivalent to a Boltzmann equation for a distribution function $f(\vec{r},\vec{p},t)$ which describes photon transport in phase space, using the correspondence $\vec{p} = \hbar\omega\vec{n}/c$ and the proportionality relation $I = \hbar c p^3 f$.

3. The quantum radiative transfer equation

The particle picture of radiation transport leads to ambiguities inherent to the definition of a photon in quantum mechanics. Symmetry considerations show that a position operator is not well defined for this particle, because it has zero mass and spin 1 [6]. Furthermore, if ever such an operator did exist, the Heisenberg uncertainty relation would prevent from knowing the position and momentum of a photon simultaneously, in contrast with that assumed in the standard radiative transfer theory. A remedy to this problem is provided by the quantum phase space formalism developed by Wigner [7]. This formalism involves "quasi-probability" distributions $W_s(\vec{r}_1,\vec{p}_1..\vec{r}_s,\vec{p}_s,t)$, which are generalizations of the s-particle phase space distribution functions used in classical kinetic theory. They account for nonclassical features such as the uncertainty principle and quantum entanglement. "Quasi-probability" means that these functions may become negative. This occurs for physical states that have no classical equivalent. A coarse graining procedure is done to avoid such a peculiarity, with a spatial scale much larger than the particle's thermal de Broglie wavelength $\hbar/\Delta p$ (where Δp is the typical momentum dispersion). For photons, the thermal wavelength denotes the coherence length $\lambda_c \sim c/\Delta\omega_{1/2}$ where $\Delta\omega_{1/2}$ is the half width at half maximum (e.g. [8]). In the following, we use an adaptation of the Wigner formalism to second quantization and focus on the one-photon distribution $W_1(\vec{r}_1,\vec{p}_1,t) \equiv W(\vec{r},\vec{p},t)$. The development is detailed in [1] and follows early works reported in [9–11]. The Wigner function is defined as the average of the phase space photon number operator $N(\vec{r},\vec{p})$:

$$W(\vec{r},\vec{p},t) = \mathrm{Tr}\{\rho(t)N(\vec{r},\vec{p})\}, \tag{4}$$

$$N(\vec{r},\vec{p}) = \sum_{\vec{\varepsilon}} \frac{1}{(2\pi\hbar)^3} \int d^3k\, a^+_{\vec{\varepsilon}}\left(\frac{\vec{p}}{\hbar} - \frac{\vec{k}}{2}\right) a_{\vec{\varepsilon}}\left(\frac{\vec{p}}{\hbar} + \frac{\vec{k}}{2}\right) e^{i\vec{k}\cdot\vec{r}}. \tag{5}$$

Here, $\rho(t)$ is the density operator of the total (radiation + matter) system, $a^{+}_{\vec{\varepsilon}}(\vec{k})$ and $a_{\vec{\varepsilon}}(\vec{k})$ are respectively the creation and annihilation operators corresponding to a photon of wavevector $\vec{k}$ and polarization $\vec{\varepsilon}$, and Tr{...} denotes a trace over the whole system's Hilbert space. The creation and annihilation operators obey the commutation rules $[a_{\vec{\varepsilon}}(\vec{k}),a^{+}_{\vec{\varepsilon}'}(\vec{k}')]=\delta_{\vec{\varepsilon}\vec{\varepsilon}'}\delta(\vec{k}-\vec{k}')$ and $[a_{\vec{\varepsilon}}(\vec{k}),a_{\vec{\varepsilon}'}(\vec{k}')]=0=[a^{+}_{\vec{\varepsilon}}(\vec{k}),a^{+}_{\vec{\varepsilon}'}(\vec{k}')]$. The sum over the polarizations performed in Eq. (5) is due to an assumption of unpolarized radiation used here. Examination of Eqs. (4) and (5) shows that $W(\vec{r},\vec{p},t)$ is normalized to the total number of photons. A closed transport equation for the one-photon Wigner function is obtained by applying the differential operator $\{\partial_t + c(\vec{p}/p)\cdot\vec{\nabla}\}\equiv D/Dt$ to each side of Eq. (4) and by using appropriate assumptions [1]. The Lagrangian derivative involves two terms of different physical origins. (i) DN/Dt denotes the time variation of the phase space number operator along a trajectory. It is identically zero in the short wavelength limit, which corresponds to propagation along defined straight rays. We will assume this condition satisfied in the following. (ii) $D\rho/Dt \equiv d\rho/dt$ denotes evolution of the Wigner distribution due to the interactions between the radiation field and the plasma, i.e., the emission and absorption of photons. This term can be described using a quantum master equation [12–14]. This treatment yields the following transport equation

$$\left(\frac{\partial}{\partial t}+c\frac{\vec{p}}{p}\cdot\vec{\nabla}\right)W(\vec{r},\vec{p},t)=S(\vec{r},\vec{p},t)-\int d^3r'\int d^3p'K(\vec{r},\vec{p},t;\vec{r}',\vec{p}')W(\vec{r}',\vec{p}',t), \tag{6}$$

where S denotes a source corresponding to spontaneous emission and the integral term describes the contribution of the absorption corrected by stimulated emission. Explicitly, S and K are given by

$$S(\vec{r},\vec{p},t)=\frac{1}{(\pi\hbar p)^3\hbar}\mathrm{Re}\int d^3r'\int d^3p'\eta_c(\vec{r}',\vec{p}',t)e^{-2i(\vec{p}-\vec{p}')\cdot(\vec{r}-\vec{r}')/\hbar}, \tag{7}$$

$$K(\vec{r},\vec{p},t;\vec{r}',\vec{p}')=\frac{c}{(\pi\hbar)^6}\mathrm{Re}\int d^3r''\int d^3p''\chi_c(\vec{r}'',\vec{p}'',t)e^{2i[(\vec{p}'-\vec{p}'')\cdot(\vec{r}-\vec{r}')-(\vec{p}-\vec{p}')\cdot(\vec{r}-\vec{r}'')]/\hbar}, \tag{8}$$

where η_c and χ_c are complex generalizations of the emission and absorption coefficients used in the radiative transfer theory. They have a structure similar to Eqs. (2) and (3). The complex line shape function is given by the following relation [2] ($\vec{n},\vec{r},t$ are not written explicitly)

$$\phi_c(\omega)=\frac{1}{\pi}\int_0^\infty d\tau C(\tau)e^{-i\omega\tau}, \tag{9}$$

with $C(\tau)$ being the atomic dipole autocorrelation function. The line shape function introduced in Sec. 2 corresponds to the real part of Eq. (9). If the Doppler effect is important, this expression is convolved with the atoms' velocity distribution function.

The transport equation (6) is more general than the standard RTE. The presence of an integral in the right-hand side denotes delocalization of the absorption and the stimulated emission processes on a phase space volume of typical extent $\hbar^3$. This result is in agreement with the Heisenberg uncertainty principle. A consequence is that the processes are delocalized spatially over a region of typical volume λ_c^3. There is also a delocalization of the emission and absorption processes with respect to the emitters' and absorbers' location, which stems from the integrals present in the source and the kernel terms [Eqs. (7) and (8)]. Equation (6) reduces to the radiative transfer equation when $\lambda_c \to 0$. In the general case, we will refer to it as a "quantum" radiative transfer equation (QRTE). The QRTE can be interpreted as a transport equation governing the evolution of quasi-particles (also: wave packets) of spatial volume λ_c^3 and with the momentum dispersion Δp^3. Figure 1 shows an illustration of the spontaneous emission and the evolution of such a quasi-particle. An ideal case is considered: we assume an isolated, motionless, and perfectly localized atom, emitting in a given direction (one-dimensional geometry). There are no absorbers, so that $K \equiv 0$ in the QRTE. In the source term, the atomic density (which appears in the complex emissivity η_c) is set equal to $\delta(z)\exp(-A_{ud}t)$, with z

being the direction under consideration. The exponential denotes depopulation due to the spontaneous emission. In a similar fashion, there is also an exponential in the dipole autocorrelation function present in the complex line shape Eq. (9) in such a way to account for natural line broadening (here: the damping factor is $A_{ud}/2$). As can be seen in the figure, the Wigner function presents a bump structure of size $\lambda_c \times \Delta p$ after a few coherence times (here the coherence time is estimated as $\tau_c = 1/A_{ud}$), which indicates the presence and the propagation of the quasi-particle. Although obtained here within a pure quantum approach, the delocalization effect induced by the spatial coherence is also expected from classical electromagnetism. A sketch of the derivation of a transport equation accounting for spatial coherence, with a classical picture, is presented in the Appendix. The result shown in Fig. 1 completes a preliminary study reported in [2]. Note, in the latter reference, an absolute value is missing on the y axis label of Fig. 2. This omission does not affect the conclusion of this work because the spread of the Wigner function over the coherence length, illustrated through the figure, does not depend on the sign of the momentum.

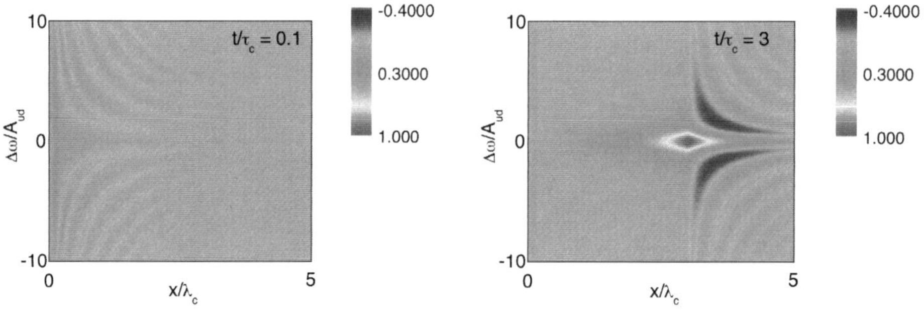

Figure 1. Phase space map of the Wigner function in the case of spontaneous emission by a motionless atom located at $z = 0$ (the correspondence between the y axis label and the momentum dispersion is done through the relation $\Delta p = \Delta\omega/c$). The Wigner function presents a bump structure of size $\lambda_c \times \Delta p$ after a few coherence times.

4. Application to the formation of spectral lines

The delocalizing integrals in Eqs. (6) – (8) are a feature of the light's spatial coherence. The latter is not retained in the standard RTE. This suggests that the usual RTE may be inaccurate in cases where the coherence length is significant with respect to the other spatial scales of interest. We have examined this point by applying the QRTE to the formation of spectral lines in one-dimensional geometry. A slab of size L containing a homogeneous hydrogen gas, between two perfectly absorbing walls at $z = \pm L/2$, is considered. We assume $p_x = 0 = p_y$, $p_z > 0$ (radiation propagating towards the $z > 0$ direction) and assume $W(-L/2) = 0$ (no incoming radiation). The Wigner function in stationary regime can be written for $z \in \,]-L/2, L/2[$ as a Fourier series

$$W(z) = \sum_{n=-\infty}^{+\infty} W_n e^{2i\pi nz/L}, \tag{10}$$

and the value at the boundary $z = L/2$ is given by

$$W(L/2) = 2\sum_{n=-\infty}^{+\infty} W_n (-1)^n, \tag{11}$$

where we have used that the periodization of W has a discontinuity at $z = L/2$ (Dirichlet's theorem). Each Fourier coefficient can be approximated as [4]

$$W_n = \frac{S_0 \delta_{n0} L - (-1)^n W(L/2)}{2i\pi n + L^2 K_{0n}}, \tag{12}$$

$$K_{0n} = \frac{1}{2L}\left[\chi_c\left(p_z - \frac{n\pi\hbar}{L}\right) + \chi_c^*\left(p_z + \frac{n\pi\hbar}{L}\right)\right]. \tag{13}$$

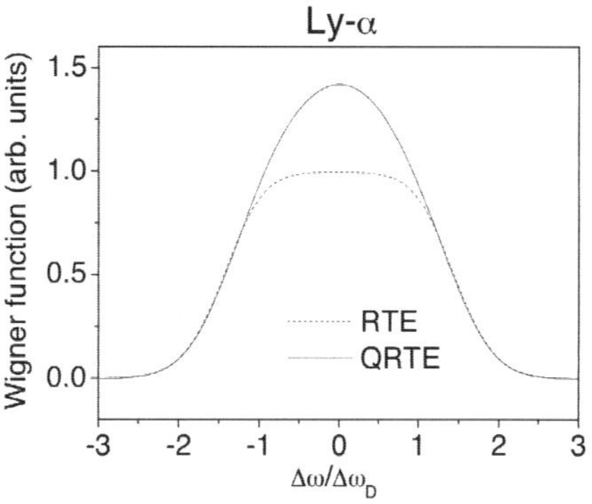

Figure 2. Plot of the Ly-α Wigner function at the outgoing radiation side, neglecting (dashed line) and retaining (solid line) coherence effects. The signal is larger than expected using the RTE.

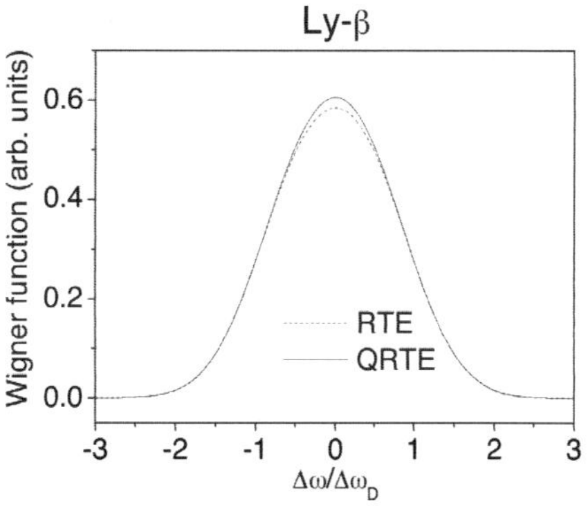

Figure 3. Plot of the Ly-β Wigner function at the same conditions as above. Here, the result of the RTE is close to that of the QRTE.

In Eq. (12), $S_0 = \eta/\hbar c p^3$ and δ_{n0} stands for the Kronecker symbol. Equations (11), (12), and (13) form a closed set, which provides the Fourier coefficients. The solution of the QRTE is obtained from solving this set for a sufficiently large number of coefficients and inserting them into Eq. (10). Figure 2 shows an application of the solution to the Lyman α line, assuming $L = 1$ mm, $N_1 = 10^{15}$ cm^{-3}, and $T = 1$ eV (with N_1, T being the density of atoms in the fundamental state and their temperature, respectively). At these conditions, the product $\lambda_c \times \chi_0$ between the coherence length [estimated as $c/\Delta\omega_D$ with $\Delta\omega_D = (\omega_0/c)\sqrt{(2T/m_p)}$ being the Doppler width] and the extinction coefficient at the line center ($\omega = \omega_0$) is of the order of unity, hence, we are in conditions such that the RTE may be inaccurate. As can be seen in the figure, a noticeable increase of the Wigner function is present at the line center. This stems from the delocalization of the photon emission and absorption processes, and it indicates that the gas seems less opaque when the delocalization is important. It should be noted that this curve does not directly represent an observable spectrum since, by construction, the Wigner function can take negative values. A proper characterization of an observable spectrum should involve a phase space integral denoting a suitable detector response. In general, the effect of spatial coherence depends of the line under consideration. This point is illustrated in Fig. 3, where a plot of the Lyman β Wigner function obtained in the same plasma conditions is shown. Here the results of the QRTE and those of the RTE are close to each other because the parameter $\lambda_c \times \chi_0$ is smaller than unity in this case. A consequence of the line dependence is that line ratios can be affected by the spatial coherence. In a diagnostic context, this means that spectra may be misinterpreted if coherence effects are not well accounted for.

5. Conclusion

The radiative transfer equation widely used in opacity problems can be derived from first principles, using the Wigner quantum phase space formalism adapted to second quantization. Such a derivation has the advantage of providing physical insight in the mechanisms involved in radiation-matter interactions. We have considered an extension of the radiative transfer equation suitable for problems involving the wave nature of light. This equation accounts for spatial coherence and reduces to the usual radiative transfer equation in the limit $\lambda_c \rightarrow 0$. An application of this equation to an ideal case has indicated significant deviations from the result of the usual radiative transfer theory. In a diagnostic context, this suggests the possibility of spectra misinterpretation if coherence effects are not well accounted for. This result is preliminary. A more rigorous treatment of the quantum radiative transfer equation, using realistic geometry, is required. Further work should also consist in comparison to experiments.

Acknowledgements

This work was carried out within the framework of the European Fusion Development Agreement and the French Research Federation for Fusion Studies. It is supported by the European Communities under the contract of Association between Euratom and CEA. The views and opinions expressed herein do not necessarily reflect those of the European Commission.

Appendix

A radiation transport equation accounting for spatial coherence can be obtained from classical electromagnetism. The classical counterpart of the Wigner function is given by

$$W(\vec{r},\vec{p},t) = \frac{1}{(2\pi\hbar)^3}\int d^3k\vec{\alpha}^*\left(\frac{\vec{p}}{\hbar}-\frac{\vec{k}}{2},t\right)\cdot\vec{\alpha}\left(\frac{\vec{p}}{\hbar}+\frac{\vec{k}}{2},t\right)e^{i\vec{k}\cdot\vec{r}}, \tag{A1}$$

where $\vec{\alpha}(\vec{k},t)$ is the normal variable of the electromagnetic field in the Fourier mode $\vec{k}$. This quantity is defined in terms of the Fourier transform of the electric and magnetic fields $\hat{\vec{E}}$, $\hat{\vec{B}}$ [12]

$$\vec{\alpha}(\vec{k},t) = \frac{-i}{2}\sqrt{\frac{2\varepsilon_0(2\pi)^3}{\hbar\omega}}\left[\hat{\vec{E}}(\vec{k},t) - c\frac{\vec{k}}{k}\times\hat{\vec{B}}(\vec{k},t)\right], \tag{A2}$$

and it obeys the following evolution equation, obtained from the Maxwell equations:

$$\frac{\partial\vec{\alpha}}{\partial t} = -i\omega\vec{\alpha} + \frac{i}{2\varepsilon_0}\sqrt{\frac{2\varepsilon_0(2\pi)^3}{\hbar\omega}}\hat{\vec{j}}. \tag{A3}$$

Here, $\omega = kc$ and $\hat{\vec{j}}$ is the transverse part (i.e., perpendicular to $\vec{k}$) of the Fourier transform of the current density associated with the atomic dipoles. A classical-harmonic-oscillator description of the dipoles (Thomson model) is suitable here. The response of a dipole $\vec{d}$ to the electric field is given by

$$\vec{d}(t) = \frac{e^2}{m_e\omega_0}\int_0^\infty d\tau e^{-\gamma\tau}\sin(\omega_0\tau)\vec{E}(\vec{R},t-\tau), \tag{A4}$$

where ω_0 is the atomic frequency, γ is proportional to the inverse lifetime of the oscillator, and $\vec{R}$ is the atom's position vector. Equation (A4) is obtained from the solution of the classical equation of motion assuming $\gamma \ll \omega_0$ (e.g. [15] for details on the classical atomic model). The exponential corresponds to the dipole autocorrelation function involved in the line shape function. If we assume non-interacting dipoles, the γ factor is identical to $A_{ud}/2$. This factor can also be written explicitly in terms of the parameters of the classical oscillator as $\gamma = r_e\omega_0^2/3c$ where r_e is the classical electron radius. The current density in real space $\vec{j}$ is obtained after deriving Eq. (A4) with respect to time and multiplying the result by the density of dipoles N:

$$\vec{j}(\vec{r},t) = \frac{e^2 N(\vec{r},t)}{m_e}\int_0^\infty d\tau e^{-\gamma\tau}\cos(\omega_0\tau)\vec{E}(\vec{r},t-\tau)\,. \tag{A5}$$

We now consider the following approximation for the electric field

$$\begin{aligned}\vec{E}(\vec{R},t-\tau) &\approx \frac{1}{(2\pi)^{3/2}}\int d^3k\hat{\vec{E}}(\vec{k},t)e^{i(\vec{k}\cdot\vec{r}+\omega\tau)} \\ &\approx \frac{1}{(2\pi)^{3/2}}\int d^3k\, i\sqrt{\frac{\hbar\omega}{2\varepsilon_0(2\pi)^3}}\vec{\alpha}(\vec{k},t)e^{i(\vec{k}\cdot\vec{r}+\omega\tau)}\end{aligned}, \tag{A6}$$

which stems from the fact that the radiation field in spectral line is nearly monochromatic. The time integral in Eq. (A5) combined with Eq. (A6) yields the complex line shape function involved in the QRTE and the current density can be written explicitly in terms of the complex extinction coefficient, using the identity $N_dB_{du} - N_uB_{ud} \equiv N \times g_u \times 2\gamma \times 4\pi^3c^2/\hbar\omega_0^3$ with $g_u = 3$:

$$\vec{j}(\vec{r},t) = \frac{\varepsilon_0 c}{(2\pi)^{3/2}}\int d^3k\, i\sqrt{\frac{\hbar\omega}{2\varepsilon_0(2\pi)^3}}\vec{\alpha}(\vec{k},t)e^{i\vec{k}\cdot\vec{r}}\chi_c^*(\vec{r},\hbar\vec{k},t)\,. \tag{A7}$$

A transport equation for the classical Wigner function is obtained by applying the Lagrangian derivative $\{\partial_t + c(\vec{p}/p)\cdot\vec{\nabla}\} \equiv D/Dt$ to each side of Eq. (A1), using Eq. (A3), the Fourier transform of Eq. (A7), and the approximation $\omega \approx \omega_0$ in the square root factors (short wavelength approximation). A calculation leads to the QRTE Eq. (6) with $S \equiv 0$, i.e. with no spontaneous emission, consistently with the classical assumption considered in this appendix.

References

[1] Rosato J 2010 *Am. J. Phys.* **78** 851–7
[2] Rosato J 2010 *AIP Conf. Proc.* **1290** 89–93
[3] Rosato J 2011 *Phys. Rev. Lett.* **107** 205001-1–4
[4] Rosato J 2012 *Transport Theor. Stat. Phys.* **41** 214–22
[5] Mihalas D 1978 *Stellar Atmospheres* (San Fransisco: Freeman)
[6] Newton T D and Wigner E P 1949 *Rev. Mod. Phys.* **21** 400–6
[7] Wigner E 1932 *Phys. Rev.* **40** 749–59
[8] Born M and Wolf E 1964 *Principles of Optics: Electromagnetic Theory of Propagation, Interference and Diffraction of Light* (Oxford: Pergamon Press)
[9] Klimontovich Iu L 1958 *Sov. Phys. JETP* **6** 753–60
[10] Osborn R K and Klevans E H 1961 *Ann. Phys. (NY)* **15** 105–40
[11] Brittin W E and Chappell W R 1962 *Rev. Mod. Phys.* **34** 620–7
[12] Cohen-Tannoudji C, Dupont-Roc J and Grynberg G 1992 *Atom-Photon Interactions: Basic Processes and Applications* (New York: Wiley-Interscience)
[13] Landi Degl'Innocenti E 1983 *Sol. Phys.* **85** 3–31
[14] Bommier V 1997 *Astron. Astrophys.* **328** 706–25
[15] Heitler W 1944 *The Quantum Theory of Radiation* (Oxford: University Press)

Virtual Laboratory Astrophysics: the STARK-B database for spectral line broadening by collisions with charged particles and its link to the European project VAMDC

Sylvie Sahal-Bréchot[1 4], Milan S Dimitrijević[2 1 3] and Nicolas Moreau[1]
[1] Laboratoire d'Étude du Rayonnement et de la Matière en Astrophysique (LERMA), Observatoire de Paris, UMR CNRS 8112, UPMC, Bâtiment Evry Schatzman, 5 Place Jules Janssen, F-92195 Meudon Cedex, France
[2] Astronomical Observatory, Volgina 7, 11060 Belgrade 38, Serbia
[3] Institute Isaac Newton of Chile, Yugoslavia Branch, Volgina 7, 11060 Belgrade 38, Serbia

E-mail: sylvie.sahal-brechot@obspm.fr

Abstract. Atomic physics in plasmas has been an essential tool for many years. Accurate spectroscopic diagnostics and modeling require the knowledge of numerous collisional line profiles. "Stark broadening" theories and calculations have been extensively developed for about 50 years. Nowadays, the access to such data via an on line database becomes essential. The aim of STARK-B is to reply to this need. It is a collaborative project between the Astronomical Observatory of Belgrade (AOB) and the "Laboratoire d'Étude du Rayonnement et de la matière en Astrophysique" (LERMA) of the Paris Observatory and CNRS. It is a database of widths and shifts of isolated lines of atoms and ions due to electron and ion impacts that we have calculated and published in international refereed journals (more than 150 papers by Dimitrijević & Sahal-Bréchot, and colleagues). It is devoted to modeling and spectroscopic diagnostics of stellar atmospheres and envelopes, laboratory plasmas, magnetic fusion plasmas, laser equipments and technological plasmas. Hence, the domain of temperatures and densities covered by the tables is wide and depends on the ionization degree of the considered ion. The STARK-B database is in free access and is is a part of VAMDC: the Virtual Atomic and Molecular Data Centre is an European Union funded collaboration between groups involved in the generation and use of atomic and molecular data. VAMDC aims to build a secure, documented, flexible and interoperable e-science environment-based interface to existing atomic and molecular data. STARK-B and its VAMDC context are presented in this paper.

1. Introduction

Pressure broadening of spectral lines arises when an atom or a molecule which emits or absorbs light in a gas or in a plasma, is perturbed by its interactions with the other particles of the medium. An atom or a molecule may be neutral as well as charged. The so-called Stark broadening is due to electron and ion colliders. The theory has been extensively developed for about 50 years and is currently used for many spectroscopic diagnostics and modeling. A number of its developments have been stimulated by the advances in Astrophysics, by the needs in laboratory and technological plasmas (tokamaks, laser produced plasmas, inertial fusion plasmas...), and also by the needs in industrial plasmas (discharge lighting). Due to the developments of the accuracy of observations in astrophysics and laboratory physics, many needs of atomic data appear at the end of the 20th century, and the needs are constantly

[4] To whom any correspondence should be addressed.

Published under licence by IOP Publishing Ltd

increasing. Hence, calculations based on a simple but enough accurate and fast methods are necessary for obtaining numerous results. Furthermore, the development of powerful computers also stimulates the development of atomic data on a large scale. Besides, the access to these atomic data via on line databases becomes essential.

In section 2, we will recall the physical conditions where Stark broadening plays a role in different plasmas. Section 3 will be devoted to a brief review of the standard impact Stark broadening theory that is at the basis of the theory, calculations and the database presented in the present paper. In fact, Dimitrijević and Sahal-Bréchot have updated and operated at a large scale the numerical code (SCP) created by Sahal-Bréchot since about thirty years. This code is based on the impact semiclassical-perturbation theory for isolated spectral lines of neutral and ionized atoms broadened and shifted by collisions with electrons and ions, [1] [2] [3] [4]. More than 150 papers are issued from the first update [5]. Then, the new need of creation of an on-line database appeared in the beginning of the 21th century, particularly in correlation with the birth and the growth of virtual Observatories and to the increasing need of exchange of interoperable data. Thus, the database STARK-B (formerly called BELDATA) was initiated in the Astronomical Observatory of Belgrade (AOB), and then a collaborative project between AOB and LERMA was born and led to the present database. STARK B, which is in free access, is currently developed at Paris Observatory and has opened on line since the end of 2008 (http://stark-b.obspm.fr [6]), and now contains the published data of all our papers obtained through the SCP theory and code. It is a part of the atomic and molecular databases of the Paris Observatory, and there is a link to the Serbian Virtual Observatory (SerVO, http://servo.aob.rs/~darko). A mirror site is planned at AOB. More details are given in Section 4, as well as the scheduled developments. STARK-B has been a database of VAMDC (Virtual Atomic and Molecular Data Centre) since the end of 2009. This FP7 European project "Research Infrastructures" was created in summer 2009 for 3.5 years. It is an interoperable e-Infrastructure for exchange of atomic and molecular data. It is an international consortium that has built an e-science interoperable platform [7] [8] (http://www.vamdc.eu, and http://portal.vamdc.eu) permitting an automated exchange of atomic and molecular data. The software infrastructure is based on an ensemble of standards and softwares ((http://www.vamdc.eu/software/). VAMDC addresses data producers and users, and gives a support for introducing new databases in the consortium, and for using standards and softwares. This will be presented in Section 5.

2. Importance of Stark broadening in plasmas

On the one hand, in astrophysics, thanks to the considerable developments of the spectral resolution and sensitivity (high *S/N*) of the recent past years, and of large ground-based telescopes and space-born missions, it becomes possible to observe very faint objects and spectra in all ranges of wavelengths (from XUV to radio) with an unequalled accuracy. For interpreting the spectra, the atomic parameters responsible for their intensities and their profiles must be known. The development of realistic models of interiors and of atmospheres of stars including stratification, and the interpretation of their evolution and the creation of elements through nuclear reactions, requires the knowledge of numerous profiles, especially for trace elements. Abundances of elements are crucial parameters to be determined. This needs an accurate interpretation of the detailed line spectra of stellar objects and thus extensive sets of atomic data, including collisional broadening and especially Stark broadening. Neglecting Stark broadening can lead to 40% error in abundances. Nowadays collisional broadening data are needed, not only for strong lines of abundant elements (H, He, C, N, O, Ne) as in the past, but also for weak lines of abundant elements, for elements of lower abundance (the Iron-peak), and then, that is more recent, for heavy elements of very low abundance which are always very weak. Even some radioactive elements, such as thorium and uranium have been recently discovered in low-metallicity stars (stars of the first or second generation), and are used for chronometric age determination. An extreme domain of temperatures (about 10^6 - 10^7 K) and electron densities (about 10^{24} cm^{-3}) concern interiors of stars. They cannot be observed but are mirrored via asteroseismology, and Stark broadening of abundant ions plays an important role for modeling and for deriving radiative opacities. In the very dense and hot atmospheres of neutron stars, the physical conditions are typical of those of stellar interiors, and highly-ionized atoms such as H and He-like Fe have been observed in X rays and the measured line strengths of lines indicate that the lines are significantly broadened by Stark effect [9], and this should provide an opportunity to determine both the mass and the radius of these exotic objects. On less extreme conditions of temperatures (10^4 to a few 10^4 K) and densities (10^{13} to 10^{15} cm^{-3}), Stark broadening is efficient for modeling and analyzing spectra of moderately hot (A), hot (B) and very hot (O) types of stars. It is dominant in comparison with thermal Doppler effect in deep layers of stellar atmospheres. In white dwarfs, collisional broadening and especially Stark broadening is dominant in all layers of the atmosphere (temperatures in the region of 10^4 K, electronic densities of the order of 10^{18}-10^{19} cm^{-3}) In stars like sun, and especially in the chromosphere which lays above the photosphere, Stark broadening can be operative for lines arising from high excited levels.

On the other hand, in laboratory and technological plasmas, considerable developments are also in progress and require new and numerous atomic data for spectroscopic diagnostics and modeling. Magnetic confinement fusion plasmas (tokamaks, such as ITER) are moderately hot (electron and ion temperatures about 0.5 to 40 keV in the core and 0.05 to 10 keV at the edge) and dense (10^{14} cm^{-3}), especially in the divertor and edge plasma regions. Expected impurities include a wide range of elements (such as tin), but some important elements are not the same as in astrophysics, with nuclear charge Z, ranging from above 70 (as tungsten) and down to 1. Light elements are also important, such as Helium-like krypton. Inertial confinement fusion plasmas (laser fusion, ion-beam fusion, such as JET, laser LMJ) are hot (several keV) and very dense (up to 10^{24} cm^{-3} for the density) and their thermodynamical conditions look like those of stellar interiors.

In addition, industrial plasmas also require atomic data for optimizing the performances of discharge lamps and lighting sources [10]. The temperatures are various: the discharge can attain 45000K, whereas the emitted white light corresponds to 3000-5000K. The electron density is rather high, so Stark broadening can play an important role in the modeling. The interesting elements are rare earth phosphors (Dy, Ho, Ce) because they are excellent radiation sources in fluorescent lamps. Temperatures of HID (High Intensity Discharge) lamps, such as metal halides, are about 1000K on the wall and 7000K on axis. The buffer gas is usually mercury vapor at high pressure. Ga and Al atomic data are needed.

3. Basis of STARK-B data: the standard theory of Stark impact broadening of isolated lines

The theory and calculation of collisional line broadening in the impact approximation follows the founding work by Baranger [11] [12] [13].

3.1. The impact approximation

First, the impact approximation is the fundamental one: the interactions are separated in time. Consequently the duration of an interaction $\tau = \rho_{typ} / v_{typ}$, where ρ_{typ} is a typical impact parameter and v_{typ} a typical relative velocity, must be much smaller than the mean time interval between two collisions ΔT, which is of the order of the inverse of the collisional line width $N v_{typ} \pi \rho_{typ}^2$, where N is the density of the perturbers. The condition of validity of the impact approximation can be written as $\rho_{typ} / v_{typ} \ll \Delta T$, or, which is equivalent, $N V_{typ} \ll 1$, V_{typ} being the collision volume [11] [12], i.e., $\rho_{typ} \ll N^{-1/3}$.

In other words, the studied radiating atom interacts with one perturber only at a time, and the perturbers are independent and their effects are additive.

3.2. The complete collision approximation

Second, the collision is assumed to be complete. This means that the atom has no time to emit or absorb a photon during the collision process. This is valid if the duration of an interaction is much smaller than the interval between two successive emissions (or absorptions) of photons, which is of the order of the inverse of the detuning, when larger than the line width. So, if the impact approximation is valid, the complete collision approximation is valid in the line center but can be invalid in the wings. In the very far wings the interactions can become "quasistatic", which means that a photon can be emitted or absorbed before the perturbers had time to move. The data of STARK-B do not apply in this case.

Consequently, within the complete collision approximation, radiative and collision processes are decoupled and the impact broadening theory becomes an application of the theory of collisions between an emitting (or absorbing) neutral (or ionized) atom and interacting particles. These two approximations are at the basis of the STARK-B data.

3.3. The case of "isolated lines"

Third, STARK-B data apply to "isolated lines" only and do not apply to "overlapping lines" [12]. This means that the levels next to the upper or lower level of the studied transition and likely to modify the broadening by introducing optical coherences do not overlap with them. So, hydrogen and hydrogenic ionic lines are excluded from the database, as well as some specific helium lines and some lines arising from Rydberg levels.

3.4. Lorentz profile

Therefore the profile of the *i-f* line emitted or absorbed between the *i* and *f* levels studied is Lorentzian, with a full width at half maximum *W* (in angular frequency units) and a shift *d* [13]. *W* can be expressed in terms of inelastic cross-sections and elastic processes as (cf. [13]):

$$W = N\int v\, f(v)\left(\sum_{i'\neq i}\sigma_{ii'}(v)+\sum_{f'\neq f}\sigma_{ff'}(v)+\sigma_{el}(v)\right),$$

where N is the density of the colliding perturbers, $f(v)$ the Maxwell distribution of the relative atom-perturber velocity v, $\sigma_{ii'}$ and $\sigma_{ff'}$ the inelastic cross-sections between the initial level i (resp. f final level) and the perturbing levels i' (resp. f') of the i-f transition. $\sigma_{el}(v)$ represents the contribution of elastic collisions and includes Feshbach resonances when ion-electron collisions are studied.

The shift is not given in the present paper, and we refer to [13] for its quantum expression.

We notice that the widths and shifts are proportional to the density N in the above expression, and depend on the temperature of the plasma through the distribution of velocities. In fact, since interactions with charged particles are concerned, at high densities, the Debye screening has to be taken into account and reduces the cross-sections. So, the widths and shifts become reduced.

Consequently, for a series of lines of a given neutral or ionized atom, STARK-B displays widths *W* and shifts *d* for a set of temperatures and densities and for various perturbers (electrons and ions). For low densities, which are not on the tables, the data, which are provided at medium densities, can be deduced through a linear extrapolation. At high densities some data are not given because the impact approximation is not valid; an asterisk, which replaces the data, indicates this. Data which are given but preceded by an asterisk mean that the impact approximation reaches its limit of validity, i.e. when $0.2 < N\, V_{typ} \leq 0.5$.

In addition, when the density increases, the lines, which are isolated at low densities, can overlap; the impact width becomes comparable to the separation $\Delta E(nl,\ nl\pm 1)$ between the perturbing energy levels and the initial or final level. The isolated line approximation becomes invalid. This is indicated in the database by a parameter *C* defined in [5] and in following papers. *C* is provided in the tables and can be used as follows: for a perturber density *N* lower than the limiting value N_{lim} (cm^{-3}) = C/W, the line can be treated as isolated even if a weak forbidden component due to the failure of this approximation remains in the wing.

3.5. Effect of fine structure and hyperfine structure on the widths and shifts

The following remark holds in LS coupling. The atomic electronic spin *S* and the orbital kinetic momentum *L* are decoupled. For electronic collisions and most often by collisions with ions, the collision time (of the order of ρ_{typ}/v_{typ}) is very much smaller than the fine structure splitting. So the electronic spin has no time to rotate during the collision and can be ignored. Thus all the fine structure components have the same widths and shifts, which are equal to those of the multiplet. This is a fortiori the same for the hyperfine components. So our calculations have most often been performed for multiplets only, and STARK-B data are most often given for multiplets. This remark does not apply for heavy atoms where departures from LS coupling can be important.

3.6. *Compared orders of magnitude of electron and ion Stark widths and shifts*

For isolated lines, the widths and shifts due to electron collisions are generally ten times higher than the widths and shifts due to ion collisions. This is due to the fact that inelastic collisions with positive ions are generally negligible, owing to the mass effect that decreases the relative velocity, and, in addition, if the radiator is an ion, the Coulomb repulsion acts. This can be not the case when the perturbing levels are very close to the levels of the studied transition. This can also be not the case at very high temperatures, and especially for ion radiators when the Coulomb repulsion becomes weak.

3.7. *Calculation of the cross-sections, and then widths and shifts*

3.7.1. The Semi-Classical Perturbation (SCP) approach. For the purposes of STARK-B, the semi-classical-perturbation treatment is adapted and gives results with a sufficient accuracy (about 20%). This is especially the case if the perturbing levels are not too far from the levels of the studied line. The basic formalism has been developed and discussed in detail in [1] [2]. Classical straight rectilinear paths for neutral radiators, and hyperbolic paths have been introduced for ion-electron and ion-ion collisions. Then the *S*-matrix has been obtained within the second order perturbation theory. The needed cross-sections are obtained through integration over the impact parameter of the transition probabilities. The needed cut-offs are determined in order to maintain the unitarity of the scattering *S*-matrix, and Debye screening is taken into account. A very fast computer code has been created. This formalism, and the computer code, have been updated and optimized several times: [3] for complex atoms, [4] for the inclusion of Feshbach resonances in the elastic ion-electron cross-sections, [5] and further papers, and [14] for transitions arising from very complex configurations.

3.7.2. The atomic structure. In the semiclassical picture, oscillator strengths and needed energy levels enter the expressions of the inelastic cross-sections. They are input data of the numerical code.

The STARK-B data issued from our oldest papers (eighties and nineties), the Coulomb approximation with quantum defect (Bates & Damgaard approximation, [15]) was used, together with measured or calculated energy levels.

In the more recent papers modern ab initio methods are used (cf. [14]). The different atomic structure computer codes or the data can be downloaded on line. Thus the calculations of widths and shifts can be made from the beginning to the end without any additional external input or experimental adjustment. The chosen atomic structure package enters our computer semi-classical code and that allows, when these methods are applicable, to obtain widths and shifts for one-two hundred of lines:

- TOPbase, the Opacity Project atomic database, contains accurate calculated oscillator strengths and energy levels for abundant neutral atoms and ions of astrophysics. They have been obtained within the close-coupling scattering theory by means of the R-matrix method with innovative asymptotic techniques [17] [18]. L-S coupling is assumed. So, TOPbase data have been especially used to light and low and moderately ionized atoms and ions.
- The Cowan code [19] is an online atomic structure package consisting of a set of computer programs for calculation of energy levels, radiative transition wavelengths and probabilities, electron impact excitation and photoionization cross sections, etc. The Hartree-Fock-Slater multi-configuration expansion method with statistical exchange is the normal option since it is most computationally efficient. The relativistic corrections are treated by perturbations. So this method is especially suited to moderately heavy atoms which are little and moderately ionized.
- SUPERSTRUCTURE (SST) [20] is well suited for computation of large quantities of atomic data for highly charged ions. The wave functions are determined by diagonalization of the nonrelativistic

Hamiltonian using orbitals calculated in a scaled Thomas-Fermi-Dirac-Amaldi potential. Relativistic corrections are introduced according to the Breit-Pauli approach. Atomic data are obtained in intermediate coupling.

Si V and Ne V line widths and shifts data have been calculated with both Bates & Damgaard and SST atomic data [21] [22]. The difference does not exceed 30%. C II widths and shifts data have been calculated with both TOPBase and Bates & Damgaard atomic data [23], and the difference does not exceed a few percent, except when configuration interaction plays an important role by making permitted a forbidden transition. This gives an idea and an order of magnitude of the importance of the chosen atomic structure on the Stark broadening data.

3.7.3. The Modified Semi-Empirical Method (MSE). A number of applications were also achieved with the Modified Semi-Empirical (MSE) method [24] [25] [26] and other papers cited in [27]. It is less accurate, though simpler to use due to the considerably smaller set of atomic data needed in comparison with the SCP theory. It can effectively replace the SCP method when this one cannot be used due to a lack of atomic data. The MSE data have not still been introduced in STARK-B.

4. Organization of the database

Since STARK-B is devoted to modeling and spectroscopic diagnostics in various plasmas, the domain of temperatures and densities covered by the tables is wide and depends on the ionization degree of the considered ion. The temperatures can vary from several thousands for neutral atoms to several millions of Kelvin for highly charged ions. The electron or ion densities can vary from 10^{12} (case of stellar atmospheres) to several 10^{22} cm^{-3} (some white dwarfs, subphotospheric layers and some laboratory and fusion plasmas). The datamodel, especially the definition of configurations, terms and levels follow the VAMDC standards, in order to allow interoperability with other atomic databases.

It is important to notice that the provided wavelengths (Å units) are most often calculated from the energy levels that are used in the calculations. Consequently, these wavelengths can be different from the measured ones, especially if energy levels originate from TOPBase, SST or Cowan code. The lines can be identified through the configurations, terms and levels. In addition, the widths and shifts data are provided in units of wavelengths (Å) and not in angular frequency units. So, if widths and shifts data are needed for measured wavelengths ($\lambda_{measured}$), or for fine structure data whereas the data are only provided for multiplets (cf. section 3.5), one has to multiply the STARK-B data by $(\lambda_{measured}/\lambda_{STARK\text{-}B})^2$.

We refer to section 3.4 for the cases of departures from the impact approximation and from the isolated line approximation at high densities.

Currently, STARK-B contains Stark widths and shifts for spectral lines of the following elements and ionization degrees:

Ag I, Al I, Al III, Al XI, Ar I, Ar II, Ar VIII, Au I, B II, B III, Ba I, Ba II, Be I, Be II, Be III, Br I, C II, C III, C IV, C V, Ca I, Ca II, Ca V, Ca IX, Ca X, Cd I, Cd II, Cl I, Cl VII, Cr I, Cr II, Cu I, F I, F II, F III, F IV, F V, F VI, F VII, Fe II, Ga I, Ge I, He I, Hg II, I I, In II, In III, K I, K VIII, K IX, Kr I, Kr II, Kr VIII, Li I, Li II, Mg I, Mg II, Mg XI, Mn II, N I, N II, N III, N IV, N V, Na I, Na X, Ne I, Ne II, Ne III, Ne IV, Ne V, Ne VIII, Ni II, O I, O II, O III (in progress), O IV, O V, O VI, O VII, P IV, P V, Pb IV, Pd I, Rb I, S III, S IV, S V, S VI, Sc III, Sc X, Sc XI, Se I, Si I, Si II, Si IV, Si V, Si VI, Si XI, Si XII, Si XIII, Sr I, Te I, Ti IV, Ti XII, Ti XIII, Tl III, V V, V XIII, Y III, Zn I.

Apart O III data, all the data corresponding to our already published papers have now been implemented.

The homepage (Fig.1) proposes several menus, among which "Introduction", "Data Description" and "Access to the Data". "Introduction" briefly recalls the approximations and methods of calculation. "Data Description" describes the data that are in the files. A graphical interface is provided in "Access to the Data" (Fig.2): first, the consumer clicks on the wished element in the Mendeleev periodic table and then on the ionization degree of interest. Yellow cells contain data, while white cells are empty. Next, with a few clicks, the user chooses the colliding perturber(s), the perturber density, the transition(s) by quantum numbers and the plasma temperature(s). The consumer can also make a query by domain of wavelengths instead by transitions. Then a table displaying the widths and shifts is generated. Bibliographic references are given and linked to the publications via the SAO/NASA ADS Physics Abstract Service [55] and/or within DOI. They can be freely downloaded if the access is not restricted. The widths and shifts data can be downloaded in ASCII and in VOTable format (XML format), adapted to Virtual Observatories.

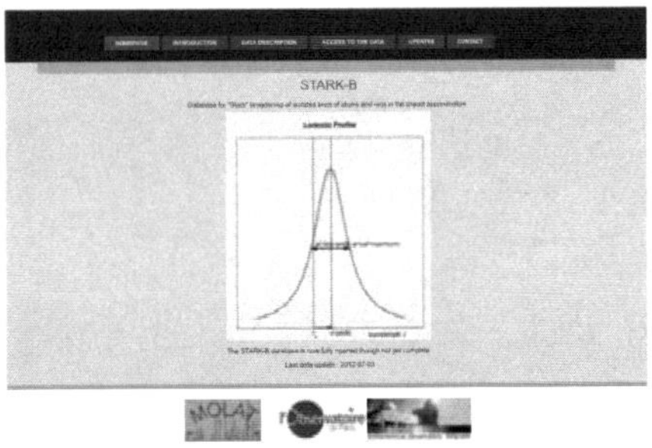

Figure1. The STARK-B home page: http://stark-b.obspm.fr

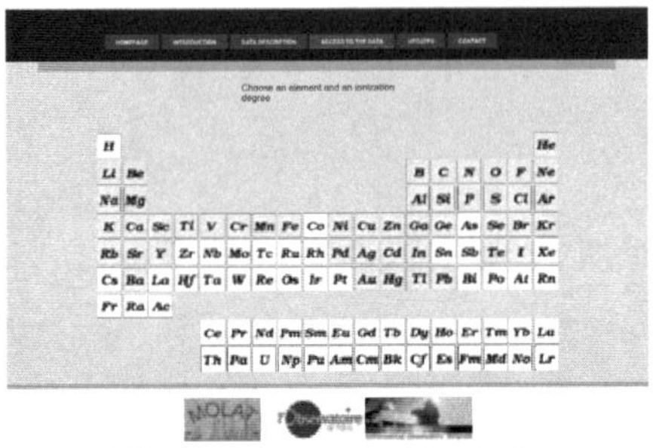

Figure 2. "Access to the data": the graphical interface

The database will be completed when new data will be calculated and published (e.g. C II results of [23]). The implementation of MSE data is planned for a next future. The implementation of our quantum data will be another future step. The further developments also concern insertion of little apps (fitting along temperatures, along principal quantum number for a given multiplet, charge of the ion collider along isoelectronic sequences, of the radiating ion, homologous ions...) by using fittings or systematic trends in order to obtain data that are missing on the database.

5. The STARK-B database and VAMDC (Virtual Atomic and Molecular Data Centre)

Many fields in astronomy, physics, energy production and industry depend on databases for atomic, molecular and particle-surface interaction processes. A reliable exchange of such data has become crucial for decades. Nowadays, many databases do exist, but they are organized and presented in different manners, with heterogeneous standards, rules and selection criteria. This is a barrier for an efficient search for data and their use. The free exchange of atomic and molecular data requires the definition both of standards which model the data structure and of tools that implement these standards and that help to carry out science using these data. Developments in computer technologies offer opportunities for new distributed tools and applications in various fields of physics. The convenient and reliable exchange of data is clearly an important component of such applications. So VAMDC [7] [8] was created in summer 2009 for 3.5 years in the framework of the FP7 "Research Infrastructures - INFRA-2008-1.2.2 - Scientific Data Infrastructures" initiative. It is an European collaboration between groups involved in the generation and use of atomic and molecular data. VAMDC involves 15 administrative partners representing 24 teams from 6 European Union member states, partners in non-EU countries (Serbia, the Russian Federation and Venezuela), and external partners in the US. It includes scientists in atomic and molecular physics with a strong coupling to the users of their data (astrochemistry, atmospheric physics, plasmas) and scientists and engineers from the Information and Communication Technologies community used to deal with interoperable e-infrastructure. The core of the VAMDC-infrastructure is founded on the databases detailed in [7]. VAMDC aims to build a secure, documented, flexible and interoperable e-science environment-based interface to existing atomic and molecular data. The VAMDC portal (http://portal.vamdc.eu/) has been released in the mid-April 2012 for the external users. It can be tested and is frequently updated from the development version that is still in progress. The results of the queries can be downloaded in XSAMS format (XML Schema for Atoms, Molecules and Solids). The XSAMS schema provides a framework for a structured presentation of atomic, molecular, and particle-solid-interaction data in an XML file. XML

(Extensible Markup Language) is the current standard for exchange of such data. It enables an automated exchange of complex contents between heterogeneous information systems through interoperability. STARK-B is one of the databases of VAMDC and participates to this undertaking.

6. Conclusion

Nowadays, numerical databases in Atomic and Molecular Physics are essential for both the modeling and the interpretation of spectra provided by observations and laboratory measurements of various plasmas. The free exchange of atomic and molecular data requires the definition both of standards which model the data structure and of tools that implement these standards and that help to carry out science using these data. The continuation of such developments and services of powerful and constantly updated online databases, like STARK-B, is crucial.

Acknowledgments

A part of this work has been supported by VAMDC. VAMDC is funded under the "Combination of Collaborative Projects and Coordination and Support Actions" Funding Scheme of The Seventh Framework Program. Call topic: INFRA-2008-1.2.2 Scientific Data Infrastructure. Grant Agreement number: 239108

References

[1] Sahal-Bréchot S 1969a *A&A* **1** 91
[2] Sahal-Bréchot S 1969b *A&A* **2** 322
[3] Sahal-Bréchot S 1974 *A&A* **35** 321
[4] Fleurier C, Sahal-Bréchot S and Chapelle J 1977 *JQSRT* **17** 595
[5] Dimitrijević M S and Sahal-Bréchot S 1984 *JQSRT* **31** 301
[6] Sahal-Bréchot S, Dimitrijević M S and Moreau N 2012 *STARK-B database, [online]. Available: http://stark-b.obspm.fr [Jul 8, 2012]. Observatory of Paris, LERMA and Astronomical Observatory of Belgrade*
[7] Dubernet ML, Boudon V et al. 2010 *JQSRT* **111** 2151, http://www.vamdc.eu/
[8] Rixon G, Dubernet M L, et al. 2011 *7th International Conference on Atomic and Molecular Data and their Applications -ICAMDATA-2010, AIP Conf. Proc.* **1344** 107
[9] Rovenskaya N I 2004 *Astrophys. Space Sci.* **291** 113
[10] Lister G and Curry J J 2011 *7th International Conference on Atomic and Molecular Data and their Applications -ICAMDATA-2010, AIP Conf. Proc.* **1344** 219
[11] Baranger M 1958a *Phys. Rev.* **111** 481
[12] Baranger M 1958b *Phys. Rev.* **111** 494
[13] Baranger M 1958c *Phys. Rev.* **112** 855
[14] Mahmoudi W F, Ben Nessib N and Sahal-Bréchot S 2008 *EPJD* **47** 7
[15] Bates D R Damgaard A 1949 *Trans. Roy. Soc. Lond. Ser. A* **242** 101
[16] Ben Nessib N 2009 *New Astron. Rev.* **53**, 255
[17] Cunto W, Mendoza C, Ochsenbein F and Zeippen CJ 1993 *A&A* **275** L5
[18] http://cdsweb.u-strasbg.fr/topbase/topbase.html
[19] Cowan RD 1981 *The Theory of Atomic Structure and Spectra* (Berkeley, CA: University of California Press)
[20] Eissner W Jones M and Nussbaumer H 1974 *Comput. Phys. Commun.* **8** 270
[21] Ben Nessib N, Dimitrijević M S and Sahal-Bréchot S 2004 *A&A* **423** 397
[22] Hamdi R, Ben Nessib N, Dimitrijević M S and Sahal-Bréchot S 2007 *ApJS* **170** 243
[23] Larbi-Terzi N, Sahal-Bréchot S, Ben Nessib N and Dimitrijević M S 2012 *MNRAS* **423** 766
[24] Dimitrijević M S and Konjević J 1980 *JQSRT* **24** 451
[25] Dimitrijević M.S and Popović L Č 2001 *J. Appl. Spectr.* **68** 893
[26] Dimitrijević M S and Kršljanin V 1986 *A&A* **165** 269
[27] Dimitrijević M S 2003 *A&A Trans.* **22** 389

Evidence of plasma polarization shift of Ti He-α resonance line in high density laser produced plasmas

F Y Khattak[*1], **O A M B Percie du Sert**[2], **F B Rosmej**[3], **and D Riley**[2]
[1]Department of Physics, Kohat University of Science and Technology, Kohat, Khyber Pakhtunkhwa, Pakistan
[2]School of Mathematics and Physics, Queen's University of Belfast, BT7 1NN, UK
[3]Sorbonne Universités, Pierre et Marie Curie, UMR 7605, LULI, case 128, 4 Place Jussieu, 75252 Paris Cedex 05, France
[4]LULI, Ecole Polytechnique, CNRS, CEA, Route de Saclay, Physique Atomique dans les Plasmas Denses - PAPD, 91128 Palaiseau Cedex, France

E-mail: fida_k@hotmail.com

Abstract. A spectroscopic study of the He-α ($1s^2\ ^1S_0$ - $1s2p\ ^1P_1$) line emission (4749.73 eV) from high density plasma was conducted. The plasma was produced by irradiating Ti targets with intense ($I \approx 1\times10^{19}$ W/cm^2), 400nm wavelength high contrast, short (45fs) p-polarized laser pulses at an angle of 45°. A line shift up to 3.4±1.0 eV (1.9±0.55 mÅ) was observed in the He-α line. The line width of the resonance line at FWHM was measured to be 12.1±0.6 eV (6.7±0.35 mÅ). For comparison, we looked into the emission of the same spectral line from plasma produced by irradiating the same target with laser pulses of reduced intensities ($\approx 10^{17}$ W/cm^2): we observed a spectral shift of only 1.8±1.0 eV (0.9±0.55mÅ) and the line-width measures up to 5.8±0.25 eV (2.7±0.35 mÅ). These data provide evidence of plasma polarization shift of the Ti He-α line.

1. Introduction

The subject of spectral line broadening and shift has been studied extensively, theoretically and experimentally both (see for example [1-22]). Its diagnostic characteristic has its relevance to plasma under extreme conditions and hence of interest to researchers actively pursuing the Inertial Confinement Fusion (ICF) scheme of harnessing the thermonuclear energy [23] and also to astrophysicists [24]. Spectral lines emitted from a system may be broadened and shifted due to interaction of the emitter with the free electrons in the system. When averaged over time, this interaction produces an excess of negative charge around the nucleus of the emitting ion. This can lead to overlapping orbits of the bound electrons of the emitter and, hence, partially shields the nuclear charge. Consequently, the energy level structures of the emitting ion in a plasma environment are shifted depending on the electron density of plasma. Such a shift is usually termed as plasma polarization shift (PPS). Recently, a generalized analytical approach (valid for any ion and quantum numbers) has been developed to calculate energy level due to plasma polarisation shifts [25].

[*] To whom any correspondence should be addressed.

Since the line shift depends strongly on the initial state of the emitter, particularly its principal quantum number, we do not find many studies about the low lying lines, for example the α and β resonance lines. Majority of the studies concentrated on low Z plasmas. With the availability of high contrast ultra-short laser pulses and the highly efficient curved crystals, however, such studies have been recently reported. Saemann *et al* [16] have reported a red polarization shift (~3.7±0.7 eV) and 20 eV for the Al Ly-α and He-β line respectively. Renner *et al* [18, 22] have presented a comprehensive study of the Al Lyman series (α-η) and He-α indicating a strong dependence of the shift on the density. The observed shifts for the Ly-α and Ly-β lines were too small to draw a definite conclusion about their agreement with the quantum mechanical impact theory [9]. Woolsey *et al* [19] have reported a wavelength shift of about 4mÅ of He-β line in Ar plasma. Their results indicate a systematic red shift with increasing density. Andiel *et al* [21] have very recently reported line shifts for Ly-α Ly-β, He-β and He-γ line in aluminium plasma generated by irradiating buried layer of Al in carbon of varying thickness using 395 nm, 150 fs pulses with plasma densities approaching $5x10^{23}$ cm^{-3}.

Despite of numerous theoretical and experimental progresses in understanding the spectral line shifts in dense plasmas, the subject remains controversy. Although the existence of x-ray line shifts in dense laser produced plasmas is now widely accepted [17, 19, 26], large red shifts at near solid density [16, 27] are sometimes questioned due to overlapping satellite transitions and uncertainties in wavelength markers [22, 28].

We report here on an experiment where we have observed a line shift of ~3 eV in He-α resonance spectral line from Ti, to our knowledge shift of the He-α line from the highest Z element reported so far, by focusing high contrast frequency doubled (400 nm), 45 fs p-polarized pulses on 12.5 micron thick high purity titanium (Ti) foils.

2. Experimental

We carried out the experiment at the Rutherford Appleton Laboratory using the terawatt short pulse laser facility ASTRA. The system is capable of delivering half a joule of infra-red (800nm) pulses with duration of 45(±5) fs. The output pulse is accompanied by a pre-pulse with a contrast of 10^{-7} reaching the target 13ns before the arrival of the main pulse. In addition, the main pulse is superimposed on a residual uncompressed CPA pedestal and ASE. We used a third order auto-correlator in 800nm wavelength to measure the pulse contrast of the beam and it was found to be 10^{-4} at 1.5 ps before the arrival of the main pulse.

We employed a 0.6 mm thick type-I KDP crystal to frequency double the 800 nm (IR) beam to a 400 nm (blue), s-polarized beam with a conversion efficiency of 20%. A beam splitter, reflecting the blue and transmitting the IR, was used at 45° to relay the blue beam onto the target. The IR part of the beam was dumped at the back of the splitter. Any IR component reflected off the beam-splitter was further reduced by two dielectric mirrors in the beam-path. With this arrangement, interaction of high contrast clean pulses with the target surface was made possible [29]. The motorized controlled target orientation mechanism ensured the interaction of p-polarized pulses with the target at different angles.

The blue beam (400nm) was focused onto the surface of the target surface at different angles from the normal to the target with the help of an f/2.5 off-axis silver-coated parabola plane. An 8-bit CCD coupled with a x40 microscope objective recorded the focal spot at different offsets using the low energy mode of the laser. The full width at half maximum (FWHM) at the best focus was measured to be ~2 μm with energy content of about 35%. A retro-alignment system was in place by collinearly injecting a diode laser via one of the dielectric mirrors that allowed us to find the best focal spot whenever the target was moved to a fresh position. A 10 μm wire was then inserted into the focal plane and the reflection of focal spot from this was transmitted back to the alignment system and an image of the focal spot formed on a CCD camera equipped with an objective. In this way, using the very small amount of backscattered light even from smooth target surfaces, the position of best focus onto the foil could be identified when the target was moved 1mm between shots.

The focal spot on target was varied by moving the parabola off the best focus position along the line-of-focus by a known amount with the help of a micro-controller, towards the target and away from the target (referred to as positive off-set and negative off-set respectively). With the negative off-set the incident beam converges inside the target while in case of the positive off-set the focus lies before the target and a divergent beam interacts with the target. A small leakage of the 400 nm light through the second dielectric mirror in the incident beam was used to monitor the energy of the blue beam in each shot. This was achieved through a photodiode coupled to an integrating sphere. An IR filter was employed to cut out 800 nm light. At the best focus position of the parabola, the intensity of the blue beam on the target reached a maximum of $\sim 10^{19}$ W/cm^2.

To ensure interaction of the beam with a fresh surface of the target, after every shot, the target was moved by 1 mm with an external computer controlled micro-drive. For every shot, before moving the parabola to the desired offset position, the beam was first imaged through the retro system at the best focus position of the parabola. The parabola was protected from the plasma debris by placing a thin glass pellicle in front of the parabola.

The time integrated single shot line emission spectra from Ti plasma, in the range of 4400 eV to 5000 eV, were recorded on individual shots with a Von- Hamos (VH) spectrometer coupled with an x-ray CCD system [30,31]. The spectrometer comprises a cylindrically curved (5cm radius of curvature), 25 mm wide and 15 mm arc-length, LiF (**200**) crystal having a 2d lattice spacing of 4.027 angstrom, on a micro-controlled base stage in a lightproof metal box. The line of sight of the spectrometer (crystal centre and the source position) made an angle of $\sim 68^\circ$ with the horizontal plane.

We used $\lambda(\text{K-}\alpha_2)_{ref}$ = 2752.195 mÅ (4504.92 eV) and $\lambda(\text{K}_\beta)_{ref}$ = 2513.96 mÅ (4931.83 eV) from the NIST site [32] as our wavelength marker. With profile fitting (see the following section), total spectral resolution of the spectrometer is estimated to be $\lambda/\delta\lambda \approx 1350$. With the given set-up, one pixel on the CCD corresponds to an interval of $\approx$ 0.65 eV ($\approx$ 0.35 m Å).

3. Results and Discussions

Figure 1a shows an integrated lineout of the spectrum obtained by an electronic sum of spectra from 13 data shots individually recorded when Ti foil is irradiated with high contrast 400 nm, 45 fs, p-polarized pulses at 45° angle of incidence. In all these shots, the parabola position is set at the best focus (with an accuracy of $\pm 15\mu$m). These shots are recorded during the same laser run without letting up the chamber. The spectral interval from the cold K-α up to K-β is shown in logarithmic intensity scale. Apart from the K-α and K-β line, the resonance line of He-α, inter-combination line Y and the K-α satellites (Li, Be, B, C, N, O, F) are also visible. With the presence of K-α and K-β line on the spectrum we can use two reference lines almost at the two extremes of the detector and check the accuracy of our wavelength calibration for the intermediate values. K-β line can be seen in single shot data taken at larger offset ($\approx$ 400micron) from the best focus position due to lower background hard x-rays. Although the level of He-α line emission in these cases was too low to be recorded, it was crucial to monitor possible changes in dispersion during the same laser session or from one session to another.

We used Voigt profile fitting to find the central pixel position for the emission lines. In figure 1b-c, K-α_1, K-α_2 and K-β are fitted by means of Voigt profiles taking the line centre position of K-α_2 and K-β as a reference line on the spectrum. The best fit for K-α_1 is obtained with a line centre position of $\lambda(\text{K-}\alpha_1)$ = 2748.7 mÅ (4510.7 eV). This value is in very good agreement (within 0.15 mÅ or 0.2 eV) with the reference value of $\lambda(\text{K-}\alpha_1)_{ref}$ = 2748.5471 mÅ (4510.899 eV) [32], indicating an appropriate use of K-α_2 and K-β as our wavelength marker to find out the intermediate dispersion with a precision well above the values of shifts and widths to be discussed below. Figure 1d shows Voigt profile fits to the experimental profile of He-α and the inter-combination line. The best fit is obtained for the He-α line centre position of at 2612.2 mÅ (4746.3 eV) and FWHM of 7.1 mÅ (12.8 eV). This indicates a red shift of 1.85 mÅ (3.4 eV) for the He-α resonance line 4749.7 eV [32]. We note that the FWHM as obtained from the fits include all sources of broadening (e.g., Stark, Doppler, opacity, line overlapping, instrumental).

Because we see in figure 1a, evidence of inner-shell K-α emission from different ion stages and because we expect fast electron resistive heating, we would like to exclude its effect on the K-α_2 spectral position. We compare single-shot spectral data from Ti foil irradiated at the optimum focus of the parabola and at an offset of 100 μm from the best focus position.

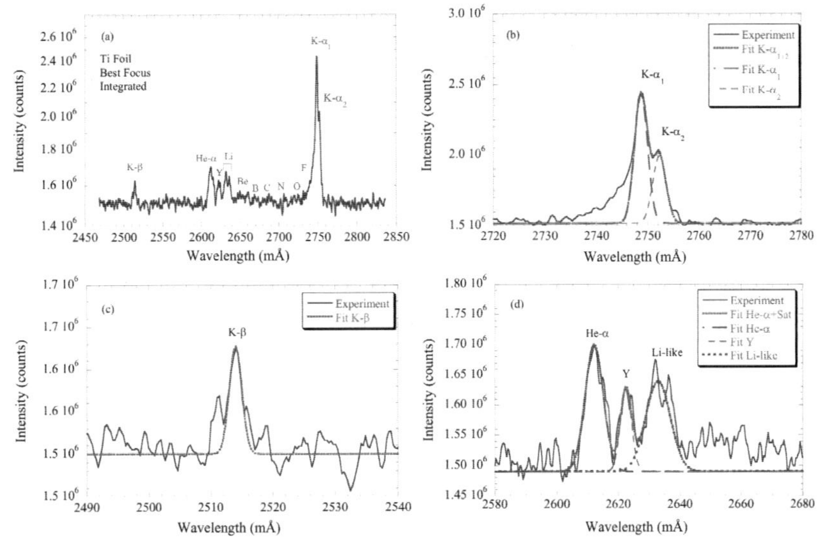

Figure 1: ***(a) An integrated line-out of the spectral data electronically summed over ~10 shots during the same session. These data are electronically corrected for the background and presentation on a log scale. The red curves show Voigt profile fits to the experimental profile of (b) K-α, (c) K-β, and (d) He-α, inter-combination line Y and Li-like satellites.***

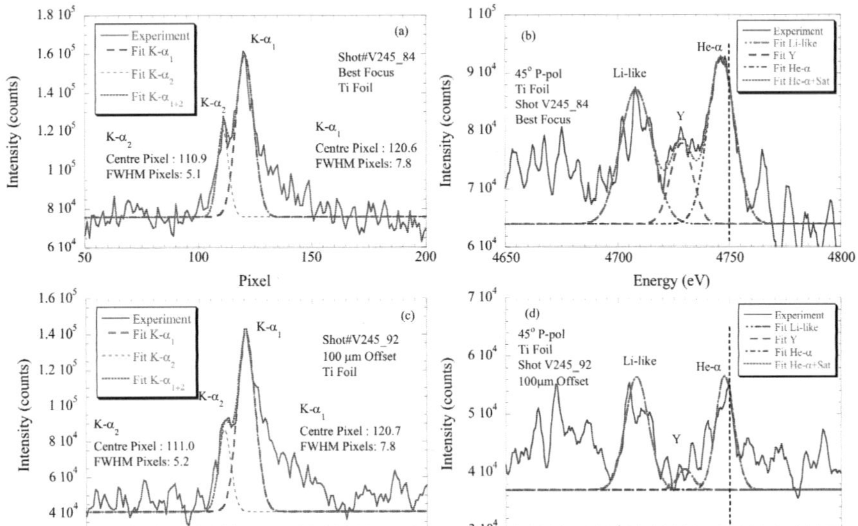

Figure 2: Spectral line out of the data recorded from single shot when Ti foil is irradiated at the optimum focus (a,b) and at an offset of 100 μm (c,d) along with the Voigt profile fits. The spectral band in the vicinity of K-α is shown in (a, c) He-α is shown in (b,d).

We compared the spectral positions and widths of the K-α and He-α resonance line emitted from Ti foil irradiated at an offset of ~100 μm (Fig 2c, d) to a shot at the best focus position (Fig 2a, b). These two shots were recorded during the same experimental session without letting-up the chamber and without disturbing the position of the spectrometer. Once again Voigt profile fitting procedure was used, as outlined above, to work out the line centre position. The K-α_2 line centre position in the best focus case is fitted at 110.9 pixel (figure 2a) whereas the 100 μm offset case gives 111.0 pixel. The difference of these values (≈ 0.1 pixel) translates into a wavelength difference of 0.04 mÅ or energy difference of 0.06 eV which is about an order of magnitude better than the desired wavelength precision. Furthermore, as can be seen from the insert in figure 2, the difference in the FWHM of the K-α emission in these two cases is also only 0.1 pixel. With two orders of magnitude difference in intensity (~10^{19} and ~10^{17} W/cm^2), our simulation (see below) show a considerable difference in temperature and density. Negligible change in the FWHM and the central pixel position of K-α emission in these two cases demonstrates that the ion stages and resistive heating have negligible effect on the observed K-α_2 wavelength position. Hence, justifying the use of $\lambda(\text{K-}\alpha_2)_{ref}$ = 2752.195 mÅ as a reference line for our wavelength calibration.

Similarly, Voigt profile fit was carried out for the He-α resonance line and inter-combination line as was done in figure 1. Figure 2b and 2d show lineouts of the spectra with Voigt fits, in the vicinity of He-α resonance line, from Ti foil irradiated at 45° with p-polarized pulses at the best focus and at an offset of 100 μm respectively. These spectra are from the same shots as above figure 2a and 2c. The vertical dotted lines indicate the un-shifted position of He-α resonance line at 4749.73 eV [33] in both cases. For the optimum focus case, we measure a spectral red shift of 3.4 eV (≈ 2 mÅ) for the He-α resonance line and FWHM of 12.1 eV (6.7 mÅ). These values are similar to the values obtained for the integrated case discussed in figure 1. For the second case of irradiation at an offset of 100 μm from best focus, we measure a spectral red shift of 1.8 eV (0.95 mÅ) and a FWHM of 5.8 eV (2.7 mÅ). These values are almost half of that measured for the optimum focus case. Our data from electronically integrated shots at best focus from Ti foil irradiated at 30° with p-polarized pulses also indicate a shift of about 3.6 eV (2.1 mÅ). Again, very similar to the data of 45° irradiation presented above. Because of the noise in the spectral data, any shift in the Li-like satellite lines could not be ascertained.

We carried out hydrodynamic simulation using HYADES [34] (the code has the facility of electromagnetic wave solver) for the Ti foil irradiated with 400 nm, 45 fs p-polarized pulses at an angle of 45° to the target normal for these two cases in order to estimate the difference in average plasma density as observed with x-ray emission spectroscopy. We observed that the density exceeded 10^{24} cm^{-3} with a peak temperature above ~3keV for the optimum focus whereas for the defocus case we noticed lower density and temperature well below 1 keV. The hydrodynamic output data from HYADES were post-processed with the collisional radiative model SOBOLEV [35] to simulate the temporal profile of the He-α emission. Our results show that for the optimum focus, the density exceeds 10^{24} cm^{-3} with a peak temperature above ~3keV whereas for the defocus case a temperature well below 1 keV is observed.

We have independently cross checked the hydro-simulations of emission averaged data with the spectral simulation code MARIA [36-38]. The fits (collisional radiative model including opacity and hot electrons, for further details see [36-38]) of the time integrated data of He-α, the intercombination line Y and the associated dielectronic satellite structure indicate a bulk electron temperature of 0.8-1.0 keV and a density of about (3-10)x10^{23} cm^{-3}. This is in reasonable agreement with the simulated =electron density of 0.5-1x10^{24} cm^{-3} and an electron temperature of 0.5-1 keV from about t=1–10 ps. Moreover, the particular spectral distribution of the Li-like 1s2l2l'-satellites indicates [36] that the electron density in the case of tight focus is near 10^{24} cm^{-3}, whereas for the case of 100 μm defocusing, the electron density is more close to 10^{23} cm^{-3}. This is also in rather good qualitative agreement with the hydrodynamic simulations.

Another possible source of line shift is the directed Doppler shift as well as differential plasma motion in optically thick plasmas. The maximum red shift due to Doppler effect in our case is ~0.3 eV (0.17 mÅ), This is far less than the observed shift. Similarly, a differential plasma motion of the order of $\delta E/E = \delta V/c \approx 0.3$ eV/4.7 keV = 6x10^{-5} will not lead to a significant shift of an optically thick line profile.

The hydrodynamic simulation shows that the density in case of the optimum focus remains about 3 times to that of the 100μm offset case for the relevant duration of the emission. As lower shifts correspond to lower plasma densities and all possible sources of line shifts are significantly below the observed ones we have a clear evidence of having observed the plasma polarisation shift of He-α of titanium, the highest Z reported so far.

4. Conclusions

To conclude, we have observed a spectral red shift of ~3.4eV for the He-α line emission from plasma generated by focussing intense high contrast 45fs, 400nm p-polarized pulses at 45° degree on Ti foil at $\sim 10^{19}$ W/cm^2. At lower irradiance of 10^{17} W/^{cm2}, this shift reduces to ~1.8eV. The difference in the peak-shift may be considered as a signature of contribution from the plasma polarization effect and our simulations of foil targets and spectral distributions support this thesis.

Acknowledgements

We would like to thank the ASTRA laser staff, target area staff and the target fabrication staff at the Rutherford Appleton Laboratory for their support. This work was supported by the UK Engineering and Physical Sciences Research Council through grant number EP/C001869/01. Support from ARCUS (grant number P003, 5.1) is also greatly acknowledged. One of the authors would like to acknowledge the support of the Higher Education Commission of Pakistan.

References

1. H. Griem *et al.*, Phys. Rev. **116**, 4 (1959)
2. H. F. Berg *et al.*, Phys. Rev. **125**, 199 (1962)
3. S. Volonté, J. Phys. B, **8**, 1170 (1975)
4. V. Demura and G. V. Sholin, J. Quant. Spectrosc. Radiat. Transfer **15**, 881 (1975)
5. M. Baker and D. D. Burgess, J. Phys. B **12**, 2097 (1979)
6. R. W. Lee, G. E. Bromage, and A. E. Richards, J. Phys. B **12**, 3445 (1979)
7. T. L. Pittman, P. Voigt, and D. E. Kelleher, Phys. Rev. Lett. **45**, 723 (1980)
8. K. G. H. Baldwin *et al.*, J. Phys. B **19**, L179 (1986)
9. H. Nguyen *et al.*, Phys. Rev. A **33**, 1279 (1986)
10. M. Koenig, P. Malnoult, and H. Nguyen, Phys. Rev. A **38**, 2089 (1988)
11. H. Griem, Phys. Fluids **4**, 346 (1992)
12. J. C. Kieffer *et al.*, Phys Fluids B **5**, 2676 (1993)
13. H. Nishimura *et al.*, Phys. Plasma, **2**, 2063 (1995)
14. Y. Leng *et al.*, Phys. Rev. E. **52**, 4328 (1995)
15. Woolsey *et al.*, Phys. Rev. E **56**, 2314 (1997)
16. A. Saemann *et al.*, Phys. Rev. Lett. **82**, 4843, (1999)
17. J. C. Junkel *et al.*, Phys. Rev. E **62**, 5584 (2000)
18. O. Renner *et al.*, J. Quant. Spectrosc. Radiat. Transfer **58**, 851 (2000)
19. N. Woolsey *et al.*, J. Quant. Spectrosc. Radiat. Transfer **65**, 573 (2000)
20. H. R. Griem, *Principles of Plasma Spectroscopy* (Cambridge, 2005)
21. U. Andiel *et al.*, Europhys. Lett. **60**, 861, 2002
22. O. Renner *et al.*, J. Quant. Spectrosc. Radiat. Transfer **99**, 523 (2006)
23. J. Lindl, Phys. Plasmas **2**, 3933 (1995)
24. K. Nazir *et al.*, Appl. Phys. Lett. **69**, 3686 (1996)
25. X. Li, and F.B. Rosmej, Europhys. Lett. 99, 33001 (2012).
26. O. Renner *et al.*, J. Phys. B: At. Mol. Opt. Phys. **31**, 1379 (1998)
27. K. Eidmann *et al.*, J. Quant. Spectrosc. Radiat. Transfer **81**, 133 (2003)
28. R. Shepherd *et al.*, J. Quant. Spectrosc. Radiat. Transfer **81**, 431 (2003)
29. D. F. Price *et al.*, Phys. Rev. Lett. **75**, 252 (1995)
30. Khattak *et al.*, Europhys. Lett. **72**, 242, (2005)
31. Khattak *et al.* J. Phys. D **36**, 2372 (2003)
32. http:/physics.nist.gov/PhysRefData/XrayTrans/Html/search.html
33. P. Beiersdorfer *et al*, Phys. Rev. A **40**, 150(1989)
34. J. T. Larsen and S. M. Jane, J. Quant. Spectrosc. Radiat. Transfer **51**, 179 (1994)
35. D. Riley, J. Quant. Spectrosc. Radiat. Transfer **60**, 221 (1998)
36. F.B. Rosmej, J. Phys. B. Lett.: At. Mol. Opt. Phys. **30**, L819 (1997).
37. F.B. Rosmej, Europhysics Letters **55**, 472 (2001).
38. F.B. Rosmej, Europhysics Letters **76**, 1081 (2006).

Journal of Physics: Conference Series **397** (2012) 012021 doi:10.1088/1742-6596/397/1/012021

Quantum-statistical line shape calculation for Lyman-α lines in dense H plasmas

S Lorenzen[1], A Wierling[1], H Reinholz[1], G Röpke[1], M C Zammit[2], D V Fursa[2], I Bray[2]

[1]Institute of Physics, University of Rostock, 18051 Rostock, Germany

[2]Institute of Theoretical Physics, Curtin University, Perth WA 6845, Australia

E-mail: Sonja.L@renzen.de

Abstract. We present results for the Lyman-α line of hydrogen in dense plasmas. Full line profiles are calculated within a quantum-statistical method, based on thermodynamic Green's functions. The contributions of plasma ions and electrons are considered separately. Linear and quadratic Stark effect as well as quadrupole effects are taken into account for ions. The model microfield method is used to include ion dynamics. The focus of this work lies on the contribution to broadening and shift by free electrons beyond the Born approximation. The effect of strong collisions can be identified as ladder-like diagrams of the electron-emitter propagator. In an effective two-particle approximation, the electronic self-energy is given in terms of scattering amplitudes, analogous to Baranger's expressions [Baranger, M 1958 *Phys. Rev.* **112** 855]. We obtained scattering amplitudes from convergent close-coupling calculations including medium effects via Debye screening. Additionally, the electronic coupling between initial and final states is taken care of by a vertex correction. In our examples, the free electron density ranges between 10^{23} and 10^{25} m^{-3} at a plasma temperature of 1 and 2 eV, respectively.

1. Introduction

Details of spectral lines like width, shift, and asymmetries can be used to determine plasma properties from an electro-magnetic spectrum [1]. The plasma can either be part of an astrophysical object or be created in arc discharges or by laser impact in the laboratory. A sound theory to calculate line profiles is needed in order to obtain accurate information about plasma parameters such as composition, temperature, and density. There exist many approaches to calculate the spectrum of bound-bound electron transitions, emitted from or absorbed by a plasma. No matter, if they use pure quantum mechanics [2], involve a semi-classical view on the perturbers, e.g. [3, 4], or depend on MD simulations, e.g. [5], they are based on the calculation of the dipole-dipole correlation. In our case, the dipole-dipole correlation – namely, the polarization function – is calculated within a many-body theory using thermodynamic Green's functions. The theory is briefly described in Sec. 2. The focus of this paper is twofold. At first, we want to improve the perturbative Born approach used so far for the perturbing electrons. It has the shortcoming to overestimate strong electron collisions. Hence, we implement an effective two-particle T-matrix approach, which accounts for strong collisions. This is done at the cost of neglecting dynamical screening for weak collisions. Instead, static screening is used in the calculation of the T-matrix in a convergent close-coupling scheme. The second focus is on the limit of the quasi-static approximation used for ion contributions so far. In this short communication, we restrict ourselves to theoretical results for the Lyman-α line of hydrogen, see Sec. 3. Lyman lines of hydrogen-like lithium and the application in plasma diagnostics will be discussed in a forthcoming paper.

2. Quantum-statistical approach to line profiles

The quantum-statistical theory for pressure broadening has been described in detail in [6–8]. Here, we give only the key formulas in atomic Rydberg units, i.e $\hbar = 2m_e = e^2/2 = 1$. These units are used throughout this paper. The emitted spectral intensity $I(\Delta\omega)$ at $\Delta\omega = \omega - \omega_0$ near the unperturbed transition frequency ω_0 is given by

$$I(\Delta\omega) = \frac{(\omega_0 + \Delta\omega)^4}{8\pi^4 c^3} e^{-\frac{\omega_0+\Delta\omega}{k_B T}} \,\mathrm{Im}\left[\sum_{ii'ff'} \{\langle i|\vec{r}|f\rangle\langle f'|\vec{r}|i'\rangle\langle i|\langle f| < U(\Delta\omega) >_{\mathrm{KP}} |f'\rangle|i'\rangle\}\right]. \quad (1)$$

The sum runs over all initial i and final f emitter states. The double sum is due to degeneracy. The contributions to the line profile are weighted with the transition probability, which is given by the dipole matrix elements $\langle i|\vec{r}|f\rangle$. In order to include ion dynamics in the theory, the model microfield method (MMM) based on a kangoroo process (KP) [9, 10] is adapted. Here, the time evolution operator $< U(\Delta\omega) >_{\mathrm{KP}}$ depends on the jump frequency $\Omega(E)$ and is given by

$$< U(\Delta\omega) >_{\mathrm{KP}} =< U(\Delta\omega|\vec{E}) >_{\mathrm{s}} + \frac{< \Omega(E)U(\Delta\omega|\vec{E}) >_{\mathrm{s}}^2}{< \Omega(E) >_{\mathrm{s}} - < \Omega^2(E)U(\Delta\omega|\vec{E}) >_{\mathrm{s}}}, \quad (2)$$

where the average $< \cdots >_{\mathrm{s}}$ is over the static ion microfield with the distribution function $W(E)$. For hydrogen, we use Hooper's low frequency tables [11] to determine $W(E)$. In our theory, the time evolution operator is constructed in the following way to include electron contributions

$$U(\Delta\omega|\vec{E}) = \{\Delta\omega - \mathrm{Re}\,[\Sigma_i(\Delta\omega) - \Sigma_f(\Delta\omega)] + i\Omega(E) + i\,\mathrm{Im}\,[\Sigma_i(\Delta\omega) + \Sigma_f(\Delta\omega)] + \Gamma_{if}(\Delta\omega)\}^{-1}. \quad (3)$$

Here, Σ_i and Σ_f are the self-energies, i.e. broadening and shift, of energy level i and f due to the surrounding plasma, respectively, and Γ_{if} is the upper-lower level coupling term. Due to different interaction time scales, the self-energy can be split into an E-field-dependent ionic part and a frequency-dependent electronic part

$$\Sigma_{i,f}(E, \Delta\omega) = \Sigma^{\mathrm{i}}_{i,f}(E) + \Sigma^{\mathrm{e}}_{i,f}(\Delta\omega). \quad (4)$$

The perturbation of the emitter by the plasma ions is mainly given by the linear and quadratic Stark effect. Furthermore, quadrupole contributions are taken into account from [12]. In the quasi-static limit, i.e. with $\Omega(E) \to 0$, the second term of Eq. (2) vanishes.

2.1. Contribution of electron-emitter collisions

Two approaches have been developed to account for the contribution by free electrons. They are either considered within a dynamically screened Born approximation (1st order) or within an effective two particle T-matrix approach. While the former includes the dynamical screening of weak collisions, it overestimates strong collisions. This can be rectified to some extend by a cut-off procedure introduced by Griem [3]. The effective T-matrix approach is a simplified version of the T-matrix approach presented in [13]. It can describe weak and strong collisions equally well, but does so far only include static screening [14]. For a non-degenerate plasma, the electronic self-energy, evaluated at $\Delta\omega = 0$, is then given by

$$\Sigma^{\mathrm{e}}_{i,f} = -\frac{2}{\pi} n_e \Lambda^3_{\mathrm{th}} \int_0^\infty \mathrm{d}k\, k^2 e^{-k^2/k_B T} f_{i,f}(0, k). \quad (5)$$

Plasma properties enter via the electron density n_e and the thermal wavelength $\Lambda_{\text{th}} = \sqrt{4\pi/k_B T}$ as well as the forward scattering amplitude $f_{i,f}(0,k)$ for elastic electron scattering at the emitter in state i and f, respectively. This expression was also found by Baranger in [2]. There, a level coupling term is derived, which is given by

$$\Gamma_{if} = \frac{2i}{\pi} n_e \Lambda_{\text{th}}^3 \int_0^\infty \mathrm{d}k\, k^3 e^{-k^2/k_B T} \int_0^\pi \mathrm{d}\theta \sin(\theta) f_f(\theta,k) f_i^*(\theta,k). \tag{6}$$

2.2. Scattering amplitudes from Debye screened convergent close-coupling

The scattering amplitudes, which enter Eqs. (5) and (6), are obtained from a convergent close-coupling calculation (CCC). It is modified to include Debye screening in the interaction potentials. Details of the CCC method and its modification can be found in [15] and [16], respectively. Here, we use 54 (Sturmian) Laguerre functions as a basis for bound and continuous emitter states. With this choice, the emitter states up to 4f are reproduced with the correct energies. The electron-electron potential is expanded in partial waves, where the first partial waves (up to 70) are considered directly, for larger numbers of partial waves extrapolation formulas are used. The coupled equations are solved in momentum space and lead to the scattering amplitude. Our method gives separate results for singlet and triplet scattering channels as well as for scattering at the emitter with initial and final states $n_i, l_i, m_i \to n_f, l_f, m_f$. Here, we consider only elastic scattering $n_i, l_i, m_i \to n_i, l_i, m_i$ for the self-energies.

3. Lyman-α line of H

To illustrate the T-matrix approach, we consider the simplest line of hydrogen, namely Lyman-α. As the lower level 1s gives a small contribution, we focus on shift and broadening of the 2p level due to electrons in the discussion. In Fig. 1, the convergence with the number of partial waves is presented for $k_B T = 1$ eV and $n_e = 2 \cdot 10^{23}$ m^{-3}, and $n_e = 1 \cdot 10^{25}$ m^{-3}, respectively. The self-energies are given for different magnetic quantum numbers m of the emitter. Singlet and triplet scattering results are averaged.

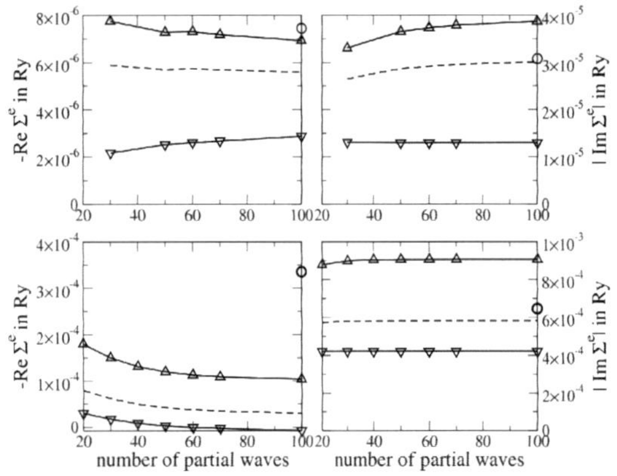

Figure 1. Convergence of the electronic contribution to the self-energy of the hydrogen 2p level with the number of partial waves. Left: real part, right: imaginary part. Dependence on the magnetic quantum number is indicated by symbols: $m = \pm 1$ ($\triangle$) and $m = 0$ ($\triangledown$), the average is (- - - -). The results are given for $k_B T = 1$ eV and $n_e = 2 \cdot 10^{23}$ m^{-3} in the upper and $n_e = 1 \cdot 10^{25}$ m^{-3} in the lower panel. For comparison, the result from the Born approximation with a cut-off procedure is given at the right ordinate axis ($\bigcirc$).

The convergence is faster for the imaginary part and at the higher density, where the screening is stronger. The plasma conditions correspond to a Debye length of 314 a_0 and 44 a_0, respectively, where a_0 is the Bohr radius. Assuming our T-matrix results as benchmark, the cut-off Born approximation overestimates the shift (real part), especially for the higher density of $n_e = 1 \cdot 10^{25}$ m^{-3}. The width (imaginary part) agrees within 10% with the average

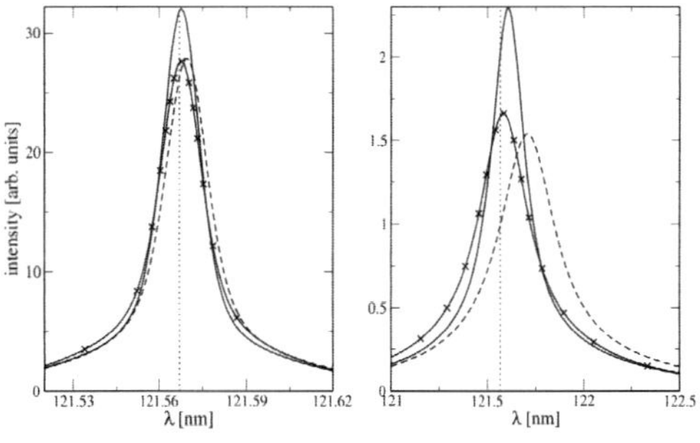

Figure 2. Hydrogen Lyman-α line calculated for quasi-static ions and electrons in either Born approximation (- - - -), Born approximation supplemented with a cut-off procedure (——) or with T-matrix approach (×——×). For $k_BT = 1$ eV and $n_e = 2 \cdot 10^{23}$ m^{-3} (left) and $n_e = 1 \cdot 10^{25}$ m^{-3} (right). All profiles are area normalized.

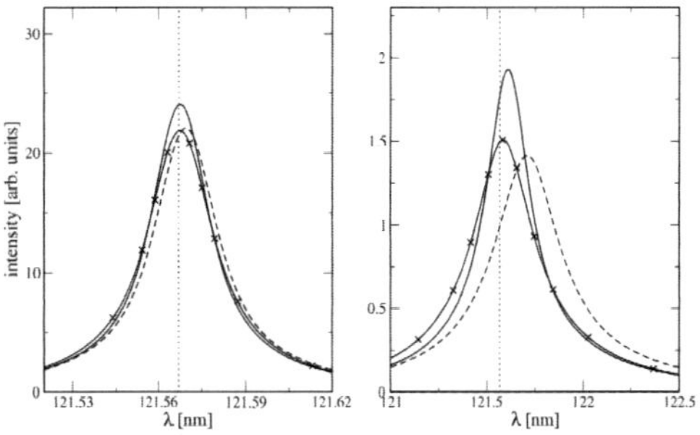

Figure 3. Same as Fig. 2 but with ion dynamics considered within the model microfield method, the scale is identical to Fig. 2.

over m. However, as the central component of the Lyman-α line is given by the transition $2p_{m=\pm1} \rightarrow 1s$, the line width is dominated by the width of the $2p_{m=\pm1}$ level. Thus, in Fig. 2, the Born approximation with cut-off underestimates the width by up to 35% compared to the T-matrix approach. There, ions are only considered quasi-statically. In Fig. 3, the lines are further broadenend by up to 50% for the lower density, and up to 30% for the higher density due to the ion dynamics (MMM). The importance of ion dynamics for H Lyman-α has been known from measurements [17] since 1977. Our full line profiles include Stark broadening by the plasma surroundings and are convoluted with a Gaussian to account for Doppler broadening.

In Tab. 1, we compare the width and shift due to electron collisions with an older close-coupling calculation [18]. There, Debye screening is included by restricting the calculation of isolated e-H scattering to a certain number of partial waves. When we average the self-energy and vertex term of different line components, our results agree within few percent with [18]. However, when we calculate the full line profile without ion effects and Doppler broadening, the line is 50% narrower and up to 40% less shifted. Thus, it is not appropriate to average width and shift over different scattering channels in line shape calculations.

Table 1. Width and shift of H Lyman-α from electron collisions compared with results from [18] for two plasmas with the same Debye length $\lambda_D = 314$ a_0. Conditions are 1) $k_BT = 1$ eV, $n_e = 2 \cdot 10^{23}$ m^{-3} and 2) $k_BT = 2$ eV and $n_e = 4 \cdot 10^{23}$ m^{-3}. All values are given in Å.

	shift:	averaged	full line	[18]	HWHM:	averaged	full line	[18]
1)		0.0087	0.0051	0.0087		0.062	0.031	0.064
2)		0.021	0.014	0.021		0.098	0.048	0.099

4. Conclusions

Full Lyman-α line profiles have been calculated for H plasmas within a quantum-statistical theory using a perturbative Born approximation as well as an effective two particle T-matrix approach for the electronic contributions. The latter needs scattering amplitudes as an input. They are obtained from convergent close-coupling calculations with Debye screening. The T-matrix approach can intrinsically treat strong electron-emitter collisions, whereas the Born approximation compensates the overestimation of strong collision contributions by an artificial cut-off procedure.

We show, that for a density of $n_e = 1 \cdot 10^{25}$ m^{-3} and a temperature of $k_B T = 1$ eV, the Born approach overestimates the shift even after cut-off by a factor of three, whereas the width agrees to the average obtained in the T-matrix approach within 10%. However, the line shape is significantly different between both approaches, because the self-energy is strongly dependent on the magnetic quantum number. This feature cannot be reproduced within the Born approximation. Ion contributions have been taken into account quasi-statically, and dynamically within the model microfield method. The ion dynamics leads to further line broadening.

Results for hydrogen Lyman-α were presented as a test case in comparison to older results. The method can be applied for other lines of hydrogen and hydrogen-like ions, as well.

5. Acknowledgement

This work was supported by the German Research Foundation DFG within SFB 652. The support of the Australian Research Council, the Australian National Computational Infrastructure Facility and its Western Australian node iVEC, are gratefully acknowledged.

References

[1] Griem H R 1997 *Principles of Plasma Spectroscopy* (Cambridge University Press, Cambridge)
[2] Baranger M 1958 *Phys. Rev.* **112** 855
[3] Griem H R 1974 *Spectral Line Broadening by Plasmas* (Academic Press, London)
[4] Sahal-Brechot S 1969 *Astron. Astrophys.* **1** 91
Sahal-Brechot S 1969 *Astron. Astrophys.* **2** 322
[5] Stamm R and Voslamber D 1979 *J. Quant. Spectros. Radiat. Transfer* **22** 599
[6] Hitzschke L, Röpke G, Seifert T and Zimmermann R 1986 *J. Phys. B* **19** 2443
[7] Günter S 1995 habilitation thesis at Rostock University, Germany
[8] Sorge S, Wierling A, Röpke G, Theobald W, Sauerbrey R and Wilhein T 2000 *J. Phys. B* **33** 2983
[9] Brissaud A and Frisch U 1971 *J. Quant. Spectros. Radiat. Transfer* **11** 1767
[10] Seidel J 1977 *Z. Naturf.* **32a** 1195
Seidel J 1980 *Z. Naturf.* **35a** 679
[11] Hooper, Jr C F 1968 *Phys. Rev.* **165** 215
Hooper, Jr C F 1968 *Phys. Rev.* **169** 193
[12] Halenka J 1990 *Z. Phys. D* **16** 1
[13] Günter S 1993 *Phys. Rev. E* **48** 500
[14] Lorenzen S, Wierling A, Reinholz H and Röpke G 2011 *Contrib. Plasma Phys.* **51** 349
Lorenzen S, Wierling A, Reinholz H, Röpke G, Zammit M, Fursa D and Bray I 2010 *AIP Conference Proceedings* vol 1290 pp 99–103 20th International Conference on Spectral Line Shapes, 6-11 June 2010, St. John's, NF, Canada
[15] Bray I and Stelbovics A T 1992 *Phys. Rev. A* **46** 6995
[16] Zammit M C, Fursa D V and Bray I 2010 *Phys. Rev. A* **82** 52705
[17] Grützmacher K and Wende B 1977 *Phys. Rev. A* **16** 243–246
[18] Unnikrishnan K and Callaway J 1991 *Phys. Rev. A* **43** 3619

Atomic Hydrogen and Molecular Carbon Emissions in Laser-Induced Breakdown Spectroscopy

Christian G. Parigger, Alexander C. Woods and Mohammad R. Rezaee

The University of Tennessee Space Institute,
Center for Laser Applications,
411 B.H. Goethert Parkway,
Tullahoma, TN 37388, USA

E-mail: cparigge@tennessee.edu

Abstract.
This article addresses hydrogen Balmer series measurements following laser-induced optical breakdown. Electron density on the order of 1 × 10^{25} m^{-3} can be inferred using H_α Stark width and shift for plasma generated in 1 to 1.3 × 10^5 Pa, gaseous hydrogen. The H_β line can be utilized for electron density up to 7 × 10^{23} m^{-3}. Laser ablation of aluminium reveals limits of application of the Balmer series. Electron excitation temperature is inferred utilizing Boltzmann plot techniques that include H_α, H_β and H_γ atomic lines. H_β and H_γ lines show presence of molecular carbon in a 2.7 and 6.5 × 10^5 Pa, expanding methane flow. Occurrence of superposition spectra in the plasma decay due to recombination or due to onset of chemical reactions necessitates consideration of both atomic and molecular emissions following laser-induced optical breakdown.

1. Introduction

In this paper we explore hydrogen plasma spectra generated by high intensity laser radiation. Investigated are the H_α, H_β, and also H_γ lines and their characteristics as well as broadening and shifts. This article elaborates on previously communicated emission spectroscopy data of the hydrogen Balmer series [1, 2, 3, 4, 5]. Laser-induced breakdown spectroscopy (LIBS) historically embraces elemental analysis, or atomic spectroscopy, and to a lesser extent molecular spectroscopy. Time-resolved diagnostics allow us to record atomic emission profiles separate from molecular recombination and/or excitation; however, atomic and molecular species can occur simultaneously within a selected spectral window, for our expanding methane experiments typically on the order of 1 to 2 micro-seconds after optical breakdown.

Following the invention of laser devices, Laser-Induced Breakdown Spectroscopy (LIBS) evolved into one of the favorite topics not only in atomic and chemical physics but also in chemistry, biology, environmental science, material engineering, and recently space exploration. Modeling of the hydrogen line shapes for the diagnostic for International Thermonuclear Experimental Reactor (ITER) is an important ongoing investigation, and it is based on the Balmer lines of the hydrogen isotopes. The use of hydrogen Balmer lines has been employed to measure the temperature and electron density in plasmas, including several theoretical approximations [6, 7, 8, 9]. In typical LIBS experiments, the main broadening mechanism

Published under licence by IOP Publishing Ltd

of spectral lines is the Stark effect. Line broadening has been studied comprehensively by H. Griem [10]. A large, linear Stark effect in hydrogen makes its lines attractive for recent research [11]. According to theory, the shapes and shifts of plasma-broadened isolated lines are mainly determined by electron impact with the radiating atom or ion, and a smaller contribution arises from the electric micro-fields generated by essentially static plasma ions. In addition, at a given electron number density of the plasma, the hydrogen lines are broader with respect to the lines of other elements. Doppler broadening has information about velocity distribution, and Stark broadening gives information on electron density [12]. These features make their use particularly interesting in LIBS experiments performed with typically 0.1-nm spectral resolution spectrometers where the instrumental broadening introduced by the spectrometer might be comparable, or even larger, than the broadening of the emission lines of elements different from hydrogen. Moreover, the Stark broadening of the emission lines is independent from the fulfillment of local thermodynamic equilibrium (LTE) conditions [13], and this aspect makes this approach for the electron density measurement an interesting and powerful tool. Measurements of line shapes are important in plasma diagnostics allowing us to investigate fundamental atomic structure as well as inferring plasma temperature and electron density. In particular, the H_β line is one of the most studied lines, and its use has become a standard technique in plasma spectroscopy [14].

In previous studies by Parigger *et al.* [15], analysis of Stark broadening and shifts in measured H_α and H_β spectra, combined with Boltzmann plots from H_α, H_β and H_γ lines to infer the temperature, have been discussed for the electron densities in the range of 10^{22} - 10^{25} m^{-3}, and for the temperatures in the range of 6,000 to 100,000 K. Lines H_α, H_β and H_γ were recorded at different delay times from optical breakdown. The electron density was evaluated using different theoretical approaches that are commonly employed for plasma diagnostics [3]. We previously reported measurements [1, 2, 3, 4, 5] and comprehensive studies of the diagnostic application of Balmer lines.

2. Experimental Details

For generation of a micro-plasma, a laboratory Nd:YAG laser operated at the fundamental wavelength of 1064-nm was used. For the hydrogen plasma investigations, a Continuum YG680S-10 Nd:YAG with 150-mJ energy per pulse and 7.5-ns pulse duration, was focused to typically 1,400 GW/cm^2 in a pressure cell that was filled with gaseous hydrogen to a pressure of 810 $\pm$ 25 Torr (1.07 $\times$ 10^5 Pa) and 1010 $\pm$ 25 Torr (1.33 $\times$ 10^5 Pa), subsequent to evacuation of the cell with a diffusion pump. Spectrometer and detector arrangements are discussed in the Parigger and Oks review article[1].

Recording of shadow graphs (not illustrated here) was accomplished with Coherent Infinity 40-100 Nd:YAG laser with 50-mJ and 300-mJ pulse energy and 3.5-ns pulse width was focused in laboratory air to an irradiance of typically 10,000 GW/cm^2. Perpendicular to the laser beam generating LIB, we typically employed pulsed laser radiation from a 308-nm XeCl excimer laser.

In the expanding methane experiments [4], we used a 1/2 m model 500 SpectraPro Acton Research Corporation spectrometer together with an intensified linear diode array (model 1460 Princeton Applied Research detector/controller optical multichannel analyzer). Averaged over 100 individual LIB events, the captured time-resolved data (measurement window of 0.1 μs) from the methane breakdown events were detector-noise/background corrected and wavelength and detector sensitivity calibrated using standard calibration lamps.

3. Experimental Results

Measurements of individual profiles of the Balmer series lines H_α, H_β and H_γ allows us to infer electron excitation temperature. However, the inferred temperature will show error-bar contributions due to the broadened atomic emissions. Presence of the molecular emissions will add to the background for the hydrogen Balmer series diagnostic. We treat the H_β and H_γ lines as background contribution to extract molecular temperature information for analysis of molecular excitations [5].

Time resolved spectroscopy measurements of hydrogen- alpha, -beta, and -gamma emissions were performed in an expanding flow of methane gas. Hydrogen Stark widths were used to characterize the plasma decay during the first 2 microseconds after laser-induced optical breakdown [4]. Analysis of the recorded hydrogen spectra include measurements of full-width-half-maximum (FWHM) and full-width-half-area(FWHA) to infer electron density N_e [2]. Temperature was determined from incomplete hydrogen Balmer Series lines [3]. Yet for time delays on the order of 1 μs, molecular emissions can be recognized from the C_2 Swan system, superposed to the H_β and H_γ Balmer series lines. C_2 emissions are clearly recognizable for time delays on the order of 2 μs. Figures 1 and 2 show these results.

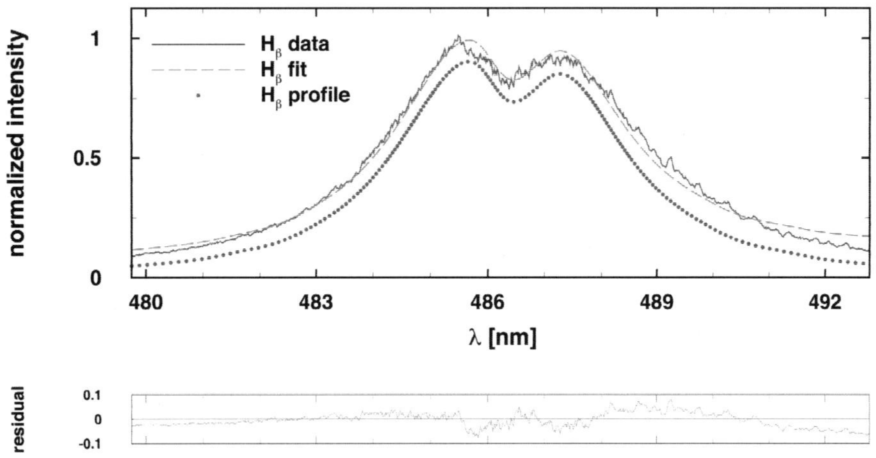

Figure 1. H_β recorded and fitted spectra. Delay timeΔ $\tau = 1.1$ μs. $N_e = 1.0 \times 10^{23}$ m^{-3}.

Figs. 1 and 2 indicate the presence of C_2 molecular emissions for a delay time ofΔ $\tau = 2.1$ μs. Temperature inferred from the fitting of the C_2 "fingerprints" revealed molecular temperatures in the range of 4000 - 6000 Kelvin [5]. The H_β profile at a delay time ofΔ $\tau = 2.1$ μs shows an electron density of 0.5×10^{23} m^{-3}. The electron excitation temperature for delay times on the order of 1.1 to 2.1 μs amounts to 1.3 to 1.1×10^4 K [3], while the molecular temperature of the residual carbon, see Fig. 2, is determined to be 0.52×10^4 K.

The H_β profiles typically show a double-peak structure consistent with computed line profiles that appear asymmetric early in the plasma decay. The major broadening mechanism here is due to the Stark effect, which should cause a Lorentzian profile, and the dip at the line center is due to absence of computed H_β shifts [16, 17] and in part due to electrostatic interactions with ions [18]. The theory for widths of the hydrogenic lines is mostly accurate and best agrees for H_β, where the overall error is about 5%. For hydrogenic systems a relatively large portion of the broadening is due to quasi-static effects.

Electron density in plasma can be found by measuring Stark broadening, and this is the most precise technique for determining the electron density N_e in plasma. This broadening will become significant for $N_e > 10^{21}$ m^{-3}. Stark broadening is due to collision between radiating species with charged species, like electrons and ions. Not only the H_β Stark width but also the

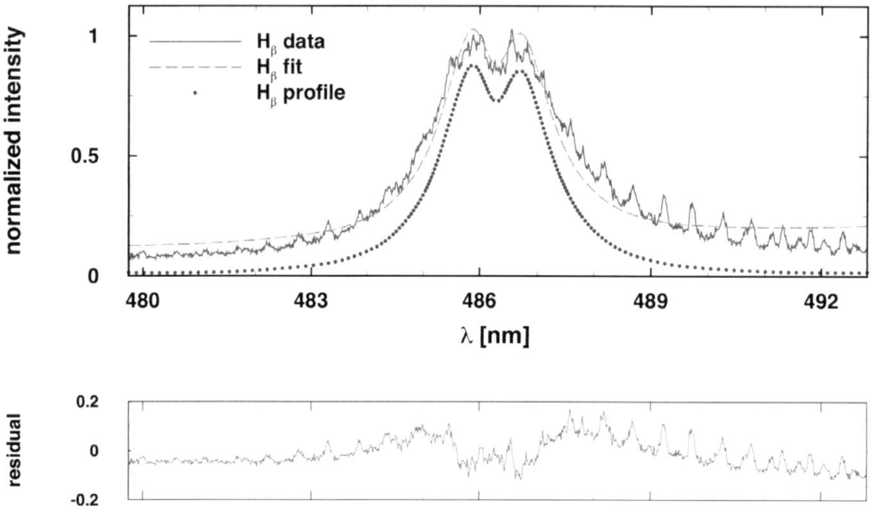

Figure 2. H_β recorded and fitted spectra. Delay time$\Delta\ \tau = 2.1\ \mu$s. $N_e = 0.5 \times 10^{23}\ m^{-3}$.

separation of the H_β double-peak structure is a measure for the free electron density [15]. Oks' theory of Stark widths incorporates several major theoretical advances and new phenomena in the Stark broadening of hydrogen lines in plasmas [19]. The Stark linewidth$\Delta\ \lambda_{FWHM}$ at full-width-half maximum (FWHM) can be extracted from the measured line width$\Delta\ \lambda_{observed}$ by subtracting instrumental$\Delta\ \lambda_{instrumental}$ and Doppler line broadening$\Delta\ \lambda_{Doppler}$ [20]. A large electron density ($N_e > 10^{23}$ m^{-3}) causes Stark broadening that is typically much larger than employed instrumental resolution and Doppler broadening. Experimental studies of aluminum laser ablation [21] within an evacuated cell reveals the limitation of Balmer series diagnostic for electron density and temperature. Early in the plasma decay, the H_α line becomes discernable at delays on the order of 0.025 to 0.050 μs after LIB, showing an electron density of 10^{24} m^{-3}. The H_β line appears recognizable at delays on the order of 0.075 to 0.100 μs, showing an electron density of 4×10^{23} m^{-3}.

4. Discussion

A plasma's temperature and density are important for its stability. For high-temperature nuclear fusion plasma, non-invasive plasma diagnostics (like optical emission spectroscopy) are necessary [22]. A typical method to infer plasma temperature is to construct a so-called Boltzmann plot [23] including consideration of non-uniformity of laser-induced plasma [24]. Relative ion populations in local thermodynamic equilibrium (LTE) can be computed using the Saha (or Saha-Eggert) equation. However, when applying the Saha equations to obtain temperature, called the temperature of the ionization equilibrium, relative intensities of lines from different ion stages of the same atom (occasionally different atoms) need to be measured [25].

The recorded emission spectra of the hydrogen Balmer series allow us to infer electron number density, using previously tabulated values. The displayed hydrogen emission profiles are also convolved with the Doppler width that amounts to 0.05 nm at 10,000 K and 0.15 nm at 100,000 K. There is a substantial red shift of the H_α line profile from its unperturbed position for N_e

on the order of 10^{24} m^{-3}. Larger red shifts in the forward region of the breakdown plasma correspond to higher temperature and also electron density, therefore, we infer that there is an electron (or excitation) temperature gradient across the length of the plasma. Temperatures on the order of 100,000 K can be found early in the plasma decay [15]. Use of Oks' theory allowed us to enhance significantly the agreement of theory and our experimental results.

Superposition spectra occur due to recombination or due to onset of chemical reactions. Analysis of both atomic and molecular emission spectra following laser-induced optical breakdown require utilizing molecular diagnostics [26] based on so-called line-strength files. The illustrated spectroscopic data display both the Balmer series hydrogen beta line and molecular C_2 Swan band presence. The molecular excitation temperature is determined using modified Boltzmann plots and fitting of spectra from selected molecular transitions [27].

5. Acknowledgements

We wish to thank for support by UTSI's Center for Laser Applications.

References

[1] Parigger C G and Oks E 2010 *Int. Rev. Atom. Molec. Phys.* **1** 13
[2] Parigger C G 2010 *Int. Rev. Atom. Molec. Phys.* **1** 129
[3] Parigger C G, Woods A C and Hornkohl J O 2011 *Int. Rev. Atom. Molec. Phys.* **2** 77
[4] Parigger C G, Dackmann M and Hornkohl J O 2008 *Appl. Opt.* **47** G1
[5] Parigger C G, Woods A C and Hornkohl J O 2012 *Appl. Opt.* **51** B1
[6] De Giacomo A, Santagata A, Legnaioli S, Lorenzetti G, Pardini L, Palleschi V and Omenetto N 2011 NASLIBS (Clearwater, FL) P20
[7] Aragón L and Aguilera J A 2010 *Spectrochim. Acta B: Atom. Spectrosc.* **65** 395
[8] El Sherbini A M, Hegozy H and El Sherbini Th M 2006 *Spectrochim. Acta B: Atom. Spectrosc.* **61** 532
[9] Detalle V, Héon R and Sabsabi M 2001 *Spectrochim. Acta B: Atom. Spectrosc.* **56** 1011
[10] Griem H 1974 *Spectral Line Broadening by Plasmas* (New York: Academic Press)
[11] Konjević N 1999 *Phys. Rep.* **316** 339
[12] Chen F F, *Introduction to Plasma Physics* (New York: Springer Verlag)
[13] Cristoforetti G, De Giacomo A, Dell'Aglio M, Legnaioli S, Tognoni E, Palleschi V and Omenetto N 2010 *Spectrochim. Acta Part B: Atom. Spectrosc.* **65** 86
[14] Djurović S, Ćirišan M, Demura A V, Demchenko G V, Nikolić D, Gigosos M A and González M Á 2009 *Phys. Rev. E* **79** 046402
[15] Parigger C G, Plemmons D H and Oks E 2003 *Appl. Opt.* **42** 5992
[16] Schrödinger E 1926 *Ann. Phys.* **385** 437
[17] Schrödinger E 1926 *Phys. Rev.* **28** 1049
[18] Laux C O 2003 *Plasma Sources Sci. Technol.* **12** 125
[19] Oks E 2006 *Stark Broadening of Hydrogen and Hydrogenlike Spectral Lines on Plasmas: The Physical Insight* (Oxford: Alpha Science International)
[20] Camacho J J, Díaz L, Santos M and Poyato J M L 2011 *Spectrochim. Acta B: Atom. Spectrosc.* **66** 57
[21] Parigger C G, Hornkohl J O and Nemes L 2007 *Appl. Opt.* **46** 4026
[22] Kim J-H and Lee H-J 2006 *J. Korean Phys. Soc.* **49** S184
[23] Aydin Ü, Roth P, Gehlen C D and Noll R 2008 *Spectrochim. Acta B: Atom. Spectrosc.* **63** 1060
[24] Gornushkin I B, Shabanov S V, Merk S, Tognoni E and Panne U 2010 *J. Anal. Atom. Spectrom.* **25** 1643
[25] Cremers D E and Radziemski L J 2006 *Handbook of Laser-Induced Breakdown Spectroscopy* (New York: John Wiley)
[26] Parigger C G and Hornkohl J O 2010 *Int. Rev. Atom. Molec. Phys.* **1** 25
[27] Hornkohl J O, Nemes L and Parigger C G 2011 *Spectroscopy of Carbon Containing Diatomic Molecules* in *Spectroscopy, Dynamics and Molecular Theory of Carbon Plasmas and Vapors, Advances in the Understanding of the Most Complex High-Temperature Elemental System* Nemes L and Irle S, Eds. (Singapore: World Scientific) chapter 4 pp 113 -169

Journal of Physics: Conference Series **397** (2012) 012023 doi:10.1088/1742-6596/397/1/012023

Evaluation of particle source rate and its influence on particle transport in fusion plasma

M Goto[1], K Sawada[2], K Fujii[3], M Hasuo[3] and S Morita[1]

[1] National Institute for Fusion Science, Toki 509-5292, Japan
[2] Department of Applied Physics, Faculty of Engineering, Shinshu University, Nagano 380-8553, Japan
[3] Department of Mechanical Engineering and Science, Graduate School of Engineering, Kyoto University, Kyoto 606-8501, Japan

E-mail: goto@nifs.ac.jp

Abstract. We have carried out numerical inversion of the Laplace transform for the observed line profile, and have derived the intensity distribution function against the atom temperature. The temperature dependence is interpreted to the spatial dependence, where other diagnostic results are used, so that the radial profile of photon emission rate is derived. The photon emission rate is then interpreted to the ionization rate with the help of collisional-radiative model calculation. From the ionization rate derived we have evaluated the particle confinement time τ_p as a function of the average minor radius, r_eff, for two discharges characterized by the different magnetic field strength. In the strong field case, a turning point is clearly seen in the τ_p profile at around $r_\mathrm{eff} = 0.6$, which corresponds to the last closed magnetic flux surface. In the weak field case, the absolute confinement time is approximately one-order smaller than that in the strong field case.

1. Introduction

The ionization rate of neutral hydrogen or, in other words, the particle source term in the continuity equation is indispensable information for the study of particle transport in the fusion plasma experiment. The conventional method of experimental determination of hydrogen atom density in the plasma has been based on spatial inversion of chord-integrated intensities of an emission line such as the Balmer-α line [1] or the Lyman-α line [2]. However, the atom density deep inside the core region tends to have a large uncertainty because the line intensity from those atoms is generally obtained after dominant fractions originated in the plasma edge region are subtracted from the measured signal.

It has been noticed that Balmer-α line profiles observed contain a considerably broad component which is thought to originate from hot atoms created by the charge exchange process [3–6]. Since hydrogen atoms are thought to have a relatively large penetration depth due to the charge exchange process [7], if that is the case, the broad line profile component may have information on the atoms or protons in the central region of the plasma. However, the analysis to date has been rather superficial and no practical interpretation of the broad component has been proposed.

Under such circumstances, it has been demonstrated that the line emissivity distribution along the line-of-sight can be derived through numerical inversion of the observed Balmer-α line

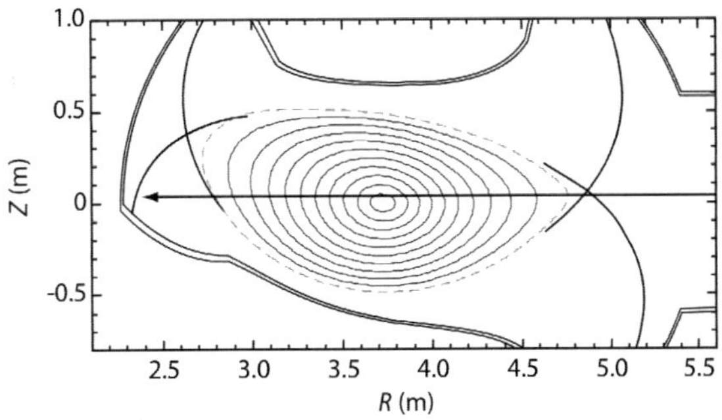

Figure 1. Poloidal cross section of the plasma with magnetic flux surfaces for the present measurement. The line-of-sight is shown with the arrow.

profile in the Large Helical Device (LHD) [8]. Furthermore, the results have been utilized to evaluate radial profiles of the ionization rate, inward neutral flux, and neutral atom density with the help of collisional-radiative model calculation.

In this paper we attempt to derive the particle confinement time as an application of the line emissivity distribution obtained with the method of line profile inversion and consider the validity of the present analysis through comparison of the results for two different discharges.

2. Experiment

The central region of a poloidal plasma cross section which is horizontally elongated is viewed with a single line-of-sight as shown in Fig. 1. The optics consists of an optical fiber with the core diameter of 100 μm and a collimator lens. The spatial resolution is approximately 30 mm. The collected light is introduced into the astigmatism-corrected UV-visible spectrometer [9] which has the focal length of 1.33 m and is equipped with a 1800 grooves/mm grating (McPherson Model 209). The reciprocal linear dispersion is 0.312 nm/mm at the wavelength of the Balmer-α line (656.280 nm). A CCD (charge coupled device) having 1024 pixels in the direction of wavelength dispersion and with 13 μm squares pixel size (Andor DV-435) is used as the detector. The absolute sensitivity of the entire system has been calibrated with a standard light source which consists of a halogen lamp and an integrated sphere (Labsphere USS-600C).

We have made analysis for two different types of discharges. One is a discharge with $R_{\rm ax} = 3.6$ m and $B_{\rm ax} = 2.75$ T, where $R_{\rm ax}$ and $B_{\rm ax}$ are the major magnetic axis radius and the magnetic field strength at the magnetic axis, respectively. The aim of the experiment is getting high stored energy with gas puff. The other is with $R_{\rm ax} = 3.59$ m and $B_{\rm ax} = 0.41$ T which aims at obtaining the highest β, the plasma pressure normalized to the magnetic pressure. The apparent difference between the two discharges is the magnetic field strength.

Figure 2 shows the Balmer-α line profiles obtained for the two discharges. It is clearly seen that the strong field case has a larger broad component than the weak field case. This is understandable because that broad component is thought to correspond to line emissions in the plasma core region and the temperature is higher in the strong field case.

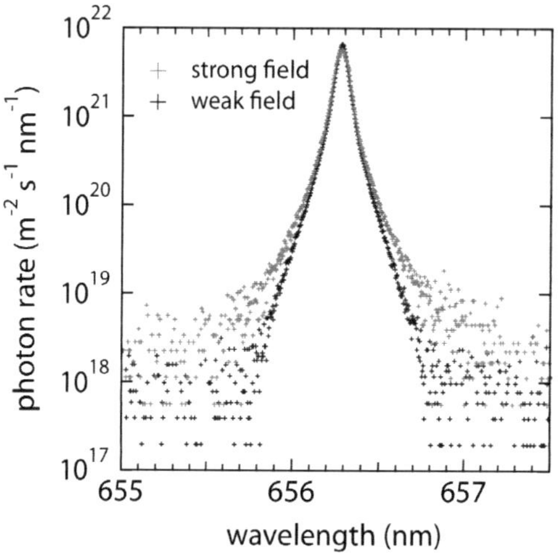

Figure 2. Balmer-α line profiles measured for the two discharges.

3. Laplace inversion of line profile

The observed line profile $I(\lambda)$ [photons m^{-2} s^{-1} nm^{-1}], where λ is the wavelength, is understood as a line-integral along the line-of-sight and can be expressed as

$$I(\lambda) = \int_{R_{\text{min}}}^{R_{\text{max}}} \eta(R) G(R, \lambda) \mathrm{d}R, \tag{1}$$

where R stands for the major radius, and R_{min} and R_{max} indicate the inboard-side and outboard-side edge radii of the line-integral, respectively. The quantities $\eta(R)$ [photons m^{-3} s^{-1}] and $G(R, \lambda)$ [nm^{-1}] are respectively the power density of line radiation at location R and the local line profile function at R that is normalized. When $G(R, \lambda)$ is the Gaussian function of λ, it is represented by the line width instead of R so that Eq. (1) can be rewritten as

$$I(\lambda) = \int_0^\infty f(w) \frac{1}{\sqrt{\pi} w} \exp\left[-\left(\frac{\lambda - \lambda_0}{w} \right)^2 \right] \mathrm{d}w, \tag{2}$$

where λ_0 is the wavelength at the line center, w is the Gaussian half width having the relation $w_{1/2} = 2\sqrt{\ln 2}\, w$, where $w_{1/2}$ is the full width at half maximum, and $f(w)$ is the contribution fraction of a Gaussian component having the width of w to the entire line profile. After replacement of variables as

$$s = (\lambda - \lambda_0)^2, \tag{3}$$

$$t = \frac{1}{w^2}, \tag{4}$$

Eq. (2) can be rewritten in the form of Laplace transform as

$$\mathscr{F}(s) = \int_0^\infty F(t) \exp(-st)\, \mathrm{d}t, \tag{5}$$

where

$$F(t) = \frac{f(t^{-1/2})}{2\sqrt{\pi} t}. \tag{6}$$

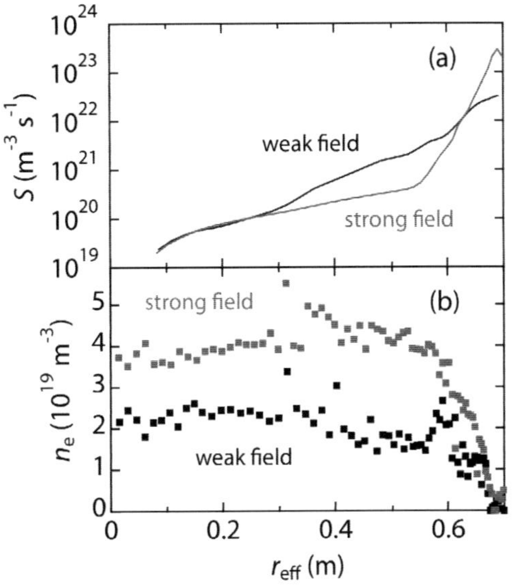

Figure 3. Profiles of the ionization rate S (a) and the electron density (b) as a function of r_{eff} for the two discharges.

From the function $\mathscr{F}(s)$, which corresponds to the line profile experimentally obtained, $F(t)$ is determined by numerical inversion of Eq. (5). The numerical inversion has been carried out with the method of Ref. [10] which is based on the Gaussian quadrature with the shifted Legendre polynomials.

The distribution function $f(w)$ is readily obtained with Eq. (6) and can be further interpreted as a function of atom temperature T_{a}, $g(T_{\text{a}})$, as

$$g(T_{\text{a}}) = f(w)\frac{\mathrm{d}w}{\mathrm{d}T_{\text{a}}}, \tag{7}$$

with

$$w = \lambda_0\sqrt{\frac{2kT_{\text{a}}}{Mc^2}}, \tag{8}$$

where M, c, and k are the atom mass, light speed, and Boltzmann constant, respectively.

4. Results and discussion

As mentioned above we assume that T_{a} is equivalent to the proton temperature T_{p} at the same location. In addition to that, since the electron density n_{e} is rather high in the present discharges, T_{p} and the electron temperature T_{e} are expected to be in equilibrium, namely, $T_{\text{p}} = T_{\text{e}}$ would be established. Under such circumstances, the T_{a} dependence in Eq. (7) is interpreted to the spatial dependence with the help of T_{e} profiles by the Thomson scattering measurement so that the radial profile of photon emission rate is derived. The photon emission rate then yields the ionization rate with the help of collisional-radiative model calculation [8]. Figure 3 (a) shows the ionization rate profiles derived, where the horizontal axis indicates the average minor radius r_{eff}. The decay of S as going deep into the core region is much faster in the strong field case

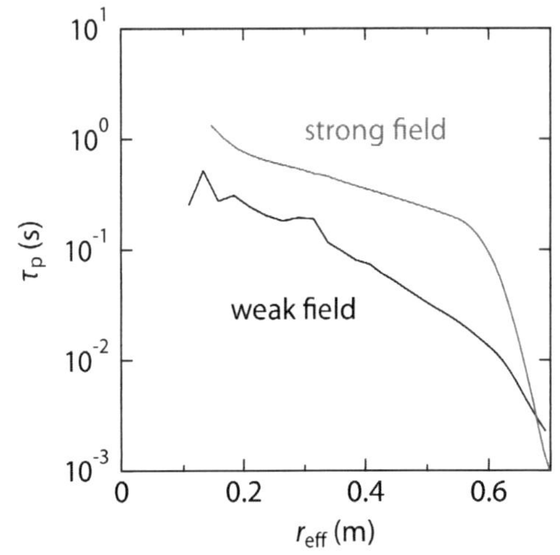

Figure 4. Particle confinement time τ_p as a function of r_{eff} for the two discharges.

than in the weak field case. This may be due to the higher electron density in the strong field case as seen in Fig. 3 (b).

The ionization rate or the particle production rate in the confined region is an important parameter to evaluate the particle confinement time, which is a measure of the performance of plasma confinement. The particle confinement time for electrons or protons inside a magnetic surface at r_{eff}, $\tau_p(r_{eff})$, is defined as

$$\tau_p(r_{eff}) = \frac{\int^{r_{eff}} n_e(\mathbf{r})\,\mathrm{d}V}{\oint^{r_{eff}} \mathbf{\Gamma}(\mathbf{r})\cdot\mathrm{d}\sigma} = \frac{\int^{r_{eff}} n_e(\mathbf{r})\,\mathrm{d}V}{\int^{r_{eff}} \nabla\cdot\mathbf{\Gamma}(\mathbf{r})\,\mathrm{d}V} \tag{9}$$

where $n_e(\mathbf{r})$ and $\Gamma(\mathbf{r})$ are the electron density and the outward flux of electrons at position $\mathbf{r}$, respectively, $\int^{r_{eff}} \cdots\, \mathrm{d}V$ means the volume integral inside the surface at r_{eff}, and $\oint^{r_{eff}} \mathbf{\Gamma}\cdot \mathrm{d}\sigma$ is the surface integral of $\mathbf{\Gamma}$ over the same surface. With the help of continuity equation, i.e.,

$$\frac{\partial n_e(\mathbf{r})}{\partial t} + \nabla\cdot\mathbf{\Gamma}(\mathbf{r}) = S(\mathbf{r}), \tag{10}$$

where $S(\mathbf{r})$ is the ionization rate of hydrogen atom, Eq. (9) is rewritten as

$$\tau_p(r_{eff}) = \frac{\int^{r_{eff}} n_e(\mathbf{r})\,\mathrm{d}V}{\int^{r_{eff}} S(\mathbf{r})\,\mathrm{d}V - \int^{r_{eff}} \frac{\partial n_e(\mathbf{r})}{\partial t}\,\mathrm{d}V}. \tag{11}$$

From the ionization rate S in Fig. 3 (a) and n_e profile by the Thomson scattering measurement, we have evaluated the three integrals in Eq. (11) and have derived $\tau_p(r_{eff})$. The results are shown in Fig. 4. In the strong field case, τ_p decreases as going outward and the decay speed is suddenly increased at around $r_{eff} = 0.6\,\mathrm{m}$. This is understandable because that location approximately corresponds to the position of the last closed magnetic flux surface. In the weak field case, the absolute confinement time is about one order smaller and its decay speed as going outward is larger than those in the strong field case. These results qualitatively agree with general understanding of the dependence on the magnetic field strength of the particle confinement time.

Journal of Physics: Conference Series **397** (2012) 012023 doi:10.1088/1742-6596/397/1/012023

References

[1] Colchin R J et al. 2000 *Nucl. Fusion* **40** 175
[2] Boivin R L et al. 2001 *Rev. Sci. Instrum.* **72** 961
[3] Stotler D P et al. 1996 *Phys. Plasmas* **3** 4084
[4] Kubo H et al. 1998 *Plasma Phys. Control. Fusion* **40** 1115
[5] Koubiti M et al. 2002 *Plasma Phys. Control. Fusion* **44** 261
[6] Iwamae A et al. 2005 *Phys. Plasmas* **12** 042501
[7] Stangeby P C 2000 *The Plasma Boundary of Magnetic Fusion Devices* (London: IOP Publishing)
[8] Goto M, Sawada K, Fujii K, Hasuo M and Morita S 2011 *Nucl. Fusion* **51** 023005
[9] Goto M and Morita S 2006 *Rev. Sci. Instrum.* **77** 10F124
[10] Bellman R E, Kalaba R E and Lockett J A 1966 *Numerical Inversion of the Laplace Transform* (New York: American Elsevier)

Changes in atomic populations due to tokamak edge plasma fluctuations

R Hammami[1], H Capes[1], F Catoire[2], L Godbert-Mouret[1], M Koubiti[1], Y Marandet[1], A Mekkaoui[3], J Rosato[1] and R Stamm[1]

[1] PIIM, Aix-Marseille Université and CNRS, centre Saint Jérôme, F-13397 Marseille Cedex 20, France.
[2] CELIA, Université de Bordeaux – CNRS - CEA, Talence, 33405, France
[3] IEK-4 Plasma Physics, Forschungszentrum Jülich GmbH, Trilateral Euregio Cluster, Euratom Association, D-52425 Jülich, Germany

E-mail: ramzi.hammami@univ-amu.fr

Abstract. The population balance of atoms and ions in a tokamak edge plasma is calculated in the presence of density and temperature fluctuations. We apply a stochastic model to simplified atomic systems and examine the sensitivity of the energy level populations to the statistical properties of the fluctuations.

1. Introduction

The energy level populations of atoms and ions are important quantities for characterizing the edge plasmas of magnetic fusion devices such as tokamaks. Radiation losses, ionization equilibrium, or the intensity of spectral lines used for diagnosing the plasma are directly related to the atomic and ionic populations. Such quantities are used to understand how the plasma interacts with the plasma facing materials. Their determination relies on the solution of an atomic population kinetic model, which is usually obtained using a stationary collisional-radiative model (CRM), assuming constant plasma parameters such as the electronic density and the temperature of each species [1]. In this work we examine the case where these parameters fluctuate as a consequence of edge plasma turbulence. In edge plasmas, density and temperature fluctuations of order unity on time scales of the order of 10 μs have been measured, and characterized with statistical quantities such as the probability density function (PDF) and the time correlation function [2]. We consider a previously developed model, which assumes that the fluctuating quantity obeys a stochastic process [5]. The density or temperature are distributed according to a prescribed PDF, and jump from one constant value to another one according to a given waiting time distribution (WTD) [5]. Assuming such a process for the fluctuations, we can solve the time dependent CRM by an analytical model or a Monte Carlo type simulation. In section 2, we briefly describe how atomic populations may be calculated in the presence of the renewal stochastic process for the fluctuations reported in [7]. Our model is applied to two ideal atomic systems in section 3, a reduced hydrogen atom, and the carbon ionization balance.

2. Modeling the plasma parameters with a renewal process

The atomic or ionic populations are described by a vector $X(t)=(x_1(t),\ldots,x_N(t))$ containing the different population x_j of each level j with an initial value $X_0 = X(t=0)$. The kinetics of the atomic populations is governed by a differential equation of the form [5]:

$$\frac{dX(t)}{dt} = M(Y(t))X(t), \tag{1}$$

where M is a time dependent matrix filled with the collisional-radiative rates, and $Y(t)$ is the fluctuating plasma parameter (electronic temperature or density in the following). We now assume that Y follows a renewal process with stepwise constant values separated by instantaneous jumps. The duration on each step obeys a prescribed waiting time distribution (WTD), and the values taken by Y are distributed along a chosen probability density function (PDF) $P(Y)$. Equation (1) is then a linear stochastic equation, which may be solved by a numerical simulation or an analytic approach. The analytic solution takes a particularly simple form if an exponential WTD $\varphi(t) = \nu\, exp(-\nu t)$ is chosen where ν the typical fluctuation frequency of Y. It may be expressed as the Laplace transform of the Green function G associated to equation 1:

$$\langle G(s)\rangle = [I - \nu G_{ST}(s)]^{-1} G_{ST}(s), \tag{2}$$

where the brackets $\langle\cdots\rangle$ imply that an ensemble average has been taken, and the subscript ST means that a static average over the PDF $P(Y)$ has been taken. It is possible to obtain the average $\langle X(t)\rangle$ of the population vector in the limit of long times by using:

$$\lim_{t\to\infty}\langle X(t)\rangle = \lim_{s\to 0^+} s\langle G(s)\rangle X_0 . \tag{3}$$

The solution of equation (1) may also be obtained by a numerical solution coupled to a simulation of the chosen stochastic process. The fluctuating plasma variable is again assumed to follow a stepwise constant stochastic process with prescribed PDF and WTD, but we now generate variable histories on the computer with pseudorandom number algorithms, associated to numerical techniques such as transformation or rejection methods [8]. In the following, the calculations are performed assuming an exponential WTD with an average time duration of steps equal to the inverse of the typical frequency ν of the fluctuations. This entails that we have chosen a process, which loses all memory of its previous states at every time. Other stochastic processes may be considered by our approach if we expect memory effects to be significant [11].

Starting with an initial population X_0 at $t=0$, we first apply during a time interval Δt_1 a constant value Y_1 of the fluctuating parameter. The population X_1 at time Δt_1 may be written as:

$$X_1 = e^{-M(Y_1)\Delta t_1} X_0 \tag{4}$$

For a given time history of the variable Y, we may then iterate this expression until we reach a time for which the value of X is constant and a stationary result has been obtained. We have averaged the values of $X(t)$ over a set of such time histories for a fixed value of ν. The number of histories required for obtaining a stable result depends on the number of time jumps required before we reach the stationary result. According to the values of the fluctuation and atomic frequencies, about 10 to 10^4 histories may be required for obtaining results being reproducible to within 1 percent. Our simulation technique, although much less computer efficient than our analytic model, is flexible since it allows to rapidly switch between different PDF and WTD already available in numerical libraries. The two approaches have been used to crosscheck several of our results.

3. Results and discussion

3.1 The gamma PDF

The three level system 1s, 2s, 2p of hydrogen has been recently analysed for the case of temperature fluctuations. Changes in the population of the 2s level depend on the typical frequency and the rate $r = \frac{\Delta T}{\langle T \rangle}$ of the electron temperature fluctuations, where $\Delta T = \sqrt{\langle T^2 \rangle - \langle T \rangle^2}$, and the average electron temperature is given by $\langle T \rangle = \int_0^\infty TP(T)dT$ with $P(T)$ the PDF of the fluctuating temperature. Assuming a gamma function for the PDF $P(T)$ is written:

$$P(T) = \frac{\alpha^\beta}{\Gamma(\beta)} T^{\beta-1} \exp(-\alpha T). \tag{5}$$

In reference [5], it has been shown that significant changes in the 2s population are observed for a fluctuating plasma temperature. Such changes depend on the typical frequency ν of the fluctuations, and reach two limits as ν is small (static case) or large (diabatic case) compared to the typical atomic frequencies given by the collisional and radiative processes. If the analysis of a specific plasma device is performed, the average frequency ν (the inverse of the turbulence correlation time) is in principle known and determined from the measure of the correlation function of the fluctuating variable.

3.2 Three level hydrogen atom

In the following we first use the same statistical properties of the fluctuating temperature, and study a three level hydrogen atom with the principal quantum levels n=2, 3 and 4. The collisional excitation rate employs the simple formula for allowed transitions found in the work of Van Regemorter [11]. That is, for the collisional frequency C_{ij} from a state i to a state j:

$$C_{ij} = N_e 2.5 \times 10^{-11} \frac{1}{E_{ij}\sqrt{T}} e^{-\frac{E_{ij}}{T}} f_{ij}, \tag{6}$$

where E_{ij} is the energy separation of the states i and j, f_{ij} is the absorption oscillator strength, and T the electron temperature in eV. A principle of detailed balance is used for obtaining de-excitation frequencies.

Our first result concerns temperature fluctuations obeying a gamma distribution with an average temperature of 2 eV, and r=0.9. The average populations (Eq. (4)) of the levels n=3, 4 plotted as a function of the fluctuation frequency ν are shown in figure 2 for a density of $N_e=10^{19}$ m^{-3}.

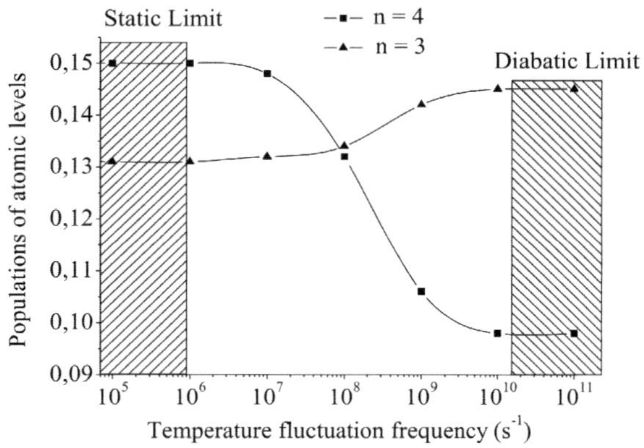

Figure 2. Populations of the n=3 and n=4 levels of a three level hydrogen atom, as a function of the fluctuation frequency of the temperature, and for a fixed electron density of $N_e=10^{19}$ m^{-3}.

This couple of density and temperature sets us in a regime for which both radiative and collisional processes contribute to the population kinetics. For frequencies ν lower than about 10^6 s^{-1}, we are in the static regime, for which $\nu<<\nu_{at}$. One enters the opposite diabatic limit where $\nu>>\nu_{at}$ for frequencies ν higher than about 10^{10} s^{-1}. In between those two asymptotic regimes, it can be seen in figure 2 that the population of the n=3 level increases with increasing frequencies by 11%, while in the same range the population of n=4 decreases by 31%. In our closed system, the population of the n=2 level changes in order to maintain a constant total population of atoms. If we consider the ratio of level populations n=3 to n=4 (the Balmer α to Balmer β intensity ratio), we find that it increases by about 70% as we vary the frequency from the static to the diabatic regime.

We now examine the effect of electronic density fluctuations which are assumed to follow a gamma distribution given by equation (6), with $r = 0.9$.

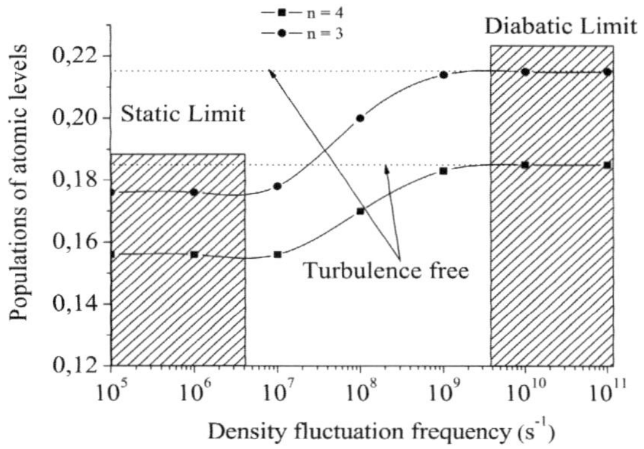

Figure 3. Populations of the n=3,4 levels of a three level hydrogen atom, as a function of the fluctuation frequency of the density, and for a fixed electron temperature of 2 eV.

In figure 3, we have plotted the stationary populations of levels n=3, 4 for an average density of 10^{19} m^{-3}, and a temperature of 2eV. A change of population is also observed between the static and diabatic limits, with a simultaneous increase of the two levels as the frequency is increased, by 19% and 22% for respectively levels n=3 and 4. Moreover, we note from figure 3, that the diabatic limit tends to the regime without plasma electron density fluctuations. Our simulation of the stochastic process is in a perfect agreement with the analytical results. Indeed, we have shown [5] that solving equation (1) for the diabatic regime, is equivalent to solve the following equation:

$$\langle M \rangle \lim_{s \to 0} s \langle \tilde{G}(s) \rangle X_0 = \langle M \rangle X_{DB} = 0, \tag{7}$$

where

$$X_{DB} = \lim_{s \to 0} s \langle \tilde{G}(s) \rangle X_0, \tag{8}$$

is the population vector for the diabatic regime. From the two cases studied, it appears difficult to predict in a general case the behaviour of a set of populations submitted to a fluctuating plasma parameter. A specific study with our model is required for each plasma condition analysed.

3.3 Application to the carbon ionisation equilibrium

Neglecting particle transport, we have studied the ionization balance of carbon in the presence of temperature fluctuations. The equation of ionization equilibrium without transport can be written for the average population of the ion with charge Z:

$$\frac{d\langle n_Z \rangle}{dt} = -\langle n_Z N_e \alpha_{Z \to Z+1} \rangle - \langle n_Z N_e \beta_{Z \to Z-1} \rangle + \langle n_{Z-1} N_e \alpha_{Z-1 \to Z} \rangle + \langle n_{Z+1} N_e \beta_{Z+1} \rangle \tag{9}$$

where $\alpha_{Z \to Z'}$ and $\beta_{Z \to Z'}$ are respectively the ionization and recombination coefficients.

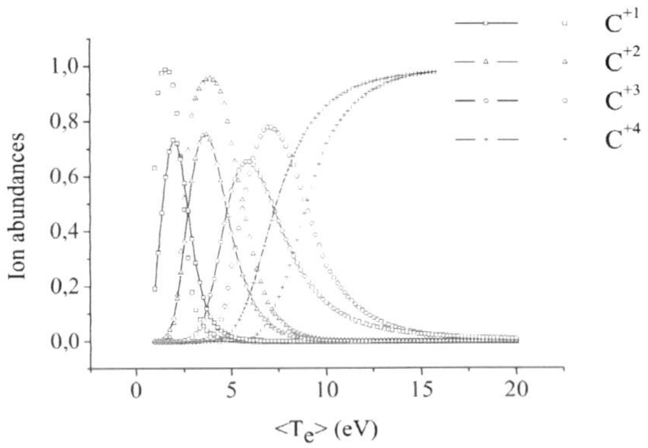

Figure 4. Different carbon ions populations (from C^{+1} to C^{+4}) as a function of the averaged temperature are plotted with (solid lines) and without (dotted lines) temperature fluctuations.

These data are taken from the ADAS database [12]. The fluctuation frequency used in our calculation is equal to $10\mu s^{-1}$ which is a typical fluctuation frequency measured in edge plasma of current magnetic fusion devices. The PDF of temperature is assumed to be a gamma distribution with a fluctuation rate $r = 0.5$. We show in Fig.4 the abundances of several carbon ions (from C^{+1} to C^{+4}) as a function of the average temperature in the range 1 eV to 20 eV, i.e. conditions which can be found in divertor plasmas. Changes in the ion populations are clearly seen, with shifts in the position of the maximum abundance modifying the ratio of the different ion species for a given temperature. Fluctuations have therefore to be retained in analysis aimed at measuring particle transport from the abundances curves, since the latter are sensitive to both of them, provided the fluctuation rate is large enough.

4. Conclusion

The effects of temperature and density fluctuations on the population of ideal atomic or ionic systems have been investigated. We have shown that significant changes occur in the populations of a three level system modeling hydrogen Balmer emission as we scan the typical frequency fluctuation of temperature or density. The ionization balance of carbon has also been studied in the presence of temperature fluctuations. The abundances of the different ions are modified in the presence of such temperature fluctuations. All these effects are significant only for fluctuation rates r larger than several tens of percent, values which can be found in tokamak edge plasmas. With the aim of achieving a fluctuating edge plasma diagnostic, further studies will concern more realistic atomic system, which will require retaining more atomic processes and a more detailed atomic structure.

Acknowledgments
This work was carried out within the framework of the European Fusion Development Agreement and the French Research Federation for Fusion Studies. It is supported by the European Communities under the contract of Association between Euratom and CEA. The views and opinions expressed herein do not necessarily reflect those of the European Commission. We also acknowledge the support of the French National Research Agency (contract ANR-11- BS09-023, SEDIBA).

References

[1] Fujimoto T 2004 *Plasma Spectroscopy*, (Oxford: Clarendon Press)
[2] Graves J P, Horacek J, Pitts R A and Hopcraft K I 2005 *Plasma Phys. Control. Fusion* **47** L1
[3] Dudson D et al. 2008 *Plasma Phys. Control. Fusion* **50** 124012
[4] Rosato J, Capes H, Catoire F, Kadomtsev M B, Levashova M G, Lisitsa V S, Marandet Y, Rosmej F B and Stamm R 2011 *J. Nucl. Mater.* **415** S617
[5] Catoire F, Capes H, Marandet Y, Mekkaoui A, Rosato J, Koubiti M and Stamm R 2011 *Phys. Rev. A.* **83** 012518
[6] Catoire F, Capes H, Rosato J, Marandet Y, Mekkaoui A, Koubiti M and Stamm R 2011 *Eur. Phys. J. D.* **65** 481
[7] Lax M and Scher H 1977 *Phys. Rev. Lett.* **39** 781
[8] Vesely F J 1994 *Computational Physics, an Introduction* (New York: Plenum Press)
[9] IMSL 2012, http://www.roguewave.com
[10] Rosato J, Catoire F, Marandet Y, Mekkaoui A, Capes H, Koubiti M, Stamm R, Kadomtsev M, Levashova M G, Lisitsa V S and Rosmej F B 2011 *Phys. Lett. A.* **375** 4187
[11] Van Regemorter H 1962 *Ap. J.* **136** 906
[12] Summers H P 2004 The Atomic Data and Analysis Structure manual, version 2.6, http://www.adas.ac.uk/

Contribution of Stark-Doppler broadening of carbon impurity lines to the analysis of tokamak divertor plasmas

M Koubiti[1], T Nakano[2], Y Marandet[1], L Mouret[1], J Rosato[1] and R Stamm[1]
[1]PIIM, UMR 7345 CNRS/Université d'Aix-Marseille, Campus de St. Jérôme, service 232, Av. Escadrille Normandie-Niemen, F-13397 Marseille Cedex 20, France
[2]Japan Atomic Energy Agency, 801-1, Mukoyama, Naka, Ibaraki, 311-0193, Japan

E-mail: mohammed.koubiti@univ-amu.fr

Abstract. We investigate the broadening mechanisms affecting carbon lines emitted from tokamak divertor plasmas with an emphasis on the C IV n=6-7 (λ=772.6 nm) line. We show how line profiles can contribute to the spatial characterization of tokamak divertors providing high-resolution line spectra along viewing chords can be measured. For simplicity, we ignore the Zeeman effect by considering only Doppler and Stark broadenings in the calculations. Therefore these calculations, which ignore the magnetic field effect, can be compared to spectra measured with the use of a linear polarizer transmitting only the π component whose polarization is parallel to the toroidal magnetic field. Comparison of theoretical profiles to high-resolution measured C IV n=6-7 line spectra allow the determination of the several plasma parameters.

1. Introduction

In magnetic fusion devices, the huge heat load escaping from the core plasma towards the walls can be damageable for their Plasma Facing Components (PFCs). To protect the PFCs, several scenarios are possible, but the most promising one consists in the radiative cooling of the divertor. In such a scenario, evaluating the power requires the knowledge of the plasma parameters with the highest possible accuracy. This can be achieved using spectroscopic measurements such as line intensities and line profiles. In several magnetic fusion devices like the JT-60U tokamak, whose wall and targets are made of carbon, radiative cooling is ensured by line radiation from carbon impurities, mainly the C^{2+} and C^{3+} ions [1-2]. Since it is focused on the broadening and shape of the C IV n=6-7 λ 772.6 nm line spectra, the work presented here is limited to magnetic fusion devices equipped with carbon targets. The aim of the paper is to show how theoretical profiles can be used to fit measured C IV n=6-7 λ 772.6 nm line spectra and contribute to the characterization of tokamak divertor plasmas. For simplicity, the Zeeman effect is ignored and only Doppler and Stark broadenings are accounted for in the calculations, in addition of course to the instrumental function. Therefore these calculations, ignoring the magnetic field effect, can be compared to spectra measured with the use of a linear polarizer transmitting only the π component whose polarization is parallel to the toroidal magnetic field. As this is the case of JT-60U, the present method was applied to measurements from this device [3-5]. The calculations of line profiles are carried out using the PPP line shape code [6]. To simplify the ion and electron temperatures are assumed equal. In addition, in the case of JT-60U data, the electron temperatures were determined from line intensities of low-resolution C IV VUV and visible spectra [1-2]. This has allowed the reduction of the number of fitting parameters as will be explained later.

2. Theoretical background and method of investigation

In this work, all synthetic profiles of the C IV n=6-7 line are calculated considering only the Stark and Doppler effects in addition to the instrumental function. For Stark broadening, we have used the standard model relying on the two well-known approaches: the impact and quasi-static approximations for the interactions of the emitter with the plasma electrons and ions respectively. The plasma conditions of the emission zone are such that there is no need to include the ion dynamics effect which can be accounted for by the PPP line shape code through its Frequency Fluctuation Model (FFM). As a consequence of ignoring the Zeeman effect in the calculations, the number of parameters governing the line profile is reduced to three (the electron density N_e, the electron and ion temperatures T_e, T_i). In these conditions, the line shape is the result of the competition between Stark and Doppler broadenings. Note that if an increase of the electron density results in the increase of Stark broadening, the consequence of an increase of the temperature on the line profile is not obvious. Indeed, by increasing the temperature we increase the Doppler effect but decrease the Stark broadening. This is shown in Figures 1 and 2, where synthetic profiles of the C IV n=6-7 (772.6 nm) accounting for Stark effect are compared to those accounting for both Stark and Doppler effects for a hydrogen plasma. Figure 1 corresponds to a hydrogen plasma with N_e=4x10^{14} cm^{-3} and T_e=T_i=5 eV. In Figure 2, in addition to the Stark-Doppler profile shown in Figure 1, we show pure Stark and Stark-Doppler profiles calculated for a higher density plasma (N_e=6x10^{14} cm^{-3} and T_e=T_i=3 eV).

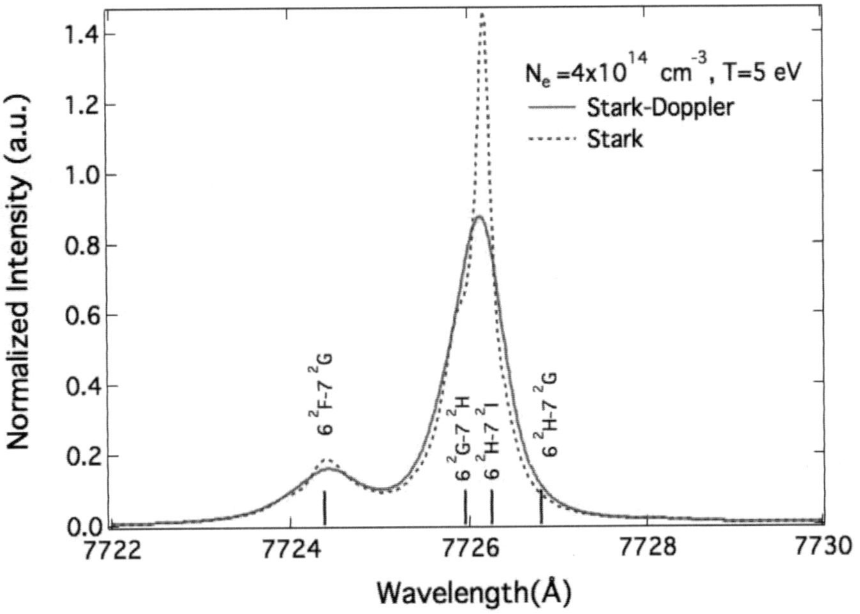

Figure 1. Comparison of calculated Stark-broadened profiles of the C IV n=6-7 line with and without the Doppler effect for typical divertor plasma parameters. Vertical short lines represent the various transitions of the line which fall in the range of interest: 7722-7730 Å. The effect of Stark broadening is visible on both peaks of the line.

In practice, the number of free parameters necessary to fit a C IV n=6-7 line spectrum with a single theoretical profile reduces to one (the electron density) if the electron and ion temperatures are equal and known. This is the case for JT-60U, where the electron temperatures are deduced from the measured intensities of low-resolution C IV line spectra coupled to a collisional-radiative model.

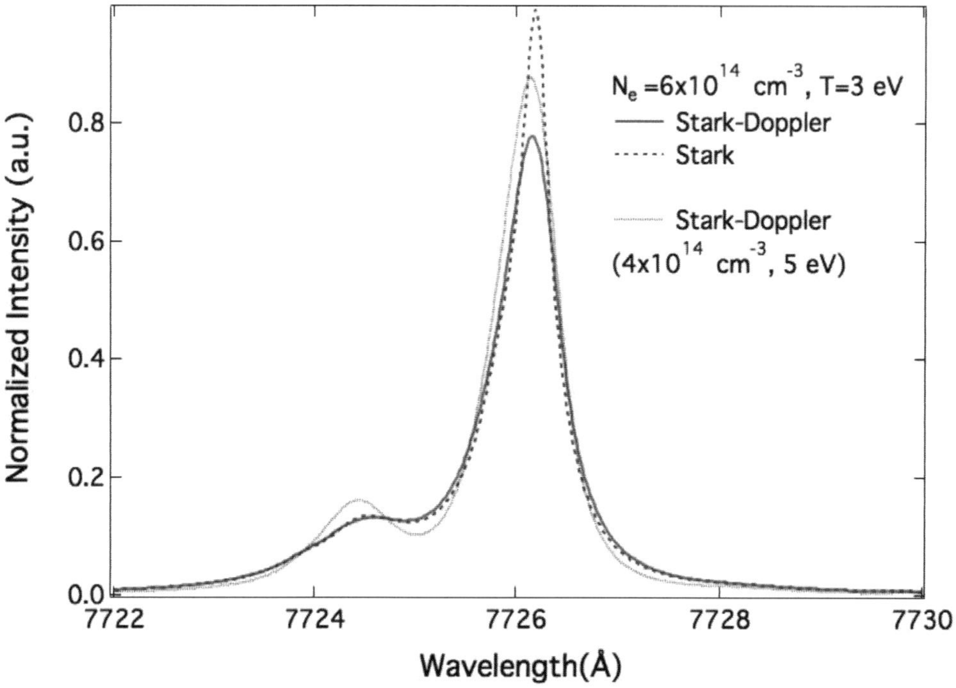

Figure 2. Same as Figure 1 but for a higher plasma density. For this high density, the main difference between Stark and Stark-Doppler profiles is visible only on the highest peak. However, the dip between the two peaks is reduced when the electron density (and therefore the Stark broadening) is increased.

3. Results and discussion

A great number of theoretical profiles of the C IV n=6-7 line were calculated as described above for a series of plasma parameters relevant to tokamak divertors, i.e., for electron densities and temperatures in the ranges $N_e=10^{14}$-10^{15} cm^{-3} and $T_e=T_i=1$-30 eV. In order to be used for comparison with JT-60U data, the appropriate instrumental function was taken into account. The results of the applicability of the method to high-resolution measurements from JT-60U were already published elsewhere [3-5] and therefore we will only recall here the main results. It has been shown that depending on the intersection of the viewing chord and the X-point, a measured spectrum is either fitted considering a single profile (a single homogeneous plasma layer) or two profiles resulting from two homogeneous plasma layers with different plasma parameters. In principle, fitting a measured spectrum with a sum of two profiles requires the determination of five parameters if no assumption is made. As this is complicated, a fitting procedure was developed and two parameters were fixed on the basis of other measurements. This has allowed the determination of the remaining plasma parameters and the relative contribution of each layer to the line emission. As it can be seen in [3-5], the use of line profiles has allowed to complete the spatial characterization of the divertor by providing electron densities which are in the range 1-8x10^{14} cm^{-3}. By considering two homogeneous plasma layers instead of one, we have somehow relaxed the assumption of the homogeneity of the emission zone. We think that our two layer model is a good approximation to a calculation with a fully inhomogeneous plasma. It should be noted that lines other than the C IV n=6-7 can be used for the same purposes if they are subject to Stark broadening, i.e., resulting from levels with a high principal quantum number, and if their emission intensity is high enough for high wavelength resolution measurement. Note that even if the calculations presented here were carried out without the Zeeman effect, it is possible to include it with the PPP line shape code. Therefore the described method can be easily extended to measurements from other machines with or without the use of polarizer provided the spectra are measured with a sufficiently high-resolution.

4. Conclusion

It was suggested to use synthetic profiles of the C IV n=6-7 line for fitting high-resolution spectra of the same line in order to infer some plasma parameters of the emission zone. A simplified case was considered where the synthetic profiles were calculated considering only the Stark and Doppler broadenings but without the Zeeman effect. Such a case is convenient if spectra are measured using a linear polarizer and a high-resolution spectrometer as it was the case of JT-60U. Although not shown here, the use of theoretical profiles has contributed to the spatial characterization of the divertor plasma of JT-60U by providing average electron densities along several viewing chords of the spectral measurements.

Acknowledgments
This work is a collaboration between PIIM laboratory (Aix-Marseille University) and JAEA (Japan). It was carried out within the framework of the European Fusion Development Agreement and the French Research Federation for Fusion Studies. It is supported by the European Communities under the contract of Association between Euratom and CEA. The views and opinions expressed herein do not necessarily reflect those of the European Commission.

References

[1] Nakano T, Kubo H, Asakura N, Shimizu K, Kawashima H and Higashijima S 2009 *J. Nucl. Mater.* **390-391** 255
[2] Nakano T, Kubo H, Asakura N, Shimizu K, Konoshima S, Fujimoto K, Kawashima H and Higashijima S 2007 *Nucl. Fusion* **47** 1458
[3] Koubiti M, Nakano T, Godbert-Mouret L, Marandet Y, Rosato J and Stamm R 2011 *J. Nucl. Mater.* **415** S1151
[4] Koubiti M, Nakano T, Capes H, Marandet Y, Mekkaoui A, Mouret L, Rosato J and Stamm R 2012 *Contrib. Plasma Phys.* **52** 455
[5] Koubiti M, Nakano T, Marandet Y, Mekkaoui A, Mouret L, Rosato J and Stamm R, spatial characterization of a divertor in the presence of an X-point MARFE by collisional-radiative modeling and line profile calculations, submitted to *J. Nucl. Mater.*
[6] Talin B, Calisti A, Godbert L, Stamm R, Lee R W and Klein L 1995 Phys Rev A **51** 1918

XXI International Conference on Spectral Line Shapes (ICSLS 2012) IOP Publishing
Journal of Physics: Conference Series **397** (2012) 012026 doi:10.1088/1742-6596/397/1/012026

Stark parameters of some asymmetrical Si II lines

B Ferhat[1] Y Azzouz[1] R Redon[2] M Ripert[2] and A Lesage[3]

[1]*LEQ,Faculté de Physique, U.S.T.H.B, BP 32 El Alia,16111, Alger, Algérie*

[2]*Université du Sud Toulon-Var, PROTEE, EA 3819, 83957 La Garde Cedex, France*

[3] *Observatoire de Paris, GEPI, 1 Place Jules Janssen 92195 Meudon, France*

E mail : belkacem.ferhat@gmail.com

Abstract. Six lines of SiII are experimentally studied in pulsed plasma generated by Nd :Yag laser breakdown on pure solid silicon target. A set of experimental Stark parameters of asymmetrical lines are measured in temperature range from 14 000 K to 18 000 K (using Boltzmann plot). Calculated values of the electron density (using Griem's formula) vary from 1.7 to 6.1 x$10^{23}$$m^{-3}$. Processed spectral lines are 333.982 nm ($3s^2 4p$ -$3s^2 6s$) and 397.746 nm, 399.177 nm, 399.801 nm, 401.622 nm (3d'$^2F^0$- 4f'^{4}G) and (3d'$^2F^0$- 4f'^{2}G) of astrophysical interest. Asymmetrical line shapes are synthesized by a sum of two semi-Lorentzian distributions. The obtained fit is in good agreement with the measured spectra.

1. Introduction

Stark parameters values of silicon lines are needed [1, 2] for the knowledge of the atmosphere physical conditions in low temperature silicon stars at temperature from 10 000 to 20 000 K. Various theories [3-10] provide calculated values of symmetrical lines of Lorentzian shape. But the presented profiles are not. That is a characteristic which has been studied before [11-15] but in the present case, up to now, there is no evident explanation.

Under the experimental conditions, almost all observed lines are of asymmetrical shape and four lines belonging to the (3d'$^2F^0$- 4f'^{4}G) and (3d'$^2F^0$- 4f'^{2}G) multiplets are studied for the first time.

2. Experimental set up

The data acquisition was performed [16] by mean of Nd:Yag laser breakdown on solid silicon (10ns pulse width, 300 mJ energy). The silicon sample of purity 99.999 % was a cylindrical rod of 5 mm in diameter and 50 mm in length. The sample, placed in the vacuum cell, was fixed to a helicoidally displacement device built up with a threaded rod, driven by a step motor (Figure. 1).

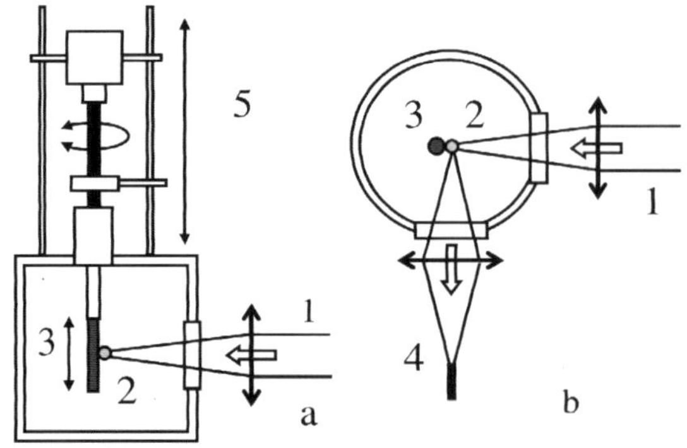

Figure.1: Helicoidally displacement device and breakdown cell:
1 - laser beam. 2 – plasma. 3 – target. 4 - optical bundle.

The rotation rate was set by a Hewlett Packard TTL pulses generator which frequency was set as a function of the repetition rate of the laser pulses. The laser beam was focused at regular intervals onto the surface of the cylindrical sample and the repeatability of the plasma controlled by observing the spectral intensity of the observed lines. A monochromator TRIAX 180 from Jobin-Yvon with a 2400 grooves per mm grating (blaze: 330nm and slit 25 μm) was coupled to an UV-visible 1024×256 ICCD camera from Jobin-Yvon. The resulting resolution was 30 pm per pixel. The bundle was built of 16 optical fibers of 200 μm, with a circular entrance aperture and a linear exit at the monochromator slit. The ICCD was intensified during 200 ns, the delay after the laser pulses was 125 ns and 295 ns for the plasma in vacuum and in Xenon, respectively. For a given set of parameters each spectrum was recorded in accumulative mode for 100 laser pulses, over the 256 lines of the ICCD. These measurements were repeated 20 times. Then, the plasma spectrum was calculated as the median of those 20 spectra.
The apparatus function was measured using Hg or Ar lamps and the thermal noise was measured at the beginning and at the end of each acquisition sequence. Few quantities of Xenon were added to determine the temperature, using Boltzmann plot of the Xe II lines. The accuracy of measured values is about 10 % due to the uncertainties induced by the spectrometric technique.

3 Reduction of the data

The unfolding procedure of Si II lines profiles was done using a Matlab program according to the algorithm described below: the two first steps constitute the spectra pretreatment and the following steps permit to calculate a modeled spectrum which is the sum of synthetic background and synthetic lines. For each studied multiplet, these contributions are depending on parameters which are optimized by the Simplex method in order to minimize the quadratic error between the real and modeled spectrum.

1. Once the wavelength calibration is performed, the program used all thermal noise files of the ICCD to calculate the median noise.
2. Then, it proceeded to the listing of all the files of the real spectra to classify them by laser beam incident energy (and therefore by plasma temperature). The adopted spectrum, for each laser energy, was the median spectrum from which the median noise spectrum must be subtracted.
3. The synthetic background is a second-degree polynomial whose coefficients are estimated from the real one. These coefficients are thereafter optimized using the Simplex method.
4. A synthetic line is made using a contribution of two semi-Lorentzian distributions : the left half profile is characterized by the half-width γ_1 and the half right profile has an half-width γ_2. This profile model successfully reproduced the observable data (best fit of the real spectrum) and has not yet theoretical justification. The half-widths of synthetic lines are optimized by the Simplex method.
5. The final synthetic spectrum is the sum of each synthetic line convoluted by the measured apparatus function added to the synthetic background.

4. Results

Series of very stable and repeatable spectra of singly ionized silicon were obtained for electron density and temperatures ranges: $N_e = 1{,}70 - 6.10^{23}\ m^{-3}$; T=14 000 K to 18 000 K.
A check for a temperature gradient along the radius of the discharge using an Abel inversion did not detect any in-homogeneity which can be, eventually, responsible of the asymmetry and no self-absorption was detected. It was noticed that the line at 399.177 nm belonging to the same multiplet was of symmetrical shape for all the temperature ranges. Due to the asymmetric shape of most of the SiII profiles the Griem's formula was used to measure the electron density, instead of the SiII line width.
Figures 2 and 3 report observed lines in a large range of wavelength where we note, particularly, the faint lines resulting from the transitions ($3d'^2F^0$- $4f'^4G$) and ($3d'^2F^0$- $4f'^2G$). Figure.4 reports the real and synthesized spectra of the doublet ($3s^24p$ -$3s^26s$) and those of the faint lines ($3d'^2F^0$- $4f'^4G$) and ($3d'^2F^0$- $4f'^2G$) are reported in the figure 5. In spite of their low intensity, they are recorded without saturation of the detector. For all multiplets, the smooth and continuous line indicates the modeled spectrum.

Table: Measured values of Stark parameters

T= 14000 °K	λ_0 (nm)	λ_m (nm)	$\Delta\lambda$(nm)	I_{max} (r.u)	γ_1 (nm)	γ_2 (nm)
	333.314	333.292	0.022	992.64	0.142	0.637
	333.982	333.917	0.065	1900.03	0.030	0.600
	397.746	397.718	0.038	555.38	0.300	0.423
	399.177	399.160	0.017	1369.51	0.282	0.282
	399.801	399.763	0.018	788.49	0.215	0.321
	401.622	401.605	0.017	570.93	0.193	0.205
T= 15000 °K	λ_0 (nm)	λ_m (nm)	$\Delta\lambda$(nm)	I_{max} (r.u)	γ_1 (nm)	γ_2 (nm)
	333.314	333.313	0.001	1110.85	0.166	0.650
	333.982	333.920	0.062	2067.71	0.033	0.651
	397.746	397.723	0.023	663.03	0.286	0.405
	399.177	399.171	0.006	1477.58	0.316	0.316
	399.801	399.720	0.079	843.57	0.120	0.420
	401.622	401.616	0.004	581.09	0.239	0.239
T= 16000 °K	λ_0 (nm)	λ_m (nm)	$\Delta\lambda$(nm)	I_{max} (r.u)	γ_1 (nm)	γ_2 (nm)
	333.314	333.339	0.025	1330.21	0.210	0.590
	333.982	333.916	0.066	2213.67	0.041	0.818
	397.746	397.718	0.028	843.64	0.309	0.464
	399.177	399.179	0.002	1906.77	0.364	0.366
	399.801	399.722	0.079	1220.95	0.105	0.424
	401.622	401.612	0.010	723.45	0.282	0.288
T= 17000 °K	λ_0 (nm)	λ_m (nm)	$\Delta\lambda$(nm)	I_{max} (r.u)	γ_1 (nm)	γ_2 (nm)
	333.314	333.334	0.020	1258.51	0.203	0.749
	333.982	333.937	0.045	2241.20	0.038	0.754
	397.746	397.736	0.010	883.98	0.419	0.482
	399.177	399.186	0.009	2070.74	0.401	0.401
	399.801	399.705	0.096	1325.40	0.080	0.487
	401.622	401.612	0.010	770.20	0.310	0.321
T= 18000 °K	λ_0 (nm)	λ_m (nm)	$\Delta\lambda$(nm)	I_{max} (r.u)	γ_1 (nm)	γ_2 (nm)
	333.314	333.429	0.115	1247.80	0.480	0.556
	333.982	333.926	0.056	2186.86	0.049	0.905
	397.746	397.766	0.020	968.62	0.365	0.443
	399.177	399.181	0.004	2490.60	0.310	0.310
	399.801	399.799	0.002	1498.47	0.252	0.348
	401.622	401.562	0.060	901.17	0.195	0.426

Measured values (with a precision of about 10%) of left (γ_1) and right (γ_2) half-widths of 6 spectral lines, as well the Stark shifts, for 5 values of the temperature are reported in table 1.
The first column indicates the central wavelengths according to Striganov and Sventitskii [18]. They correspond, in a more satisfactory way, to the observable values than those brought back from the NIST [19] which, moreover, even mention the spectral characteristics of the transitions (3d' $^2F^0$- 4f' ^4G) and (3d' $^2F^0$- 4f' ^2G) (neither energy levels, neither spectral terms, nor statistical weights etc…)
The second column brings back the wavelengths of maximum observed for the transitions considered, while $\Delta\lambda$ indicates the Stark shift calculated by subtraction between the values of the first and the second columns.

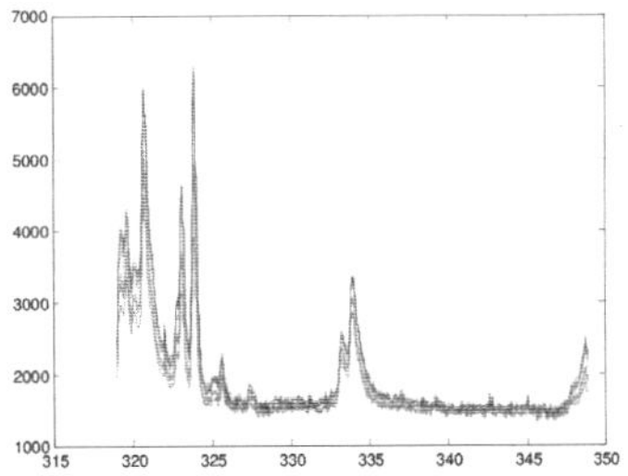

Fig. 2: Observed lines of the doublet $3s^24p$ -$3s^26s$ in large range of wavelength

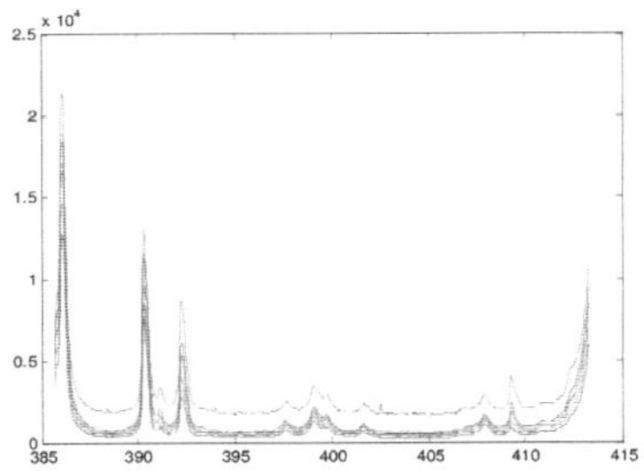

Fig. 3: Observed lines of $3d'^2F^0$- $4f'^4G$ and $3d'^2F^0$- $4f'^2G$ transitions in large range of wavelength

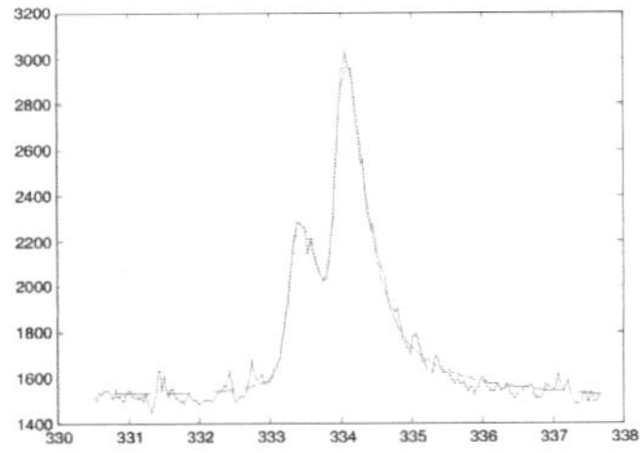

Fig. 4: Real and synthesized spectra of the doublet $3s^24p$ -$3s^26s$

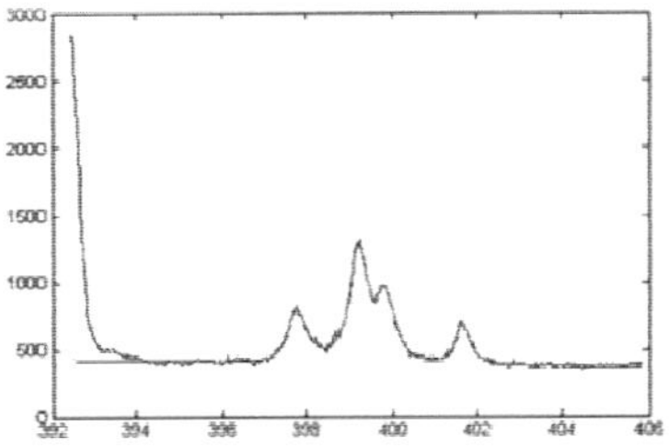

Fig. 5: Real and synthesized spectra of the doublets $3d'^2F^0$- $4f'^4G$ and $3d'^2F^0$- $4f'^2G$

The intensities of the lines are reported in columns 4 and 5 while γ_1 and γ_2 are the two half-widths of the lines according to the above mentioned method of calculation. The numerical values of the half-widths at half maximum indicate that lines are asymmetrical except the line 399.177 nm which is symmetrical for all the values of temperature and the line 401.622 nm asymmetric for T=18000°K only. The Stark shift is very weak for the line 399.177 nm. To our knowledge asymmetry for the presented transitions, neither experimental nor theoretical, has been, up to now, reported.

References

[1] Lanz T and Artru M C 1985 *Physica Scripta* **32** 115
[2] Lesage A and Redon R 2004 *A&A* **478** 765
[3] Griem H R 1968 *Phys.Rev* **165** 258
[4] Griem H R 1974 *Spectral line broadening by plasmas* (New York, Academic press)
[5] Sahal-Brechot S 1969a *A&A* **1** 91
[6] Sahal-Brechot S 1969b *A&A* **2**,322
[7] Lanz T Dimitrievic M S and Artru M C 1988 *A&A* **192** 249
[8] Ralchenko Yu V Griem H R and Igor Bray 2003 *JQSRT* **81** 371
[9] Ralchenko Yu V Griem H R Igor Bray and Dimitri V Fursa 1999 *Phys. Review A* **59** 3
[10] Alexiou S 1997 in *13th Int. Conf. on Spectral Line Shapes*, ed. M Zoppi and L Ulivi
[11] Ciurylo R *Physical Review A* **58**, Issue 2, pp. 1029-1039
[12] Torres J Palmarès J M Gigolos M A Gamero A Sola A and Van der Mullen J J A M 2008 *Spectrochimica Acta Part B: Atomic Spectroscopy* **63 (9)** 939
[13] Nikolic D. Djurovic S Mijatovic Z Kobilarov R Vujiic B and Irian M 2004 *JQSRT* **86** (3) 285
[14] Günter S and Könies A 1994 *JQSRT* **52** (6) 819
[15] Szudy J and Baylis W E 1977 *JQSRT* **17** (5) 681
[16] Matheron P Escarguel A Redon R Lesage A and Richou J 2001 *JQSR* **69** 535
[17] Deron C Perrin M-Y and Soufiani A. 2005, in *6th World Conference on Experimental Heat Transfer, Fluid Mechanics and Thermodynamics* Matsushima Miyagi Japan
[18] Striganov A R and Sventitskii N S 1968, in *Tables of Spectral lines of Neutral and Ionized Atoms* IFI/Plenum
[19] Ralchenko Yu Kramida A E Reader J and NIST ASD Team 2008 *NIST Atomic Spectra Database* (version 3.1.4) [Online] National Institute of Standards and Technology Gaithersburg MD

Generalized energy approach in an electron-collisional spectroscopy of multicharged ions in plasma in the Debye approximation

A V Loboda and T B Tkach

Odessa State University – OSENU, P.O. Box 24a, Odessa-9, SE-65009, Ukraine

E-mail: andmolec@mail.ru , quantper@mail.ru

Abstract. The generalized relativistic energy approach is used for the studying the spectra of plasma of the multicharged ions and calculation of electron-ion collision strengths, cross-sections in Ne-like ions. The approach is based on the Debye shielding model relativistic Dirac Hamiltonian for electron-nuclear and electron-electron potentials and energy approach. The relativistic many-body perturbation theory with using gauge-invariant scheme of generation of the optimal one-electron representation is used. The measured and calculated electron-collisional excitation cross-sections for Ne-like barium for value of incident electron energy 8.20keV and the effective collision strengths of the Ne-like krypton excitation states for temperature $T=5\times10^6$K and electron density $n_e==10^{14}cm^{-3}$ are presented and analysed.

1. Introduction

The properties of laser-produced hot and dense plasmas, known as laser plasma, have drawn considerable attention over the last two decades through the recent laser-fusion studies [1,2]. Similar interest is also stimulated by importance of this information for correct determination of the characteristics for plasma in thermonuclear (tokamak) reactors, searching new mediums for X-ray range lasers. In recent years the X-ray laser problem has stimulated a great number of papers devoted to the development of theoretical methods for modelling the elementary processes in collisionally pumped plasma (c.f. [1-25]). The laser effects have been discovered on the transitions of Li-, Ne-, Ni-like and other ions. Very useful data on the X-lasers problem are firstly received and collected in the papers by Ivanova et al [2,13-16]. Calculation of emission spectra of the plasma ions based in the precise theoretical techniques is practical tool, which may be used instead of very expensive sophisticated experiments. An important application of the theory of atomic spectra in plasma is search of the optimum plasma excitation condition for lasing and discovery of new pumping approaches. In addition, these investigations are important to understand the plasma processes themselves. Different atomic levels are populated in laboratory plasma by different physical processes. This results in a different dependence of each line intensity on the plasma parameters. In recent years the X-ray laser problem has stimulated a great number of papers devoted to the development of theoretical methods for modelling the elementary processes in collisionally pumped plasma. There are constructed first lasers with using plasma of Li-, Ne-, Ni-like ions as an active medium. Two principal theoretical problems must be solved in order to develop a special code adequate to predict the plasma parameters needed to generate a soft-X-ray or VUV lasing: i). accurate calculation of electron-collisional strengths and rate coefficients for processes in plasma; ii). kinetics calculations to determine level inversions, intensities, gain coefficients at definite plasma parameters.

In this paper the generalized relativistic energy approach is used for the studying the spectra of plasma of the multicharged ions and calculation of electron-ion collision strengths, cross-sections in Ne-like ions. The approach is based on the Debye shielding model relativistic Dirac Hamiltonian for electron-nuclear and electron-electron potentials and energy approach [2,3,6-8,22]. The relativistic many-body perturbation theory (PT) with using the gauge-invariant scheme of generation of the optimal one-electron representation is used. The measured and calculated electron-collisional excitation cross-sections for Ne-like barium for value of incident electron energy 8.20keV and the effective collision strengths of the Kr^{26+} Ne-like ion excitation states for temperature $T=5\times10^6$K and electron density n_e==$10^{14}cm^{-3}$ are presented and analysed.

2. Energy approach in scattering theory

In the theory of the non-relativistic atom a convenient field procedure is known for calculating the energy shifts ΔE of degenerate states. This procedure is connected with the secular matrix M diagonalization [2,3,16]. In constructing M, the Gell-Mann and Low adiabatic formula for ΔE is used. A similar approach, using the Gell-Mann and Low formula with the QED scattering matrix, is applicable in the relativistic atom theory [2-6]. In contrast to the non-relativistic case, the secular matrix elements are already complex in the PT second order (the first order of the inter-electron interaction). Their imaginary parts are connected with the radiation decay possibility. The total energy shift of the state is usually presented in the form:

$$\Delta E = \mathrm{Re}\Delta E + \mathrm{i}\ \mathrm{Im}\Delta E \tag{1a}$$

$$\mathrm{Im}\ \Delta E = -\Gamma/2 \tag{1b}$$

where Γ is interpreted as the level width.

The whole calculation of the energies and decay probabilities of a non-degenerate excited state is reduced to calculation and diagonalization of the complex matrix M .and definition of matrix of the coefficients with eigen state vectors $B^{IK}_{ie,iv}$ [2-5]. To calculate all necessary matrix elements one must use the basis's of the 1QP relativistic functions. In many calculations of characteristics of the atomic elementary processes it has been shown that adequate description of these characteristics requires using the optimized basis's of wave functions. In ref. [3] it has been proposed "ab initio" optimization principle for construction of cited basis's. There is used the minimization of the gauge dependent multielectron contribution of the lowest QED PT corrections to the radiation widths of atomic levels. In the fourth order of QED PT there appear diagrams, whose contribution into the ImδE accounts for the polarization effects. This contribution describes collective effects and it is dependent upon the electromagnetic potentials gauge (the gauge non-invariant contribution ΔE_{ninv}). The minimization of the functional $Im\ \Delta E_{ninv}$ leads to the integral differential equation, that is numerically solved. In result one can get the optimal one-electron basis of the PT [2,5,6].

The energy approach has been generalized to cover the problems of scattering theory (c.f.[2-5]). Here we briefly outline the main idea considering the collisional de-excitation of the Ne-like ion:

$$((2j_{iv})^{-1}3j_{ie}[J_iM_i],\ \varepsilon_{in})\to(\Phi_o,\varepsilon_{sc}).$$

Here Φ_o is the state of the ion with closed shells (ground state of the Ne-like ion); J_i is the total angular moment of the initial target state; indices iv, ie are related to the initial states of vacancy and electron; indices ε_{in} and ε_{sc} are the incident and scattered energies, respectively to the incident and scattered electrons. The initial state of the system "atom plus free electron" can be written as

$$|I>=a^+_{in}\sum_{m_{iv},m_{ie}} a^+_{ie}a_{iv}\Phi_o C^{J_i,M_i}_{m_{ie},m_{iv}} \tag{2}$$

Journal of Physics: Conference Series **397** (2012) 012027 doi:10.1088/1742-6596/397/1/012027

Here $C^{J_i,M_i}_{m_{ie},m_{iv}}$ is the Clebsh-Gordan coefficient. Final state is as follows: $|F>= a^{+}_{sc}\Phi_o$, where Φ_o is the state of an ion with closed electron shells (ground state of Ne-like ion), $|I>$ represents three-quasiparticle (3QP) state , and $|F>$ represents the one-quasiparticle (1QP) state. The justification of the energy approach in the scattering problem is in details described in refs. [2,3,5,15]. For the state (2) the scattered part of energy shift $Im\ \Delta E$ appears first in the second order of the atomic perturbation theory.

The collisional de-excitation cross section is defined as follows [2,5]:

$$\sigma(IK \to 0) = 2\pi \sum_{j_{in},j_{sc}} (2j_{sc}+1)\{\sum_{j_{ie},j_{iv}} < 0 \mid j_{in}, j_{sc} \mid j_{ie}, j_{iv}, J_i > B^{IK}_{ie,iv}\}^2 \tag{3}$$

The amplitude like combination in (3) has the following form:

$$< 0 \mid j_{in}, j_{sc} \mid j_{ie}, j_{iv}, J_i >= \sqrt{(2j_{ie}+1)(2j_{iv}+1)}(-1)^{j_{ie}+1/2} \times \sum_{\lambda}(-1)^{\lambda+J_i} \times$$
$$\times \{\delta_{\lambda,J_i} /(2J_i+1) Q_{\lambda}(sc,ie;iv,in) + \begin{bmatrix} j_{in}\ldots j_{sc}\ldots J_i \\ j_{ie}\ldots j_{iv}\ldots\ldots\lambda \end{bmatrix} Q_{\lambda}(ie;in;iv,sc)\}, \tag{4}$$

where $Q_\lambda = Q^{\mathrm{Qul}}_{\lambda} + Q^{\mathrm{Br}}_{\lambda}$ is, as a rule, the sum of the Coulomb and Breit matrix elements. The details of their calculating can be found in [2-4,13-15,19,20].

3. The Debye shielding model

As it is well known (c.f.[17,22] and refs. therein) in the classical theory of plasmas developed by Debye and Hückel, the interaction potential between two charged particles in a plasma is modelled by a Yukawa - type potential as follows:

$$V(r_i, r_j) = (Z_a Z_b / |r_a - r_b|) exp(-\mu \cdot |r_a - r_b|), \tag{5}$$

where r_a, r_b represent respectively the spatial coordinates of particles A and B and Z_a, Z_b denote their charges. A difference between the Yukawa type potential and standard Coulomb potential is in account for the effect of plasma, which is modeled by the shielding parameter μ [17]. In fact, this parameter describes the shape of the long-rang potential. The parameter μ is connected with the plasma parameters such as the temperature T and the charge density n as follows:

$$\mu \sim \sqrt{e^2 n / k_B T} \tag{6}$$

Here, as usually, e is the electron charge and κ_B is the Boltzman constant. The density n is given as a sum of the electron density N_e and the ion density N_k of the k-th ion species having the nuclear charge q_k :

$$n = N_e + \sum_k q_k^2 N_k \tag{7}$$

Under typical laser plasma conditions of T~1keV and n~10^{22} cm^{-3} the parameter μ is of the order of 0,1 in atomic units [17]. By introducing the Yukawa - type electron-nuclear attraction and electron-

electron repulsion potentials, the electronic Hamiltonian for a N-electron multicharged ion in a plasma is given in atomic units as follows:

$$H = \sum_i [\alpha cp - \beta mc^2 - Z\exp(-\mu r_i)/r_i] + \sum_{i>j} \frac{(1-\alpha_i\alpha_j)}{r_{ij}} \exp(-\mu r_{ij}) \tag{8}$$

A difference between our model Hamiltonian (8) and analogous model Hamiltonian with the Yukawa potential of ref. [17] (c.f. refs there too) is in using the relativistic approximation, which is obviously necessary for adequate description of such relativistic systems as the multicharged ions of the neon isoelectronic sequence.(c.f. details in refs. [13,14]). Further in the calculation procedure we introduce also the model electron-core potential of Ivanova et al [13,15] with parameter, calibrated within the special ab initio procedure within the relativistic energy approach [3].

4. Results

We applied our approach to calculate the electron collisional excitation cross-sections, strengths and rate coefficients for electron-collisional excitation for some Ne-like ions. To test our theory we compare our calculations results on the collisional cross-sections for Ne-like iron with the known theoretical data [2,5,6,11,12,21]. In tables 1 and 2 we present the measured (plasma) electron-collisional excitation cross-sections σ for Ne-like barium for two values of incident electron energy 5.69keV and 8.20 keV [11] and the theoretical data: our calculation data and results of four other theoretical papers [2,5,6,12,13,22]. It should be noted that the experimental information about the electron-collisional cross-sections for high-charged Ne-like ions is very scarce and is extracted from indirect observations. Such experimental information for a few collisional excitations of the Ne-like barium ground state has been presented in Ref.[11]. Let us note that in ref. [2] one-electron basis of PT was calibrated using the experimental one-electron energies. Here we use ab initio one-electron basis. The PT first order correction is calculated exactly, the high-order contributions are taken into account for effectively: polarization interaction of two above-core quasi-particles and an effect of their mutual screening (internal correlation effects) [13-16].

Table 1. Comparison of measured and calculated electron-collisional excitation cross-sections for Ne-like barium for incident electron energy 5,69 keV (units are 10^{-21}cm^2).

Level	J	Measured Marrs etal [11]	Calculated Ivanov etal [2,13]	Calculated Zhang etal [12]	Calculated Reed [21]	Calculated Glushkov et al [5]	Our paper
Sum (J=0)		2.50±0.35	2.48	2.58	2.60	2.51	2.63
$2p_{3/2}3d_{5/2}$	1	3.98±0.56	3.20	3.44	3.56	3.25	3.61
$2p_{1/2}3d_{3/2}$	1	2.12±0.30	1.78	2.42	2.00	1.84	2.02

Table 2. Comparison of measured and calculated electron-collisional excitation cross-sections for Ne-like barium for incident electron energy 8,20 keV (units are 10^{-21}cm^2).

Level	J	Measured Marrs etal [11]	Calculated Ivanov et al [2,13]	Calculated Zhang etal [12]	Calculated Reed [21]	Calculated Glushkov et al [5]	Our paper
Sum (J=0)		2.27±0.32	1.83	1.89	1.94	1.86	1.97
$2p_{3/2}3d_{5/2}$	1	3.30±0.46	2.87	2.99	3.23	2.93	3.27
$2p_{1/2}3d_{3/2}$	1	1.82±0.25	1.64	2.10	1.82	1.69	1.84

In table 3 we present the theoretical data on the effective collision strengths of the Kr^{26+} Ne-like ion excitation states for the temperature T=5·10^6 K and the electron density n_e=10^{14} cm^{-3} (the typical electron density of the tokamak plasma). The R-matrix data by DF Griffin et al (RM) [23] are listed for comparison too. It should be noted that strong compensation of different PT terms is a characteristic feature of the states with vacancies in the core. This is one of the main reasons for the fact that the accuracy of conventional a priori calculations of such states does not always satisfy the requirements arising in many applications. Summation over j_{in},j_{sc} in (18) spreads over the range 1/2-23/2. The convergence of this sum has been numerically investigated. It should be noted that the higher partial-wave contribution is less than 0,8% for all states considered. For some levels the corrections due the correlation effects change the results by a factor of 2-3,5. Our final electron-collisional excitation cross-sections appear to be typically larger than those of Ref.[2,5,12,21], where the plasma shielding effects are not directly included. But, in whole there is physically reasonable agreement between theory and experiment.

Table 3. The effective collision strengths of the Kr^{26+} Ne-like ion excitation states for the temperature T=5·10^6 K and electron density n_e=10^{14} cm^{-3} (see text).

Term	RM [23]	Our paper
$2p^53s\ (3/2,1/2)_2$	8.29(-3)	8.13(-3)
$2p^53s\ (3/2,1/2)_1$	9.36(-3)	9.19(-3)
$2p^53p\ (3/2,1/2)_1$	3.49(-3)	3.38(-3)
$2p^53p\ (3/2,1/2)_2$	4.30(-3)	4.18(-3)
$2p^53s\ (1/2,1/2)_0$	1.32(-3)	1.21(-3)
$2p^53s\ (1/2,1/2)_1$	7.69(-3)	7.56(-3)
$2p^53p\ (3/2,3/2)_3$	4.03(-3)	3.89(-3)
$2p^53p\ (3/2,3/2)_1$	3.14(-3)	3.01(-3)
$2p^53p\ (3/2,3/2)_2$	3.36(-3)	3.12(-3)
$2p^53p\ (3/2,3/2)_0$	8.67(-3)	8.49(-3)
$2p^53p\ (1/2,1/2)_1$	2.69(-3)	2.54(-3)
$2p^53p\ (1/2,3/2)_1$	2.80(-3)	2.72(-3)
$2p^53p\ (1/2,3/2)_2$	3.27(-3)	3.16(-3)
$2p^53d\ (3/2,3/2)_0$	1.24(-3)	1.13(-3)
$2p^53p\ (1/2.1/2)_0$	1.71(-2)	1.58(-2)
$2p^53d\ (3/2,3/2)_1$	3.45(-3)	3.31(-3)
$2p^53d\ (3/2,3/2)_3$	3.80(-3)	3.67(-3)
$2p^53d\ (3/2,5/2)_2$	4.13(-3)	3.96(-3)

Using of the Debye shielding approach and an account for the highly-lying excited states is quantitatively important for the adequate description of the collision cross-sections. In conclusion let us note that in any case our calculation encourages us to believe that using QED energy approach with the optimal one-electron PT basis and Debye shielding model block is quite consistent and effective one from the point of view of the theory correctness and results exactness. This fact was surely confirmed by calculations of the oscillator strengths, radiative widths, hyperfine structure constants in atoms, multicharged ions (c.f. [3,19,20]). Our ab initio approach can be used in calculations of the cross sections of other collisional processes and, in general speaking, other systems [24,25]. Its using is especially perspective when the experimental information about corresponding properties and systems is very scarce.

References

[1] Nakamura N, Kavanagh A P, Watanabe H, Sakaue H A, Li Y, Kato D, Curell F J and Ohtani S 2007 *Journ. Phys. Conf. Ser.* **88** 012066

[2] Ivanov L N, Ivanova E P and Knight L 1993 *Phys. Rev.* A **48** 4365
[3] Glushkov A V and Ivanov L N 1992 *Phys. Lett.* A **170** 33
[4] Glushkov A V and Ivanov L N 1993 *J. Phys.* B: *At. Mol. Opt. Phys.* **26** L379
[5] Glushkov A V, Ambrosov S V, Loboda A V, Gurnitskaya E P and Prepelitsa G P 2005 *Int. Journ. Quant.Chem.* **104** 562
[6] Glushkov A V, Malinovskaya S V, Loboda A V, Shpinareva I M, Gurnitskaya E P and Korchevsky D A 2005 *Journ. Phys. Conf. Ser.* **11** 188
[7] Glushkov A V, Malinovskaya S V, Loboda A V, Shpinareva I M and Prepelitsa G P 2006 *Journ. Phys. Conf. Ser.* **35** 420
[8] Glushkov A V, Malinovskaya S V, Chernyakova Yu G and Svinarenko A A 2004 *Int. Journ. Quant. Chem.* **99** 889
[9] Glushkov A V, Ambrosov S V, Ignatenko A V and Korchevsky D A 2004 *Int. Journ. Quant. Chem.* **99** 933
[10] Glushkov A V, Khetselius O Yu and Malinovskaya S V 2008 *Europ. Phys. Journ* ST **160** 195
[11] Marrs R, Levine M, Knapp D and Henderson J 1988 *Phys. Rev. Lett.* **60** 1715
[12] Zhang H, Sampson D, Clark R and Mann J 1988 *At. Dat. Nuc. Dat. Tabl.* **37** 17
[13] Ivanov L N, Ivanova E P, Glushkov A V and Kramida A E 1985 *Phys. Scripta* **32** 513
[14] Ivanova E P and Glushkov A V 1986 *J. Quant. Spectr. Rad. Transf.* **36** 127
[15] Ivanova E P and Gulov A V 1991 *Atom. Data. Nucl. Data. Tabl.* **49** 1
[16] Ivanov L N, Ivanova E P and Aglitsky E V 1988 *Phys. Rep.* **166** 315
[17] Okutsu H, Sako T, Yamanouchi K and Diercksen G H F 2005 *J. Phys.* B*: At. Mol. Opt. Phys.* **38** 917
[18] Badnell N R 2007 *Journ. Phys. Conf. Ser.* **88** 012070
[19] Khetselius O Yu 2009 *Int. Journ. Quant. Chem.* **109** 3330
[20] Khetselius O Yu 2009 *Phys. Scripta* T **135** 014023
[21] Reed K J 1988 *Phys. Rev.* A **37** 1791
[22] Malinovskaya S V, Glushkov A V, Khetselius O Yu, Loboda A V, Lopatkin Yu M, Svinarenko A A, Nikola L V and Tkach-Perelygina T B 2011 *Int. Journ. Quant. Chem.* **111** 288
[23] Griffin D C, Balance C P, Mitnik D M and Berengut J C 2008 *J. Phys.* B: *At. Mol. Opt. Phys.* **41** 21520
[24] Glushkov A V, Khetselius O Yu and Svinarenko A A 2011 *Advances in the Theory of Quantum Systems in Chemistry and Physics, Series: Frontiers in Theoretical Physics and Chemistry* vol 22, ed P Hoggan, E Brandas, G Delgado-Barrio and P Piecuch (Berlin: Springer) p 51
[25] Loboda A V, Prepelitsa G P and Svinarenko A A 2011 *Photoelectronics* **20** 40

High precision line shape studies in low pressure ammonia for an accurate determination of the Boltzmann constant

C Lemarchand, M Triki, B Darquié, P L T Sow, S Mejri, C Chardonnet, Ch J Bordé and C Daussy

Université Paris 13, Sorbonne Paris Cité, Laboratoire de Physique des Lasers, CNRS, UMR 7538, F-93430, Villetaneuse, France

E-mail: christophe.daussy@univ-paris13.fr

Abstract. In this paper, we present the progress performed in an experiment dedicated to the determination of the Boltzmann constant by an accurate measurement of the Doppler absorption profile of a line of ammonia in the gas phase in thermal equilibrium. We have recorded the hyperfine structure of the ν_2 saQ(6,3) rovibrational line by saturation spectroscopy and determined very precisely the induced broadening of the absorption line in linear absorption. In addition, we present an accurate analysis of the shape of the probed self-broadened line. Experimental spectra are confronted to various models that take into account Dicke narrowing or speed-dependent effects. Our results show a deviation from the Voigt profile dominated by speed-dependent broadening and shift. We also theoretically demonstrate that there is strictly no additional broadening from finite transit time across the laser beam. Finally, from the reported analysis, we anticipate that a first optical determination of the Boltzmann constant with a competitive accuracy of 1.8 parts per million is now reachable.

1. Introduction

A renewed interest in the Boltzmann constant, k_B, is related to the possible redefinition of the International System of Units (SI) [1-4]. A new definition of the kelvin would fix the value of the Boltzmann constant to a value determined by The Committee on Data for Science and Technology (CODATA) [5]. Since 2004 we have developed a new approach based on laser spectroscopy for the Boltzmann constant determination. The principle [6, 7] consists in recording the Doppler profile of a well-isolated absorption line of a molecular gas in thermal equilibrium in a cell. This profile reflects the Maxwell-Boltzmann distribution of velocities along the laser beam axis. In a first experiment we demonstrated the potential of this approach [8-12] on an ammonia rovibrational line. We were soon followed by at least four other groups who started similar experiments on CO_2, acetylene, H_2O and rubidium [13-17].

In this paper, we present the spectrometer developed to record the ν_2 saQ(6,3) rovibrational line of ammonia both by linear and saturated absorption spectroscopy. We give an evaluation of the effect of the hyperfine structure on the Doppler width measurement accuracy. We also present an accurate line shape analysis of the probed self-broadened absorption transition, recorded by linear absorption laser spectroscopy. Spectra are analyzed using various models which enable to clearly identify the profile which best matches the data as the speed-dependent Voigt profile (SDVP). We present a determination of the impact of collisional effects on the Doppler width measurement accuracy. Finally, we theoretically demonstrate that there is strictly no additional broadening from finite transit time across the laser beams. The molecular recoil and the second-order Doppler effect are included in this analysis following a complete quantum treatment.

2. The spectrometer

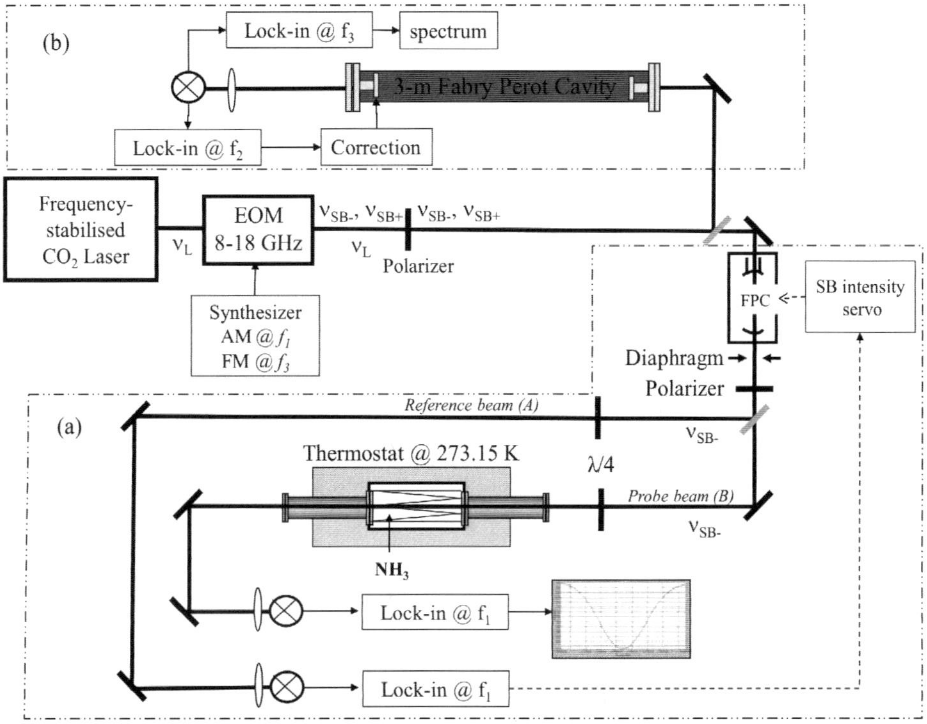

Figure 1. Experimental setup for (a) linear absorption spectroscopy and (b) saturated absorption spectroscopy (AM: amplitude modulation, FM: frequency modulation, EOM: electro-optic modulator, FPC: Fabry Perot cavity, SB: sideband, Lock-in: lock-in amplifier).

The spectrometer (figure 1) is based on a CO_2 laser source which operates in the 8-12 μm range. The laser frequency stabilisation scheme is described in reference [18]: a sideband generated with a tunable electro-optic modulator (EOM) is stabilized on an OsO_4 saturated absorption line detected on the transmission of a 1.6-m long Fabry-Perot cavity (FPC). Since its tunability is limited to 100 MHz, our CO_2 laser source is coupled to a second EOM which generates two sidebands SB- and SB+ on both sides of the fixed laser frequency. Their frequency is tunable from 8 to 18 GHz with respect to the carrier frequency. The intensity ratio between these two sidebands and the carrier is about 10^{-4}. After the EOM, a grid polarizer attenuates the carrier by a factor 200 but not the sidebands which are cross-polarized. Figure 1.b represents the saturated absorption spectrometer used for recording the hyperfine structure of the rovibrational line. The modulation f_2 is used to stabilize the resonator frequency and can be applied either on one mirror mounted on a piezoelectric transducer or directly on the sideband frequency via the synthesizer that drives the 8–18 GHz EOM. The hyperfine components of the molecular line are detected in transmission of a 3-m long Fabry Perot cavity after demodulation at f_3, a modulation applied on the sideband frequency via the EOM. Figure 1.a represents the linear absorption spectrometer. An FPC is used to drastically filter out the residual carrier and the unwanted

SB+ sideband and to stabilise the intensity of the transmitted sideband SB-. A 50/50 beamsplitter splits the transmitted beam in two parts. Beam A intensity is compared and locked to a very stable voltage reference by acting on the length of the FPC, which results in beam B probing NH_3 in the 37-cm long absorption cell with a constant incident laser power. Both the reference beam (A) and the probe beam (B) are amplitude-modulated at f_l via the 8-18 GHz EOM for noise filtering and signals are obtained after demodulation at f_l. The sideband is tuned close to the desired molecular resonance and scanned to record the Doppler profile.

3. The hyperfine structure analysis

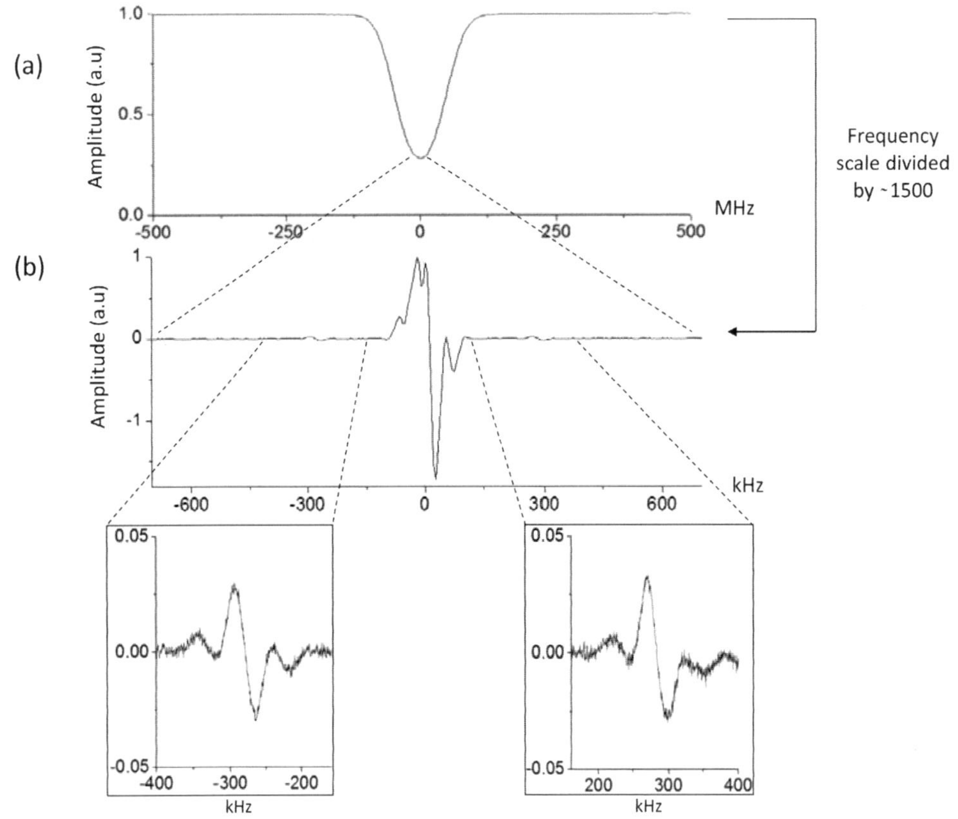

Figure 2. $^{14}NH_3$ saQ(6,3) absorption line recorded by linear absorption (a) and at higher resolution by saturated absorption spectroscopy (b). At about 300 kHz on both sides of the central components Doppler-generated level crossings are observed.

The selected absorption line is the ν_2 saQ(6,3) rovibrational line of the ammonia molecule $^{14}NH_3$ at the frequency ν = 28 953 693.9(1) MHz. This well-isolated transition has been chosen to avoid any deformation of the profile due to line mixing with neighbouring lines. Owing to the non-zero spin values of the nitrogen and hydrogen nuclei, an unresolved hyperfine structure is present in the Doppler profile of this rovibrational line. Figure 2 compares (a) the ν_2 saQ(6,3) linear absorption signal recorded in transmission of the 37-cm long absorption cell (see figure 1.a) and (b) the saturated absorption signal recorded in transmission of the 3-m long Fabry-Perot cavity (see figure 1.b). Saturated absorption spectroscopy displays Doppler-generated level crossings, clearly visible around ± 300 kHz from the central components.

Combined with microwave spectroscopy data, an analysis of saturated absorption spectra led to an accurate determination of the hyperfine constants of the ν_2 saQ(6,3) line [19] and to the estimation of the impact of the hyperfine structure on the Doppler width measurement. The global spread of the overall hyperfine structure and the relative intensities of hyperfine components of the linear absorption spectrum (and the respective uncertainties) result in a very precise determination of the correction to be applied on the value of the Doppler width: -4.356(13) parts per million, leading to an uncertainty of 0.026 part per million on k_B.

4. The self-broadened line shape analysis

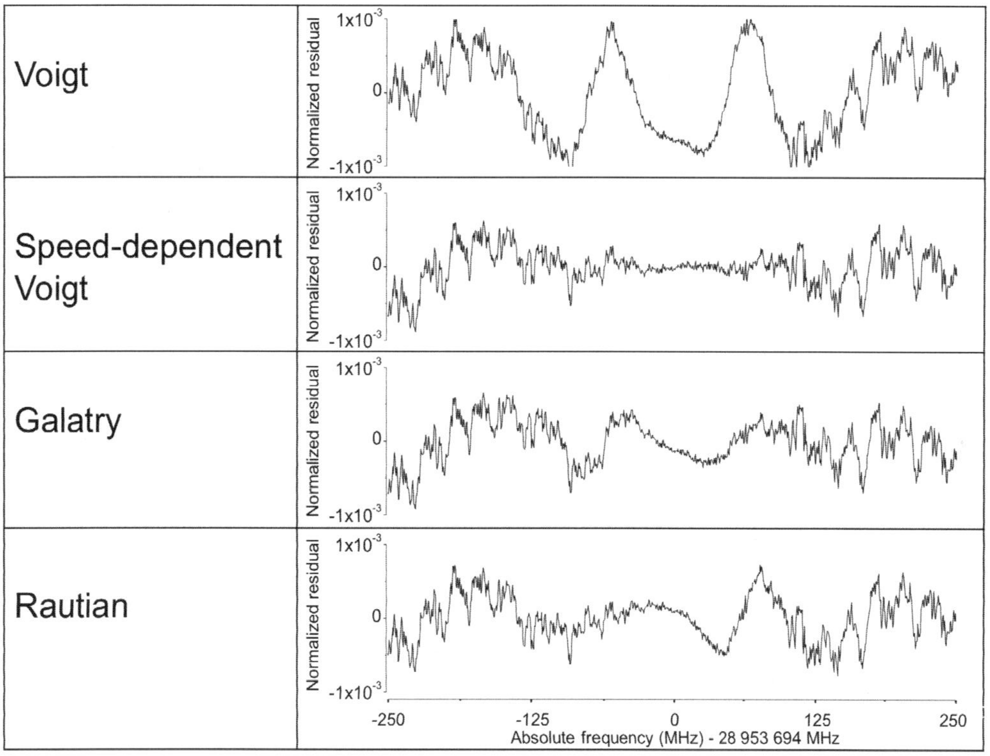

Figure 3. Normalized residuals for non-linear least-squares fits of spectra recorded at 17.3 Pa, with a VP (Voigt Profile), SDVP (Speed-Dependent Voigt Profile), GP (Galatry profile) and RP (Rautian Profile). The frequency scale is offset by 28 953 694 MHz.

In the very low pressure regime explored, the line shape can be described simply by a Voigt profile (VP). With respect to the VP, the line shape can be narrowed either by speed-dependent or Lamb-Dicke-Mössbauer (LDM) effects. On the one hand, the SDVP takes into account a dependence of the relaxation mechanism on molecular speed through a speed-dependent collisional shift and broadening.

On the other hand the LDM effect results in a reduction of the Doppler width due to velocity changing collisions. Assuming soft – resp. hard – collisions between molecules then leads to the Galatry – resp. Rautian – profile (GP – resp. RP) [20, 21].

Spectra recorded around 17 Pa have been fitted with a VP, SDVP, GP and RP. Residuals obtained are displayed on figure 3. For the VP, GP and RP, residuals clearly indicate that these profiles do not match the experimental data. By contrast, analysis of spectra with the SDVP clearly demonstrates a good agreement between this model and experimental line shapes. Based on our analysis (reported in detail in ref [22]) we can state that the deviation from a VP of the ν_2 saQ(6,3) line of NH_3 is most probably caused by speed-dependent effects and not LDM effects, in the explored pressure range.

The impact of collisional effects on the Doppler width measurement has been estimated. The associated uncertainty is only limited by our knowledge on the SDVP collisional parameters [22]. In order to reduce the impact of this source of uncertainty, the measurement of the Boltzmann constant is performed at low pressure, below 2 Pa. Simulated spectra corresponding to these experimental conditions have been fitted taking into account uncertainties on the SDVP parameters. Following the line-absorbance based analysis recently proposed by A. Castrillo et al. [23] an experimental uncertainty of 0.9 part per million is obtained on the Doppler width measurement, leading to an uncertainty of 1.8 parts per million on k_B.

5. The theory of linear absorption line shape: case of the speed-dependent Voigt profile

The purpose of this paragraph is to outline the derivation given in [21] of a line shape including the modal structure of the laser beams, with an explicit introduction of a speed-dependent broadening and shift (besides the second-order Doppler shift), in order to investigate possible contributions of transit-time effects. The starting point is the expression for the dimensionless absorbance in the momentum representation:

$$A_\nu = 2\pi\alpha n^{(0)} d_{ba}^2 L \frac{e^{-E_a\beta}}{Z_{int}} \frac{\omega}{V} \int d^3k\, \alpha^*(\vec{k})\, \alpha(\vec{k}) \overline{f}_{ba}(\vec{k},\omega) + c.c. \tag{1}$$

where α is the fine structure constant, $n^{(0)}\frac{e^{-E_a\beta}}{Z_{int}}$ the population of the lower level, d_{ba} the dipole moment of the transition divided by the elementary charge, L the absorption length, $V = LS$ the laser mode volume. The normalized line shape for each plane wave component $\alpha(\vec{k})$ of the laser mode is the velocity average $\overline{f}_{ba}(\vec{k},\omega) = \int f_{ba}(\vec{v},\vec{k},\omega)\, d^3v$ where $f_{ba}(\vec{v},\vec{k},\omega)$ is proportional to the first-order off-diagonal density matrix element $\rho_{ba}^{(1)} = i\Omega_{ba} f_{ba}$ through the Rabi frequency Ω_{ba} and satisfies the integral equation:

$$\begin{aligned}\left[i\left(\omega - \omega_{ba}(1 - v^2/2c^2) - \Delta_{ba}(v) - \vec{k}.\vec{v} - \delta\right) + \gamma_{ba} + \Gamma_{ba}(v)\right] f_{ba}(\vec{v},\vec{k},\omega) \\ = F_M(\vec{v}) + \int d^3v' W_{ba}(\vec{v}' \to \vec{v}) f_{ba}(\vec{v}',\vec{k},\omega)\end{aligned} \tag{2}$$

which includes the second-order Doppler shift, the recoil shift δ, a velocity-dependent shift $\Delta_{ba}(v)$ and broadening $\Gamma_{ba}(v)$, the natural linewidth γ_{ba}, and where $F_M(\vec{v})$ is the Maxwell-Boltzmann distribution of velocities and $W_{ba}(\vec{v}' \to \vec{v})$ is the collision kernel describing the probability for a velocity change from $\vec{v}'$ to $\vec{v}$.

The wave vector appears in this equation only through the projection $\vec{k}.\vec{v}$ so that, thanks to the isotropy of the velocity distribution and of collisions, there is no privileged direction (except for that of gravity whose influence needs to be carefully evaluated). Thus $\overline{f}_{ba}$ cannot depend on the direction of $\vec{k}$ but only on its modulus $|\vec{k}|$ and finally, due to the dispersion relation $k = \omega/c$, $\overline{f}_{ba}$ results in a function of ω only. The mode volume V cancels with $\int d^3k\, \alpha^*(\vec{k})\, \alpha(\vec{k})$ and any dependence on the laser mode content disappears. This gives for the absorbance:

$$A_\nu = 4\pi\alpha n^{(0)} d_{ba}^2 L \frac{e^{-E_a\beta}}{Z_{int}} \omega Re\overline{f}_{ba}(\omega) \tag{3}$$

which is independent of the structure of the light beam and hence has strikingly no additional transit-time broadening as we shall illustrate below in a specific case.

In this paper we consider more specifically the case where velocity-changing collisions are ignored and hence the integral term of equation (2) is neglected:

$$2Re\bar{f}_{ba}(\omega) = 2\frac{\sqrt{\pi}}{\Delta\omega_D} Re\, z(\omega) \tag{4}$$

where

$$z(\omega) = \int d^3v \frac{F_M(\vec{v})}{i(\omega-\omega_{ba}(1-v^2/2c^2)-\Delta_{ba}(v)-\vec{k}.\vec{v}-\delta)+\gamma_{ba}+\Gamma_{ba}(v)}$$
$$= \frac{4}{\sqrt{\pi}\Delta\omega_D}\int_0^{+\infty} d\theta\theta e^{-\theta^2}\tan^{-1}\left(\frac{i\theta}{\zeta(\theta)}\right) \tag{5}$$

is a generalized speed-dependent Voigt profile which, thanks to the isotropy of $F_M(\vec{v})$, depends on $\vec{k}$ only through the Doppler width $\Delta\omega_D = ku$ ($u = \sqrt{2k_BT/m_{NH_3}}$ is the most probable speed of molecules) and the recoil shift and hence has been expressed as a single quadrature on $\theta = v/u$, with $\zeta(\theta) = \left(\omega - \omega_{ba}(1-\theta^2u^2/2c^2) - \Delta_{ba}(\theta) - \delta + i(\gamma_{ba} + \Gamma_{ba}(\theta))\right)/\Delta\omega_D$.

In section 4, we followed the formalism developed by Berman, Ward and Pickett for the SDVP leading to the collisional parameters [24-26]:

$$\Gamma_{ba}(\theta) = \frac{\Gamma_{ba}}{2^{m/2}}\, {}_1F_1\left(-\frac{m}{2},\frac{3}{2},-\theta^2\right) \text{ and } \Delta_{ba}(\theta) = \frac{\Delta_{ba}}{2^{n/2}}\, {}_1F_1\left(-\frac{n}{2},\frac{3}{2},-\theta^2\right) \tag{6}$$

where ${}_1F_1(a,b,z)$ is the Kummer confluent hypergeometric function, and $m = 0.360(7)$ and $n = -3.8(3)$ are parameters related to the intermolecular potential [22].

More generally equation (2) cannot be solved without a numerical calculation of the collision kernels [27, 28]. There are however simple cases where it can be solved analytically e.g. when the integral term in (2) is a convolution product [29, 30]. Another example is the model of Rautian, Nelkin and Ghatak [31, 32] in which each collision is assumed to be strong enough to redistribute the velocities according to the Maxwell-Boltzmann distribution:

$$W_{ba}(\vec{v}' \to \vec{v}) = \nu_c F_M(\vec{v}) \tag{7}$$

where ν_c is an adjustable collision frequency. We obtain the generalized speed-dependent Rautian profile:

$$2Re\bar{f}_{ba}(\omega) = 2\frac{\sqrt{\pi}}{\Delta\omega_D} Re \frac{z(\omega)}{1-\nu_c(\sqrt{\pi}/\Delta\omega_D)z(\omega)} \tag{8}$$

If velocity-changing collisions are neglected, $\nu_c = 0$ and the generalized speed-dependent Rautian profile reduces to the generalized speed-dependent Voigt profile which is considered in this paper. Other models for the collision kernels are presently under study but as long as these are isotropic no transit-time effect will occur.

6. Conclusion

This paper presents a study of the influence of the hyperfine structure on the measurement of the Doppler broadening of the ν_2 saQ(6,3) rovibrational line of ammonia close to 273.15 K. Its role cannot be neglected at a level of a few parts per million but it can now be taken into account with a negligible residual uncertainty. We have reported on an accurate analysis of self-broadened ν_2 saQ(6,3) line shape of NH_3. The deviation from a Voigt profile of the present line shape seems dominated by speed-dependent effects. We have also demonstrated the absence of any additional transit broadening whatever the optical quality of the laser beam wave fronts and effective diameters, even in presence of relaxation effects dependent on molecular speed. Finally, from the accurate knowledge of the absorption line shape we anticipate that a first determination of k_B with a competitive accuracy of 1.8 parts per million by measurement of the Doppler width in ammonia is now reachable.

Acknowledgments
This work is funded by CNRS, the Laboratoire National de Métrologie et d'Essais and by European Community (EraNet/IMERA).

References

[1] Mills I M, Mohr P J, Quinn T J, Taylor B, N, and Williams E R 2005 *Metrologia* **42** 71
[2] Mills I M, Mohr P J, Quinn T J, Taylor B, N, and Williams E R 2006 *Metrologia* **43** 227
[3] BIPM 2010 *http://www.bipm.org/utils/common/pdf/si_brochure_draft_ch2.pdf*
[4] Bordé C J 2005 *Phil. Trans. Roy. Soc A* **363** 2177
[5] Mohr P J, Taylor B N and Newell D B 2012 *arXiv:*1203.5425
[6] Bordé C J, Conference given at the Symposium to celebrate the 125th Anniversary of the meter Convention Paris 2000
[7] Bordé C J 2002 *Metrologia* **39** 435
[8] Daussy C, Briaudeau S, Guinet M, Amy-Klein A, Hermier Y, Bordé C J and Chardonnet C 2005 in *Laser Spectroscopy* edited by A. F. ed. by E. Hinds, E. Riis (World Scientific, Cairngorms National Park, Scotland)
[9] Daussy C, Guinet M, Amy-Klein A, Djerroud K, Hermier Y, Briaudeau S, Bordé C J and Chardonnet C 2007 *Phys. Rev. Lett.* **98** 250801
[10] Djerroud K, Daussy C, Lopez O, Amy-Klein A, Briaudeau S, Hermier Y and Chardonnet C 2007 *Ann. Phys. Fr.* **32** 175
[11] Djerroud K *et al.* 2009 *C. R. Physique* **10** 883
[12] Lemarchand C, Djerroud K, Darquié B, Lopez O, Amy-Klein A, Chardonnet C, Bordé C J, Briaudeau S and Daussy C 2010 *Int. J. Thermophys.* **31** 1347
[13] Casa G, Castrillo A, Galzerano G, Wehr R, Merlone A, Di Serafino D, Laporta P and Gianfrani L 2008 *Phys. Rev. Lett.* **100** 200801
[14] Yamada K M T, Onae A, Hong F-L, Inaba H, Matsumoto H, Nakajima Y, F. I and Shimizu T 2008 *J. mol. spec.* **249** 95
[15] De Vizia M D, Rohart F, Castrillo A, Fasci E, Moretti L and Gianfrani L 2011 *Phys. Rev. A* **83** 052506
[16] Sun Y R, Pan H, Cheng C F, Liu A W, Zhang J T and Hu S M 2011 *Optics Express* **19** 19993
[17] Truong G-W, May E F, Stace T M and Luiten A N 2011 *Phys. Rev. A* **83** 033805
[18] Bernard V, Daussy C, Nogues G, Constantin L, Durand P E, Amy-Klein A, van Lerberghe A and Chardonnet C 1997 *IEEE Journal of Quantum Electronics* **QE-33** 1282
[19] Lemarchand C, Triki M, Darquie B, Borde C J, Chardonnet C and Daussy C 2011 *New Journal Of Physics* **13** 073028
[20] Hartmann J-M, Boulet C and Robert D 2008 *Collisional Effects on Molecular Spectra: Laboratory Experiments and Models, Consequences for Applications* (Elsevier Science)
[21] Bordé C J 2009 *C. R. Physique* **10** 866
[22] Triki M, Lemarchand C, Darquié B, Sow P L T, Roncin V, Chardonnet C and Daussy C 2012 *Phys. Rev. A* **85** 062510
[23] Castrillo A, De Vizia M D, Moretti L, Galzerano G, Laporta P, Merlone A and Gianfrani L 2011 *Phys. Rev. A* **84** 032510
[24] Berman P R 1972 *Journal Of Quantitative Spectroscopy & Radiative Transfer* **12** 1331
[25] Ward J, Cooper J and Smith E W 1974 *Journal Of Quantitative Spectroscopy & Radiative Transfer* **14** 555
[26] Pickett H M 1980 *Journal Of Chemical Physics* **73** 6090
[27] Avrillier S, Bordé C J, Picard J and Tran Minh N 1981 edited by B. Wende and W. de Gruyter (Berlin, New York) p 1161
[28] Avrillier S, Bordé C J, Picard J and Tran Minh N 1982 *Journal de Physique* **43** 1213
[29] Bordé C J, Avrillier S and Gorlicki M 1977 *Journal de Physique Lettres* **38** 249
[30] Bordé C J, Avrillier S and Gorlicki M 1979 *Journal de Physique Lettres* **40** 35
[31] Nelkin M and Ghatak A 1964 *Phys. Rev. A* **135** 4
[32] Rautian S G and Sobel'man I I 1967 *Sov. Phy. Usp.* **9** 701

Highly-accurate line shape studies in the near-IR spectrum of $H_2{}^{18}O$: Implications for the spectroscopic determination of the Boltzmann constant

Livio Gianfrani
Dipartimento di Matematica e Fisica,
Seconda Università degli Studi di Napoli, 81100 Caserta, Italy

E-mail: livio.gianfrani@unina2.it

Abstract. The outcomes of a specific study on line shape deviations from the usual Voigt profile, in the near-IR spectrum of water, will be discussed. Particularly, a great attention is paid to the role of speed-dependence of collisional broadening. From the experimental point of view, high spectral fidelity has been reached by using the technique of offset-frequency locking between two extended-cavity diode lasers. In such a system, the master laser is frequency stabilized against a narrow sub-Doppler molecular line, while the slave laser is actively controlled so that its emission frequency maintains a given offset with respect to the frequency of the master. This method allows one to perform highly accurate and reproducible frequency scans of the slave laser around a given vibration-rotation line. I will present recent findings on the speed dependence of collision parameters, influencing the shape of $H_2{}^{18}O$ absorption lines. Finally, I will discuss the implications for the spectroscopic determination of the Boltzmann constant.

1. Introduction

The General Conference on Weights and Measures (CGPM) agreed at its 24th meeting in October 2011 on new definitions of the kilogram, ampere, kelvin, and mole in terms of fixed numerical values of the Planck constant (h), elementary charge (e), Boltzmann constant (k_B), and Avogadro constant (N_A), respectively [1]. The changes proposed for the International System of units (SI) will be adopted only after a further refinement of the experimental results on the new definitional constants. Meanwhile, the CGPM encourages scientists to continue their work and efforts for the aims of a better determination of the h, e, k_B, and N_A constants. The new definition for the unit kelvin would require precision measurements of k_B by at least three fundamentally different methods. Presently, the most accurate way to access the value of the Boltzmann constant is from measurements of the speed of sound in a noble gas. In fact, acoustic gas thermometry has recently provided a k_B determination with a relative standard uncertainty of 1.24 parts in 10^6 [2]. Another approach is known as dielectric constant gas thermometry and it is based upon the Clausius-Mossotti equation. A relative standard uncertainty of 9.2 parts in 10^6, at the temperature of the triple point of water, has been recently demonstrated by using this method [3]. Other upcoming primary thermometry techniques, which are currently at the stage of further development and optimization, include Doppler broadening thermometry (DBT) and quantum voltage calibrated electrical noise thermometry [4].

Based upon precision laser spectroscopy in the linear regime of interaction, DBT consists in retrieving the Doppler width from the highly-accurate observation of the absorption profile corresponding to a

given atomic or molecular line in a gaseous sample at the thermodynamic equilibrium. In addition to a very well designed experiment allowing for extremely high fidelity in observing the absorption profile of the selected line, the spectroscopic determination of the thermodynamic temperature requires a very reliable physical understanding of line shape that may present quite strong departures from the usual Voigt profile. Besides the Doppler effect, other mechanisms can influence the line profile and they all contribute to the overall width [5]. Collisions between particles increases the relaxation rate (thus giving rise to the collision broadening), but they also have an averaging effect on the velocity of the particles, leading to a narrowing of the line, usually referred to as the Dicke effect. Collisional narrowing and collisional broadening effects can be correlated, as velocity-changing collisions and internal state perturbations may occur simultaneously. Furthermore, speed dependence can be evidenced, consisting in the fact that the relaxation rate of a given energy level presents a dependence on the relative velocity of the absorber-perturber system. In some cases, it is also necessary to consider collision time asymmetry, because of the finite duration of each collision [5].
As a result, the retrieval of the Doppler width from high quality absorption spectra is far from being trivial. A true calculation of the spectral line shape would require the density-matrix formalism of quantum mechanics that leads to a Boltzmann-like transport/relaxation equation [6]. A similar approach is still prohibitively complex for molecular systems. As a consequence, one should resort to semiclassical models, whose physical interpretation is often dubious. Since a perfect representation of reality is not available yet, there is presently a wide consensus on the fact that retrieval of the Doppler width at a given gas temperature requires spectra acquisitions at different gas pressures, with the aim of determining a possible dependence on the pressure and, eventually, extrapolating the zero-pressure value of the Doppler width [7].
Alternatively, instead of probing weakly absorbing molecules in a diffusive regime, it is possible to use a low-pressure atomic vapor system, like Rb [8]. In this case, being the atomic motion effusive, collisions are extremely rare so that pressure-induced variation of the absorption line shape can be completely neglected. This removes the need to extrapolate results to an equivalent zero-pressure value. On the other hand, Doppler-width determinations from atomic transitions are significantly affected by systematic deviations due to effects such as magnetic sensitivity and optical pumping.
In reason of that, the most accurate determinations of the Boltzmann constant by means laser spectroscopy have been performed on molecular systems, like ammonia and carbon dioxide, despite the difficulties in modeling the line shape [9, 10].
Another interesting molecular target for DBT experiments is water, due to the occurrence of relatively strong absorption features in the spectral window around 1.4 μm [11]. This is a telecom wavelength whose technical benefits can be fully exploited for the aims of an optimum implementation of DBT. Similarly, it is worth noting the extremely high linearity in the response of standard InGaAs photodiodes, which is a further important ingredient for a highly-accurate spectroscopic measurement of the Boltzmann constant [12].

In this paper, I will resume the outcomes of a recent investigation on the spectral profile of individual vibration-rotation lines of the $H_2{}^{18}O$ molecule at 1.4 μm. High spectral fidelity was ensured by the use of a dual-laser absorption spectrometer, expressly designed and developed with the ambitious goal of reaching the target uncertainty of 10^{-6}, in the spectroscopic determination of k_B. Pushing the experimental precision to extreme levels, it was possible to provide new insights towards a better physical understanding of the line shape problem in the near infrared spectrum of water.

2. Experimental details

Figure 1 shows a block diagram of the laser absorption spectrometer, which has been already described elsewhere [13]. It basically consists of an extended cavity diode laser with an emission wavelength in the range between 1.38 and 1.41 μm, a frequency control system, an intensity stabilization feed-back loop, and an isothermal cell. Precise control, stabilization and synchronization of the laser frequency is achieved by using the technique of offset frequency locking, in which the probe laser (usually referred to as the slave laser) is forced to maintain a precise frequency-offset from a reference laser (namely, the master laser). This offset is provided by a radio-frequency (rf) synthesizer, which in turn is phase-locked to an ultra-stable Rb oscillator. The master laser presents a subkilohertz absolute stabilization of its central frequency, being locked to an ultra-narrow saturated absorption signal, resulting from the nonlinear interaction that occurs inside a high-finesse optical resonator, in coincidence with a particular vibration-rotation line of a water isotopologue. For this purpose, the optical cavity is filled with a water-vapor sample at very low pressures, either enriched in ^{17}O and ^{18}O, or characterized by a natural abundance, depending on the selected reference line. Hence, the beat note between master and slave lasers is first detected by means of a fast photodiode, properly amplified, scaled in frequency, and subsequently compared with the rf signal, so as to produce an error signal driving a double-servo system, which in turn acts on the diode injection current and on the extended-cavity length of the slave laser. By tuning the rf frequency, it is possible to perform continuous, highly linear and highly accurate frequency scans of the slave laser around a given center frequency.

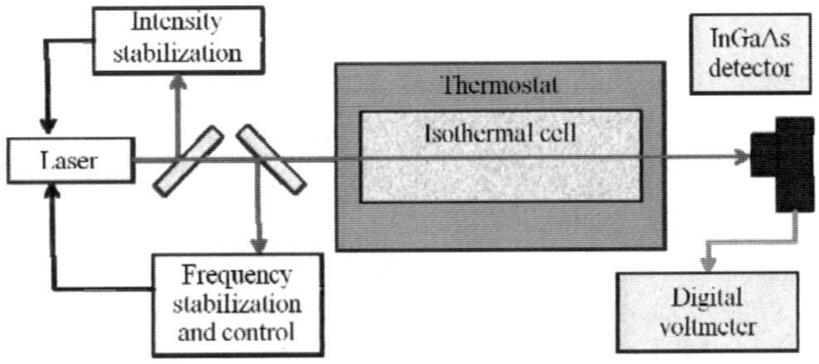

Figure 1. Sketch of the absorption spectrometer which has been developed for a highly-accurate spectroscopic determination of k_B.

The most subtle source of systematic deviations, namely, amplitude variation in the background baseline, is essentially removed by implementing an intensity control feedback loop, which effectively compensates for any power variation associated to a laser frequency scan, as well as for the periodic modulation of the power arising from spurious etalon effects [14].

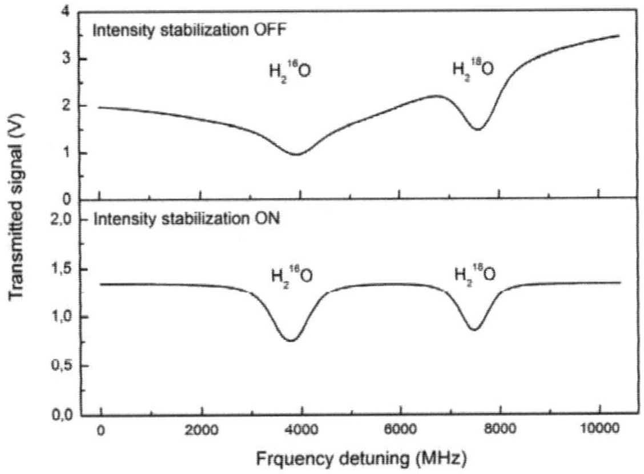

Figure 2. Correction of the laser power variation associated with a laser frequency scan around a pair of $H_2{}^{16}O$ and $H_2{}^{18}O$ lines.

This technical development represents a further element of novelty, as compared to previous works dealing with line shape studies by means of diode laser absorption spectroscopy. It also cancels out any influence of absorption from water molecules along the path of the beam in air. This is clearly shown in figure 2, where a typical absorption spectrum is shown without (upper trace) and with (lower trace) the action of the servo loop. Intensity stabilization at the level of 10^{-4} is achieved by using an AOM as an actuator, thus controlling continuously the amount of laser power that is deflected from the primary beam to the first diffracted order, which is used for the spectroscopic experiment. This method of background flattening is more effective compared to other approaches based on background subtraction procedures or the use of balanced dual-beam detection strategies. In fact, the optoelectronic control of the laser intensity ensures a better stability of the background signal and, consequently, a better experimental reproducibility and accuracy, because of the active nature of its operation, in a relatively large electronic bandwidth (~50 kHz).

Equipped with a pair of antireflection-coated BK7 windows, the isothermal cell is 150-mm long and it is referenced to the triple point of water (TPW), with temperature accuracy, uniformity and stability at the level of 0.1 mK, as extensively described elsewhere [15].

3. Line shape investigations

Speed-dependent (SD) effects may take place regardless of the gas pressure and usually lead to an absorption line shape that is asymmetric and narrower than the Voigt profile. Characterized by very different physical origins, the two narrowing effects, namely, optical diffusion and speed-dependence of collisional broadening, can occur simultaneously. Past experiments have shown that SD effects are particularly important in the case of large perturber-to-absorber mass ratios [16].

A recent experimental and theoretical analysis of the $H_2{}^{18}O$ line shapes at 1.38 μm has revealed that SD effects in self-colliding water molecules cannot be ignored [17]. More specifically, it has been demonstrated that the molecular confinement alone is unable to explain entirely the departures from the Voigt profile and that the speed dependence of pressure-induced broadening and shifting has to be considered, even in the case of pure water samples at relatively small pressures (between 10 and 500 Pa). In fact, absorption spectra were successfully interpolated using the uncorrelated version of the speed-dependent Galatry profile, with a hypergeometric dependence on the absorber speed for both pressure broadening and shifting parameters, thus reaching an agreement between theory and experiment at the level of 5×10^{-5}. This outcome agrees with what we expect when comparing the Lenard-Jones radius with the collision diameter retrieved from the relaxation parameter. It is worth noting that the hypergeometric speed dependence results from the application of the formalism developed by Berman and Pickett [18, 19].

Since the confluent hypergeometric function entails long numerical calculations for line fitting, a quadratic approximation is often employed, leading to the so-called quadratic speed-dependent models [20]. In this approximation, the collisional width is given by:

$$\Gamma(v_a) = \Gamma_0\left[1 + a_W\left(\left(\frac{v_a}{v_{a0}}\right)^2 - \frac{3}{2}\right)\right],$$

where $v_a = |\vec{v}_a|$ is the speed of the absorbing molecule, v_{a0} is the most probable speed, $\Gamma_0 = \langle\Gamma(v_a)\rangle$ is the average collisional width over molecular speeds, and a_W is a coefficient characteristic of the speed dependence, which is independent on the pressure. A similar equation holds for the pressure shifting parameter. Figure 3 shows the absolute residuals resulting from line fitting to the quadratic and hypergeometric speed-dependent Voigt (SDVP-Q and SDVP-HG) and speed-dependent Galatry profiles (SDGP-Q and SDGP-HG) of an experimental spectrum as acquired in coincidence with the $2_{2,1}$->$2_{2,0}$ line of the $H_2{}^{18}O$ $\nu_1+\nu_3$ band, at 1.3846 μm. The gas pressure was 500 Pa, while the temperature was kept at the constant value of the triple point of water. It is clear that the quadratic approximation, either used alone or considered in conjunction to the Dicke narrowing effect, fails to reproduce the experimental spectrum. In particular, the best agreement between theory and experiment was found when using the speed-dependent Galatry profile, with a hypergeometric dependence (SDGP-HG).

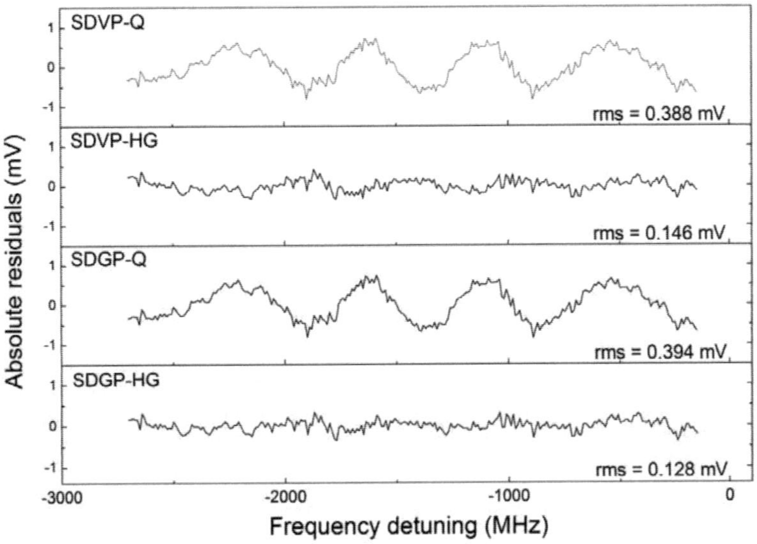

Figure 3. Results of nonlinear least-squares fits of an absorption profile to the quadratic and hypergeometric versions of the speed-dependent Voigt and Galatry profiles. The signal-to-noise ratio was about 15000, while the empty-cell transmission signal was ≈2 V.

This failure can be ascribed to the fact that the perturber-to-absorber mass ratio, α, is equal to exactly 1, the perturber being identical to the absorber. In fact, Ciurylo and co-workers, in a previous work, argued that the quadratic approximation, in a model describing simultaneously the two narrowing effects, is compatible with a soft-collision assumption, which implies $\alpha << 1$ [21].

4. Consequences for the spectroscopic determination of the Boltzmann constant

In proof-of-principle experiments, performed on ammonia and carbon dioxide molecules, respectively at ~10.3 and ~2 μm, spectroscopic determinations of the Boltzmann constant were performed with a combined relative uncertainty of 1.9 10^4, for the NH_3 spectrometer, and 1.6 10^4, for the CO_2 spectrometer [22, 10]. At that time, spectral analysis was performed by using Gaussian and Voigt profiles, respectively. In the last few years, with the ambitious goal of approaching the target accuracy of one part over 10^6, technical improvements of the experimental setups have been accompanied by more and more refined modeling of the line shapes, also involving other molecular targets, like water and oxygen [23, 24]. As a consequence, it has been possible to look for speed-dependent collisional effects even in those cases in which their influence on the experimental line shapes is expected to be rather small. This is the case of self-colliding molecules in the very low pressure regime (namely, $p<100$ Pa), as already discussed in the previous section. In this respect, I should also mention the recent work performed on ammonia by Triki *et al.* [25].

Coming back to water, the implications for the spectroscopic determination of the Boltzmann constant have been deeply investigated both theoretically and experimentally. A careful analysis of numerically simulated spectra has demonstrated that the choice of the line shape model is a key factor to reach low uncertainties in DBT [23]. A large number of simulations were done by accurately calculating the SDGP-HG function through numerical integration, adopting the Berman-Pickett model for a quadrupole-quadrupole interaction. The narrowing parameter per unit pressure was fixed at a reasonable value, taking into account the upper limit to the effective frequency of velocity changing collisions, provided by diffusion theory. A random noise was added to the simulated spectra so as to

reproduce the observed signal-to-noise ratios. A non-linear least-squares fitting procedure was applied to the simulated spectra, using Voigt, Galatry and speed-dependent Voigt models, in order to retrieve the Doppler width as a function of the simulated $H_2{}^{18}O$ pressure, at the temperature of the triple point of water. In any case, the Doppler width values exhibited a clear pressure dependence, which could be modeled by using a second-order polynomial function, at the aims of extrapolating the zero-pressure value ($\Delta\nu_0^D$). It turned out that, for Voigt and Galatry fits, $\Delta\nu_0^D$ underestimates the real Doppler width, with a relative systematical deviation of about 10^{-4}, for a signal-to-noise ratio of 10000 (calculated with respect to the maximum transmitted signal of the simulated spectra). The SDVP-HG model did much better in the fits of simulated spectra. More particularly, as a result of the analysis of 1050 simulated spectra, characterized by a signal-to-noise ratio of 50000, the extraordinary accuracy of 3 10^{-7} was demonstrated [23].
Experimental results seem to confirm the outcomes of the simulations. Figure 4 shows the Doppler widths as retrieved from the analysis of 150 spectra, recorded in coincidence with the $2_{2,1}\rightarrow 2_{2,0}$ line of the $H_2{}^{18}O$ $\nu_1+\nu_3$ band, as a function of the gas pressure, at the temperature of the triple point of water.

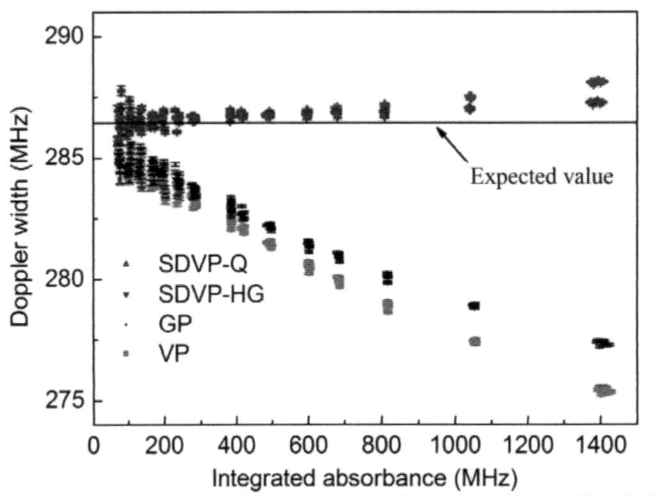

Figure 4. Examples of Doppler-width retrieval by means of non-linear least-squares fits of absorption spectra at different pressures (between 10 and 400 Pa). The black line represents the expected value (286.4530 MHz).

Of the tested models, the speed-dependent Voigt profile with hypergeometric dependence on both pressure broadening and shifting coefficients gives the best results, being the extrapolated zero-pressure value of the Doppler width equal to 286.469 (22) MHz, which differs from the expected one by less than 6 parts in 10^5 (in terms of relative difference between measured and expected values). Instead, both Voigt and Galatry profiles (VP and GP) give smaller values (as compared to the expected one), with relative deviations of the order of 10^{-3}. It is worth noting that the quadratic model, even doing better than VP and GP, does not reach the performance of the SDVP-HG model.

5. Conclusions

Recent advances in precision molecular spectroscopy enable one to reach unprecedented fidelity in the observation of the shape associated to a given vibration-rotation transition. Consequently, interesting investigations can be carried out, aimed at a better understanding of the physical mechanisms determining the spectral profiles. This is of utmost importance for improving the capabilities of quantitative laser absorption spectroscopy, for a variety of applications, including primary gas thermometry. I have shown and discussed the outcomes of recent studies on the $H_2{}^{18}O$ near-IR spectrum. The experimental accuracy is so high that it is possible, for the first time, to recognize the influence of a restricted number of molecules, moving at the highest velocities, for which the quadratic approximation departs from the actual behavior of the relaxation rate as a function of the molecular speed. Experimental tests of DBT at the triple point of water confirm that the SDVP-HG model is superior to the others. This is an important step-forward towards a spectroscopic determination of the Boltzmann constant with a combined uncertainty at the ppm (part per million) level.

Acknowledgments
I am indebted to past and present PhD students and post-doctoral fellows who contributed at this work, as well as my colleagues, Giovanni Casa, Antonio Castrillo and Luigi Moretti. I also thank Francois Rohart for fruitful discussions.

References

[1] Resolutions adopted by the 24th meeting of the General Conference on Weights and Measures: http://www.bipm.org/en/CGPM/db/24/
[2] Pitre L, Sparasci F, Truong D, Guillon A, Risegari L, Himbert M E 2011 *Int. J. Thermophys.* **32** 1825
[3] Fellmuth B, Fischer J, Gaiser C, Jusko O, Priruenrom T, Sabuga W, Zandt T 2011 *Metrologia* **48** 382
[4] Benz S P, Pollarolo A, Qu J F, Rogalla H, Urano C, Tew W L, Dresselhaus P D, White D R 2011 *Metrologia* **48** 142
[5] Ciurylo R 1998 *Phys. Rev. A* **58** 1029
[6] Shapiro D A, Ciurylo R, Drummond J R, May A D 2001 *Phys. Rev. A* **65** 012501
[7] Yamada K M T, Onae A, Hong F-L, Inaba H, Shimizu T 2009 *C. R. Phys.* **10** 907
[8] Truong G-W, May E F, Stace T M, and Luiten A N 2011 *Phys. Rev. A* **83** 033805
[9] Lemarchand C, Triki M, Darquié B, Bordé Ch J, Chardonnet C, Daussy C 2011 *New J. Physics* **13** 073028
[10] Casa G, Castrillo A, Galzerano G, Wehr R, Merlone A, di Serafino D, Laporta P, Gianfrani L 2008 *Phys. Rev. Lett.* **100** 200801
[11] Castrillo A, Casa G, Merlone A, Galzerano G, Laporta P, Gianfrani L 2009 *C. R. Phys.* **10** 894
[12] Yoon H W, Butler J J, Larason T C, Eppeldauer G P 2003 *Metrologia* **40** S154
[13] Castrillo A, Fasci E, Galzerano G, Casa G, Laporta P, Gianfrani L 2010 *Opt. Express* **18**, 21851
[14] Casa G, Parretta D A, Castrillo A, Wehr R, Gianfrani L 2007 *J. Chem Phys* **127** 084311
[15] Merlone A, Moro F, Castrillo A, Gianfrani L 2010 *Int. J. Thermophys.* **31** 1360
[16] Lisak D, Hodges J T, Ciurylo R 2006 *Phys. Rev. A* **73**, 012507
[17] De Vizia M D, Rohart F, Castrillo A, Fasci E, Moretti L, Gianfrani L 2011 *Phys. Rev. A* **83** 052506
[18] Berman P R 1972 *J. Quant. Spectrosc. Radiat. Transfer* **12** 1331
[19] Pickett H M 1980 *J. Chem. Phys.* **73** 6090
[20] Rohart F, Mäder H, Nicolaisen H-W 1994 *J. Chem. Phys.* **101**, 6475
[21] Ciuryło R, Jaworski R, Jurkowski J, Pine A S, Szudy J 2001 *Phys. Rev. A* **63**, 032507
[22] Daussy C, Guinet M, Amy-Klein A, Djerroud K, Hermier Y, Briaudeau S, Bordé C J, Chardonnet C 2007 *Phys. Rev. Lett.* **98** 250801
[23] De Vizia M D, Moretti L, Castrillo A, Fasci E, Gianfrani L 2011 *Mol. Phys.* **109** 2291
[24] Cygan A, Lisak D, Trawiński R S, Ciurylo R 2010 *Phys. Rev. A* **82** 032515
[25] Triki M, Lemarchand C, Darquié B, Sow P L T, Roncin V, Chardonnet C, Daussy C 2012 *Phys. Rev. A* **85** 062510

Optical and RF Spectroscopy of Spin Noise

E B Aleksandrov [1] and V. S. Zapasskii [2]
[1] Ioffe Physico-Technical Institute, Saint-Petersburg, Russia
[2] Saint-Petersburg State University, Spin Optics Laboratory, Saint-Petersburg, Russia

E-mail: zap@vz4943.spb.edu

Abstract. In the lecture, we briefly outline basic ideas underlying the method of spin noise spectroscopy and consider informative capabilities of the optical and RF spectroscopy of spinnoise power detected by means of magneto-optics.

1. Introduction

The method of spin noise spectroscopy (SNS) proposed in early 80s [1] and substantially developed during the last decade [2, 3] is considered primarily as a specific optical technique of magnetic resonance. Spectral shape of the RF "noise signal", in the vicinity of the Larmor frequency of the spin-system, is determined by its magnetic characteristics and, in fact, provides magnetic resonance spectrum of the system. Behavior of the other degree of freedom of the Faraday-rotation-based SNS, related to its optical channel, is controlled not only by magneto-optical characteristics of the system, but also, in a rather sophisticated way, by interplay of optical transitions associated with different spin-systems of the paramagnet. [4].

2. Basics of the SNS

The basic idea of the spin noise spectroscopy (SNS) is very simple. Magnetization of a spin-system in an external magnetic field is known to be the result of summation of a multitude of projection of magnetic moments, each of them being a random quantity. The equilibrium nonzero magnetization is created due to the difference in probabilities for each particular spin to be oriented parallel or anti-parallel to the applied field. As a result, the field dependence of the equilibrium magnetization of the spin-system is described by the Brillouin-type function, and, for an isotropic paramagnet, the spin-system magnetization, in a fixed magnetic field, has a definite fixed value and is directed exactly along the applied field.

This is correct, however, only to within magnetization fluctuations, which are inevitable for a paramagnet with a finite number of spins. If we look at the tip of the magnetization, so to say, through a microscope, we will see that it experiences fluctuations both in magnitude and in direction. These two types of magnetization fluctuations – along and across the applied magnetic field – appear to be essentially different (Fig. 1).

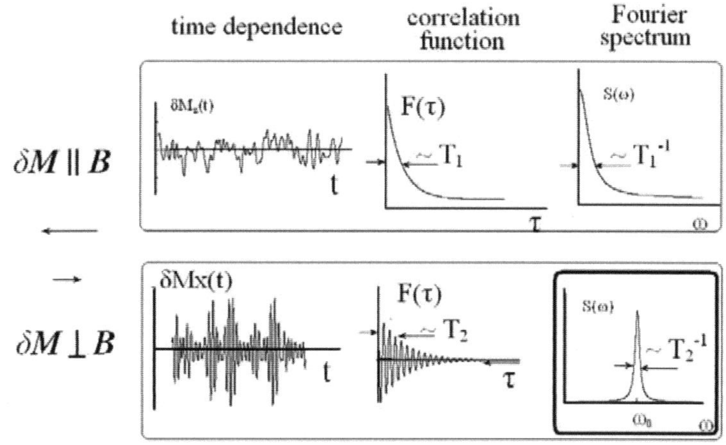

Fig. 1. Time dependences, correlation functions, and spectra of magnetization fluctuations in the longitudinal (δ**M** ∥ **H**) and transverse (δ**M** ⊥ **H**) magnetic field.

Characteristic correlation time of the longitudinal magnetization fluctuations is controlled by the so-called longitudinal relaxation time of the spin-system (T_1), so that the correlation function of the process has an exponential shape with the decay constant ~ T_1. In conformity with the Wiener-Khinchin theorem, spectrum of this magnetization noise, represented by the Fourier transform of the correlation function, is shaped as a Lorentzian with the width ~T_1^{-1} centered at zero frequency.

An essentially different behavior is revealed by fluctuations of the transverse magnetization. Any random nonzero transverse component of the magnetization will evidently exhibit Larmor precession around the applied magnetic field. This precession will last during the time of about T_2 (dephasing time or transverse relaxation time) and then will be replaces by another realization of transverse random component with different magnitude and phase. As a result, the transverse magnetization has to reveal an excess noise at the frequency of Larmor precession. Correspondingly, the correlation function of this process will have an oscillatory behavior with the decay time ~ T_2, while its spectrum will be presented by a Lorentzian with the width T_2^{-1} centered at Larmor frequency of the spin-system (Fig.1). In other words, the transverse magnetization noise reveals the spectrum which we usually refer to as the magnetic resonance (EPR) spectrum. So, the paramagnet in an external magnetic field, under equilibrium conditions, should reveal its EPR spectrum without any external perturbation.

All this looks good on paper. In reality, for a macroscopic paramagnet, these thermodynamic fluctuations appear to be extremely small, and the experimental problem of their detection may seem, at first sight, hardly solvable. Fortunately, there exist polarimetric methods of magneto-optics that allow one to detect tiny variation of magnetization of a paramagnet practically with no perturbation of the spin-system.

This experimental technique is based on the fact that the magnetic circular anisotropy of a paramagnet (circular birefringence, responsible for the Faraday rotation, and circular dichroism, responsible of the field-induced ellipticity) contains the so-called "paramagnetic" contribution which is directly proportional to spin-system magnetization [5]. The most convenient method of optical detection of magnetization implies the use of the Faraday rotation (FR), which can be observed in the region of transparency where the spin-system can be probed in a perturbation-free way. Sensitivity of the FR technique to small variations of the magnetization is determined not only by the proportionality factor controlling conversion of magnetization to FR (which depends on the value of spin-orbit interaction in the system and on the light wavelength), but also by polarimetric sensitivity of the experimental setup. The possibility of detecting spontaneous noise of magnetization of real macroscopic spin-systems became clear after achieving shot-noise-limited polarimetric sensitivity lying, for the laser power of several tens of mW, in the range of $10^{-8} - 10^{-9}$ rad (see [6] and references therein).

The first experiment on detection of magnetic resonance in the FR noise spectrum [1] was performed with sodium atoms at a pressure of ~ 10 Torr. Schematic of the experimental setup is shown in Fig. 2. As a light source, we used a cw dye laser tuned to close vicinity of the D_1 or D_2 line of Na. To decrease the number of atoms contributing to the FR noise and thus to increase the level of the detected noise, the laser beam was focused in the cell. Excess noise of the laser source was suppressed by the balanced detector which subtracted correlated intensity fluctuations in two arms of the polarization beamsplitter and summed up unticorrelated intensity fluctuations related to polarization fluctuations of the incident beam. The cell with sodium vapor was placed into a transverse magnetic field created by a pair of Helmholz coils. The output signal of the balanced detector was selectively amplified at a frequency of 1.3 MHz, then quadratically detected (rectified), lock-in amplified at the field modulation frequency, and recorded as a function of the applied magnetic field. The signal of spontaneous magnetic resonance was recorded, in these experiments, with the signal-to-noise ratio (a sort of "noise-to-noise" ration) was about 100:1 for the accumulation time ~ 2 s. This was the first experimental demonstration of feasibility of the spin noise spectroscopy.

Until the beginning of our century, however, the experiment [1] remained more or less unnoticed. A considerable impact to development of the SNS was given by publication of Crooker with coauthors [2], in which the experiment [1] was extended to some other alkali atoms (Rb and K) and by the experiment of Oestrrich et al. [7], in which the authors have managed to apply this technique to solid semiconductor systems. A real technical breakthrough occurred in the SNS in 2007 [8], when the sweeping spectrum analyzed, in the system of data acquisition, was replaced by the one with the real-time fast Fourier transform (FFT) processing system. As a result, the sensitivity of the technique was increased by more than two orders of magnitude, and SNS turned into a real practical tool of the magnetic resonance spectroscopy.

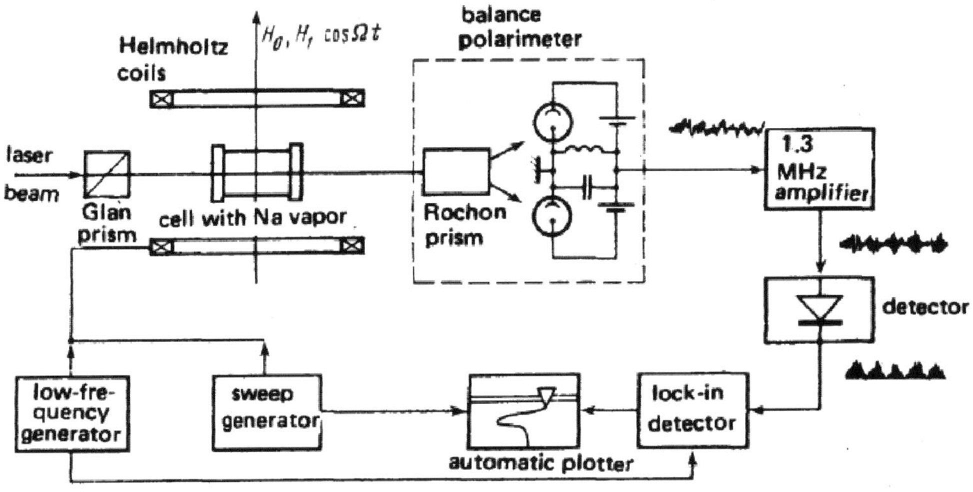

Fig. 2. Schematic of the first experiment on detection of magnetic resonance in the FR noise spectrum.

The use of optical field for monitoring magnetic structure of the spin-system gives birth to two spectroscopic aspects of the technique, related, respectively, to the RF spectrum of the spin noise (magnetic resonance spectrum) and to optical spectrum (wavelength dependence) of the spin noise power (Fig. 3). These two aspects of the SNS are its complementary sides which enrich each other and make the technique much more informative. Consider them successively.

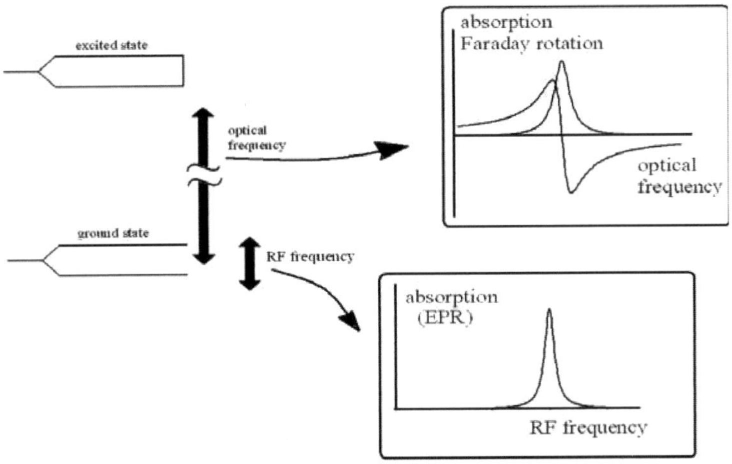

Fig.3. Two spectroscopic aspects of the FR-based SNS.

3. RF spectroscopy of spin noise

Initially, SNS was considered just as a perturbation-free method of detection of magnetic resonance. In other words, all the information was supposed to be obtained from RF spectrum of the spin noise, with the role of the optical channel being restricted by providing maximum value of the signal. Specific properties of the RF spin-noise resonance, whose lineshape and position obey standard laws of the magnetic resonance spectroscopy, are (i) absence of any ac-field-induced distortion of the resonance line (in view of absence of any ac field) and (ii) independence of the spin noise (magnetic resonance) signal of the population difference between magnetic sublevels of the spin-system (no magnetic polarization of spin-system is needed to detect the spin-resonance signal). In addition, the FR-based SNS reveals the typical for optical methods high special resolution achieved under condition of tight focusing of the probe beam (which is anyway needed to increase sensitivity of the technique). For localized spins, this fe4ature allows one to study relief of inhomogeneity of the sample at high special frequencies and may considerably mitigate requirements to the magnetic field homogeneity. For non localized spins, as, e.g., in atomic vapors or in semiconductors with free carriers, there arises an addition "time-of-flight" broadening, which depends on the light beam cross section and can be used to get information about kinetic characteristics of the spin ensemble. In particular in [9-11], this opportunity was used to distinguish between localized and delocalized conduction-band electrons in n-GaAs and to detect the effect of Brownian motion of the electrons in the conduction band on the spin noise line width.

4. Optical spectroscopy of spin noise

Magnetization of a spin-system is converted into the FR in accordance with the spectral behavior of the latter. This spectrum, in turn, for an isolated absorption band, has a dispersion-like shape as shown in Fig. 4a Accordingly, the FR noise power (integrated over the RF spectrum) is expected to be proportional to the FR (or Verdet constant) squared, vanishing at the center of the band where the FR turns into zero (Fig. 4b). This is, however, not necessarily the case [4].

When an isolated absorption band (or a more complicated optical spectrum) is related to a single spin-system or, in other words, the absorption spectrum of the ensemble coincides with that of a single spin-system, then, indeed, dependence of the spin noise power on optical frequency is described by FR squared (Fig. 4,b), and the behavior of the spin noise spectrum with optical frequency will vary in magnitude as shown in Fig. 4,c. If, however, the absorption spectrum comprises several components related to different spin-systems, the situation changes.

Journal of Physics: Conference Series **397** (2012) 012030 doi:10.1088/1742-6596/397/1/012030

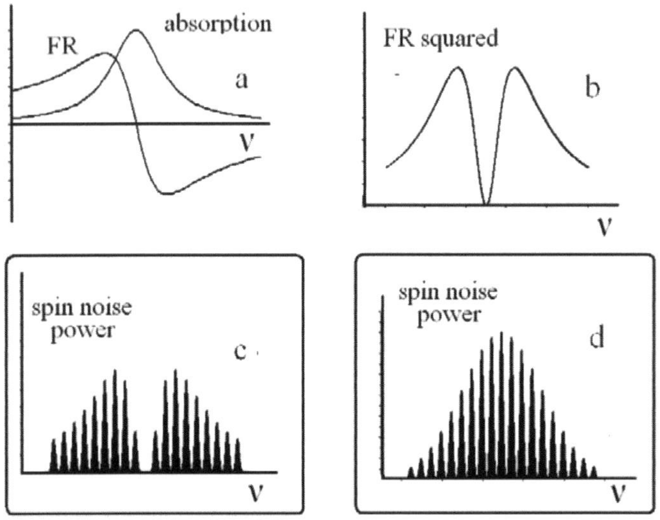

Fig. 4. Optical spectra of absorption and FR (a), FR squared (b), spin noise resonance for the case of homogeneous (c) and strongly inhomogeneous (d) broadening.

Consider, as an example, an optical spectrum comprised of two closely spaced components (Fig. 5a). The FR spectrum of the system will look as shown in Fig. 5b regardless of whether these components are related to one and the same spin-system or not. At the same time, spectral behavior of the FR noise (spin noise power) will be essentially different for these two cases. While in the case of a single spin-system fluctuations of the FR of the two components appear to be correlates, and the FR noise is described by the total FR squared, in the second case, fluctuations of the FR in two components are uncorrelated, and the optical spectrum of the FR noise will be described by the sum of squared of the two partial contributions of the components. As a result, optical spectra of the spin noise, in these two cases appear to be strongly different (Fig. 5, c, d).

The most spectacular example of this difference is given by the case of inhomogeneously broadened spectral bands, when a great multitude of spectral components comprising the band provide independent contributions to the FR noise. In this case, the FR noise of the inhomogeneously broadened band looks as shown in Fig. 4, d, with no dip in the center of the band, where the FR proper vanishes. As the ratio of inhomogeneous broadening to homogeneous increases, the depth of the dip in the optical spin noise spectrum decreases and, what is highly important, the magnitude of the noise signal increases (Fig. 6). This fact provides additional advantages of the spin noise spectroscopy with respect to inhomogeneously broadened systems. These conclusions have been confirmed experimentally in [4].

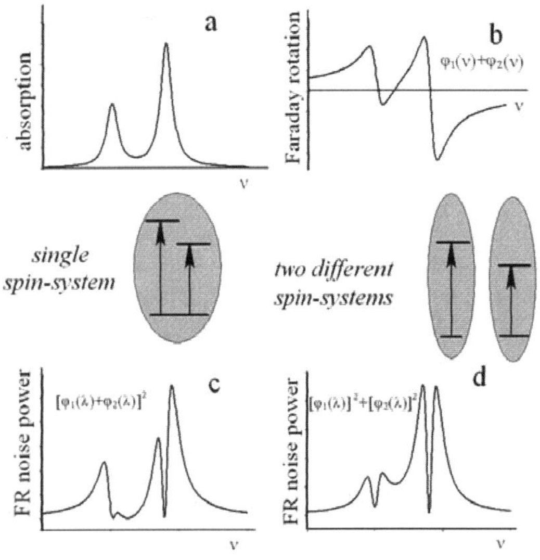

Fig. 5. An example of two closely spaced components of the optical spectrum. The absorption spectrum (a), FR spectrum (b), and optical spectra of spin noise power for the case when two components of the optical spectrum are related to a single spin-system (c) or to different ones (d).

5. Conclusions

The spin noise spectroscopy which is getting more and more popular for studying diverse paramagnets possesses two channels of spectroscopic information, corresponding to the RF and optical frequency range and contained primarily in the lineshape of the RF and optical resonances of spin noise power. These two channels in combination considerably widen experimental potential of this promising tool of the EPR spectroscopy providing it with new sources of information.

Journal of Physics: Conference Series **397** (2012) 012030 doi:10.1088/1742-6596/397/1/012030

References

[1] E. B. Aleksandrov and V. S. Zapasskii, JETP **54**, 64 (1981)
[2] S. A. Crooker, D. G. Rickel, A. V. Balatsky, and D. Smith, Nature, **431**, 49 (2004)
[3] G. M. Muller, M. Oestreich, M Romer, and J. Hubner, Physica E, **43**, 569 (2010)
[4] V. S. Zapasskii, A. Greilich, S. A. Crooker, Y. Li, G. G. Kozlov, D. R. Yakovlev, D. Reuter, A. D. Wieck, and M. Bayer, 2012, arXiv:1206.4168v1 [cond-mat.mes-hall]
[5] V. S. Zapasskii, "Optical Detection of Spin-System Magnetization in Rare-Earth activated Crystals and Glasses", Spectroscopy of Solids Containing Rare-Earth Ions, Eds. A. A. Kaplyanskii and M. F. Macfarlane, Elsevier Science Publishers B.V., 1987, pp. 674-811.
[6] V. S. Zapasskii, J. Appl. Spectroscopy, **37**, 181-196 (1982)
[7] M. Oestreich, M. Romer, R. G. Haug, and D. Hagele, Pys. Rev. Lett., **95,** 216603 (2005)
[8] M. Römer, J. H. Hubner, and M. Oestreich, Rev. Sci. Instrum., **78,** 103903 (2007).
[9] S. A. Crooker, L. Chrng, and D. L. Smith, Phys. Rev. B, **79**, 035208 (2009)
[10] G. M. Muller, M. Romer, D. Schuh, W. Wegscheider, J. Hubner, and M. Oestreich, Phys. Rev. Lett., **101**, 206601 (2008)
[11] G. Muller, D. Schuh, J. Hubner, M. Oestreich, Phys. Rev. **B 81**, 075216 (2010)

Quasimolecular absorption and emission near the forbidden atomic line Cd $\left(5^3P_2 - 5^1S_0\right)$ in Cd + Ar and Kr collisions

O S Alekseeva[1,2,4], **A Z Devdariani**[2,3], **M G Lednev**[1], **A L Zagrebin**[1,2]

[1]Department of Physics, Baltic State Technical University, St. Petersburg, 190005, Russia
[2]Institute of Physics, St. Petersburg State University, Peterhof, 198504, Russia
[3]Department of Theoretical Physics, Herzen State Pedagogical University of Russia, St. Petersburg, 191186, Russia

E-mail: o.alek@rambler.ru

Abstract. We report on the theoretical study of qusimolecular absorption and emission near the forbidden atomic line Cd $\left(5^3P_2 - 5^1S_0\right)$ in a mixture of cadmium vapor with inert gases (argon and krypton). By using the semiempirical method of quasimiolecular term analysis and available experimental potential curves the interaction potentials of the excited atoms Cd^* + Kr and Cd^* + Ar and the probabilities of the radiative transitions were calculated. Based on them the processes of quasimolecular absorption and emission in mixtures of Cd with Ar and Kr atoms were considered and the absorption coefficients and emission spectra were calculated. Also the rate constants of radiative decay of the metastable state of Cd atoms in collisions with Kr and Ar were calculated.

1. Introduction

In this paper the collision-induced absorption and emission near the forbidden line Cd $\left(5^1S_0 - 5^3P_2\right)$ in mixtures of cadmium vapor with krypton and argon are investigated theoretically.

The calculation of spectral characteristics of these processes requires that the potential curves of the interaction of atoms in the ground and the excited states and the square of the dipole moment of the transitions between these states as a function of the interatomic distance be known. Therefore at first the interaction potentials for $Cd\left(5\ ^3P_2\right)$-RG in $\Omega = 0^-,1,2$ states and the probabilities of qasimolecular radiative transitions $1\left(^3P_2\right) - 0^+\left(^1S_0\right)$ were obtained from the avaible experimental interaction potentials for $1(^{1,3}P_1)$ and $0^+(^{1,3}P_1)$ states [1-5] by making use of the semiempirical method of quasimolecular term analysis [6]. Based on these results the processes of quasimolecular absorption and emission in Ar and Kr in collisions with Cd atoms were considered and the absorption coefficients and emission spectra were calculated.

[4] To whom any correspondence should be addressed.

2. Potential curves and probabilities of quasi-molecular radiative transitions

To calculate the potentials of interatomic interaction and the probabilities of quasi-molecular radiative transitions we used the effective Hamiltonian method [7] in the formulation [8].The nonzero matrix elements of the effective Hamiltonian in the basis $\left|{}^{1,3}P_J\Omega_{ic}\right\rangle$ of diabatic quasimolecular wave functions of coupling case c, which are the products of the atomic wave fuctions

$$\left|{}^{1,3}P_J\Omega_{ic}\right\rangle = \left|\mathrm{Cd}\left(5^{1,3}P_J\Omega\right)\right\rangle_{ic}^{at}\left|\mathrm{RG}\left({}^1S_0\right)\right\rangle \tag{1}$$

($\left|\mathrm{Cd}\left(5^{1,3}P_J\Omega\right)\right\rangle_{ic}^{at}$ - is the atomic wave function for intermediate coupling case), are presented in [6]. The potential curves for the excited states were obtained within the framework of the semiempirical method of quasimiolecular term analysis [6]. The method is based on the comparison of four eigenvalues of the matrix of the effective Hamiltonian with interaction potentials for the states $1\left({}^3P_1\right)$, $1\left({}^1P_1\right)$, $0^+\left({}^3P_1\right)$ and $0^+\left({}^1P_1\right)$ reconstructed from experimental data. The point is, all matrix elements of the effective Hamiltonian can be expressed through atomic constants and the quantities ${}^{1,3}H_\sigma(R)$, ${}^{1,3}\mathrm{H}_\pi(R)$, which are represent the potentials of the interaction in ${}^{1,3}\Sigma$ and ${}^{1,3}\Pi$ states without regard for the spin–orbit interaction in the Cd atom. Thus the functions ${}^{1,3}H_\sigma(R)$, ${}^{1,3}\mathrm{H}_\pi(R)$ can be determined from the available experimental potentials for four states and then the potential curves for the remaining four states can be found from these functions by diagonalization of the matrix of the effective Hamiltonian.

Using the quantities ${}^{1,3}H_\sigma(R)$ and ${}^{1,3}\mathrm{H}_\pi(R)$ calculated, the probability $\Gamma\left(1\left({}^3P_2\right),R\right)$ of the quasimolecular radiative transitions $1\left({}^3P_2\right)\rightarrow 0^+\left({}^1S_0\right)$ per unit time at a fixed interatomic distance R can be determined. The adiabatic quasimolecular wave functions $\left|\Omega\left({}^3P_J\right)\right\rangle$ (the eigenfunctions of the effective Hamiltonian) can be presented as linear combinations of the diabatic functions. In particular,

$$\left|1\left({}^3P_2\right)\right\rangle = c_1(R)\left|{}^1P_1 1_{ic}\right\rangle + c_2(R)\left|{}^3P_2 1_{ic}\right\rangle + c_3(R)\left|{}^3P_1 1_{ic}\right\rangle, \tag{2}$$

where the coefficients $c_i(R)$ are determined by diagonalizing the matrix of the effective Hamiltonian. The atomic wave functions for the intermediate coupling case in turn can be presented as linear combination of the wave functions for the LS coupling case with amplitudes a and b:

$$\left|\mathrm{Cd}\left({}^1P_1\Omega\right)\right\rangle_{ic}^{at} = a\left|\mathrm{Cd}\left({}^1P_1\Omega\right)\right\rangle_{LS}^{at} + b\left|\mathrm{Cd}\left({}^3P_1\Omega\right)\right\rangle_{LS}^{at}, \tag{3}$$

$$\left|\mathrm{Cd}\left({}^3P_1\Omega\right)\right\rangle_{ic}^{at} = -b\left|\mathrm{Cd}\left({}^1P_1\Omega\right)\right\rangle_{LS}^{at} + a\left|\mathrm{Cd}\left({}^3P_1\Omega\right)\right\rangle_{LS}^{at}. \tag{4}$$

In the calculations the approximate values of the amplitudes a = 0.998 and b = - 0.062 were used, which, as in [9], were determined from the energies of atomic levels using the semiempirical method of the atomic spectra analysis [10].

To calculate the probabilities $\Gamma\left(\Omega\left({}^3P_J\right),R\right)$ of the quasimolecular radiative transitions in adiabatic wave functions we isolated the contribution from the diabatic functions $\left|{}^1P_1 0^+_{LS}\right\rangle$ and $\left|{}^1P_1 1_{LS}\right\rangle$, since in the diabatic molecular basis $\left|{}^{1,3}P_J\Omega_{LS}\right\rangle$ corresponding to the *LS* coupling case, the dipole moments of the transitions to the ground state are nonzeros only for the ${}^1P_1 0^+_{LS}$ and ${}^1P_1 1_{LS}$ states. In particular,

$$\left\langle {}^1P_1 1_{LS} \middle| 1\left({}^3P_2\right)\right\rangle = ac_1 - bc_3. \tag{5}$$

The dipole moment $\left\langle {}^1P_1 1_{LS} \middle| \hat{d} \middle| {}^1S_0 0\right\rangle$ is associated with the dipole moment $\left\langle {}^3P_1 1_{ic} \middle| \hat{d} \middle| {}^1S_0 0\right\rangle$ of an observed atomic transition. Thus the probability $\Gamma\left(1\left({}^3P_2\right),R\right)$ of the quasi-molecular radiative transition can be expressed in terms of the probability $\Gamma\left({}^3P_1\right)$ and frequency $\omega\left({}^3P_1\right)$ of the atomic intersystem crossing transition and the calculated amplitudes a, b, c_1, c_3:

$$\Gamma\left(1\left({}^3P_2\right),R\right) = \Gamma\left({}^3P_1\right)\left(\frac{\omega\left(1\left({}^3P_2\right),R\right)}{\omega\left({}^3P_1\right)}\right)^3\left(\frac{a}{b}c_1 - c_3\right)^2. \tag{6}$$

Figures 1 and 2 display calculated values of reduced widths

$$\gamma\left(1\left({}^3P_2\right),R\right) = \frac{\Gamma\left(1\left({}^3P_2\right),R\right)}{\Gamma\left({}^3P_1\right)}\left(\frac{\omega\left({}^3P_1\right)}{\omega\left(1\left({}^3P_2\right),R\right)}\right)^3. \tag{7}$$

They are almost independent of the quasi-molecular ground state, which affects the magnitude of $\omega(R)$, and characterize the square of the ratio of the dipole moment of the quasi-molecular transition $1\left({}^3P_2\right) - 0^+\left({}^1S_0\right)$ to the dipole moment of the ${}^3P_1 - {}^1S_0$ atomic transition.

The interaction potentials for the ground U_0 and excited U^* states and the difference potentials ΔU are shown also in Figures 1 and 2. For the ground state the potentials determined in studies [1, 6] have been used.

3. Quasi-molecular absorption and emission near the forbidden line Cd $\left(5^1\mathrm{S}_0 - 5^3\mathrm{P}_2\right)$

For a mixture of cadmium vapor with inert gases, the optical absorption and emission spectrum near the forbidden line Cd $\left(5^1\mathrm{S}_0 - 5^3\mathrm{P}_2\right)$ are determined by the processes

$$\mathrm{Cd}\left(5s^{2\,1}\mathrm{S}_0\right) + \mathrm{RG} + \hbar\omega \rightarrow \begin{cases}\mathrm{Cd}\left(5s5p\,{}^3\mathrm{P}_2\right) + \mathrm{RG}, \\ \mathrm{CdRG}\left(1\left({}^3\mathrm{P}_2\right)\right)\end{cases} \quad \text{(absorption)}$$
$$\mathrm{Cd}\left(5s5p\,{}^3\mathrm{P}_2\right) + \mathrm{RG} \rightarrow \begin{cases}\mathrm{Cd}\left(5s^{2\,1}\mathrm{S}_0\right) + \mathrm{RG} + \hbar\omega, \\ \mathrm{CdRG}\left(X\,{}^1\Sigma\right) + \hbar\omega\end{cases} \quad \text{(emission)} \tag{8}$$

and

$$\mathrm{CdRG}\left(X\,{}^1\Sigma\right)+\hbar\omega \rightarrow \begin{cases} \mathrm{Cd}\ \left(5s5p\,{}^3\mathrm{P}_2\right)+\mathrm{RG}, \\ \mathrm{CdRG}\left(1\left({}^3\mathrm{P}_2\right)\right) \end{cases} \quad \text{(absorption)}$$
$$\mathrm{CdRG}\left(1\left({}^3\mathrm{P}_2\right)\right) \rightarrow \begin{cases} \mathrm{Cd}\ \left(5s^2\,{}^1\mathrm{S}_0\right)+\mathrm{RG}+\hbar\omega, \\ \mathrm{CdRG}\left(X\,{}^1\Sigma\right)+\hbar\omega \end{cases} \quad \text{(emission)} \tag{9}$$

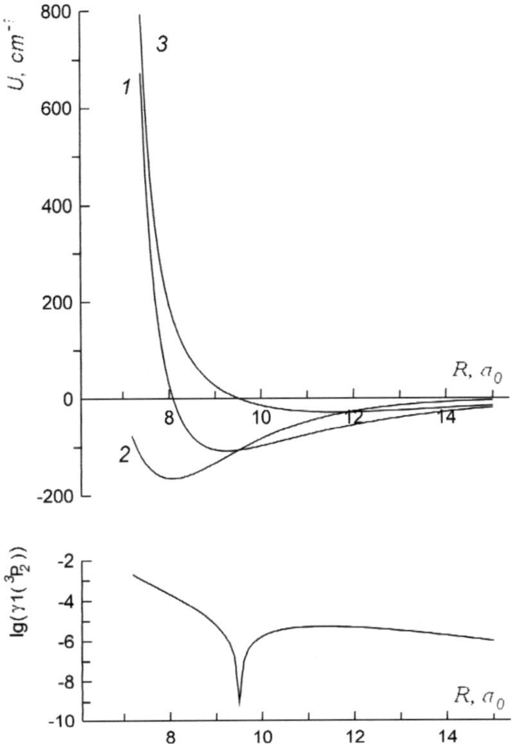

Figure 1. The interaction potentials (1) U^*, (2) U_0, and (3) ΔU (upper) and the reduced probability γ (lower) for the Cd + Kr system

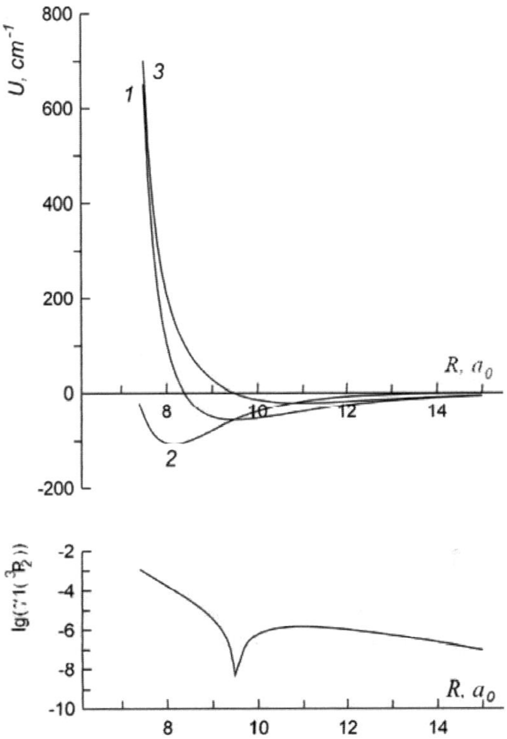

Figure 2. The interaction potentials (1) U^*, (2) U_0, and (3) ΔU (upper) and the reduced probability γ (lower) for the Cd + Ar system

The first of them correspond to free-free and free-bound transitions and the second corresponds to bound-free ana bound-bound transitions.

The spectral distribution of absorption coefficient of a mixture of cadmium vapor with an inert gas in the short-wavelength (with respect to the forbidden atomic line) spectral region, where the role of bound-bound transitions is insignificant, and the spectral distribution of photons emitted in the processes (10) and (11) normalized by the condition $\int I\left(T,\Delta\omega\right)d\Delta\omega=1$, are determined within the framework of the well-known quasi-static approximation [11]

$$K_{abs}(T,\lambda) = \frac{\lambda^2}{8\pi}\frac{g^*}{g_0}\frac{4\pi R_C^2\Gamma(R_C)}{\left|\frac{d\Delta U(R)}{d\hbar R}\right|_{R=R_C}}\exp\left(-\frac{U_0(R_C)}{kT}\right), \tag{10}$$

$$I(T,\Delta\omega) = \frac{2}{5}\frac{4\pi R_C^2\Gamma(R_C)}{K(T)\left|\frac{d\Delta U(R)}{d\hbar R}\right|_{R=R_C}}\exp\left(-\frac{U^*(R_C)}{kT}\right), \tag{11}$$

where λ is the wavelength of quasi-molecular absorption, T is the temperature of the gas mixture, k is the Boltzmann constant, $\frac{g^*}{g_0} = 2$ is the ratio of the statistic weights of the electron states $1(^3P_2)$ and $X\,^1\Sigma$, U_0 and U^* are the interaction potential for atoms on the ground and excited states respectively, ΔU is the difference potential, $\Gamma(R)$ is the probability of the quasi-molecular radiative transition $1(^3P_2) \rightarrow X\,^1\Sigma$, R_C is the Condon point, and the quantity

$$K(T) = \frac{2}{5}4\pi\int_0^\infty \Gamma(R)\exp\left(-\frac{U^*(R)}{kT}\right)R^2dR \tag{12}$$

characterizes the integrated (over the spectrum) rate constant of radiative decay of the metastable state.

The values of $K(T)$ for T = 300 K and T = 700 K are presented in Table 1.

Table 1. Integrated (over the spectrum) rate constants $K(T)$ (in 10^{18} cm^3s^1) of radiative decay of the metastable state in the processes (8) and (9) for CdKr and CdAr

	T, K	
	300	700
CdKr	2.4	3.3
CdAr	1.1	1.9

The results of calculations of the spectral distributions of the absorption coefficients $K_{abs}(T,\Delta\omega)$ and emission spectrum of mixtures of cadmium vapor with krypton and argon for T = 300 K and T = 700 K are presented in the figures 3 and 4.

As follows from the calculation results the position of the maximum of the spectral distibution of the absorption coefficient is shifted toward the shorter wavelenghts slowly enough and the magnitude of the maximum decreases with increasing temperature. The width of the spectral distribution grows with increasing temperature. The maximum of the emission spectrum moves also rather slowly to the short-wavelength side with increasing temperature, whereas the distribution is broadened.

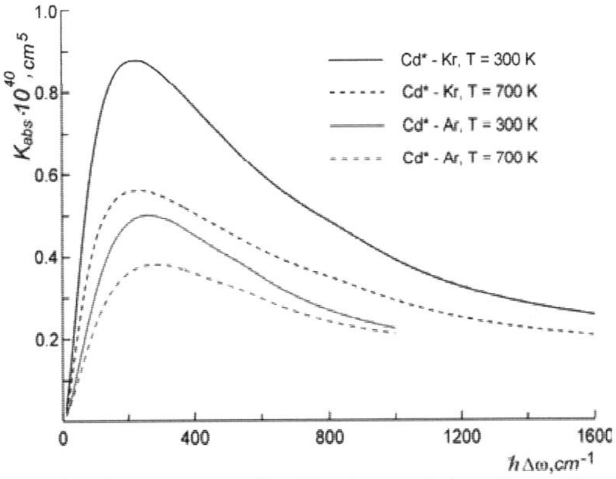

Figure 3. The spectral distributions of the absorption coefficients of mixtures of Cd vapor with Kr and Ar for $T = 300$ K and $T = 700$ K

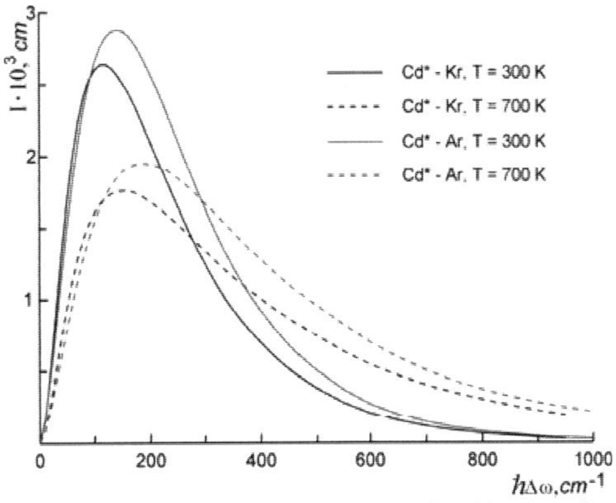

Figure 4. The normalized spectral distribution for the quasimolecular emission of mixtures of Cd vapor with Kr and Ar for $T = 300$ K and $T = 700$ K

4. Conclusion

The emission spectra represent continuous bands shifted to the short-wavelength region with respect to the position of the forbidden atomic line. For $T \geq 300$ K the depth of the potential well of the state $1\left({}^3P_2\right)$ is smaller than kT. Therefore the relative population of bound states is small and states of continuum are preferentially populated. The major contribution to the emission of mixtures of cadmium vapor with Kr and Ar near the forbidden atomic line is made by quasi-molecular radiative transitions in the region of the nearest approach of atoms in the collision process, i.e. the spectrum is formed by the radiative-collisional quenching of the metastable state.

Absorption spectra also represents continuous bands. As follows from the behavior of the interaction potentials and the reduced width the process of absorption is most efficient in the short-wavelength (with respect to the forbidden line) spectral region. The main result of the process of the absorption is the formation of metastable atoms $Cd(^3P_2)$ (long-lived excited molecules CdRG in the state $1\left({}^3P_2\right)$ can be formed only for relatively small shifts), therefore the processes of optical absorption lead to selective population of the metastable state $Cd(^3P_2)$.

References

[1] Funk D J, Kvaran A, Breckenridge W H 1989 *J. Chem. Phys.* **90** 2915
[2] Ruszczak M, Strojecki M, Koperski J 2005 *Chem. Phys. Lett.* **416** 147
[3] Kvaran A, Funk D J, Kowalski A, Breckenridge W H 1989 *J. Chem. Phys.* **89** 6069
[4] Koperski J, Kiełbasa Sz M, Czajkowski M 2000 *Spectrochim. Acta* **56A** 1613
[5] Koperski J, Łukomski M, Czajkowski M 2002 *Spectrochim. Acta* **58A** 2709
[6] Alekseeva O S, Devdariani A Z, Lednev M G, Zagrebin A L 2011 *Russian Journal of Physical Chemistry* B **5** 946
[7] Nikitin E E and Umanskii S Ya 1980 *Semiempirical Methods of Calculation of Interatomic Interaction* (Moscow)
[8] Zagrebin A L and Lednev M G 1994 *Opt. Spectrosc.* **77**, 481
[9] Zagrebin A L, Lednev M G, Tserkovnyi S I 1993 *Opt. Spectrosc.* **74**, 24
[10] Sobel'man I I 1977 *Vvedenie v Teoriyu Atomnykh Spectrov (Introduction to the Theory of Atomic Spectra)* (Moscow)
[11] Gallagher A 1984 *Topics in Applied Physics* **30** 139

Statistical models of scalar collisional interference incorporating phase shifting

John Courtenay Lewis[1] **and Roger M. Herman**[2]
[1]Memorial University of Newfoundland, St. John's NL Canada A1B 3X7
[2]Department of Physics, Pennsylvania State University
E-mail: john.lewis@mun.ca

Abstract. In this paper we include phase shifting in the simple statistical models of collision sequence interference developed earlier by the authors. The Lindholm model of shifting and broadening is used.

1. Scalar collisional interference

Scalar collisional interference is a phenomenon known principally in HD and it's mixtures with hydrogen and rare gases, though it should also be found in HT and DT.

HD has a small permanent dipole moment and small transition moments. These are due to small breakdowns in the Born–Oppenheimer separation.

The permanent or transition moments interact electromagnetically with the (also small) collision–induced dipole moment.

Perhaps the clearest of the HD−He spectra are shown in ref. [1]. Fig. 1 of [1] shows the overall spectrum around the fundamental band with the P, Q, and $R(0)$ and $R(1)$ branches clearly shown, and Fig. 2 of [1] shows the pressure dependence of the shape of the $R(0)$ band with a constant HD density of 14.5 amagat and He density varying from 29.4 amagat to 205 amagat. The inversion of the line shape with increasing He density can clearly be seen.

The"scalar interference" contribution to the R and P transitions is actually a combination, based upon two effects: an interference between the scalar-modulated internal dipole induced in collisions and the small permanent dipole allowed through a breakdown of the Born-Oppenheimer approximation in the mass-asymmetric HD molecule; and an interference between successive scalar–modulated collisionally induced internal dipole transition moments. The designation "scalar collisional interference" refers to the scalar nature of functions of intermolecular separation (which resemble the intermolecular energy function) that modulate the internal dipole. The collision dynamics of importance are those that influence these functions as the HD molecule experiences them throughout collision sequences. The internal dipole then contributes to$\Delta J = \pm 1$ transitions, and can interfere with any other scalarly modulated internal transition operator having dipole symmetry, including the permanent dipole itself.

2. Theoretical and computational studies

The observed line shapes are well fit by a dispersion function [2, 3]:

$$A(\nu) = D_0 \frac{\left(1 + 2\Delta\nu/(q\Gamma)\right]^2}{\left(1 + (2\Delta\nu/\Gamma)^2\right]}.$$

Herman and coworkers [2, 3] argued that

$$q^{-1} = \rho_{FG}\,\Delta'' I / \left(1 + \rho_{FG}\Delta' I\right)$$

with total intensity going as

$$\tilde{\alpha}_1(\rho_{FG}) = \tilde{\alpha}_1(0)\left(1 + 2\rho_{FG}\,\Delta' I + \rho_{FG}^2\left(\Delta'^2 - \Delta''^2\right) I^2\right]$$

and this fits the observational data of ref. [1] and other studies within observational error, over the range of densities used.

Herman and coworkers provided little insight into the various parameters used in the above expressions. Extensive theoretical studies were carried out by J. Cooper and George Tabisz and co–workers (references can be found in the bibliography of [4]) and a thorough computational study was performed by Magnus Gustafsson and Lothar Frommhold [5]. These have provided considerable understanding of the phenomenon at the level of few–body interactions.

3. On the usefulness of simple models

Simple soluble models have proved of inestimable value in the development of statistical mechanics. The first of these was the two–dimensional Ising model with zero field, which was solved by Lars Onsager in 1944 [6].

The study of exactly soluble models, and models which were numerically tractable, was an essential component of the development of the study of critical phenomena in the 1960s and 1970s.

The study of line shapes, in contrast, has moved from simple (simplistic) models to more detailed and realistic ones, to good effect.

The reason is no mystery: correlated collective effects are not very important in typical spectra. Modelling atomic and molecular interactions in pairs and other small groups in great detail is, on the other hand, very important.

Collision–sequence interference offers one partial exception, because in collision–sequence interference the collision–induced spectra show (usually) sharp features which reflect correlations extending over several and in some instances many collisions.

4. "Simple stochastic models"

"Simple stochastic models" of collision–sequence interference are time series in which the actual molecular interactions are replaced with point pulses of integrated intermolecular force (*i.e.* impulses) and dipole moments.

Typically only one particle is followed. After each "interaction" its velocity is changed at random. The induced dipole moments or induced changes are represented as powers of the impulse.

The first and simplest of these stochastic models for collision–sequence interference was described in ref. [7]. That model was two–dimensional, with isotropic scattering, constant speed, and with fixed time steps between successive collisions, and looked only at vector collision–sequence interference.

Scalar interference was treated in the stochastic models in 2008 [8].

It proved relatively easy to generalize to Gaussian velocity distributions, but always with zero persistence of velocity: the distributions of velocity before and after the collision are uncorrelated Gaussians, with the same temperature parameter. It also was straightforward to introduce simple Poisson distributions of intervals between collisions [9]. Collision time distributions do not confirm perfectly to simple Poisson distributions but are close to Poissonian [10, 11].

These models allowed estimation of the infilling of the vector interference dip by differences between the induced dipole moment and the intermolecular force, which is otherwise very difficult to obtain.

They also are useful for developing the numerical methods used to analyse molecular dynamics simulations which involve collision–sequence interference [12].

In these conference proceedings two enhancements of the models are described. The first is the extension from zero persistence of velocity to arbitrary persistence of velocity ($0 \leq \Delta < 1$). This is largely the work of H.T. Wheeler, and is reported elsewhere in these conference proceedings. The results are interesting but not unexpected – primarily a narrowing of the vector interference dip, predicted a long time ago [13].

The second enhancement is the incorporation of a line shifting mechanism into the model. In essence Lindholm broadening is used [14]. H_2 and HD are small nearly spherical molecules and a more elaborate mechanism is not justified in a first step.

5. Assumptions of the present model

It is assumed that N collisions lie in the finite large time interval $(0, T)$,

$$0 \leq t_1 \leq \cdots \leq t_{N-1} \leq t_N \leq T\,.$$

These times are assumed to form a simple Poisson process: The intervals between collisions are assumed independent of velocity.

The velocity after collision k is $\boldsymbol{v}_{k+1}$. The impulse on collision k (mass is unity) is

$$\boldsymbol{f}_k = \boldsymbol{v}_{k+1} - \boldsymbol{v}_k\,.$$

6. Scalar modulation

The scalar modulation of the dipole moment absent shifting is given by

$$\mu(t) = \left[A + \sum_k \mu_k\, \delta(t - t_k)\right] e^{-\iota\omega_0 t}$$

In these models, μ_k is expressed in terms of the impulse $\boldsymbol{f}_k$. In full generality,

$$\mu_k = \mu_k\left[\boldsymbol{f}_k\right]$$

but so far we have usually taken proportionality constant equal to unity:

$$\mu_k = f_k\,.$$

7. Phase shifting

During a collision the frequency of oscillation of the molecule shifts.

In more vigorous collisions, a nonradiative transition may take place, a possibility which can be ignored for simple models of the hydrogens.

We assume that during a collision k the phase of the oscillation changes from $-\iota\omega_0\, t$ to $-\iota\omega_0\, t - \iota\eta_k$.

In the spirit of the simple stochastic models it is assumed that the shift in phase induced in a collision k is given by $\eta_f = \eta\left[\boldsymbol{f}_k\right]$. Usually we take

$$\eta_f = \eta f_k\,.$$

where η is a real number.

Then the time dependence of the allowed part of the induced dipole moment modulation is given by

$$\mu_{\mathrm{A}}\left(t\right) = Ae^{-\iota\omega_0 t-\iota\eta\,\varphi(t)}$$
$$\varphi\left(t\right) = \sum_{k=1}^{N} f_k\, h\left(t-t_k\right)$$

where h is a Heaviside step function.

The induced part of the scalar modulation with phase shifting is

$$\mu_{\mathrm{I}}\left(t\right) = \sum_{k=1}^{N} f_k\,\delta\left(t-t_k\right)e^{-\iota\omega_0 t-\iota\eta\,\varphi(t)}$$

The total scalar modulation of the dipole moment becomes

$$\mu\left(t\right) = \left[A+\sum_{k=1}^{N} f_k\,\delta\left(t-t_k\right)\right]e^{-\iota\omega_0 t-\iota\eta\,\varphi(t)}\,.$$

The Fourier transform of $\mu\left(t\right)$ is $a_T\left(\omega\right)$.

We define

$$\begin{aligned}\tilde{a}\left(\tilde{\omega}\right) = {}& \left(e^{\iota\tilde{\omega}t_1}-1\right)+e^{-\iota\eta f_1}\left(e^{\iota\tilde{\omega}t_2}-e^{\iota\tilde{\omega}t_1}\right)\\ &+e^{-\iota\eta(f_1+f_2)}\left(e^{\iota\tilde{\omega}t_3}-e^{\iota\tilde{\omega}t_2}\right)+\dots\\ &+e^{-\iota\eta(f_1+\ldots+f_{N-1})}\left(e^{\iota\tilde{\omega}t_N}-e^{\iota\tilde{\omega}t_{N-1}}\right)\\ &+e^{-\iota\eta(f_1+\ldots+f_N)}\left(e^{\iota\tilde{\omega}T}-e^{\iota\tilde{\omega}t_N}\right)\end{aligned}$$

and the closely related function

$$\begin{aligned}\hat{a}\left(\tilde{\omega}\right) = {}& e^{\iota\tilde{\omega}t_1}\left(1-e^{-\iota\eta f_1}\right)+e^{-\iota\eta f_1}e^{\iota\tilde{\omega}t_2}\left(1-e^{-\iota\eta f_2}\right)\\ &+e^{-\iota\eta(f_1+f_2)}e^{\iota\tilde{\omega}t_3}\left(1-e^{-\iota\eta f_3}\right)\\ &+\ldots+e^{-\iota\eta(f_1+\ldots+f_{N-1})}e^{\iota\tilde{\omega}t_N}\left(1-e^{-\iota\eta f_N}\right)\end{aligned}$$

The unaveraged periodogram can be written in terms of $\hat{a}$ and $\tilde{a}$:

$$\begin{aligned}S_T\left(\omega\right) &\equiv \frac{1}{T}\left|a_T\left(\omega\right)\right|^2\\ &= \frac{1}{T}\left|A\right|^2\frac{\left|\tilde{a}\left(\tilde{\omega}\right)\right|^2}{\tilde{\omega}^2}+\frac{2}{T\eta}\,\mathfrak{Re}\,A\frac{\hat{a}\left(\tilde{\omega}\right)\tilde{a}\left(\tilde{\omega}\right)^*}{\tilde{\omega}}+\frac{1}{T\eta^2}\left|\hat{a}\left(\tilde{\omega}\right)\right|^2\end{aligned}$$

8. A surprising fact

From the definition of $\hat{a}$ it can be seen that

$$\hat{a}\,(0) = 1 - e^{-\iota\eta(f_1+\ldots+f_N)} = \mathcal{O}\,(1)$$

whereas, for $\widetilde{\omega} \neq 0$,

$$\hat{a}\,(\widetilde{\omega})_{\widetilde{\omega}\neq 0} = \mathcal{O}\left(\sqrt{N}\right)$$

The remarkable consequence of this is that for $\widetilde{\omega} = 0$, that is for $\omega = \omega_0$, the pure induced spectrum goes to zero:

$$S_{\mathrm{II}}\,(\omega = \omega_0) = \lim_{T\to\infty} \frac{1}{T\eta^2} \left\langle |\hat{a}\,(0)|^2 \right\rangle = 0\,.$$

9. Simulations

This system is easy to simulate. Normal deviates can easily be generated, and these give the velocities. As for the intervals between collisions, generating exponentially distributed random variables is easy but it is easier still to use a uniform random number generator on $(0\,,T)$ to generate N points. If N is large enough (and we use 1000 to 10000 points in a run) the distribution is very close to Poisson. We used the Gnu Scientific Library, which has several random number generators, and most of the simulations were run on a laptop. Each run was repeated about 200 times with the same parameters, and the results were averaged.

The figure Fig. 1 below shows the allowed component for one set of parameters ($A = 20$, $\eta = 0.1$). Shifting as well as broadening is clearly evident.

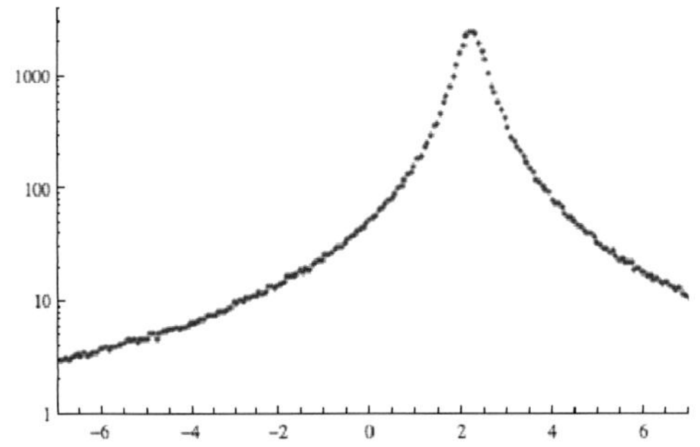

Figure 1. Allowed spectrum, log intensity against $\widetilde{\omega} = \omega - \omega_0$.

The pure induced component is shown in Figs. 2 and 3 below. It is strongly asymmetric and goes (almost) to zero for $\widetilde{\omega} = 0$, as predicted. This is much more clearly seen when log intensity is plotted against $\widetilde{\omega}$. For reasons unknown the cross spectrum is small in this example.

We have not endeavoured to obtain the spectra for this model analytically. However, Roger Herman suggested a simplification, in which the phase shift is identically $\eta =$ constant in every collision. The resulting expressions for the components of the spectrum can then be evaluated, and the total spectrum is found to be

$$\begin{aligned} S\,(\omega) &= S_{\mathrm{AA}}\,(\omega) + S_{\mathrm{AI}}\,(\omega) + S_{\mathrm{II}}\,(\omega) \\ &= \bar{\nu}\,(1-\cos\eta)\,\frac{2A^2 + 2A\,\widetilde{\omega}\cos\eta + \frac{\widetilde{\omega}^2}{\eta^2}}{4\,\bar{\nu}^2\sin^4\frac{\eta}{2} + (\widetilde{\omega} - \bar{\nu}\sin\eta)^2}\,. \end{aligned}$$

A number of changes can be rung on these models but we will leave them for now.

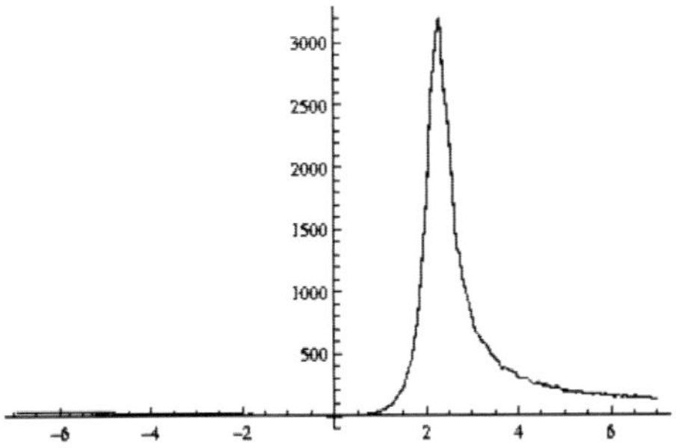

Figure 2. Pure induced spectrum, intensity against $\widetilde{\omega} = \omega - \omega_0$.

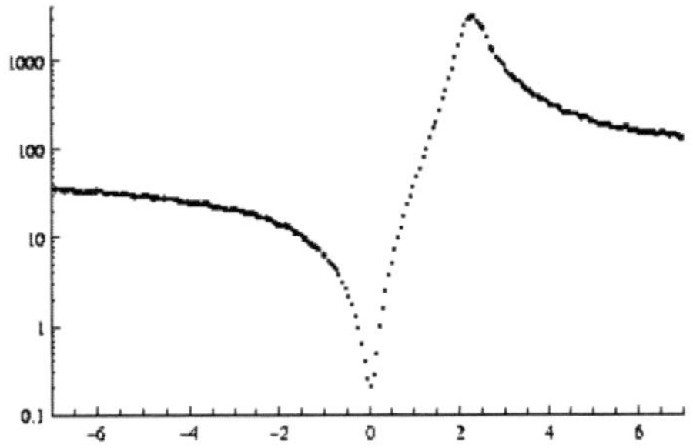

Figure 3. Pure induced spectrum, log intensity against $\widetilde{\omega}$.

10. Conclusions

The inclusion of phase shifting into the stochastic model drastically modifies the resulting spectra. These become strongly asymmetric, and in fact develop a deep minimum which may actually be zero.. However, they not replicate the observed behaviour of actual gas phase HD spectra, though they do seem to resemble the spectra of solid HD. The reasons for this are obscure at present.

References

[1] McKellar A R W and Rich N, 1984 *Can. J. Phys.* **62** 1665–72
[2] Herman R M 1979 *Phys. Rev. Lett.* **42** 1206–9
[3] Herman R M, Tipping R H and Poll J D 1979 *Phys. Rev.* A **20** 2006–12
[4] Lewis J C and Herman R M 2011 *International Review of Atomic and Molecular Physics* **2** 25
[5] Gustafsson M and Frommhold L 2001 *Phys. Rev.* A **63** 052514
[6] Onsager L 1944 *Phys. Rev.* **65** 117 – 49
[7] Lewis J C 2001 *Proc. of the 15th Int. Conf. on Spectral Line Shapes* ed Seidel J (AIP Conference Proceedings 559) 394–6
[8] Lewis J C 2008 *Phys. Rev.* A **77** 062702
[9] Lewis J C 2009 *International Journal of Spectroscopy* 561697
[10] Chapman S and Cowling T G 1964 *Mathematical Theory of Non-Uniform Gases* (Cambridge University Press)
[11] J. C. Lewis 1980 *Chem. Phys. Lett.* **76** 96–100
[12] Lewis J C 2002 *Proc. of the 16th Int. Conf. on Spectral Line Shapes* ed Back C (AIP Conference Proceedings 645) 394–400
[13] Lewis J C and van Kranendonk J 1972 *Can. J. Phys.* **50** 2902–13
[14] Lindholm E 1942 *Über die Verbreiterung und Verschiebung von Spektrallinien: experimentelle und theoretische Beiträge* thesis Stockholms högskola

Spectral line broadening of atoms by noble gases

A V Demura[1], S Ya Umanskii[2], A V Scherbinin[3], A V Zaitsevskii[4]

[1]Tokamak Physics Institute, National Research Center "Kurchatov institute", Kurchatov Sq. 1, Moscow 123182, Russia
[2]N.N. Semenov Institute of Chemical Physics of Russian Academy of Sciences, Kosigina Street 4, Moscow 117992, Russia
[3]Faculty of Chemistry, Moscow State University, Leninskie Gory, Moscow 119991, Russia
[4]Kurchatov NBIC center, National Research Center "Kurchatov institute", Kurchatov Sq. 1, Moscow 123182, Russia

E-mail: **Alexander.Demura@hepti.kiae.ru**

Abstract. The complex universal approach for calculations of the binary broadening of atomic spectral lines by buffer noble gas atoms is presented. The interatomic interaction is described on the basis of the Buckingham potential, parameterized with the help of ***ab initio*** electron modeling. The Unified Frank-Condon Theory is used for the calculations of the total line profile. For the first time the impact limit is determined exactly using the Van der Waals long-range tail of the potential and taking into account the degeneracy of quasimolecular states but neglecting non-adiabatic effects. The general asymptotic wings dependences of the profile are established for the Buckingham parameterized potential. The receipts of the wings cut-offs are formulated for the approximate but sufficiently accurate evaluation of the total line profile using only the Van der Waals part of potential to be applied in large scale calculations of radiative transfer.

1. The impact Van der Waals broadening with account of degeneracy

The impact width of the spectral line corresponding to the dipole transition of the compound system of the radiator atom **A** and buffer gas atom **X** is related to the elements of scattering matrix, expressed in the form (compare with [1])

$$\sigma^{(1)}_{\gamma_i j_i;\gamma_f j_f}(\varepsilon_t) = \mathrm{Re}\left(\frac{\pi}{k^2}\sum_{J_i J_f}\sum_{ll'}(-1)^{l-l'}(2J_i+1)(2J_f+1)\begin{Bmatrix} J_f & J_i & 1 \\ j_i & j_f & l \end{Bmatrix}\begin{Bmatrix} J_f & J_i & 1 \\ j_i & j_f & l' \end{Bmatrix}\left[\delta_{ll'} - S^{J_i,\gamma_i j_i}_{l';l}(\varepsilon_t)\cdot\right.\right.$$
$$\left.\left.\cdot S^{J_f,\gamma_f j_f}_{l';l}(\varepsilon_t)\right]\right). \qquad (1)$$

Here $\hbar\mathbf{J}_i$, $\hbar\mathbf{J}_f$ are total (electronic + nuclear) angular momenta of the diatom **A–X**, $\hbar\mathbf{l}$, $\hbar\mathbf{l}'$ are angular momenta of the relative motion of **A** and **X**, **k** is the wave number of the relative motion of **A** and **X** at infinite separation, $S^{J_i,\gamma_i j_i}_{l';l}(\varepsilon_t)$ and $S^{J_f,\gamma_f j_f}_{l';l}(\varepsilon_t)$ are scattering matrix elements for $\mathbf{A}(\gamma_i \mathbf{j}_i)$–**X** and $\mathbf{A}(\gamma_f \mathbf{j}_f)$–**X** collisions at relative kinetic energy $\varepsilon_t = \hbar^2 k^2 / 2\mu$, μ is the **A–X** reduced mass, γ stands for the corresponding sets of quantum numbers for the initial **i** and final states **f** of **A**, while the curly brackets designate **6-j** symbols [2]. Since the impact broadening corresponds to elastic distant

Published under licence by IOP Publishing Ltd

collisions the main contribution to the above sum is given by **J**$_i$**, J**$_f$ >> 1. At the same time **j**$_i$, **j**$_f$ ~ **1** as mostly lower excited levels are of importance for problem under consideration. Then **ℓ, ℓ'** may differ from **J**$_i$, **J**$_f$ by the value of the order of 1, and the difference between **J**$_i$ and **J**$_f$ is also of the order of **1**. Keeping this in mind it is suitable to introduce the new summation variables $J=(J_i+J_f)/2$, $\rho=J_i-J_f$, $m=J_i-\ell$, $m'=J_i-\ell'$. As **J >> 1** the asymptotic of **6-j** symbols could be used (see, e.g. [3]), and the expression for the broadening cross section $\sigma^{(1)}_{\gamma_i j_i;\gamma_f j_f}(\varepsilon_t)$ could be simplified as follows

$$\sigma^{(1)}_{\gamma_i j_i;\gamma_f j_f}(\varepsilon_t)=\mathrm{Re}\left\{\frac{\pi}{k^2}\sum_J 2J\sum_{mm'\rho}\frac{1}{3}(-1)^{m-m'}\begin{bmatrix} j_i & j_f & 1\\ m & \rho\text{-}m & m\end{bmatrix}\begin{bmatrix} j_i & j_f & 1\\ m' & \rho\text{-}m' & m'\end{bmatrix}\right.$$
$$\left.\left[\delta_{mm'}-e^{-i\pi\rho}S^{J+(\rho/2),\gamma_i j_i}_{m';m}(\varepsilon_t)S^{J-(\rho/2),\gamma_f j_f}_{m'-\rho;m-\rho}(\varepsilon_t)\right]\right\}, \quad (2)$$

where square brackets designate **3-j** symbols [2]. Since the distant collisions correspond to the Van der Waals tail of real potential, the matrix elements of interaction in the Hund case (**c**) could be represented in the collisions frame in the form

$$\langle\gamma jm'|V^{(d)}(R)|m''j\gamma\rangle=-\frac{C_6^{(0,j\gamma)}}{R^6}\left[\delta_{m'm''}-B_6^{(2,j\gamma)}F(j,m',m'')\right], \quad (3)$$

$$C_6^{(0,j\gamma)}=\frac{1}{\pi\,(2j+1)^{1/2}}\int_0^\infty\langle\gamma j\,\|\alpha_A^{(0)}(\omega)\|\,\gamma j\rangle\langle{}^1S\,\|\alpha_X^{(0)}(\omega)\|\,{}^1S\rangle d\omega, \quad (4)$$

$$B_6^{(2,j\gamma)}=\frac{\int_0^\infty\langle\gamma j\,\|\alpha_A^{(2)}(\omega)\|\,\gamma j\rangle\langle{}^1S\,\|\alpha_X^{(0)}(\omega)\|\,{}^1S\rangle d\omega}{\sqrt{2}\int_0^\infty\langle\gamma j\,\|\alpha_A^{(0)}(\omega)\|\,\gamma j\rangle\langle{}^1S\,\|\alpha_X^{(0)}(\omega)\|\,{}^1S\rangle d\omega}, \quad (5)$$

$$F(j,m',m'')=\sum_{\tilde\Omega=-j}^{j}D^j_{\tilde\Omega m'}(0,\pi/2,\pi)*\begin{bmatrix} j & 2 & j\\ \tilde\Omega & 0 & \tilde\Omega\end{bmatrix}D^j_{\tilde\Omega m''}(0,\pi/2,\pi), \quad (6)$$

where **R** is the **A-X** internuclear distance; $\alpha_A^{(0)}(\omega),\alpha_A^{(2)}(\omega),\alpha_X^{(0)}(\omega)$ are reduced components in the $\{\boldsymbol{\kappa},\mathbf{q}\}$ representation of the irreducible tensor of dynamical polarizability of radiator atom **A** and buffer gas atom **X**; $D^j_{\tilde\Omega m''}(\alpha,\beta,\gamma)$ are the Wigner rotation matrices [2]. The quantities **ħm**, **ħm'** are the projections of the atom **A** electronic angular momentum upon the direction of the total (electronic + nuclear) angular momentum of the diatom **A–X**, that is perpendicular to the collisions frame. Using semiclassical approximation and assuming rectilinear trajectories, it is possible to obtain analytical result

$$\sigma^{(1)}_{\gamma_i j_i;\gamma_f j_f}(\varepsilon_t)=2\pi\int_0^\infty b\,db\sum_{m,\rho}\frac{1}{3}\begin{bmatrix} j_i & j_f & 1\\ m & \rho\text{-}m & m\end{bmatrix}^2\times\left\{1\text{-}\cos\left[\frac{3\pi\left(C_6^{(0,j_i\gamma_i)}-C_6^{(0,j_f\gamma_f)}\right)}{8\hbar b^5u}\cdot\Gamma^{(j_i\gamma_i;j_f\gamma_f)}_{m;\rho}\right]\right\}, \quad (7)$$

$$\Gamma^{(j_i\gamma_i;j_f\gamma_f)}_{m;\rho}=1+\frac{1}{C_6^{(0,j_i\gamma_i)}-C_6^{(0,j_f\gamma_f)}}\left[C_6^{(0,j_f\gamma_f)}B_6^{(2,j_f\gamma_f)}F(j_f,m-\rho,m-\rho)-C_6^{(0,j_i\gamma_i)}B_6^{(2,j_i\gamma_i)}F(j_i,m,m)\right]. \quad (8)$$

In (5) **b** is the impact parameter and **u** is the relative velocity of motion. It was assumed also that the angular momentum of the relative nuclear motion is **ħJ** = $\mu u b$. Performing in (7) integration over impact parameters **b** and averaging over thermal velocities **u**, assuming Maxwell distribution, one could obtain at last for the impact halfwidth $\Delta\omega_{1/2}^{(\mathrm{imp})}$

$$\Delta\omega_{1/2}^{(\mathrm{imp})} = \left(\frac{1}{2^{16/5}}\cdot\left[3^{2/5}\cdot(1+\sqrt{5})\cdot\pi^{7/5}\cdot\Gamma\left(\frac{3}{5}\right)\right]\cdot\left(\frac{4}{\pi}\right)^{1/5}\cdot\Gamma\left(\frac{9}{5}\right)\right)\cdot n_X\cdot\langle u\rangle\left(\frac{\Delta C_{6,\mathrm{eff}}^{(\mathrm{imp})}}{\hbar\langle u\rangle}\right)^{2/5} \approx 3.9504\cdot n_X\cdot\langle u\rangle\cdot\left(\frac{\Delta C_{6,\mathrm{eff}}^{(\mathrm{imp})}}{\hbar\langle u\rangle}\right)^{2/5}, \tag{9}$$

$$\Delta C_{6,eff}^{(imp)}(\gamma_i j_i;\gamma_f j_f) = \left|C_6^{(0,j_i\gamma_i)} - C_6^{(0,j_f\gamma_f)}\right|\left[\sum_{m\rho}\frac{1}{3}\begin{bmatrix} j_i & j_f & 1 \\ m & \rho\text{-}m & m\end{bmatrix}^2\left|\Gamma_{m;\rho}^{(j_i\gamma_i;j_f\gamma_f)}\right|^{2/5}\right]^{5/2}. \tag{10}$$

Here **<u>** is the mean velocity, $\mathbf{n_X}$ is the density of buffer gas atoms, $\mathbf{\Gamma(z)}$ is the gamma function. The obtained result takes into account strong Coriolis coupling between degenerate quasimolecular states, but the non-diagonal matrix elements between different "m" are neglected. At large distances the Coriolis interaction is larger in absolute value than the splitting between different molecular terms, arising due to the Van der Waals interaction between **A(j)** (with **j>0** or ½) and **X**. That is why the adiabatic approach to the description of Van der Waals broadening is quite adequate, as well as assumption of the rectilinear trajectories of relative **A** and **X** motion and implementation of the quasiclassical theory of atomic collisions (see [4]).

2. Parameterizing of modified Buckingham potential

The study of interatomic interactions showed that they are well approximated by modified Buckingham type of potentials [5]. These potentials have typical repulsive exponential short-range core and the attractive Van der Waals long-range tail

$$U_B(r) = A\exp(-\alpha r) - \frac{C_6}{(r^2+r_0^2)^3}. \tag{11}$$

Although this approximation could not mimic all characteristics of the real potential curves, it possesses by all main properties regarding the behavior of real potentials first of all at small and large interatomic distances. The parameters in (11) for the particular states of the radiator **A** and the buffer gas atom **X** could be found either experimentally or from the quantum chemical calculations (**QCh**). The value of $\mathbf{C_6}$ could be calculated as well in more universal manner using the perturbation theory in molecular basis and confronted with the values given by **QCh**. As an illustration of this approach the **QCh** calculations were performed for the potential curves $\mathbf{X^2\Sigma_{1/2}}$, $\mathbf{A^2\Pi_{3/2}}$ and $\mathbf{B^2\Sigma_{1/2}}$ of the **Na-Xe** system, and the parameters of modified Buckingham potential were retrieved [6]: $\mathbf{X^2\Sigma_{1/2}}$ **, A = 116.869 eV,** $\boldsymbol{\alpha}$ **= 1.654** $\mathbf{Å^{-1}}$**,** $\mathbf{C_6}$ **= 946.502 eV×**$\mathbf{Å^6}$**,** $\boldsymbol{r_0}$ **= 1.504 Å;** $\mathbf{A^2P_{3/2}}$ **, A = 3214.79 eV,** $\boldsymbol{\alpha}$ **= 3.185** $\mathbf{Å^{-1}}$**,** $\mathbf{C_6}$ **= 22.774 eV×**$\mathbf{Å^6}$**,** $\boldsymbol{r_0}$ **= 1.960 Å;** $\mathbf{B^2\Sigma_{1/2}}$ **, A = 102.829 eV,** $\boldsymbol{\alpha}$ **= 1.256** $\mathbf{Å^{-1}}$**,** $\mathbf{C_6}$ **= 3408.543 eV×**$\mathbf{Å^6}$**,** $\boldsymbol{r_0}$ **= 1.566 Å.** The comparison with experimental data and results of other calculations demonstrates that both could be accurately approximated by (11). The coincidence of retrieved from **QCh** calculations $\mathbf{C_6}$ values with ones, obtained within the perturbation theory is also quite reasonable.

3. Total profile within Unified Frank-Condon Theory and wing asymptotic

Basing on the knowledge of parameterized Buckingham potential for the particular quasimolecular states, involved in the certain radiative transition of compound system **A-X**, and having in the pocket the result for the impact width, obtained in the previous section, it is possible to describe the total profile **F(Δω)** of spectral line within Unified Frank-Condon Theory (**UFCT**) [7] in the binary approximation

$$\mathrm{F}(\Delta\omega) = \frac{1}{\pi}\frac{\Phi(\Delta\omega)}{(\Delta\omega)^2+(\Delta\omega_{1/2}^{(\mathrm{imp})})^2},\quad \Phi(\Delta\omega) = \sum_{if} g(i\to f)\,\Phi_{if}(\Delta\omega),\quad g(i\to f) = \frac{N(i\to f)_{\mathrm{AX}}}{N(i_{\mathrm{A}}\to f_{\mathrm{A}})}, \tag{12}$$

$$\Phi_{if}(\Delta\omega) = \pi\, n_X\langle u\rangle\int_0^\infty\frac{\varepsilon\,d\varepsilon}{(k_BT)^2}\exp\left(-\frac{\varepsilon}{k_BT}\right)\times\int_0^\infty 2b\,db\,G_{if}(b,\varepsilon,\Delta\omega)^2, \tag{13}$$

$$G_{if}(b,\varepsilon,\Delta\omega) = \frac{2}{\hbar}\int_0^{\infty} \mathrm{d}t\ \Delta U_{if}\left[R(t,b,\varepsilon)\right]\cos\left[\Delta\omega t - \eta_{if}(t,b,\varepsilon)\right], \tag{14}$$

$$\eta_{if}(t,b,\varepsilon) = \frac{1}{\hbar}\int_0^{t} \Delta U_{if}\left[R(t',b,\varepsilon)\right]\mathrm{d}t', \qquad \Delta U_{if}(r) = U_i(r) - U_f(r). \tag{15}$$

where **$\Delta\omega = (\omega-\omega_0)$**, ω_0 is the frequency of transition in the free atom **A**, ω is the radiation frequency, $G_{if}(b,\varepsilon,\Delta\omega)$ is the reduced Frank-Condon factor, $\eta_{if}(t,b,\varepsilon)$ is the phase shift versus the relative **A-X** motion along the trajectory **$R(t,b,\varepsilon_i)$** with the initial kinetic energy ε_i and the impact parameter **b**, **T** is the temperature, **k_B** is the Boltzmann constant. In (12) $g(i \to f)$ stands for the statistical weight, and $N(i \to f)_{AX}$ is the number of the allowed dipole transitions with **$\Delta\Omega$=0,±1** (**Ω** is the projection of the quasi-molecular electronic angular momentum on internuclear axis) between quasi-molecular electronic states corresponding to the potential curves $U_i(r)$, and $U_f(r)$, while $N(i_A \to f_A)$ is the total number of the allowed dipole transitions $i_A \to f_A$ with **m_i-m_f= 0,±1** (**m_i, m_f** are magnetic quantum numbers of **A**) in a free radiating atom **A**. The dipole moment of quasi-molecular transition is assumed to vary very slowly versus **r** and cancelled during normalization of the profile. It will be assumed for all quasimolecular potentials of interest (**γ=*i*, *f***), that the temperature **T** is sufficiently high, so, that the inequality takes place

$$\mathbf{k_B T \gg D_{e,\gamma}}, \tag{16}$$

where **$D_{e,\gamma}$** is the depth of the well of the potential **$U_\gamma(r)$**.

The construction of the profile is accomplished by splitting it in the two parts, one of which corresponds to the short-range interactions $\Phi^{(1)}_{if}(\Delta\omega)$ and the other $\Phi^{(2)}_{if}(\Delta\omega)$ to the long-range one $\Phi_{if}(\Delta\omega) = \Phi^{(1)}_{if}(\Delta\omega) + \Phi^{(2)}_{if}(\Delta\omega)$. Accordingly the different types of trajectories correspond to the different ranges of the impact parameter variation in (12)-(15). For large enough **b** the trajectories are assumed to be rectilinear (the second type) and do not depend explicitly on the potential, but for small **b** it is necessary to introduce the physically reasonable effective "mean" potential **$U_0(r)$** like

$$U_0(r) = A_0\exp(-\alpha_0 r);\ A_0 = \sqrt{A_i A_f},\ \alpha_0 = (\alpha_i + \alpha_f)/2. \tag{17}$$

The trajectory of the first type in the potential **$U_0(r)$** then could be found analytically using the Takayanagi modified wave number approximation [8], while the characteristic value of the impact parameter **R_0**, separating the two regions of its variation, could be found from the value of diffusion coefficient [9], obtained in the first order Chapman-Enskog approximation for the core part of potential, and equal to: $R_0 \approx \frac{1}{\alpha_0}\ln\left(\frac{A_0}{k_B T}\right)$.

The derivation of the asymptotic dependences of line profiles in the wings are accomplished on the basis of analysis of the behavior of **$G_{if}(b,\varepsilon,\Delta\omega)$** on $\Delta\omega$ as a function of the Massey parameter [1] **$\xi(\Delta\omega,b,\varepsilon)$ = $|\Delta\omega|\tau(b,\varepsilon)$**, where **$\tau(b,\varepsilon)$** is the characteristic variation time of **$\Delta U_{if}[R(t,b,\varepsilon)]\exp[i\eta_{if}(t,b,\varepsilon]$**. The Lorentz center of the profile corresponds to $\Delta\omega$ small enough, where **$\xi(\Delta\omega,b,\varepsilon) < 1$**. It is determined by the Van der Waals interaction, because the Weisskopf radius [4] **$b_W > R_0$**, where $b_W = (3\pi|C_{6,i} - C_{6,f}|/8\hbar < u >)^{1/5}$ according to (7). The wings correspond to large $|\Delta\omega|$, where **$\xi(\Delta\omega,b,\varepsilon)$** > 1, for the wide range of impact parameters and $\varepsilon \sim$ **$k_B T$**. If **$\xi(\Delta\omega, b, \varepsilon)$** > 1, the stationary phase approximation could be applied to evaluate the integral, which value is determined in the vicinity of the stationary phase point **$t_c(\Delta\omega,b,\varepsilon)$**, being the root of the equation **$\Delta\omega = 2\pi\Delta U_{if}[R(t,b,\varepsilon)]/h$**. In fact there is superposition of contributions from different types of trajectories in the red and blue wings and both $\Phi_{ij}^{(1,2)}(\Delta\omega)$ have to be analyzed.

For certainty but without loss of generality consider the argument of cosine in (7) as negative. Then as the minimum of $\mathbf{\Delta U_{if}(r)}$ practically always takes place at $\mathbf{r > R_0}$ and hence $\mathbf{b > R_0}$ the trajectories, which give the main contribution to $\mathbf{\Phi_{if}(\Delta\omega)}$ at negative $\Delta\omega$ (*the red wing of the line*) are rectilinear as soon as $2\pi\mathbf{\Delta U_{if,min}(r)/h} > |\Delta\omega| > \mathbf{<u>/b_W}$, and this part of the red wing is described by the asymmetric classical statistical Van der Waals **(V-W)** distribution [4]

$$I_{red}^{(V-W)}(|\Delta\omega|)=\frac{2\pi n_X\,|\Delta C_{6,if}\,|}{|\Delta\omega|^{3/2}}\exp\left(-\frac{4\pi^3}{9}|\Delta C_{6,if}\,|\frac{n_X^2}{|\Delta\omega|}\right),$$
$$\Delta C_{6,if}<0,\ \Delta\omega=\omega-\omega_{if}<0,\ \ |\Delta C_{6,if}\,|\approx\left|C_6^{(0,j_i\gamma_i)}-C_6^{(0,j_f\gamma_f)}\right|. \tag{18}$$

Remind that at $|\Delta\omega| < \mathbf{<u>/\ b_W}$ the profile has the symmetric Lorentz shape, and within **UFCT** for the scalar **V-W** interaction it gradually transforms around the corresponding Weisskopf frequency $\mathbf{\Omega_W} = \mathbf{<u>/b_W}$ into the classical statistical **V-W** line shape that according to (17) for the red wing signifies the deceleration of its power decrease of intensity from the impact index **2** to the statistical index **3/2**. At $\Delta\omega = \mathbf{\Delta U_{if,min}/\hbar}$ the red satellite appears. For the negative $\Delta\omega < \mathbf{2\pi\Delta U_{if,min}(r)/h}$ no real roots exist for equation $\Delta\omega = \mathbf{2\pi\Delta U_{if}[R(t,b,\varepsilon)]/h}$, and $\mathbf{\Phi_{if}(\Delta\omega)}$ decrease exponentially $\sim \mathbf{exp[-\xi(\Delta\omega,b,\varepsilon)]}$ with increasing $|\Delta\omega|$. It is to be noted that this part of the red wing is determined by the contribution of the second type of trajectories. Thus the line profile should exhibit the strong exponential decrease in the red wing at detuning $\Delta\omega < \Delta\omega_{c,red} = \mathbf{2\pi\Delta U_{if,min,eff}/h}$, where $\mathbf{\Delta U_{min,eff}}$ is some average minimum value of the potential differences of initial and final states. The value of $\Delta\omega_{c,red}$ is specific for the each **A**-X pair and could be estimated from the approximate potential curves, including attractive and repulsive parts. For the systems of interest $\Delta\omega_{c,red}$ may vary in the wide range from hundreds to thousands of cm^{-1}.

For the positive $\Delta\omega$ (*the blue wing of the line*) the situation is different. At $|\Delta\omega| < \mathbf{<u>/\ b_W}$ the blue wing has the Lorentz part symmetric with respect to the red wing, and within **UFCT** for the second type of trajectories. But it gradually vanishes around the corresponding Weisskopf frequency $\mathbf{\Omega_W} = \mathbf{<u>/b_W}$, according to [4], if $\mathbf{\Delta C_{6,if} < 0}$. Depending on the specific interaction potentials the range of positive $\Delta\omega$ for which equation $\Delta\omega = \mathbf{2\pi\Delta U_{if}[R(t,b,\varepsilon)]/h}$ has real solution may be practically unrestricted or restricted from above. If the maximum of $\mathbf{\Delta U_{if}(r)}$ exists then for $\Delta\omega > \mathbf{2\pi\Delta U_{if,max}/h}$ the roots are complex, and corresponding $\mathbf{G_{if}(b,\varepsilon,\Delta\omega)}$ decreases exponentially $\sim \mathbf{exp[-\xi(\Delta\omega,b,\varepsilon)]}$ with increasing $\Delta\omega$. Since the interaction $\mathbf{\Delta U_{if}(r)}$ usually takes positive values at $\mathbf{r < R_0}$, which correspond to dominance of the repulsive core, the Massey parameter is of the order of $\mathbf{\Delta\omega/\alpha_0<u>}$. If $\mathbf{\Delta U_{if}(r)}$ have maximum, the blue satellite will appear at $\Delta\omega = \mathbf{\Delta U_{if,max}}$. It is to be noted that this satellite is usually very far from the center of the line (up to several thousands of cm^{-1}). With increasing $\Delta\omega$ the equation $\mathbf{2\pi\Delta U_{if,max}/h} = \Delta\omega$ will have real roots for increasing energy ε, that will lead to exponential decrease of $\Phi_{if}^{(1)}(\Delta\omega)$ with increasing $\Delta\omega$ due to thermal average in (13). Therefore the strong exponential decrease of the blue wing should occur at detunings $\Delta\omega < \Delta\omega_{c,blue}=\alpha_0\langle u\rangle$ that effectively is shown up as its cutting. Typically α_0 is about 2-3 $\mathring{A}^{-1}$.

Thus the described procedure allows to obtain the total profile with its smooth transition from the impact to the quasistatic region of the detuning from line center $\Delta\omega$ and to take properly into account the characteristic repulsive core of the interaction potential that leads to the exponentially decreasing of its far wings.

The difficulties of performing explicitly such kind of calculations and especially lacking of the accurate independent information about diatomic potential curves pushed earlier to search for more simple procedures of calculations that also would contain the impact and quasistatic contributions in the total line profile. And indeed, Stormberg [10] proposed to construct the approximate total profiles on the basis of convolution of the purely impact and purely quasistatic contours due to Van der Waals broadening [4] that allowed to arrive to closed analytical expression in terms of the Faddeev function [10]. This approximate method became very popular in the applied spectroscopy, although it gives wrong behaviour of the far profile wings due to the appearance there the impact contribution. It was recognized that using such kind of profiles leads to the dramatic overestimation of the radiation transfer in the wings, and the various empirical procedures of far profile wings artificial cutting were proposed, that alas had no theoretical justification background. Now the presented above study allows to provide physically justified simple receipts of wing cutting procedure, that would improve the wings behaviour of convolved total profile [10]. Indeed, as follows from the above consideration the total profile [10] could be corrected in the following way $I(\Delta\omega) = I_{conv}(\Delta\omega)\cdot c(\Delta\omega)$, where $c(\Delta\omega)$ is the cutting function, defined as $c(\Delta\omega)=\sum_{if} g(i\to f)c_{if}(\Delta\omega)$ and in the red wing $\Delta\omega < 0$ represented by the expressions

$$c_{if}(\Delta\omega)=\begin{cases} 1, \text{ if } \Delta\omega<0,\ \Delta\omega_{c,if}^{(red)} > |\Delta\omega| > 0, \\ \exp\left[-\kappa_{if}(|\Delta\omega| - \Delta\omega_{c,if}^{(red)})\right], \text{ if } \Delta\omega<0,\ |\Delta\omega| > \Delta\omega_{c,if}^{(red)} > 0, \end{cases} \tag{19}$$

$$\Delta\omega_{c,if}^{(red)} = \frac{\Delta U_{if,\min}}{\hbar},\ \mathit{and}\ \kappa_{if} = \left(\frac{1}{6}\frac{\kappa_B T}{\mu}\left[\left.\frac{d^2\Delta U_{if}(R)}{\hbar\, d^2R}\right|_{R=R_{\min}}\right]\right)^{-1/3}. \tag{20}$$

Here $\mathbf{\Delta U_{if,min}=U_i(R)-U_f(R)}$ is the minimum of the difference of potentials at $\mathbf{R=R_{min}}$. And in the blue wing the partial cutting function $\mathbf{c_{if}(\Delta\omega)}$ could be presented in the form

$$c_{if}(\Delta\omega)=\begin{cases} 1, \text{ if } \ \Omega_{W,if} > \Delta\omega > 0, \\ 0, \text{ if } \ \Delta\omega > \Omega_{W,if} > 0, \end{cases} \tag{21}$$

where the Weisskopf $\mathbf{\Omega_{W,if}}$ frequency is marked also by the labels of the upper and lower levels of radiative transition. Thus the receipts (19)-(21) correct the behavior of the approximate total profile [10].

Resuming it could be stated: * for the binary case the impact limit of Van der Waals broadening was found exactly taking into account the degeneracy of quasimolecular states; ** the general asymptotic dependences of the red and blue wings of the profile were established on the basis of parameterized with the help of ***ab initio*** calculations Buckingham potential and **UFCT** methods; *** the receipts of cutting procedure for improving the line wings behaviour of the approximate total profiles, obtained by convolution of the Van der Waals pure impact and pure statistical profiles are formulated.

References

[1] E. E. Nikitin, S. Ya. Umanskii, *Theory of Slow Atomic Collisions,* Springer, Heidelberg, 1984.
[2] A. R. Edmonds, *Angular Momentum in Quantum Mechanics,* Princeton Press, New Jersey, 1957.
[3] P. J. Brussard, H. A. Tolhoek, Physica **23**, 955 (1957).
[4] I. I. Sobelman, *Introduction to the Theory of Atomic Spectra*, Pergamon Press, Oxford – New York, 1972.
[5] R. A. Buckingham, Proc. Roy.Soc., London. Series A, **168**, 264-283 (1938).
[6] V. A. Astapenko, A. V. Demura, G. V. Demchenko, B. V. Potapkin, A. V. Scherbinin, S. Ya. Umanskii, A. V. Zaitsevskii, 20th International conference on Spectral Line Shape, eds. John Lewis, Adriana Predoi-Cross, AIP CP v. 1290, NY, 2010, p. 39.
[7] J. Szudy, W.E. Baylis, JQSRT **15**, 641 (1975); *ibid* **17**, 681 (1977).
[8] K. Takayanagi, J. Phys. Soc. Japan **14**, 75 (1959).
[9] A. V. Eletskii, L. A. Palkina, B. M. Smirnov, *Transport Phenomena in Weakly Ionized Plasmas,* Atomizdat, Moscow, 1975 (in Russian).
[10] H.-P. Stormberg, J. Appl. Phys. **51**, 1963 (1980).

The probabilities of the $\nu'\ 1\left({}^3P_2\right)-\nu''\ 0^+\left({}^1S_0\right)$ transitions and the radiative lifetimes of the $\nu'\ 1\left({}^3P_2\right)$ states of the CdAr and CdKr molecules

O S Alekseeva[1,2,4], A Z Devdariani[2,3], M G Lednev[1], A L Zagrebin[1,2]

[1]Department of Physics, Baltic State Technical University, St. Petersburg, 190005, Russia
[2]Institute of Physics, St. Petersburg State University, Peterhof, 198504, Russia
[3]Department of Theoretical Physics, Herzen State Pedagogical University of Russia, St. Petersburg, 191186, Russia

E-mail: o.alek@rambler.ru

Abstract. The probabilities of the $\nu'\ 1\left({}^3P_2\right)-\nu''\ 0^+\left({}^1S_0\right)$ transitions and the radiative lifetimes of the metastable $1\left({}^3P_2\right)$ states of CdKr and CdAr molecules as functions of the vibrational excitation degree are calculated.

1. Introduction

The radiative characteristics of the excimer states of Cd-RG (RG – rare-gas) molecules are considered. The basis of the calculation of the radiative lifetime of the excimer $1\left({}^3P_2\right)$ state are the potential of the interatomic interaction for this state and the dipole moment of the molecular radiative transition $1\left({}^3P_2\right)-0^+\left({}^1S_0\right)$, where $0^+\left({}^1S_0\right)$ is the ground molecular state.

2. Potential curves and probabilities of quasi-molecular radiative transitions

To calculate the potentials of interatomic interaction and the probabilities of quasi-molecular radiative transitions we used the effective Hamiltonian method [1] in the formulation [2].The nonzero matrix elements of the effective Hamiltonian in the basis $\left|{}^{1,3}P_J\Omega_{ic}\right\rangle$ of diabatic quasimolecular wave functions of coupling case *c* are presented in [3].

The potential curves for the excited states were obtained within the framework of the semiempirical method of quasimiolecular term analysis [3]. The method is based on the comparison of four eigenvalues of the matrix of the effective Hamiltonian with interaction potentials for the states

[4] To whom any correspondence should be addressed.

$1(^3P_1)$, $1(^1P_1)$, $0^+(^3P_1)$ and $0^+(^1P_1)$ reconstructed from experimental data [4-8]. The point is, all matrix elements of the effective Hamiltonian can be expressed through atomic constants and the quantities $^{1,3}H_\sigma(R)$, $^{1,3}H_\pi(R)$, which are represent the potentials of the interaction in $^{1,3}\Sigma$ and $^{1,3}\Pi$ states without regard for the spin–orbit interaction in the Cd atom. Thus the functions $^{1,3}H_\sigma(R)$, $^{1,3}H_\pi(R)$ can be determined from the available experimental potentials for four states and then the potential curves for the remaining four states can be found from these functions by diagonalization of the matrix of the effective Hamiltonian.

The obtained semiempirical and experimental interaction potentials for the triplet states are shown in Figures 1 and 2. The Morse potential parameters for the obtained semiempirical potentials and for the experimental potentials [1-5] are listed in Table 1.

Table 1. Parameters for the Morse potentials for the $1(^3P_2)$ and $0^+(^1S_0)$ states of the CdAr and CdKr molecules

	CdKr		CdAr	
	$1(^3P_2)$	$0^+(^1S_0)$ [8]	$1(^3P_2)$	$0^+(^1S_0)$ [4]
R_e, a.u.	9.25	8.07	9.48	8.14
D_e, cm^{-1}	107.86	165.00	56.87	106.80
ω_e, cm^{-1}	10.7	18.1	11.8	25.3
$\omega_e x_e$, cm^{-1}	0.3	0.5	0.6	1.5

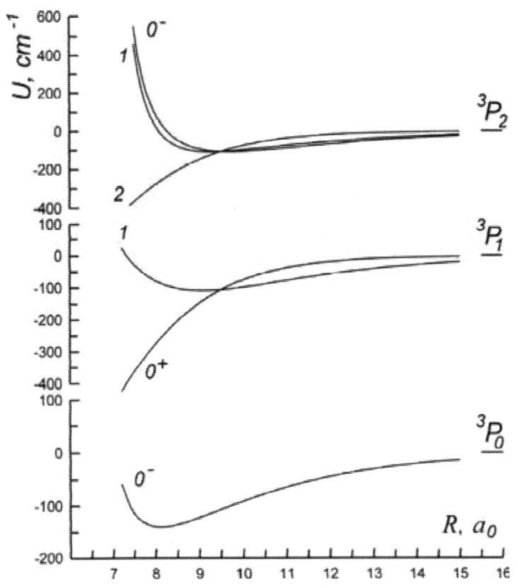

Figure 1. The interaction potentials of the triplet terms for CdKr

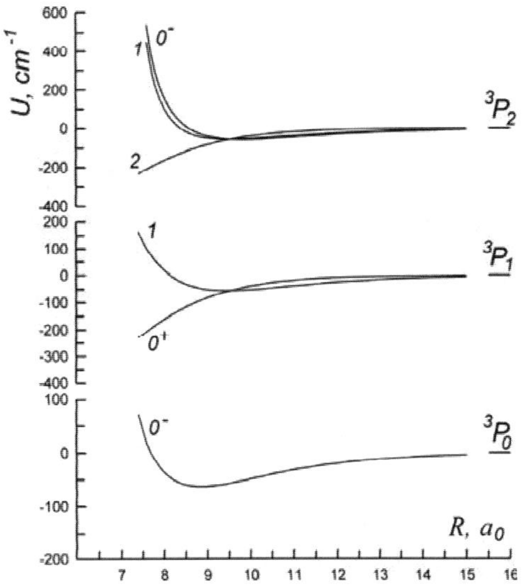

Figure 2. The interaction potentials of the triplet terms for CdAr

The adiabatic quasimolecular wave functions $|\Omega(^3P_J)\rangle$ (the eigenfunctions of the effective Hamiltonian) can be presented as linear combinations of the diabatic functions. In particular,

$$\left|1\left({}^3P_2\right)\right\rangle = c_1(R)\left|{}^1P_1 1_{ic}\right\rangle + c_2(R)\left|{}^3P_2 1_{ic}\right\rangle + c_3(R)\left|{}^3P_1 1_{ic}\right\rangle, \quad (1)$$

where the coefficients $c_i(R)$ are determined by diagonalizing the effective Hamiltonian matrix. To calculate the probabilities $\Gamma\left(\Omega\left({}^3P_J\right),R\right)$ of the quasimolecular radiative transitions in adiabatic wave functions we isolated the contribution from the diabatic functions $\left|{}^1P_1 0^+_{LS}\right\rangle$ and $\left|{}^1P_1 1_{LS}\right\rangle$, since in the diabatic molecular basis $\left|{}^{1,3}P_J \Omega_{LS}\right\rangle$ corresponding to the *LS* coupling case, the dipole moments of the transitions to the ground state are nonzeros only for the ${}^1P_1 0^+_{LS}$ and ${}^1P_1 1_{LS}$ states. The dipole moment $\left\langle {}^1P_1 1_{LS}\left|\hat{d}\right|{}^1S_0 0\right\rangle$ can be expressed through the dipole moment $\left\langle {}^3P_1 1_{ic}\left|\hat{d}\right|{}^1S_0 0\right\rangle$ of an observed atomic transition. Thus the probability $\Gamma\left(1\left({}^3P_2\right),R\right)$ of the quasi-molecular radiative transition can be expressed in terms of the probability $\Gamma\left({}^3P_1\right)$ and frequency $\omega\left({}^3P_1\right)$ of the atomic intersystem crossing transition, the calculated amplitudes c_1, c_3 and the the amplitudes a and b of the expansion of atomic eigenfunctions for the intermediate coupling case in the wavefunctions for the *LS* coupling case (in the calculations the approximate values of the amplitudes a = 0.998 and b = - 0.062 were used, which, as in [9], were determined from the energies of atomic levels using the semiempirical method of the atomic spectra analysis [10]):

$$\Gamma\left(1\left({}^3P_2\right),R\right) = \Gamma\left({}^3P_1\right)\left(\frac{\omega\left(1\left({}^3P_2\right),R\right)}{\omega\left({}^3P_1\right)}\right)^3 \left(\frac{a}{b}c_1 - c_3\right)^2. \quad (2)$$

3. The radiative lifetimes of the $\mathrm{v}'\ 1\left({}^3P_2\right)$ states and the probabilities of the $\mathrm{v}'\ 1\left({}^3P_2\right)-\mathrm{v}''\ 0^+\left({}^1S_0\right)$ transitions

Given the proximity of the frequencies of the transitions $\mathrm{v}'\ 1\left({}^3P_2\right)-\mathrm{v}''\ 0^+\left({}^1S_0\right)$ to the frequency of the forbidden transition $5\,{}^3P_2 - 5\,{}^1S_0$, the probability of the molecular transitions can calculated without regard for the dependence of the transition frequency on the interatomic distance, by setting it constant and equal to the frequency of the $5\,{}^3P_2 - 5\,{}^1S_0$ transition. In this approximation, the lifetime of the state $\mathrm{v}'\ 1\left({}^3P_2\right)$ is given by

$$\tau^{-1}(\mathrm{v}') = \left\langle \mathrm{v}'\right|\Gamma\left(1\left({}^3P_2\right),R\right)\left|\mathrm{v}'\right\rangle, \quad (3)$$

where $\left|\mathrm{v}'\right\rangle$ is the wave function of the vibrational state, which set identical to the corresponding Morse function.

The probability of the radiative transition $\mathrm{v}'\ 1\left({}^3P_2\right)-\mathrm{v}''\ 0^+\left({}^1S_0\right)$ is defined as

$$A(\mathrm{v}',\mathrm{v}'') = \frac{\omega\left({}^3P_2\right)}{\omega\left({}^3P_1\right)}\Gamma\left({}^3P_1\right)\left|\left\langle \mathrm{v}'\right|\frac{ac_1}{b} - c_3\left|\mathrm{v}''\right\rangle\right|^2. \quad (4)$$

The calculated lifetimes and transition probabilities are listed in Table 2. The calculations were performed for lower vibrational states for which the approximation of the potential by the Morse function is rather accurate.

Table 2. The probabilities A(v', v'') (in s^{-1}) of the $v'\ 1\left({}^3P_2\right)-v''\ 0^+\left({}^1S_0\right)$ transitions and the radiative lifetimes $\tau(v')$ (in 10^{-6} s) of the $v'\ 1\left({}^3P_2\right)$ states of the CdAr and CdKr molecules

RG	v' \ v''	0	1	2	3	4	5	$\tau(v')$ 0	$\tau(v')$ 1
Ar	0	74	102	4	51	36	0	3700	1580
	1	408	750	237	0.7	2.4	74		
Kr	0	110	321	242	19	28	41	1300	615
	1	408	750	237	0.7	2.4	74		

4. Conclusions

Thus, we expressed the *R*-dependent probabilities $\Gamma\left(\Omega\left({}^3P_J\right),R\right)$ of the quasimolecular radiative transitions $\Omega\left({}^3P_J\right)-0^+\left({}^1S_0\right)$ through the corresponding amplitudes of the expansions of adiabatic wave functions and the probabilities of atomic transitions. This made it possible to establish a direct relationship with the known probabilities of the single-atom transitions and readily calculate the lifetimes of the $v'\ 1\left({}^3P_2\right)$ vibronic quasimolecular states and the probabilities per unit time of the $v'\ 1\left({}^3P_2\right)-v''\ 0^+\left({}^1S_0\right)$ transitions.

References

[1] Nikitin E E and Umanskii S Ya 1980 *Semiempirical Methods of Calculation of Interatomic Interaction* (Moscow)
[2] Zagrebin A L and Lednev M G 1994 *Opt. Spectrosc.* **77**, 481
[3] Alekseeva O S, Devdariani A Z, Lednev M G, Zagrebin A L 2011 *Russian Journal of Physical Chemistry* B **5** 946
[4] Funk D J, Kvaran A, Breckenridge W H 1989 *J. Chem. Phys.* **90** 2915
[5] Ruszczak M, Strojecki M, Koperski J 2005 *Chem. Phys. Lett.* **416** 147
[6] Kvaran A, Funk D J, Kowalski A, Breckenridge W H 1989 *J. Chem. Phys.* **89** 6069
[7] Koperski J, Kiełbasa Sz M, Czajkowski M 2000 *Spectrochim. Acta* **56A** 1613
[8] Koperski J, Łukomski M, Czajkowski M 2002 *Spectrochim. Acta* **58A** 2709
[9] Zagrebin A L, Lednev M G, Tserkovnyi S I 1993 *Opt. Spectrosc.* **74**, 24
[10] Sobel'man I I 1977 *Vvedenie v Teoriyu Atomnykh Spectrov (Introduction to the Theory of Atomic Spectra)* (Moscow)

Comparative Study of Emission Spectra of He(3S)-He(2P) at 706 and 728 nm Due to the Triplet and Singlet Transitions.

N F Allard[1,2], A Monari[3], B Deguilhem [3], and F X Gadéa[3]

[1] Observatoire de Paris, GEPI, UMR8111, CNRS, 61, Avenue de l'Observatoire, F-75014, Paris, France

[2] Institut d'Astrophysique, UMR7095, CNRS, Université Paris VI, 98bis Boulevard Arago, F-75014, Paris, France

[3] Laboratoire de Chimie et Physique Quantique, UMR5626, CNRS, Université Paul Sabatier, 118 route de Narbonne, F-31400 Toulouse, France

E-mail: `nicole.allard@obspm.fr`

Abstract. This study of the self-broadening of the 728 nm He line is performed in order to establish similarities and differences to the self-broadening of the triplet 3*s*-2*p* line at 706 nm. In a previous work we presented the first calculations of He-He collisional profiles at 706nm due to the triplet 3*s*-2*p* transition in a unified line shape semi-classical theory using *ab-initio* molecular potentials. The excellent agreement between experimental and theoretical determinations of the near wing of the line profiles established the accuracy of the interaction potentials. New *ab-initio* potentials for the singlet 3*s*-2*p* transitions allow us to extend our previous study to the line at 728 nm.

1. Potentials

The interatomic interactions are the main physical quantities needed for a good understanding of collisional processes. The *ab initio* computation of the adiabatic potential energy curves of He_2 have been carried out using the MOLPRO 2009 package [1]. A huge MultiReference Configuration Interaction (MRCI) has been performed starting from a very large Multi Configuration Self Consistent Field (MCSCF). The accuracy of the approach is below the wavenumber compared to full CI approaches [1], as we have checked for the quintet states. Moreover errors due to BSSE are also far below the wavenumber thanks to the huge basis set involved. Therefore an accuracy about the wavenumber is expected. However, non adiabatic and relativistic corrections have been ignored. The *ab initio* energy curves for the He_2 line at 728 nm are shown in figure 1. They are input data for a unified spectral line shape evaluation of the complete He-He collisional line profiles [2].

2. Line profile

In radiative collision transitions it is the difference potential ΔV between the final and initial states that determines the frequency and the energy emitted or absorbed by a single photon.

[1] http://www.molpro.net

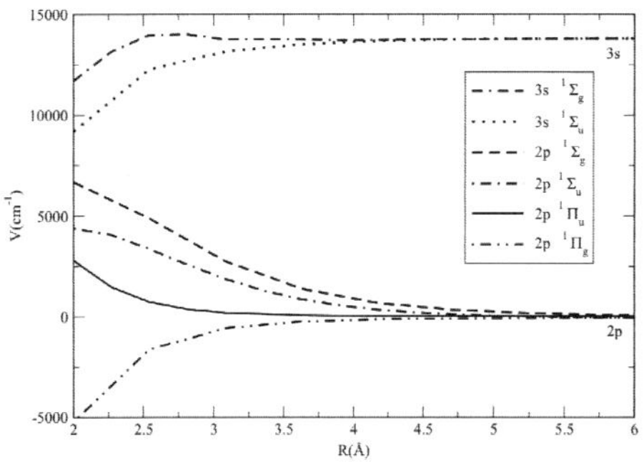

Figure 1. *Ab initio* potential energies for the singlet line.

The difference potential maxima (figures 2 and 3) lead to blue satellite bands in proximity to the atomic lines. Figure 4 shows the asymmetrical shape of the singlet and triplet lines in agreement with experimental spectra reported in [3, 4]. The blue asymmetry is due to a blend of four quasi-molecular satellites. Because of the blending, the line profile can be accurately described *only* with a *unified* spectral line shape theory that is valid from the center to the far wing.

The emission spectrum produced at 300 K in dense plasmas of helium excited by a corona discharge shows an asymmetric atomic line at 706nm. Theoretical calculations of the line profiles have been performed for the experimental physical conditions, T=300 K and $n_{He}= n_{He}(T,P)$ for the triplet line. The agreement for a pressure of 1 and 1.6 MPa is remarkable [4]. We emphasize that no free potential parameters are used to readily fit the measured line shape. Such an agreement has *only* been possible because of the accuracy of the *ab initio* potentials.

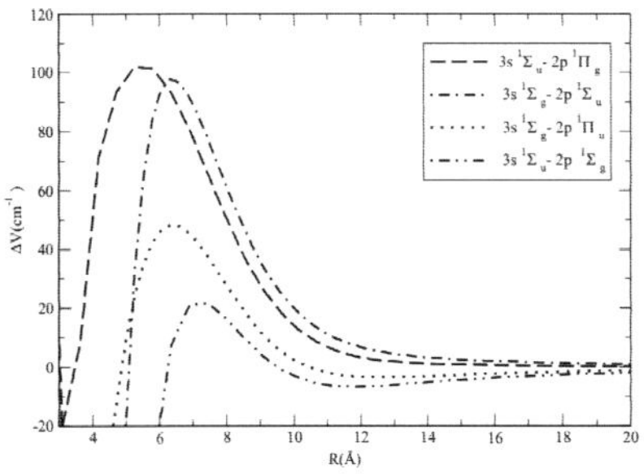

Figure 2. ΔV for the 4 transitions contributing to $3s\ ^1S \rightarrow 2p\ ^1P$ line.

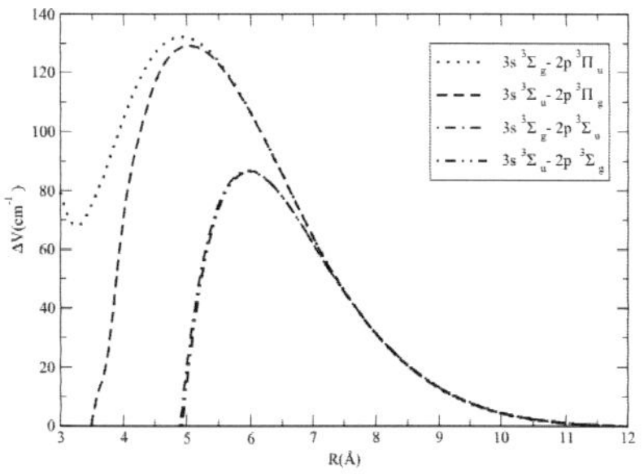

Figure 3. ΔV for the 4 transitions contributing to $3s\ ^3S \rightarrow 2p\ ^3P$ line.

Line shapes show a strong pressure dependence. Important features of the 728 nm line is the development of blue satellites already observed in the 706 nm line. Increasing the He density results in a shift of the whole profile towards the position of the satellite band (figure 6). When the He density reaches 10^{21} cm^{-3} the satellite band is more intense than the main line for the singlet line. Whereas it has to 1.2× 10^{21} cm^{-3} for the triplet line as the line satellite is farther in the wing [4]. We can, then, predict that the dependence in perturber density on the line parameters of the singlet line, as done in [5, 6], will show a greater slope, as the associated line satellites are closer to the parent line.

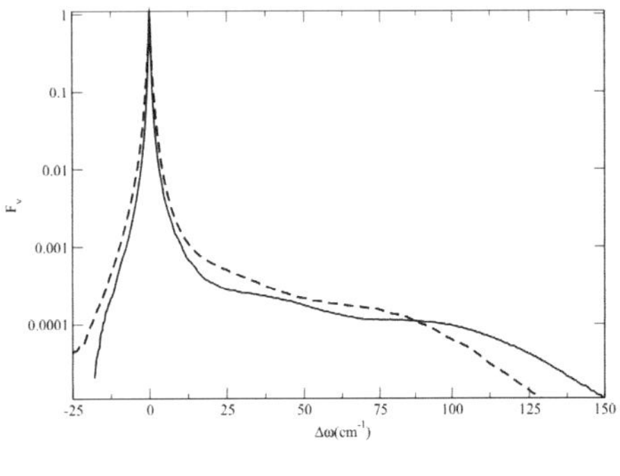

Figure 4. Unified profiles for singlet (dashed lines) and triplet (full line). T=300 K and n_{He} $=10^{19}$ cm^{-3}.

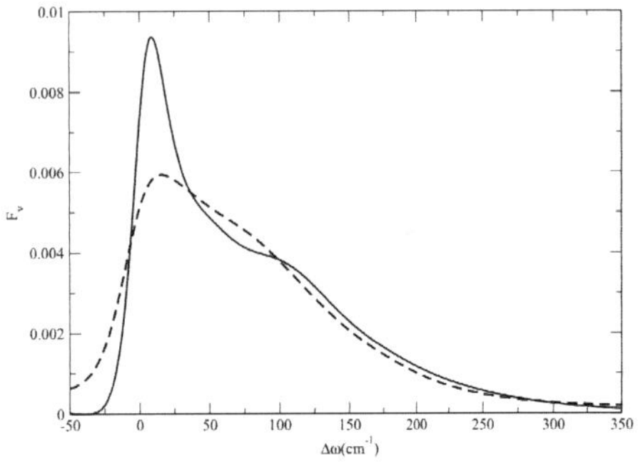

Figure 5. Unified profiles for singlet (dashed lines) and triplet (full line). T=300 K and n_{He} $=7\times10^{20}$ cm^{-3}.

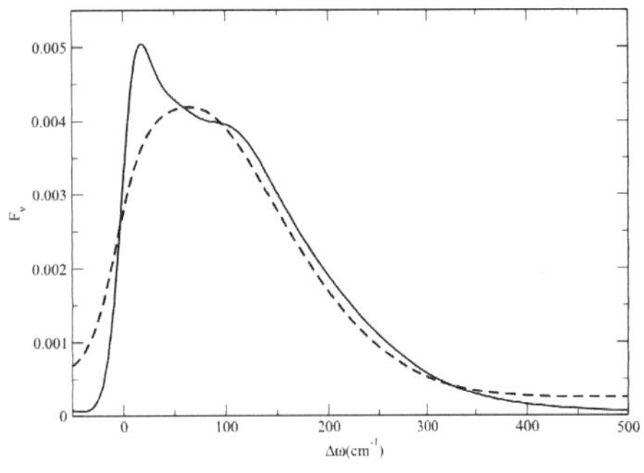

Figure 6. Unified profiles for singlet (dashed lines) and triplet (full line). T=300 K and n_{He} $=10^{21}$ cm^{-3}.

Acknowledgments
One of us (NFA) would like to acknowledge PNPS for the financial support which allows the collaboration with the Laboratoire de Chimie et Physique Quantique in Toulouse.

References

[1] Bonifaci N, Aitken F, Atrazhev V M, Fiedler S L and Eloranta J 2012 *Phys. Rev. A* **85** 042706
[2] Allard N F, Royer A, Kielkopf J F and Feautrier N 1999 *Phys. Rev. A* **60** 1021
[3] Li Z 2008 *Ph. D. Thesis* (Grenoble, France: University of J. Fourier)
[4] Allard N F, Deguilhem B, Gadéa F X, Bonifaci N and Denat A 2009 *Europhysics Letters* **88** 53002
[5] Allard N F, Bonifaci N and Denat A 2011 *European Physical Journal D* **61** 365–372
[6] Allard N F 2012 21th *ICSLS St Petersburg, Russia, June 3-10, 20012* (JPCS)

Formation of Xe_2 molecules in glow gas discharge

E Yu Kleymenov[1], S A Klemeshev, P A Saveliev and N A Kryukov[2]

Institute of Physics, St. Petersburg State University, Ulianovskaja 1, St. Petersburg, 198904 Russia

E-mail: nkryukov@yandex.ru

Abstract. Rate constants of formation of Xe_2 excimers in the first excited 0^+_u and 1_u molecular states were measured in low-presser (5-12 Torr) low-current (3 mA) glow discharge in neat xenon. Concentrations of the lowest excited 0^+_u and 1_u molecular states were measured with vacuum ultraviolet emission spectroscopy in the wavelength region 147-220 nm. Concentrations of the respective excited atomic states were estimated with infrared absorption at the Xe electronic transition wavelengths 881.9 nm and 828.0 nm. The formation rate constants $k_f\,(0^+_u) = 4 \cdot 10^{-32}$ cm^{6}s^{-1} and $k_f\,(1_u) = 0.3\ 10^{-32}$ cm^{6}s^{-1} were obtained from a pressure dependence of the atomic and molecular concentrations.

1. Molecular state formation and decay processes

In the present publication we discuss the formation of Xe_2 molecules in the lowest excited molecular states under the gas discharge conditions. In the vicinity of the equilibrium internuclear distances and at longer distances, Xe_2 molecular states are usually described with the Hund's angular momentum coupling case (c) nomenclature $\Omega^{p'}_p$ [1]. The lowest excited 0^+_u and 1_u Xe_2 molecular states have dissociation limits Xe($5p^5 6p\ ^3P_1$) + Xe($5p^6\ ^1S_0$) and Xe($5p^5 6p\ ^3P_2$) + Xe($5p^6\ ^1S_0$), respectively. The molecular states are attractive, with the potential well depth D_e=3700 cm^{-1} and 3200 cm^{-1}, respectively, and the equilibrium internuclear distance R_e=3.0 Å [1]. The ground 0^+_g molecular state is formed by two Xe($5p^6\ ^1S_0$) ground-state atoms. The potential well depth of the ground state is R_e=194 cm^{-1} and the equilibrium internuclear distance is 4.4 Å [2], i.e. at the room temperature the state can be considered as repulsive. The VUV spectrum of xenon gas discharge contains two well distinguishable features known as the first and second continua (Fig. 1), which correspond to transitions from the low and high vibration levels, respectively. The groups of high and low vibrational levels will be further denoted with superscripts "h" and "l". At low pressures, the first continuum contains two peaks, which are formed by the low vibrational levels of 0^+_u and 1_u.

Table 1 summarizes the main molecular state formation and decay processes. The generally accepted mechanism of Xe_2 molecule formation under the gas discharge conditions is three-body recombination (1) [6-11], which involves one excited Xe atom, and two Xe atoms in the ground state. The excited atom is reacting with a ground-state atom, while another ground-state atom carries out the excess of kinetic energy. Thus, a Xe_2 excimer in a high vibrational state is formed. The reverse process, impact dissociation (2) results in two ground-state atoms and one excited atom. Apart from the dissociation, there is three decay mechanisms of the radiative states: collisional vibration

[1] Current affiliation: Institute of Robotics and Intelligent Systems (IRIS), ETH Zurich, 8092 Zurich, Switzerland.
[2] Corresponding author.

Table 1. Processes of formation and decay of Xe_2 molecular states and respective rate constants or radiative decay rates.

Process		Rate constant or radiative decay rate		Reference
		$\Omega=0^+_u, j=1$	$\Omega=1_u, j=0$	
(1) Three-body recombination	k_{rec}	$(4 \pm 4)\ 10^{-32}\ cm^6s^{-1}$	$(0.3 \pm 0.1)\ 10^{-32}\ cm^6s^{-1}$	This work
$Xe(^3P_j) + 2\ Xe \rightarrow Xe_2(\Omega^h) + Xe$		$(4.4\ ..\ 21)\ 10^{-32}\ cm^6s^{-1}$	$(7.7\ ..\ 9.0)\ 10^{-32}\ cm^6s^{-1}$	[5-10]
(2) Impact dissociation	k_{diss}	$3\ 10^{-9}\ cm^3s^{-1}$	$5\ 10^{-10}\ cm^3s^{-1}$	This work
$Xe_2(\Omega^h) + Xe \rightarrow Xe(^3P_j) + 2\ Xe$				
(3) Vibrational relaxation by collision	k_{rel}	$7\ 10^{-11}\ cm^3s^{-1}$		[5]
$Xe_2(\Omega^h) + Xe \rightarrow Xe_2(\Omega^l) + Xe$		$(3.9\pm0.6)\ 10^{-10}\ cm^3s^{-1}$		[4]
(4) Radiative decay				
$Xe_2(\Omega^h) \rightarrow 2\ Xe + h\nu$	Γ^h	$3.5\ 10^8\ s^{-1}$	$2.4\ 10^8\ s^{-1}$	[4]
$Xe_2(\Omega^l) \rightarrow 2\ Xe + h\nu$	Γ^l	$3.0\ 10^8\ s^{-1}$	$1.0\ 10^8\ s^{-1}$	[4]
		$(2.2 \pm 0.1)\ 10^8\ s^{-1}$	$(1.01 \pm 0.02)\ 10^7\ s^{-1}$	[5]
(5) Mixing of 0^+_u and 1_u				
$Xe_2(0^+_u) + Xe \rightarrow Xe_2(1_u) + Xe$	k_{01}	$(1.3 \pm 0.4)\ 10^{-13}\ cm^3s^{-1}$		[5]
$Xe_2(1_u) + Xe \rightarrow Xe_2(0^+_u) + Xe$	k_{10}	$(4.6 \pm 0.5)\ 10^{-15}\ cm^3s^{-1}$		[5]

relaxation (3), radiative decay (4), and mixing of the 0^+_u and 1_u states (5). At low pressures, the mixing of the 0^+_g and 1_u states can be neglected. Therefore, the main decay processes are the vibrational relaxation, radiative decay, and dissociation. From kinetic equations for processes (1)-(4) (for details, see [4]), one can obtain:

$$F([Xe]) \equiv \frac{[Xe(^3P_j)][Xe]^2}{[Xe_2(\Omega^h)]} = \frac{k_{diss}+k_{rel}}{k_f}[Xe] + \frac{\Gamma^h}{k_f} \qquad (6)$$

The formation constants k_f and k_{diss} can be estimated from the slope and intercept term of the linear regression of concentration dependence of the left part in Eq. (6).

2. Experiment and data processing

The glow discharge in neat xenon was induced in a glass cell with an inner diameter of 1 cm by applying a high voltage to cylindrical electrodes in the cell (Fig. 2). The temperature of the discharge cell walls was always equal to the room temperature, 300 K. The length of the positive discharge column was 16 cm. The emitted VUV radiation was detected using a Rowland-scheme vacuum monochromator with a cylindrically bent 1200 lines/mm diffraction grating with a radius of curvature of 1000 mm. The monochromator volume was separated from the discharge cell by an MgF window. The entrance and exit slits were set to 10 and 200 um, respectively. The signal was detected with a photo-multiplier tube (PMT) detector in the analog mode. The spectral resolution of the VUV spectrometer was 0.08 nm at 185 nm. The sensitivity of the spectrometer was measured with a hydrogen-deuterium lamp calibrated with synchrotron radiation. The geometry of the discharge was taken into account for estimation of the power density of VUV radiation.

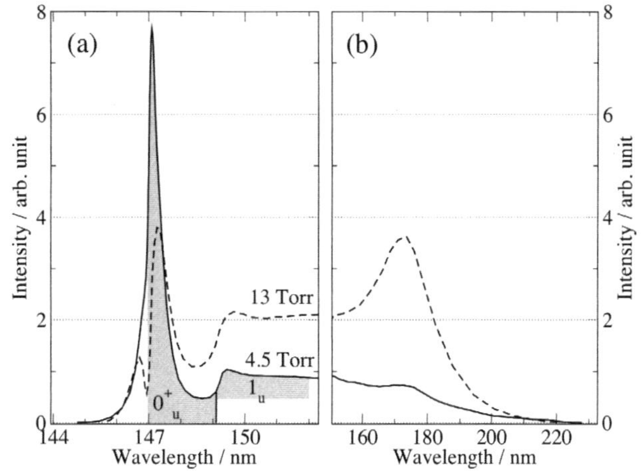

Figure 1. Spectrum of Xe_2 at 1 mA, 4.5 Torr (solid line) and 13 Torr (dashed line). (a) First continuum; (b) Second continuum.

Concentrations of the molecular Xe_2 excited states were measured with VUV emission spectroscopy and that of the atomic $Xe(^3P_j)$ states with IR absorption. The concentration of excimers was calculated as a ratio of the integral VUV intensity and radiative decay rate:

$$[Xe_2(\Omega^h)] = \frac{1}{hc\,\Gamma^h}\int_{\lambda_1}^{\lambda_2} I(\lambda)d\lambda \qquad (7)$$

The integration wavelength region (λ_1, λ_2) was (147.0 nm, 149.1 nm) for 0^h and (149.1 nm, 152 nm) for 1^h (Fig. 1). The method of background subtraction for these regions is shown on Fig. 1.

Concentrations of excited Xe atoms were measured with IR absorption spectroscopy at the Xe $5p^56s$ ($^3P_2$) -> $5p^56p$ ($^3D_2$) (λ=881.9 nm) and Xe $5p^56s$ ($^3P_1$) -> $5p^56p$ ($^3P_0$) (λ=828.0 nm) transitions. The absorption was measured perpendicular to the discharge cell. The radiation of a hollow-cathode xenon lamp transmitted through the discharge cell was dispersed with a compact monochromator and detected with a PMT. The resolution of the monochromator was 1 nm at 880 nm, which was sufficient to separate the emission lines. No self-absorption was observed in the lamp radiation spectrum recorded in a separate experiment. The concentration of the atoms was calculated numerically using the line intensities and absorption coefficients for the hyperfine structure components [12] and taking

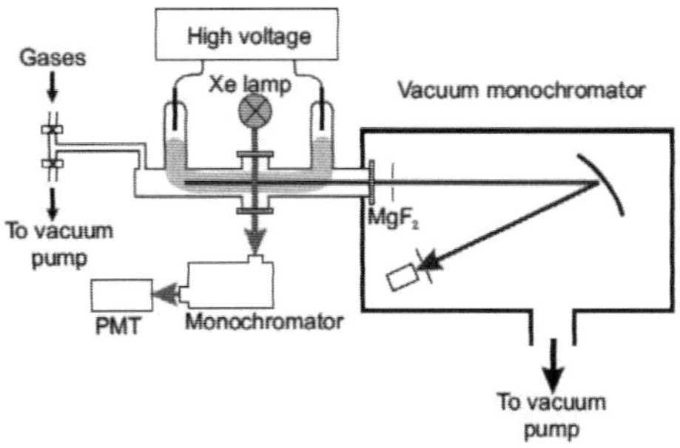

Figure 2. Experimental setup.

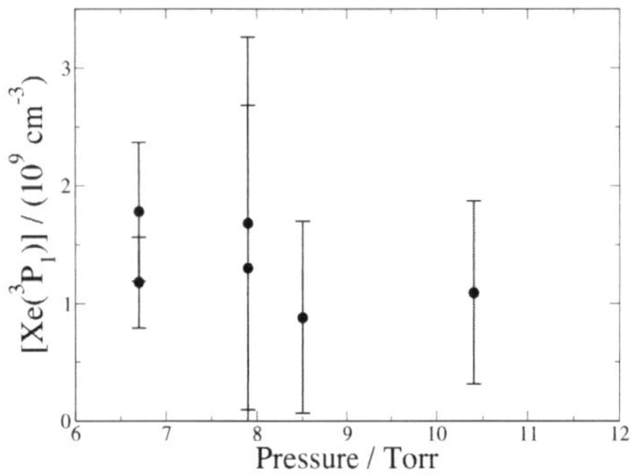

Figure 3. Concentration of Xe($^3P_1$) .

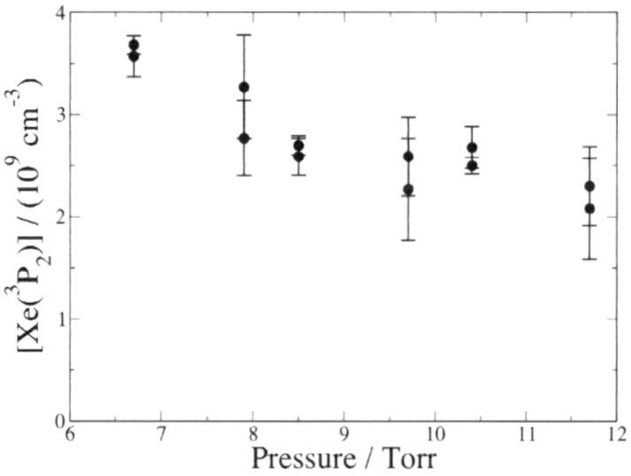

Figure 4. Concentration of Xe($^3P_2$) .

into account the Doppler broadening of the hyperfine components of the absorption and emission lines.

3. Results and discussion

Concentrations of the excited atoms were proven to depend linearly on the discharge current in the range from 1 to 16 mA. The results discussed here were obtained at a discharge current of 3 mA. At that current the discharge was diffuse at pressures up to 12 Torr. The pressure dependence of excited atom concentrations is shown on Figs. 3 and 4. The error bar represents the experimental uncertainty, which is the standard deviation of multiple measurements. The absorption at 881.9 nm is five times weaker than that at 828.0 nm, which results in a much larger uncertainty of the Xe($^3P_1$) atomic concentrations compared to that of Xe($^3P_2$). The concentrations of excited atoms were decreasing with pressure, which can be addressed to the collisional de excitation.

Concentrations of the molecular states (not shown here) were in the range (1-2) 10^3 cm^{-3} and (7-9) 10^3 cm^{-3} for 0^h and 1^h states, respectively. The concentrations were increasing with pressure. The resulting function F([Xe]) (see Eq. (2)) calculated with the detected atomic and molecular concentrations is shown on Figs. 5 and 6. The linear regression of the data has the slope (0.8 ± 0.4)

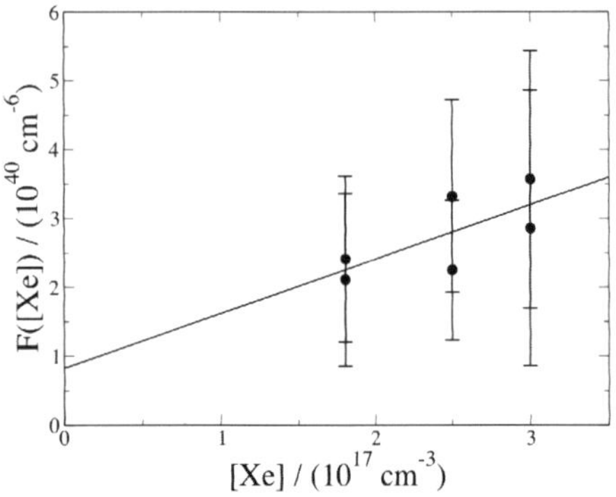

Figure 5. F(0^+_u) function.

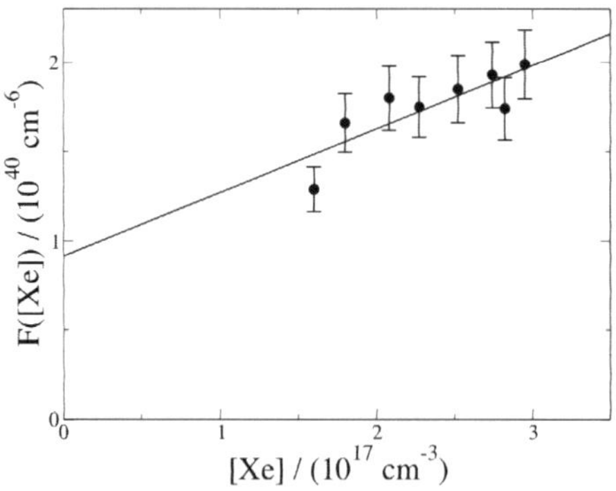

Figure 6. F(1_u) function.

10^{23} cm^{-3}, (0.4 ± 0.1) 10^{23} cm^{-3} and intercept (0.8 ± 1.0) 10^{40} cm^{-6}, (0.9 ± 0.2) 10^{40} cm^{-6} for 0^+_u and 1_u, respectively. The root mean square (rms) deviation (square root of reduced chi squared) of the linear regression is 0.42 and 0.96 respectively. The rms value of the 1_u data fit is close to one, which is the optimal value indicating a good fit of the experimental data. It supports the hypothesis of the three-body recombination as the main mechanism of the excimer formation. Measurement of a temperature dependence of the formation constant would be a direct proof of the hypothesis. The rms value of the 0^+_u data fit is significantly lower than one, which indicates that the experimental uncertainty of the data needs to be improved. The linear regression coefficients combined with the radiative decay rates listed in Tab. 1 yield the rate constants of molecular state formation: k_f (1_u) = (0.3 ± 0.1)·10^{-32} cm^6s^{-1} and k_f (0^+_u) = (4 ± 4) · 10^{-32} cm^6s^{-1}. The formation constants are significantly lower than these obtained by multiple research groups [5-10] with time-resolved measurements of xenon radiation in the gas discharge afterglow. Using the formation constants and slope of the linear regressions, the dissociation constants k_{diss} (1_u) = 5 10^{-10} cm^3s^{-1} and k_{diss} (0^+_u) = 3 10^{-9} cm^6s^{-1} can be obtained, although their uncertainties are 100% and 200%, respectively. Decreasing the experimental uncertainties is possible if using a longer absorption path for the measurement of the excited atom concentrations.

References

[1] Herzberg G 1989 *Molecular Spectra and Molecular Structure* vol 1 (Malabar: Krieger Publishing Company)
[2] Messing I, Eckstrom D J and Lorents D C 1990 *J. Chem Phys.* **93** 34-41
[3] Aziz R A and Slaman M J 1986 *Mol. Phys.* **57** 825-40
[4] Zagrebin A L, Kryukov N A and Savel'ev P A 1999 *Optics and Spectroscopy* **86** 678-85
[5] Bonifield T, Rambow F, Walters G, McCusker M, Lorents D and Gutcheck R 1980 *J . Chem. Phys.* **72** 2914-24
[6] Sewraj N, Gardou J P, Salamero Y and Millet P 2000 *Phys. Rev. A* **62** 052721
[7] Leichner P, Palmer K, Cook J and Thieneman M 1976 *Phys. Rev. A* **13** 1787–92
[8] Millet P, Birot A, Brunet H, Galy J, Pons-Germain B and Teyssier J L 1978 *J. Chem. Phys.* **69** 92
[9] Salamero Y, Birot A, Brunet H, Galy J and Millet P 1984 *J. Chem. Phys.* **80** 4774
[10] Ledru G, Marchal F, Sewraj N, Salamero Y and Millet P 2006 *J. Phys. B* **39** 2031
[11] Marchal F, Sewraj N, Jabbour G, Rodriguez Akerreta P and Ledru G 2010 *J. Phys. B* **43** 235210
[12] Bochkova O and Oshemkova V 1977 *Vestnik LGU (Bulletin of Leningrad State University) series 4: Physics and Chemistry* **4** 385–9 (in Russian)

Collisional shift of the heavy atoms hyperfine lines in an atmosphere of the inert gas

T A Florko, S V Ambrosov, A A Svinarenko and T B Tkach

Odessa State University – OSENU, P.O. Box 24a, Odessa-9, 65009, Ukraine

E-mail: quantflo@mail.ru

Abstract. New relativistic approach, based on the relativistic many-body perturbation theory (PT) with using the optimized wave functions sets , is applied to calculating the potentials and hyper fine structure collision shift for heavy atoms in atmosphere of inert gases. Data for the collisional shifts of the Cs, Tl and Yb atoms in atmosphere of the inert gases are presented.

1. Introduction

Studying the collisional shifts and broadening of the hyperfine structure lines for heavy elements (alkali, alkali-earth, lanthanides, actinides and others) in an atmosphere of inert gases is one of the important and actual topics of the modern quantum chemistry, atomic optics and spectroscopy [1-11]. It should be mentioned also that the heavy atoms are interesting from the point of view of studying a role of weak interactions in atomic optics and heavy-elements chemistry. Besides, calculation of the hyperfine structure line shift and broadening allows one to check the quality of wave functions and study a contribution of relativistic effects in two-center (multi-center) atomic systems. From the applied science point of view, the mentioned physical effects form a basis for creating an atomic quantum measure of frequency [8]. For a long time the corresponding phenomenon for thallium atom attracted a special attention because of possibility to create the thallium quantum frequency measure. Alexandrov and co-workers [8] have realized the optical pumping of the thallium atoms on the line of 21GHz, which corresponds to transition between the components of hyperfine structure for the ground state, and have measured the collisional shift of this line due to buffer (bath) gas. Naturally, the inert buffer gases (He, Ar etc.) were used. The detailed non-relativistic theory of the collisional shift and broadening of the hyperfine structure lines for simple elements (light alkali elements etc.) has been developed by many authors (see discussions in refs. [1-9]). However, consideration of heavy elements faces serious difficulties related with account for the relativistic and correlation corrections. By the way, in last years the lanthanide and actinide elements attract a great interest because of their perspectives in many applications, including the sensor physics and atomic optics devices (see refs. [2,10,11] and references there).

It is very curious that until now a consistent, accurate quantum mechanical approach for calculating main characteristics of the collisional processes was not developed though many different simplified models have been proposed (see, for example [2-11]). The most widespread approach is based on the calculation of the corresponding collision cross-section, in particular, in a case of the van der Waals interaction between colliding particles. However, such an approach does not factually define any difference between the Penning process and resonant collisional one and gives often non-correct results for cross-sections. More consistent method requires data on the process probability $G(R)$ as a function of inter nuclear distance. It should be noted that these data are practically absent at present time. In this paper a new relativistic optimized approach, based on the gauge-invariant perturbation theory (PT) with using the optimized wave functions basis's, is applied to calculating the inter atomic potentials, hyper fine structure collision shift for alkali and ytterbium atoms in atmosphere of inert gases. The basic expressions for the collision shift and broadening of the hyperfine structure spectral lines are taken from the kinetic theory of spectral lines [6,7,11]. The exchange perturbation theory (the modified version EL-HAV) has been used for calculating the corresponding potentials (see details in [1]). New data for the collisional shifts of the Tl and Yb atoms in atmosphere of the inert gases are presented.

2. Hyperfine lines shift for system "heavy atom- inert gas"

To calculate the collision shift of hyperfine structure spectral lines one could use the following expression known from kinetic theory of spectral line form (see [5-7,11]):

$$f_p = \frac{\Delta}{p} = \frac{4\pi N_b}{p} \int_0^{\infty} e^{-U(R)/kT} \left[1 + g(R)\right] \delta\omega(R) R^2 dR, \tag{1}$$

where $U(R)$ is the effective potential of the interatomic interaction, which has a central symmetry in a case of the systems A-B (in our case, for example, B=He; A=Yb); T is a temperature, w_0 is a frequency of the hyperfine structure transition in the isolated active atom; $d\omega(R)=Dw(R)/w_0$ is the relative local shift of the hyperfine structure lines, which is arisen due to the disposition of the active atoms (say, atom of ytterbium Yb and helium He) on a distance R, $\{1+g(R)\}$ is a temperature factor. To calculate an effective potential of the interatomic interaction a method of the exchange perturbation theory (the modified version EL-HAV) [1] has been used. Within exactness to second order terms on potential of Coulomb interaction of the valent electrons and atomic cores one can write:

$$\delta\omega(R) = \frac{S_0}{1-S_0} + \Omega_1 + \Omega_2 - \frac{C_6}{R^6}\left(\frac{2}{\bar{E}_a} + \frac{1}{\bar{E}_a + \bar{E}_B}\right), \tag{2}$$

where C_6 is the van der Waals constant for interaction A-B (*e.g.*, a pair of Yb-He; look below); I, $E_{Ia,b}$ are the ionization potential and excitation energy on the first level for atoms A, B correspondingly; S_0 is the overlapping integral. The value of $\bar{E}_{\alpha,b}$ can be simply defined as follows [7]:

$$\bar{E}_{\alpha b} = (I_{a,b} + E_{1a,b})/2,$$

The values Ω_1, Ω_2 in the expression (2) are the non-exchange and exchange non-perturbation sums of the first order correspondingly, which are defined as follows:

$$\Omega_1 = \frac{2}{N(1-S_0)\rho_0} \sum_k \frac{\langle \Phi_0'(1) | H_{CT}' | \Phi_k'(1) \rangle V_{k0}}{E_0 - E_k},$$

$$\Omega_2 = \frac{2}{N(1-S_0)\rho_0} \sum_k \frac{\langle \Phi_0'(1) | H_{CT}' | \Phi_k'(1) \rangle U_{k0}}{E_0 - E_k}$$

$$\rho_0 = \langle \Phi_0^1(1) | H_{CT}' | \Phi_0'(1) \rangle / \langle \Phi_0^1(1) | \Phi_0'(1) \rangle$$

where H'_{CT} is the operator of hyperfine interaction, N is the total number of electrons taken into account in calculation; E_{k}, $\Phi'_k(1) = F'_{k_a}(1)\varphi_{k_b}(2...N)$ — energy and non-symmetrized wave function of state $k = \{k_a, k_b\}$ for isolated atoms A and B.

The non-exchange matrix element of the Coulomb interatomic interaction is as follows:

$$V_{ko} = < \Phi'_k(1) \mid V(1) \mid \Phi'_0(1) >.$$

Correspondingly the exchange matrix element is as follows:

$$U_{k0} = \sum_{i=2}^{N} \langle \Phi'_k(1) | V(i) | \Phi'_0(i) \rangle$$

Operator $V(i)$ is, for example, in a case of system Yb-He, as follows:

$$V(i) = U_{SCF}(r_{a3}) + U_{SCF}(r_{a4}) - 2U_{SCF}(R) + \frac{1}{r_{bi}}, \tag{3}$$

where $U_{SCF}(r)$ is the self-conjunctive field, created by the ytterbium atomic core.

Let us return to consideration of the van der Waals constant C_6 for the interatomic A-B interaction. As a rule, one could use the approximate values for the van der Waals constant C_6 etc. Often the sufficiently great mistake in determination of the van der Waals constants provides non-high accuracy of the inter-atomic potentials calculation and further inaccuracies. The van der Waals constant may be written as follows [10]:

$$C_6(L, M) = C_{6,0}(L) - \frac{3M^2 - L(L+1)}{(2L-1)(2L+3)} \cdot C_{6,2}(1) \ , \tag{4}$$

where $C_{6,0}$ (L) is the isotropic component of the interaction and $C_{6,2}$ (L) is the component corresponding to the $P_2(\cos\theta)$ term in the expansion of the interaction in Legendre polynomials, where the angle specifies the orientation in the space-fixed frame.

The dispersion coefficients $C_{6,0}$ (L) and $C_{6,2}$ (L) may be expressed in terms of the scalar and tensor polarizabilities $\alpha_0(L;iw)$ and $\alpha_2(\mathrm{L;iw})$ evaluated at imaginary frequencies [7,11]. In particular, for the helium case one may write:

$$C_{6,0}(L) = \frac{3}{\pi} \cdot \int_0^{\infty} \alpha_0(L;iw)\overline{\alpha}_{He}(iw)dw, \tag{5}$$

where $\overline{\alpha}_{He}$ is the dynamical polarizability of helium. The polarizabilities at imaginary frequencies are given in atomic units as follows:

$$\alpha_{\parallel}(L, M; iw) = 2\sum_{\gamma, M_\gamma} \frac{(E_\gamma - E_L) |< LM \mid \hat{z} \mid L_\gamma M_\gamma >|^2}{(E_\gamma - E_L)^2 + w^2} \tag{6}$$

where E_γ is the energy of the electronically excited state $|L_\gamma M_\gamma>$ and the z axis lies along the internuclear axis.

Usually (see [2]) the non-relativistic Hartree-Fock sets of the wave functions are used. More sophisticated approach is based on using the relativistic Dirac-Fock wave functions (first variant) [5-7,11-15]. Another variant is using the relativistic wave functions as the solutions of the Dirac equations with different density functionals (the Kohn-Sham DFT theory) and effective potentials [2,10,12,16-19]. In this paper we have used the set of the relativistic functions, generated by the Dirac equation with the optimized Ivanov-Ivanova model potential [20]. The detailed approbation of this model potential in studying spectra and radiative characteristics of the ytterbium and thallium atoms is given in refs. [17-19]. In a number of papers it has been rigorously shown that using the optimized sets in calculating the atomic electron density dependent properties has a decisive role (see discussions in refs. [21,22]). Here we will not in details discuss this question.

3. Results and conclusions

In table 1 we present our theoretical results for the line shift f_p (1/Torr) for the Cs-He pair. The observed value of the line shift (T=323K) and other theoretical results for f_p are given in table 1 too. Other theoretical data are obtained on the basis of the exchange PT with using the wave functions sets in Clementi et al and Hartree-Fock approximations [6,7,23]. The important feature of our scheme is a correct account of the correlation and polarization effects with using special effective functionals from [17,21]. A difference between our theoretical data and other calculation data is explained by using the different sets of wave functions and different PT schemes. For other temperatures there are no quite precise data. It is obvious that using the gauge-invariant optimized sets of the wave functions and correct version of the exchange PT will be necessary in a case of the alkali elements in an atmosphere of more heavy inert gases.

Table 1. Observed f_p (10^{-9} 1/Torr) shifts for systems Cs-He and theoretical data.

System	Cs-He	Cs-He	Cs-He	Cs-He	Cs-He
T, K	Experiment	Our theory	Theory [3]	Theory [6]	Theory [6]
323	135	134	137	126	109
423	—	121	123	111	96
523	—	109	112	100	85
623	—	103	105	94	78

In table 2 we present our theoretical results for the shift f_p (Hz/Torr) for the Tl-He, Ar, Xe pairs. The observed value of the line shift (T=700K) and other theoretical results with using the Dirac-Fock and optimized Dirac-Fock methods are listed in table 2 too [3,7].

Table 2. The shift f_p of the Tl hyperfine line for the pairs: Tl-He, Kr, Xe (T=700 K).

System	Tl-He	Tl-Ar	Tl-Xe
Experiment	130 ± 30	-490±20	-1000±80
Theory [7]	155.0	-850.0	-1420.0
Theory [5]	137.2	-504	-1052
Our theory	135.4	-501	-1044

The ground configuration for ytterbium is: $[Xe]4f^{14}6s^2$ (term: ^1S). First of all, let us present our results for the scalar static polarizability α_0 (in units of a_0^3, a_o is the Bohr radius) and isotropic dispersion coefficient $C_{6,0}$ (in units of $E_H \cdot a_0^6$, E_H is the Hartree unit of energy). Our results are as follows: $C_{6,0}$= 45, α_0=168. For comparison let us present the data (from ref. [10]) by Dalgarno et al: $C_{6,0}$= 39.4, α_0=157.3 and by Buchachenko et al: $C_{6,0}$= 44,5. In table 3 we present our preliminary calculation results for the observed f_p (Hz/Torr) shift for the system Yb-He. It is obvious that the

Table 3. The observed f_p (Hz/Torr) shift for the system Yb-He.

T, K	f_p
700	146.8
750	144.1
800	141.3
850	139.0
900	136.2

pair Yb-He is significantly more complicated system. Until now there are no any experimental or theoretical data for this system. Thus, our data may be considered as the first preliminary reference.

References

[1] Kaplan I G 1985 *Theory of intermolecular interactions* (Moscow: Nauka)
[2] Freeman A J and Frankel R H 1987 *Hyperfine interactions* (New York: Plenum)
[3] Malinovskaya S V, Glushkov A V, Khetselius O Yu, Mischenko E V, Loboda A V and Svinarenko A A 2009 *Int. J. Quant. Chem.* **109** 3031
[4] Glushkov A V, Khetselius O Yu, Gurnitskaya E P, Loboda A V and Mischenko E V 2009 *Theory and Applications of Comp. Chem.* (AIP) **1102** 172
[5] Khetselius O Yu, Glushkov A V, Gurnitskaya E P, Loboda A V, Mischenko E V, Florko T A and Sukharev D E 2008 *Spectral Line Shapes* (AIP) **15** 231
[6] Batygin V V, Gorny M B and Gurevich B M 1978 *J. Techn. Phys.* **48** 1097
[7] Batygin V V and Sokolov I M 1983 *Opt. Spectr.* **55** 30
[8] Alexandrov E B, Popov V I and Yakobson N N 1979 *Opt. Spectr.* **46** 404
[9] Bichutskaya E N, Devdariani A Z, Zagrebin A L and Sebyakin Y N 1999 *Opt. Spectr.* **87** 197
[10] Chi X, Dalgarno A and Groenenborn G C 2007 *Phys. Rev.* A **75** 032723
[11] Sobel'man I I 1977 *Introduction to theory of atomic spectra* (Moscow: Nauka)
[12] Wilson S 2007 *Recent Advances in Theoretical Physics and Chemistry Systems, Series: Progress in Theoretical Chemistry and Physics* vol 16, ed J Maruani, S Lahmar, S Wilson and G Delgado-Barrio (Berlin: Springer) p 11
[13] Glushkov A V, Ambrosov S V, Gurnitskaya E P, Loboda A V and Khetselius O Yu 2006 *Recent Advances in the Theory of Chemical and Physical Systems, Series: Progress in Theoretical Chemistry and Physics* vol 15, ed P Julien, J Maruani, D Mayou, S Wilson and G Delgado-Barrio (Amsterdam: Springer) p 285
[14] Glushkov A V, Khetselius O Yu, Gurnitskaya E P, Loboda A V, Florko T A, Sukharev D E and Lovett L 2008 *Frontiers in Quantum Systems in Chemistry and Physics, Series: Progress in Theoretical Chemistry and Physics* vol 18, ed S Wilson, P J Grout, J Maruani, G Delgado-Barrio and P Piecuch (Berlin: Springer) p 505
[15] Khetselius O Yu 2009 *Int. J. Quant. Chem.* **109** 3330
[16] Glushkov A V 1998 *Opt. Spectr.* **84** 670
[17] Ivanova E P, Ivanov L N, Glushkov A V and Kramida A E 1985 *Phys. Scripta* **32** 512
[18] Vidolova-Angelova E, Ivanov L N and Angelov D 1988 *J. Phys. B: At. Mol. Opt. Phys.* **21** 3877
[19] Vidolova-Angelova E, Ivanova E P and Ivanov L N 1981 *Opt. Spectr.* **50** 243
[20] Ivanov L N and Ivanova E P 1979 *Atom. Data Nucl. Data Tabl.* **24** 95
[21] Glushkov A V and Ivanov L N 1992 *Phys. Lett.* A **170** 33
[22] Glushkov A V, Malinovskaya S V, Svinarenko A A and Chernyakova Yu G 2004 *Int. J. Quant. Chem.* **99** 889
[23] Clementi E, Roothaan C C J and Yoshimine M 1962 *Phys. Rev.* **127** 1618

Electronic radiative transition in He($2^{1,3}$S)-Ne weakly bound molecules. Temperature dependence

A K Belyaev [1,3], A Z Devdariani[1,2], V S Rybak[1] and I A Zlatkin[2]

[1] Department of Theoretical Physics, Herzen University, St. Petersburg 191186, Russia.

[2] Department of Optics and Spectroscopy, St. Petersburg University, St. Petersburg 198904, Russia.

E-mail: belyaev@herzen.spb.ru

Abstract. The metastable states $He(1s2s2\,^{1,3}S)$ are the champions in terms of radiative lifetime of neutral atomic states, $\sim 20\cdot 10^{-3}s$ and $\sim 7\cdot 10^{3}s$ for the single and triplet states respectively. However in numerous applications one deals with a metastable atom surrounded by buffer gas atoms. In the case of low temperatures the processes of recombination lead to the formation of clusters and dimers containing exited He^{*} atoms. The present calculations reveal that lifetimes of quasi-bound molecules $He(2^{1,3}S)Ne$ are lowered by a factor equal to $\sim 10^{-9}$ - 10^{-7} in comparison with free metastable atoms. Temperature dependences of lifetimes are presented as well as averaged spectral line shapes. It is shown that radiative transitions mainly lead to the formation of free He and Ne atoms in their ground-state with the kinetic energy of relative motion in the range of 0-0.03 eV.

1. Introduction

Knowledge of the $He(2^{1,3}S)$ state lifetimes is important owing to the role that metastable atoms play in numerous applications. The extremely long lifetimes in atomic scale, $\tau_1^0 \approx 20\cdot 10^{-3}s$ for the singlets [1] and $\tau_3^0 \approx 7\cdot 10^{3}s$ for the triplets [2], arise from the fact that direct photon decay of the $2^{1,3}S$ states to the 1^1S ground state is forbidden by selection rules $\Delta S = 0, \Delta L = \pm 1$. However, in numerous applications, e.g., in ionosphere and discharge plasmas, one deals with a metastable atom surrounded by buffer gas atoms. Thus, one needs to account for the influence of the interaction between atoms on radiative properties of metastable atoms including lifetimes and spectral profiles. The issue is of topical in the case of low temperatures when the processes of recombination lead to the formation of clusters and dimers containing excited atoms. The present work focuses on the radiative transitions in

[3] To whom any correspondence should be addressed.

the weakly bound dimers. In the case of $He(2^{1,3}S)-Ne$ the transitions can lead either to the formation of free $He(1^1S)$ and $Ne(1^1S)$ atoms:

$$He(2^{1,3}S)-Ne(1^1S) \Rightarrow He(1^1S)+Ne(1^1S)+\hbar\omega \quad (1)$$

or to the formation of bound molecules in ro-vibrational states of the ground molecular state:

$$He(2^{1,3}S)-Ne(1^1S) \Rightarrow HeNe(v,j)+\hbar\omega \quad (2)$$

with $\hbar\omega \sim 166300 cm^{-1}$. The physical reason of the reactions (1)-(2) is related to the optical transition dipole moment of the $He(2^{1,3}S)-Ne$ dimers. An atomic interaction destroys the spherical symmetry of the helium atom in the metastable state via admixture of p-atomic states of helium with the zero projection of the electric moment on the molecular axis and takes off the selection rules [3].

The present paper accounts for both reactions (1,2) in performing the calculations of the averaged lifetimes and spectral profiles produced by optical transitions from four vibrational states of the weakly bound molecule $He(2^{1,3}S)-Ne$. The calculations performed are based on an improved version of the recent *ab initio* quantum chemical data [4].

2. Quantum chemical input

The starting point of the present calculations of lifetimes and spectral profiles is a quantum chemical input that includes potential energy curves and electronic dipole transition moment. As the previous calculations [5] for the singlet case, the present calculations are mainly based on the *ab initio* results [4] for the potential energy curves of the initial excited molecular states 0^+ and 1 that correlate at large interatomic distances with the atomic states $He(2^{1,3}S)$ correspondingly and for the potential energy curve of the ground molecular state 0^+. Thus obtained potentials allowed us to calculate both the positions of vibrational levels and the nuclear wave functions by numerical solution of the time-independent Schrodinger equation by means of the expansion of nuclear wave functions in the Discrete Variable Representation (DVR) basis functions sinc(R). The results for the adiabatic potential energies together with positions of vibrational levels and wave functions are shown in figure 1 for the ground and excited 0^+ states in the case of $He(1^1S), He(2^1S)-Ne$ dimers. The dipole moment for the parallel radiative transition [4] is shown in figure 1c. As discussed above, this information was used afterwards for the calculation of the lifetime.

2.1. Excited state

The excited molecular state produced by the interaction of the metastable atom $He(2^1S)$ with the $Ne(1^1S)$ atom gives rise to the adiabatic potential energy curve with two wells at $R_m^i = 5.1a_0$ and $R_m^e = 10.6a_0$ separated by a hump at the interatomic distance $R_b = 7.4a_0$; see figure 1a. As discussed in [4], the asymptotic value of the calculated potential should be lowered in order to match the experimental data for the separated atoms. The new fitting of the $He(2^1S)-Ne(1^1S)$ potential performed in the present study leads to the correction of 469 cm-1. The ambiguity in the asymptotic value for the potential energy curve with two potential wells implies an ambiguity in the number of vibrational states. The ground vibrational state, whose wave function is shown by the thin solid line in figure 1a, is entirely localized in the interior well. This well is about 150 cm-1 in depth with respect to the atomic asymptote (260 cm-1 to the top of the barrier), but rather narrow, so that the ground vibrational level is located roughly 86 cm-1 below the dissociation limit. The external well can be regarded as a van der Waals one, although its depth is about 38 cm-1. The calculations in [4] mainly concentrated on the relatively short distances in the 3–8 a0 range. Therefore the present value 38 cm-1 should be treated as an estimate only, due to the problems with the multireference CI approach. This problem is not crucial for the present lifetime calculation of the ground vibrational state (the vibrational quantum number v = 0), as the corresponding vibrational wave function is localized entirely at short internuclear distances, $R < 7a_0$, see figure 1a. It is important for the vibrational states with v = 1, 2, 3, whose levels lie above the minimum of the external well and whose vibrational wave functions are localized completely in this well; so that their vibrational level values should be considered as the first-order estimates only.

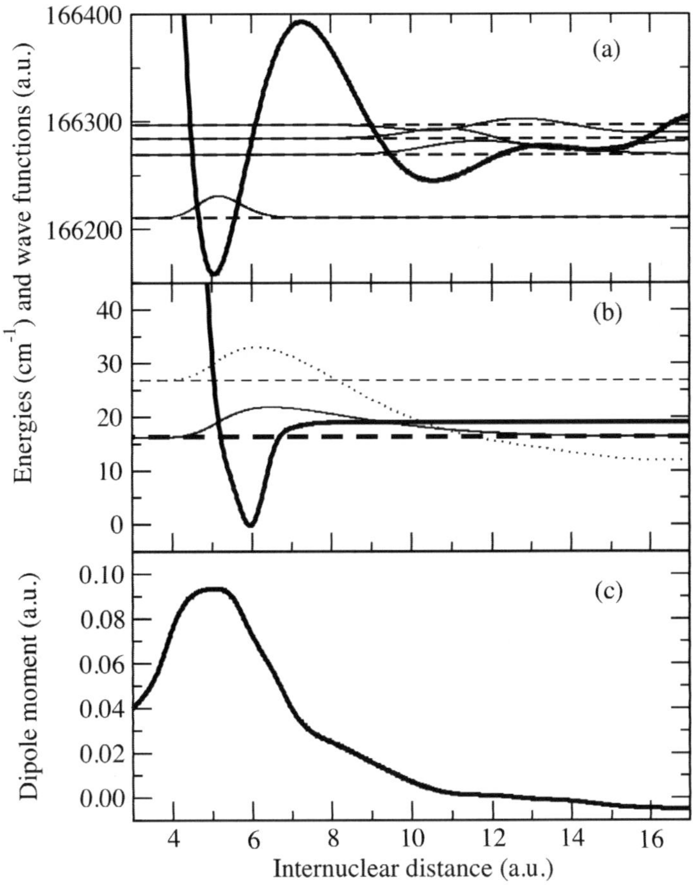

Figure 1. (a) The excited state 0^+ adiabatic potential (the thick solid curve) for the $He(2^1S)-Ne$ dimer. The dashed lines refer to the positions of the vibrational levels. The thin lines show the corresponding wave functions.
(b) The ground-state 0^+ adiabatic potential for the $He(1^1S)-Ne$ dimer molecules, the notations are similar to the ones on figure1a. The thin dashed line and the dotted line present an energy in the continuum and the corresponding wave function.
(c) The electric transition dipole moment between the states under consideration.

2.2. Ground state

The calculated potential for the ground state, the thick solid line in figure 1b, is in good agreement with the previous results [6]. The key parameters of the potential are the well depth $D_m = 20cm^{-1}$ and the equilibrium distance $R_m = 5.7a_0$. The present calculation shows that the ground-state potential well supports one vibrational level with the dissociation energy $D_m = 3.6cm^{-1}$, see figure 1b, the thick long-dashed line.

2.3. Electric transition dipole moment

The electric transition dipole moment for transitions $He(2^1S)-Ne(1^1S) \Rightarrow He(1^1S)-Ne(1^1S)$ as a function of internuclear distance is shown in figure 1c. Dipole moment has two important features: a sharp maximum in the region of strong configuration interaction at $R \approx 5a_0$ and a node at $R \approx 13a_0$. In general, the existence of a node is important in calculations of radiative properties, especially of spectral profiles, as it leads to the formation of a minimum in the radiative width and the spectral profiles.

3. Results and conclusions

The present calculation of the $He(2^{1,3}S)-Ne$ radiative lifetime is based on the following formula (in atomic units) [7,8]:

$$\tau^{-1} = \frac{4}{3}\sum_f \left(\alpha\omega_{if}\right)^3 \times \left[\frac{j+1}{2j+1}\langle ij|d(R)|f(j+1)\rangle^2 + \frac{j}{2j+1}\langle ij|d(R)|f(j\ \ 1)\rangle^2\right] \quad (3)$$

where α is the fine structure constant; ω_{if} is the angular frequency of the emitted photon, which means that $\hbar\omega_{if} = E_i - E_f$ in the case of a bound-bound transition or $\hbar\omega_{if} = E_i - T_f$ in the case of a bound-free transition, whereby the energies of the bound states E_{if} and the kinetic energy T_i are taken with respect to the ground-state asymptotic value; $d(R)$ is the electric transition dipole moment calculated between the initial and the final electronic adiabatic states at fixed nuclei; $|ij\rangle$ and $|fj\rangle$ are the nuclear radial wave functions for the initial (excited) and the final (ground) molecular states, respectively, with the orbital quantum number j. In the present case, the initial states are the electronically excited molecular states $He(2^{1,3}S)-Ne$ in allowed vibrational levels v. Table 1 shows the lifetime values calculated by equation (3) for four vibrational states of the excited adiabatic term $He(2^1S)-Ne$. The results for the average lifetime under the Boltzmann distribution are shown in figure 2. The thick solid curve referred to the case when four allowed vibrational levels are populated, dashed line and dotted line refer to the cases of three and two vibrational levels respectively, and finally, thin line depicts the case when only the ground vibrational level is populated.

Table 1. Calculated molecular lifetimes, $He(2^1S)-Ne$.

v	$\tau(s)$
0	1.35×10^{-8}
1	2.11×10^{-5}
2	1.00×10^{-5}
3	1.30×10^{-5}

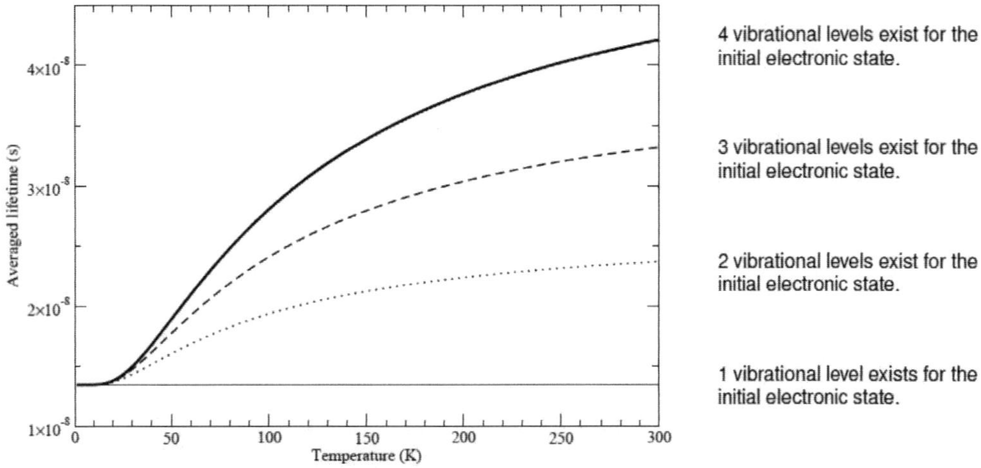

Figure 2. Temperature dependence of averaged lifetimes for the $He(2^1S)-Ne$ molecules.

The square of the calculated averaged dipole moment matrix elements $\langle i0|d|f1\rangle^2$ is shown as a function of the emitted photon energy in figure 3. It is worth mentioning that the bound-bound transition contributes 16% to the inverse lifetime in equation (3), the rest is coming from the bound-free transitions. Figure 3 clearly shows that the radiative transitions mainly lead to the formation of the ground-state He and Ne atoms with the kinetic energy of relative motion in the range of $T_f = 0 - 0.03eV$. The spectral profiles of the bound-free transitions averaged over the Boltzmann distribution are shown in figure 4.

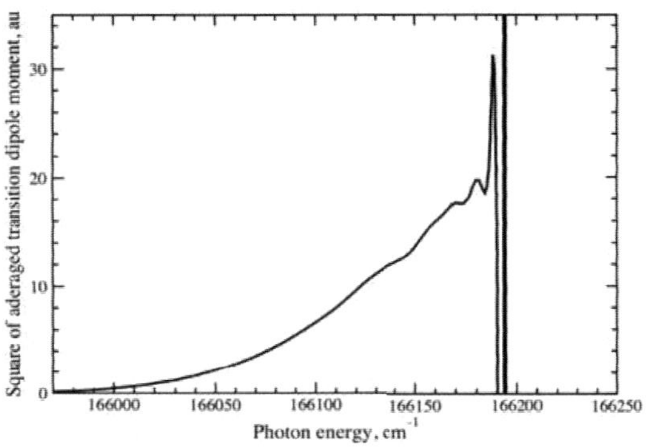

Figure 3. The square of the transition dipole moment averaged over the initial and the final wave functions in the case of the $He(2^1S)-Ne$ molecules as a function of photon energy, which determines the kinetic energy of relative motion of free He and Ne atoms. Vertical solid line at 166194 cm^{-1} shows the position of the bound–bound transitions.

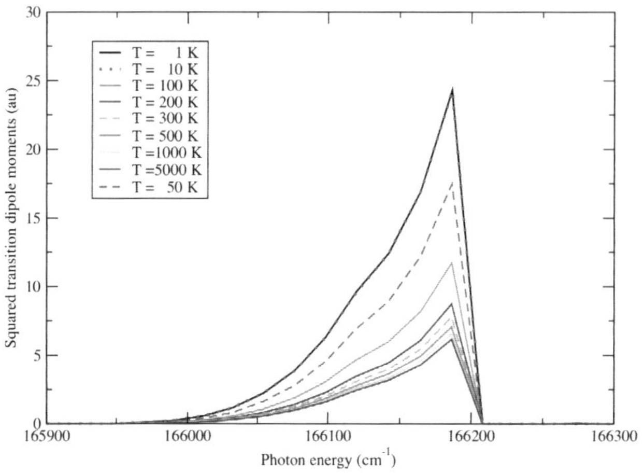

Figure 4. The temperature dependences of the averaged spectral profiles for bound-free transitions in the radiative decay the $He(2^1S)-Ne$ molecules.

Thus, the interaction of atoms lead to the pronounced decreasing of lifetimes of the metastable atomic states, namely $\tau_1 \approx 6.7\cdot 10^{-7}\tau_1^0$ for the singlet state and $\tau_3 \approx 1.3\cdot 10^{-9}\tau_3^0$ for the triplet one. Although $He-Ne$ gas mixtures and, in particular, radiative emission of the excited He atoms have been intensively studied since the advent of the first gas laser it would be interesting to study such transitions in the $He-Ne$ gas mixtures under temperatures below 100 K, when the process of the excitation transfer is negligible [9] but recombination processes should produce a large number of the $He(2^1S)-Ne$ molecules. It is expected that the increase of surrounding buffer gas atoms will decrease the lifetimes. The same is true for experiments in cold matrices.

References

[1] Dyck S van, Johnson C E and Shugart H A 2007 *Phys. Rev.* A **4** 1327

[2] Hodgman S S, Dall R G, Byron L J, Baldwin K G H, Buckman S J and Truscott A G 2009 *Phys. Rev. Lett.* **103** 053002

[3] Devdariani A Z, Zagrebin A L and Blagoev K B 1989 *Ann. Phys. Fr.* **14** 467
Devdariani A Z, Zagrebin A L and Blagoev K B 1992 *Ann. Phys. Fr.* **17** 365

[4] Buenker R J, Liebermann H-P and Devdariani A Z 2007 *J. Phys. Chem.* A **111** 1307

[5] Devdariani A, Belyaev A K, Alekseyev A B, Lieberman H-P and Buenker R J 2010 *Mol. Phys.* **108** 757
Devdariani A Z, Belyaev A K, Alekseev A B, Liebermann H-P and Buenker R J 2011 *Russ. J. Phys. Chem.* B **5** 952

[6] Keil M, Danielson L J and Dunlop P J 1991 *J. Chem. Phys.* **96** 296

[7] Huestis D L 1996 Radiative transition probabilities *Atomic, Molecular & Optical Physics Handbook* ed Gordon W F Drake (New York: AIP Press) chapter 32 pp 394-410

[8] Devdariani A Z and Chesnokov E A 2005 *Optics and Spectroscopy* **99** 866

[9] Devdariani A Z and Buenker R 2004 *Khim. Fyz.* **23** 61

XXI International Conference on Spectral Line Shapes (ICSLS 2012) IOP Publishing
Journal of Physics: Conference Series **397** (2012) 012039 doi:10.1088/1742-6596/397/1/012039

Study of absorption on the wings of sodium resonance line perturbed by molecular hydrogen

N F Allard[1,2], V A Alekseev[3] and F Spiegelman[4]

[1] Observatoire de Paris, GEPI, UMR8111, CNRS, 61, Avenue de l'Observatoire, F-75014, Paris, France
[2] Institut d'Astrophysique, UMR7095, CNRS, Université Paris VI, 98bis Boulevard Arago, F-75014, Paris, France
[3] Institute of Physics, St.Petersburg State University, Uljanovskaja St.1, Peterhof, 198504 St. Petersburg, Russia
[4] Laboratoire de Chimie et Physique Quantique, Université de Toulouse (UPS) and CNRS, 118 route de Narbonne, Toulouse, France

E-mail: `nicole.allard@obspm.fr`

Abstract. The unified theory line profiles for neutral Na perturbed by H_2 are obtained using new set of *ab initio* potentials calculated for several collision geometries. The results are of interest for modeling exoplanet and brown dwarf atmospheres.

1. Introduction

The construction of model atmospheres and synthetic spectra for cool brown dwarfs and extrasolar giant planets is necessary to derive reliable atmospheric parameters and the surface chemical composition for these objects. The studies of observed L and T dwarfs by [1, 2] showed clearly the importance of extended wings of both sodium and potassium doublets, and they pointed out the need for accurate line profile calculations. A first improvement was done by [3] using multiconfiguration self-consistent field Hartree-Fock potentials, considering the angular dependence Θ =90, 60, 45, 30 and 0°, where Θ is the angle between the direction of the molecular axis and the interaction axis. They calculated the far wing line profiles of the 0.589μm sodium D and 0.77μm potassium doublets using the *uniform approximation* to the wing profile in the [4, 5] theory. This is a major improvement compared to the unrealistic use of a Lorentzian so far in the wings. Also, in contrast to a simple quasistatic model, this formula has no singularity and should in principle predict the line satellite shapes. In Allard *et al.* 2003 [6] we presented absorption profiles of sodium and potassium perturbed by helium and molecular hydrogen. The line profiles were calculated in a unified line shape semi-classical theory (Allard *et al.* 1999 [7]) using Rossi and Pascale 1985 [8] (hereafter labeled RP85) pseudo-potentials. The pressure broadening calculations were extended to the case of the other alkalies: lithium, rubidium and cesium, where calculations were done for the $D1$ and $D2$ lines from $T_{\rm eff}$ = 500 K to 3000 K with a step size of 500 K [9, 10]. Tables of absorption coefficients are now provided in an on-line database for convenient use in atmosphere modeling [1].

[1] http://mygepi.obspm.fr/ allard/

More recently we presented in [11, 12] unified line profiles for the 3s-3p and 3p-4s transitions using *ab initio* NaH_2 potentials of one of us (hereafter labeled S11). Calculations were performed for two symmetries C_{2v} (Θ =90°) and $C_{\infty v}$ (Θ =0). In the present paper we study the $3s$-$3p$ line profile for an extended set of geometries.

2. Theory

A unified theory of spectral line broadening has been developed to calculate neutral atom spectra given the interaction and radiative transition moments for relevant states of the radiating atom with other atoms in its environment. Complete details and the derivation of the theory are given by [7]. In our theoretical work, the spectral line is computed from the Fourier transform (FT) of a dipole autocorrelation function. A pairwise additive assumption allows us to calculate the total profile $I(\Delta\omega)$ when all the perturbers interact as the FT of the Nth power of the autocorrelation function $\phi(s)$ of a unique atom-perturber pair. We obtain for a perturber density n_p

$$\Phi(s) = e^{-n_p g(s)}, \tag{1}$$

where decay of the autocorrelation function with time leads to atomic line broadening. For a transition $\alpha = (i, f)$ from an initial state i to a final state f, we have

$$\begin{aligned} g_\alpha(s) \;&= \frac{1}{\sum_{e,e'}{}^{(\alpha)} |d_{ee'}|^2} \sum_{e,e'}{}^{(\alpha)} \int_0^{+\infty} 2\pi\rho d\rho \int_{-\infty}^{+\infty} dx\; \tilde{d}_{ee'}[\,R(0)\,] \\ & [e^{\frac{i}{\hbar}\int_0^s dt\, V_{e'e}[R(t)]}\; \tilde{d}^*{}_{ee'}[R(s)] - \tilde{d}_{ee'}[R(0)]\,]. \end{aligned} \tag{2}$$

The e and e' label the energy surfaces on which the interacting atoms approach the initial and final atomic states of the transition as $R \to \infty$ (R denotes the internuclear distance between the radiator and the perturber). $\Delta V(R)$, the difference potential, is given by

$$\Delta V(R) \equiv V_{e'e}[R(t)] = V_{e'}[R(t)] - V_e[R(t)]\;, \tag{3}$$

and represents the difference between the electronic energies of the quasi-molecular transition. The interatomic interactions are the main physical quantities needed for a good understanding of collisional processes.

In [7] we define $\tilde{d}_{ee'}(R(t))$ as a *modulated* dipole

$$D(R) \equiv \tilde{d}_{ee'}[R(t)] = d_{ee'}[R(t)]e^{-\frac{V_e[R(t)]}{2kT}}\;, \tag{4}$$

V_e is the lower state potential when we consider absorption profiles, or the upper state for the calculation of a profile in emission. Note that over regions where $V_e(R) < 0$, the factor $e^{-\frac{V_e(R)}{2kT}}$ accounts for bound states of the radiator-perturber pair, but in a classical approximation wherein the discrete bound states are replaced by a continuum; thus any band structure is smeared out.

3. *Ab initio* Na-H_2 potentials

Potentials in the present study (hereafter labeled A12) and transition dipole moments were calculated using 2010 version of MOLCAST package [13]. Calculations were performed at the CASPT2 level of theory using atomic natural orbital type basis sets [14]. Adiabatic energies with inclusion of the spin-orbit interaction and transition dipole moments from the ground Na(3s)+H_2 to Na(2p)+H_2 states were obtained using RASSI-SO code in MOLCAS [15]. As an illustration of accuracy, the CASPT energies of Na 2p states calculated using ANO-RCC-VQZP basis set is approximately 150 cm^{-1} below the experimental values.

The Na-H_2 interaction potentials significantly depend on the orientation of the H_2 axis relative the collision axis. Recent multireference MRPT2 configuration interaction calculations for NaH_2 were restricted to Θ =90, and 0° geometries [11, 12]. Present calculations were performed for four geometries Θ =90, 60, 30 and 0°. The H-H bond length was kept fixed at the $H_2(X)$ state equilibrium value r_e = 1.401 au.

As an example of present results, left part of figure 1 shows potentials for the Θ =90° geometry calculated using ANO-RCC-VQZP. The results from [11, 12] are shown for comparison. The right part of figure 1 compares the difference potentials for the *B-X* transition. The agreement with S11 is reasonably good, with the difference of computed energies not exciding 200 cm^{-1}. Similar result was found from comparison of potentials for the Θ =0° geometry.

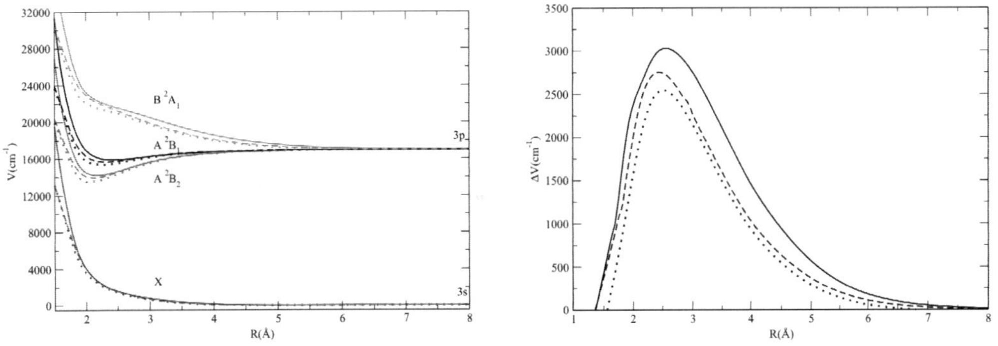

Figure 1. Left: Potential curves for the 3*s* and 3*p* states of NaH_2 molecule for the symmetry C_{2v} .
Right: Difference potentials ΔV for the *B-X* states.
A12 (dashed lines) compared to S11 (dotted lines) and RP85 (solid lines).

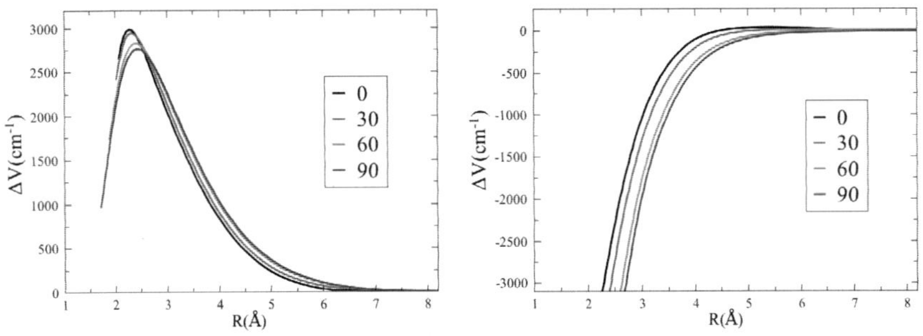

Figure 2. Difference potentials ΔV of *B-X* (left) and $A\ ^2\Pi_{1/2}$-*X* (right) for the four geometries Θ =0, 30, 60 and 90°.

4. Absorption cross sections on the red and blue wing

While *B* states radiates in the blue wing, *A* states radiate in the red wing. We present in figure 3 the variation of the semi-classical absorption line wing at various orientation angles, Θ, of H_2.

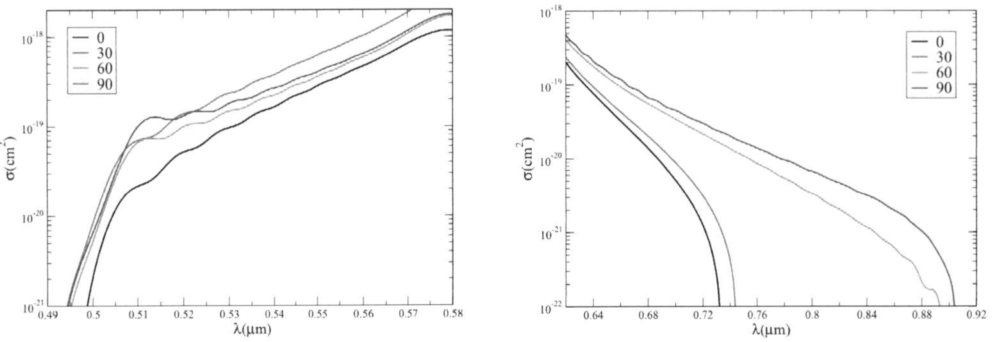

Figure 3. Contribution of the various geometries to the blue wing (left) and red wing (right). The red wing is a sum of weighted contributions from transitions to the $^2\Pi_{1/2}$ and $^2\Pi_{3/2}$ states.

The line wings after averaging over the different angles is shown in figure 4, it is compared to line profiles obtained with RP85 and S11 potentials taking into account the symmetries C_{2v} and $C_{\infty v}$. We assume an equal weight for each orientation.

Figure 4 emphasizes also the non-Lorentzian behavior of the D2 component of the Na–H_2 resonance line. The Lorentzian profile is calculated using the line widths based on RP85 presented in table 2 of [16].

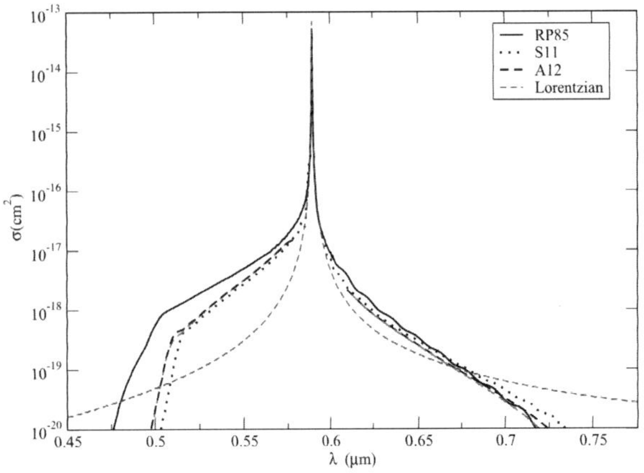

Figure 4. Absorption cross sections of the 3*s*-3*p* D2 line component taking into account the symmetries C_{2v} and $C_{\infty v}$. Red dashed curves take into account the four geometries Θ =0, 30, 60 and 90° of A12. Comparison to the corresponding Lorentzian profile. $n_{H_2} = 10^{20}$ cm^{-3} at 1000 K.

Acknowledgments
One of us (NFA) would like to acknowledge PNPS for the financial support which allows the collaboration with the Laboratoire de Chimie et Physique Quantique in Toulouse.

References

[1] Liebert J, Reid I, Burrows A, Burgasser A, Kirpatrick J and Gizis J 2000 *ApJ* **533** L155

[2] Burrows A, Hubbard W B, Lunine J I and Liebert J 2001 *Rev. Mod. Phys.* **73** 719
[3] Burrows A and Volobuyev M 2003 *ApJ* **583** 985
[4] Szudy J and Baylis W 1975 *J. Quant. Spectrosc. Radiat. Transfer* **15** 641
[5] Szudy J and Baylis W 1996 *Physics Reports* **266** 127
[6] Allard N F, Allard F, Hauschildt P H, Kielkopf J F and Machin L 2003 *A&A* **411** 473
[7] Allard N F, Royer A, Kielkopf J F and Feautrier N 1999 *Phys. Rev. A* **60** 1021
[8] Rossi F and Pascale J 1985 *Phys. Rev. A* **32** 2657
[9] Allard N F, Allard F and Kielkopf J F 2005 *A&A* **440** 1195
[10] Allard N F and Spiegelman F 2006 *A&A* **452** 351
[11] Allard N F, Kielkopf J F, Spiegelman F, Tinetti G and Beaulieu J P 2012 Collisional line profiles of sodium perturbed by H_2 from the optical to the infrared spectral range *ECLA-2011: Proceedings of the European Conference on Laboratory Astrophysics, European Astronomical Society* ed C Stehlé, C Joblin and L d'Hendecourt in press
[12] Allard N F, Kielkopf J F, Spiegelman F, Tinetti G and Beaulieu J P 2012 *A&A* **543** A159
[13] Aquilante F, De Vico L, Ferré N, Ghigo G, Malmqvist P-Å, Neogrády P, Pedersen T B, Pitoňák M, Reiher M, Roos B O, Serrano-Andrés L, Urban M, Veryazov V and Lindh R 2010 *J. Comput. Chem.* **31** 224
[14] Roos B O, Lindh R, Malmqvist P-Å, Veryazov V and Widmark P O 2004 *J. Phys. Chem. A* **108** 2851
[15] Malmqvist P-Å, Roos B O and Schimmelpfennig B 2002 *Chem. Phys. Lett.* **357** 230
[16] Allard N F, Kielkopf J F and Allard F 2007 *EPJ D* **44** 507

Journal of Physics: Conference Series **397** (2012) 012040 doi:10.1088/1742-6596/397/1/012040

$H^- - H$ Collision Induced Radiative Transitions

A V Dadonova[1,3], **A Z Devdariani**[1,2]

[1] Department of Theoretical Physics, Herzen University, St. Petersburg 191186, Russia
[2] Department of Optics and Spectroscopy, St. Petersburg University, St. Petersburg 198904, Russia.

E-mail: alladadonova@mail.ru

Abstract. Exchange interaction leads to the formation of gerade and ungerade states of temporary molecules (quasimolecules) formed during the $H^- + H$ slow collisions. The work deals with the radiation produced by optical transitions between those states. The main characteristics involved in the description of optical transitions in quasimolecules, i.e., energy terms, an optical dipole transition moments, have been calculated in the frame of zero-range potentials model. The main feature of calculations is that the results can be expressed analytically in closed forms via the Lambert W function.

The present work is aimed at the theoretical study of radiating transitions produced in slow collisions of negative ions with neutral atoms. The process of radiating is treated as a dipole transition between gerade and ungerade quasimolecular states that is as a transition between two states of the temporary molecule formed during the collisions. The concept of quasi-molecular transitions is widely used to describe various features of the spectral profiles [1]. More specifically the process of radiating followed by charge-exchange is considered :

$$H^- + H \rightarrow H + H^- + \hbar\omega . \tag{1}$$

The spectral profile produced by radiating transitions can be described in the quasi-static approximation by the formula:

$$I = 4\pi R_c^2 \left(\frac{A(R_c)}{d\Delta E / dR_c} \right) \exp\left(-U_i(R_c)/(kT)\right), \tag{2}$$

where R_c - the position of the Condon point, i.e. the interatomic distance at which the transition takes place in accordance with the Franck-Condon principle that is the distance at which the frequency of radiation $\omega = \Delta E$, where ΔE - the quasimolecular potential energy curve difference,

$$A(R_c) = (4/3g)\omega^3 \alpha^3 |D(R_c)|^2 \tag{3}$$

---

[3] To whom any correspondence should be addressed.

is the probability of radiation or the Einstein coefficient for spontaneous emission, g – statistical factor equal 2 in our case, $\alpha = 1/137$, $U_i(R_c)$ - the adiabatic energy of the initial ungerade state, D is the electric dipole moment for transitions between the two states involved [1]. Thus, all we need to calculate spectral profiles in the quasi-static approximation is the $\Delta E, U_i, D$ distance dependencies.

It is known [2] that the main atomic characteristics of the negative ion H^- as well as the collision processes involving H^- can be described in the frame of the zero-range potential model (ZRP model). In the ZRP model the quasimolecular potential energy curve difference can be found as the solution of a transcendental equation [2]. However, it turns out that not only ΔE but other characteristics of the process (1) can be expressed in terms of the Lambert function $W(z)$ which is a solution of the equation

$$z = W(z)\exp[W(z)] \tag{4}$$

elaborated in [3]. Namely,

$$\Delta E = \left(\chi_-^2 - \chi_+^2\right)/2, \tag{5}$$

where

$$\begin{cases} \chi_- = \beta + W[-\exp(-\beta R)]/R_c \\ \chi_+ = \beta + W[\exp(-\beta R)]/R_c \end{cases} \tag{6}$$

$\varepsilon = -\beta^2/2$ is the bound energy of the extra electron in a negative ion H^- and $R_c = W[\beta/\omega]/\beta$. The adiabatic energy of the initial term is given in the form:

$$U_i(R_c) = -\frac{1}{2}\left\{W^2[-\exp(-\beta R_c)]/R_c^2 + 2\beta \cdot W[-\exp(-\beta R_c)]/R_c + \beta^2\right\}. \tag{7}$$

In the frame of the LCAO approach the dipole moment is the equal to:

$$D = -R_c\Big/\left(2\sqrt{1-S^2}\right), \tag{8}$$

where S is an overlapping integral between the wave functions of the gerade and ungerade quasimolecular states.

Figure (1) shows the results for the spectral profiles averaged over Maxwell's distributions in relative units as a function of frequency in atomic units for two temperatures. The maximum intensity decreases with increasing temperature and is shifted towards higher frequencies with increasing temperature.

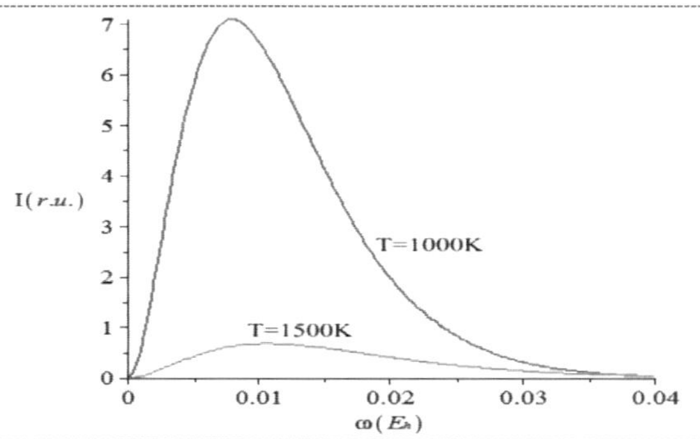

Figure 1. The averaged spectral profile produced in $H^{-}+H$ collisions

It should be stressed that the maximum of the intensity is formed by electron transitions at 7-9 a_0. It means that the radiating transitions are followed by the formation of unstable molecules H_2^- in high rotational states.

References

[1] Devdariani A Z 1999 Emission of quasimolecules *Optics and spectroscopy* **86** 954-959

[2] Demkov Yu N and Ostrovsky V N 1988 *Zero-Range Potentials and Their Application in Atomic Physics* (New York: Plenum Press) p 288

[3] Corless R M, Connet G H, Hare D E G, Jeffrey D J and Knuth D E 1996 *Adv. Comput. Math.* **5** 329-359.

XXI International Conference on Spectral Line Shapes (ICSLS 2012) IOP Publishing
Journal of Physics: Conference Series **397** (2012) 012041 doi:10.1088/1742-6596/397/1/012041

Temperature dependence of the pressure broadening of spectral lines

G D Roston and M S Helmi

Department of Physics, Faculty of Science, Alexandria University, Egypt

E mail: dr.gamal_daniel@yahoo.com

Abstract. The aim of this work is to obtain a formula relating the pressure broadening coefficient of the spectral line β with the temperature T, when the difference potential ΔV(R) between the upper and lower states of the emitting atom is represented by (Lennard – Jones) potential, The obtained formula is a power index law of β on T. This formula is applied for calculating β for different interactions of Ar, Ne, TI, Hg, Cd and Zn with the inert gases (Xe, Kr, Ar, Ne and He) at different temperatures. The results of these calculations are in good agreement with the corresponding values obtained before numerically. The obtained formula is considered very important in astrophysical problems.

1. Introduction

In some applications there is a tendency to represent the dependence of the pressure broadening coefficient $\beta = \gamma / N$ on the temperature T in the form of a power law:

$$\beta \approx T^k \qquad (1)$$

where γ is the half width of the spectral line and k is the temperature dependence index [1, 2]. γ in the impact region is proportional to the density N of the perturbing atoms. The power dependence can be derived theoretically from the classical phase – shift theory of line broadening in the impact limit if the interatomic potentials, V(R) for the upper and lower levels of the emitting atom are approximated by $V(R) = C_p R^{-p}$, where R is the interatomic separation and C_p is a constant. In such a case the impact theory predicts for the index k a formula given by [3]:

$$K = (p - 3) / 2(p - 1) \qquad (2)$$

This means that k is dependent only on the power p characterizing the type of interaction. Bielski, et al, [4], obtained the power law dependence of β for a Lennard – Jones through the function B(α), which was obtained by Hindmarsh, et al, [5]. The obtained formula by Bielski, et al, [4] did`t give any explicit form for the power law dependence of β on temperature. Moreover, the calculation of β in [4] is performed numerically. The authors [6-8] proposed algebraic formulas for the broadening ρ_{ob} and shift ρ_{os} impact parameters of a spectral line based on the assumption that the ranges of the interaction respectively for the broadening and shift are different. In this work the algebraic formula for the broadening is solved and the broadening coefficient β is obtained as a power index law on the temperature T.

2. Mathematical treatment

2.1. Power Law Dependence For a Lennard – Jones Potential

If *V'(R)* and *V''(R)* represent respectively the interaction potentials for the lower and upper states of the emitting atom, are given by Lennard – Jones potential, then the difference potential $\Delta V(R) = V''(R) - V'(R)$ can be written as

$$\Delta V(R) = \hbar(\Delta C_{12} R^{-12} - \Delta C_6 R^{-6}) \tag{3}$$

where (ΔC_6 and ΔC_{12}) are respectively the interaction parameters between the colliding particles in case of the long range attractive and repulsive van der Waals potentials. In previous papers [6-8], the authors obtained a formula relating the impact broadening parameter ρ_{ob} with temperature T using the potential equation 3 as:

$$(\alpha_{12}\Delta C_{12}/\sqrt{8K/\pi\mu})\rho_{ab}^{-11} - (\alpha_6 \Delta C_6/\sqrt{8K/\pi\mu})\rho_{ab}^{-5} = 0.63\ T^{1/2} \tag{4}$$

where α_p is a constant depending on the number p ($\alpha_p = \sqrt{\pi}\Gamma((p-1)/2)/\Gamma(p/2)$), and μ is the reduced mass between the interacting particles. The parameter ρ_{ob} is related to the pressure broadening coefficient β, and the relative velocity □ in angular frequency unit by the relation:

$$\beta = \bar{v}\rho_{ob}^{\ 2}/c = \sqrt{8kT/\pi\mu}\ \rho_{ob}^2/c \approx T^K \tag{5}$$

C here is the velocity of light. The first and second terms on the left hand side of equation 4 represent respectively the phase shifts of the broadening $\eta_p = (\alpha_p \Delta C_p/\check{v})\rho_{ob}^{-(p-1)}$] due to the repulsion [$\Delta C_6$=0] and attraction [$\Delta C_{12}$= 0]. Comparing equations.1 and 5, and using equation 2 the parameter ρ_{ob} is related to T by the relation:

$$\rho_{ob} \approx T^{(2k-1)/4} = T^{-1/(2(p-1)}} \tag{6}$$

It is seen from equation 2 that for the long range attractive van der Waals potential (p = 6) and the pure repulsive form of the potential equation 3 (p =12), the temperature index k are respectively equal to 0.3 and 9/22. These are the cases of equation 6 when the formula 4 is used in its attractive form (ΔC_{12}= 0), or in its repulsive form (ΔC_6= 0).

2.2. The Fitting formula

To solve equation (4) for ρ_{ob}, we put:

$$(\alpha_{12}\Delta C_{12}/\sqrt{8K/\pi\mu})\rho_{ob}^{\ -11} = A\rho_{ob}^{\ -11} = x\sqrt{T} \tag{7a}$$

and

$$(\alpha_6\Delta C_6/\sqrt{8K/\pi\mu})\rho_{ob}^{\ -5} = B\rho_{ob}^{\ -5} = (x-0.63)\sqrt{T} \tag{7b}$$

$$\rho_{ob} = (A/B)^{1/6}(x/(x-0.63))^{-1/6} \tag{8}$$

and

$$T = (B^{11}/A^5)^{1/3}(x^{5/11}/(x-0.63))^{11/3} = (B^{11}/A^5)^{1/3}F_1(x) \tag{9}$$

where

$$F_1(x) = (x^{5/11}/(x-0.63))^{11/3} \tag{10}$$

The function $F_1(x)$ was fitted to the power law of X, so that.

$$F_1(x) = L_1\ x^{-L_1'} \tag{11}$$

The values of the constants L_1 and L_1` depend on the ranges of x as shown in table 1

Table 1. The Fitting parameters for the two Functions $F_1(x)$ and $F_2(x)$ in the different ranges of X

Range of X		Values of $F_1(x)$		Parameters of $F_1(X)$		Values of $F_2(x)$		Parameters of $F_2(X)$	
From	To	From	To	L1	L`1	From	To	L2	L`2
0.66	0.75	191923.71	1472.6658	0.03	- 36.3	122.6963394	17.9641291	0.238	- 15.03
0.76	1.2	1122.6138	10.643740	46.77	- 10.5	16.46814133	2.13084863	4.820	- 4.477
1.21	2.5	10.125263	0.4639603	23.96	- 4.53	2.492365681	0.57127846	3.670	- 2.030
2.6	10	0.4091784	0.012695	5.06	- 2.64	0.580905335	0.110253	1.940	-1.262
10.1	25	0.0124140	0.0017570	1.772	-2.15	0.10905300	0.041561	1.2637	-0.6727

From this, we conclude that, if the temperature T and the interaction parameters (ΔC_6 and ΔC_{12}) are known, then $F_1(x)$ is obtained and consequently x from the table 1 and the relation 11. As x is known, then ρ_{ob} can be obtained using equation 8, and consequently β from equation 5.

2. 3. The broadening parameters β

We have from equation 7a

$$\rho_{ob} = A^{1/11} x^{-1/11}\ T^{-1/22} \tag{12}$$

Then equation 5 will take the form

$$\beta = D\rho_{ob}^2\sqrt{T} = DA^{2/11}[x^{-4/9}T]^{9/22} = DA^{2/11}(T_x)^{9/22} \tag{13}$$

where

$$D = \sqrt{8k/\pi\mu}/C \tag{14}$$

Substituting for T in T_x from equation 9 , we have

$$(T_x)^{9/22} = [x^{-4/9}T]^{9/22} = B^{3/2}\ A^{-15/22}\ (x^{1/3}/(x-0.63))^{(3/2)} \tag{15}$$

$$\beta = DB^{3/2}\ A^{-1/2}(x^{1/3}/(x-0.63))^{3/2} = \beta_o\ F_2(x) \tag{16}$$

where

$$\beta_0 = DB^{3/2}\ A^{-1/2} \tag{17}$$

and

$$F_2(x) = (x^{1/3}/(x-0.63))^{3/2} \tag{18}$$

2.4. The relations between the two functions $F_1(x)$ and $F_2(x)$

We have from equation 13

$$T_x = x^{-4/9}T \tag{19}$$

We put, instead of T, its expression from equation 9

$$T_x = x^{-4/9}(B^{11}/A^5)^{1/3}F_1(x) \tag{20}$$

Substituting in equation 13, we have

$$\beta = DA^{2/11}(x^{-4/9}(B^{11}/A^5)^{1/3}F_1(x))^{9/22} \tag{21}$$

$$\beta = D(B^3/A\ )^{1/2}x^{-2/11}(F_1(x))^{9/22} \tag{22}$$

Comparing equations 16 and 22, we have

$$F_2(x) = x^{-2/11}(F_1(x))^{9/22} \tag{23}$$

From the fitting formula 11, equation 22 takes the form

$$F_2(x) = L_1^{9/22}\ x^{-9/22\,(L_1'+4/9)} = L_2 x^{-L_2'} \tag{24}$$

where $L_2 = L_1^{9/22}$ and $L_2' = (9/22)(L_1'+4/9)$ (25)

The values of the constants L_2 and $L`_2$ are listed in table 1 depending on the ranges of the values of the function $F_2(x)$.

2.5. The power dependence of β on the temperature T

From equation 13 we have

$$\beta = DA^{2/11}\ x^{-2/11}\ T^{9/22} \tag{26}$$

Introducing instead of x, its value from equations (9) and (11), we have

$$F_1(x) = L_1\ x^{-L_1'} = B^{-11}A^5T \tag{27}$$

$$x = L_1^{1/L_1'} B^{11/3L_1'} A^{-5/3L_1'} T^{-1/L_1'} \tag{28}$$

Substituting equations 28 in 26, we have

$$\beta = DA^{2/11} L_1^{-2/11L_1'} B^{-2/3L_1'} A^{10/33L_1'} T^{2/11L_1'} T^{9/22} \tag{29}$$

$$\beta = DL_1^{-2/11L_1'} B^{-2/3L_1'} A^{2/11(1+5/3L_1')} T^{9/22(1+4/9L_1')} = \beta_0' \, T^k \tag{30}$$

where $\beta_0' = DL_1^{-2/11L_1'} B^{-2/3L_1'} A^{2/11(1+5/3L_1')}$ and $k = \frac{9}{22}(1+4/9L_1')$ (31)

From equation 30, we see that by applying the (L-J) potential, the broadening parameter β is a power law of temperature as in cases when the difference potential Δ (R) for the upper and lower levels of the emitting atom is given by $\Delta(R) = C_p R^{-P}$.

3. Results

The broadening coefficients were calculated for different interactions of Ar, Ne, TI, Hg, Cd and Zn atoms with the inert gases (Xe, Kr, Ar, Ne and He). The results are illustrated in the table 2 with the corresponding Hindmarsh and Roston and Helmi values β_H and β [7, 8] with the temperature dependence index k and the values of β_0'

Table 2. The calculated values of pressure broadening coefficient β and $\beta_0^`$ with the corresponding Hindmarsh values β_H and β [7,8] in units 10^{-20} cm^{-1}/atom cm^{-3} for Ar, Ne, TI, Hg, Cd and Zn perturbed by inert gases. ΔC_6 in units 10^{-32} cm^6 rad s^{-1} and ΔC_{12} in units 10^{-74} cm^{12} rad s^{-1} are taken from the labelled references.

Perturber	ΔC_6	ΔC_{12}	$F_1(x)$	k	β_0'	β	β_H	β [7,8]
Ar line 703 nm , T = 330 K								
Ar[9]	531.8	83750	20.6022	0.4264	0.58942	6.987	7.040	6.930
Ne[9]	130.0	55870	2696.02	0.4141	0.7388	8.155	8.250	8.170
He[9]	70.50	48190	72810.28	0.4141	1.24391	13.73	13.89	13.75
Ar + Ar interaction T = 330 (λ in nm)								
λ =603.2 [10]	1114	2.248x106	329.43	0.4264	1.1247	13.33	13.35	13.27
λ =649.6[10]	2551	1.011x108	8981.1	0.4141	2.485	27.43	27.62	27.3
λ =522.1[10]	4993	2.733x109	186391	0.4141	4.595	50.72	50.84	50.27
Ne line 540 nm , T = 330 K								
Ne[11]	5.400	1.2400	7.2480	0.4492	0.0866	1.172	1.153	1.170
He[11]	2.900	0.8800	120.82	0.4264	0.1557	1.846	1.870	1.870
TI line 377.68 nm , T = 860 K								
Ar[12,13]	58.70	15.250	0.0593	0.4779	0.05124	1.295	1.900	1.949
Ne[12,13]	14.30	6.5800	4.6626	0.4492	0.0864	1.798	1.640	1.800
Hg line 253.6 nm , T = 860 K								
Xe[14]	0.2970	0.0010	0.6975	0.4492	0.0089	0.186	0.207	0.202
Kr[15]	0.7920	0.0080	0.8218	0.4492	0.014806 416	0.308	0.304	0.314
Ne[16]	0.7450	0.0640	106.08	0.4264	0.0473	0.843	0.850	0.847
Cd line 326.1 nm , T = 860 K								
Xe[17]	1.3900	0.0110	0.172	0.4779	0.0116	0.292	0.359	0.370
Kr[18]	0.2190	0.0127	242.12	0.4264	0.024	0.429	0.428	0.437
Ar[17]	0.2197	0.0306	1720.8	0.4141	0.0387	0.636	0.549	0.510
Ne[17]	0.3740	0.0388	616.46	0.4264	0.0457	0.816	0.775	0.753
He[17]	0.3170	0.0106	576.56	0.4264	0.0663	1.184	1.190	1.200
Zn line 307.6 nm , T = 860 K								
Zn[19]	1.7875	0.0199	0.3431	0.4779	0.0174	0.441	0.491	0.441
Ar[20]	0.1650	0.0072	524.58	0.4264	0.0290	0.518	0.514	0.515
Ne[21]	0.1337	0.0030	417.99	0.4264	0.0298	0.531	0.529	0.531

4. Conclusions

From the foregoing discussions and relations we conclude the following:

1- If T and (ΔC_6 , ΔC_{12}) are known, then the broadening parameters β can be obtained using equations 9, 8 and 5.

2- If β and (ΔC_6 , ΔC_{12}) are known, then $F_2(x)$ is obtained using equation 16, which leads to the function $F_1(x)$ from equation 24 and then T using equation 9

3- If β and T are known, then ΔC_6 and ΔC_{12} can be obtained using equations 9 and 16.

4- The broadening parameters β were calculated for different interactions. These calculations are illustrated with the corresponding values of Hindmarsh and the previous work of the authors [7,8] as in table 2. It is seen that our calculating results for β are in good agreement with the values obtained before especially for interactions containing light particles. The discrepancy between the values of β for heavy particles can be attributed to the change of the sign of the phase value ± 0.63.

5- The obtained analytical formulas can be used simply to obtain information about the far source parameters "astrophysical problems".

6- The values of the temperature dependence index k for all interactions are very near to the value (9/22=0.4091) when the repulsion part from the potential only is applied.

References

[1] Lwin A, Mc Cartan D D and Levis E L 1977 *J. Astrophys.* **213** 599
[2] Spielfiedel A and Roueff E 1981 *J. Phys.* B. **14** L795
[3] Peach G 1981 *Adv..Phys.* **30** 367
[4] Bielski A, Bobkowski R, and Szudy J 1989 *Astrom. Astrophys.* **208** 357
[5] Hindmarsh W R, Petford A D and Smith G 1967 *Proc. Roy.Soc.*.London. **A297** 296
[6] Helmi MS 1994 *Pak. J. Sci. Ind. Res.* **37** 12 499
[7] Helmi M S and Roston G D 2000 *Physica Scripta* **62** 36
[8] Roston G D , Ghatass Z F and Obaid F S 2009 *JQSRT.* **110** 175 183
[9] Bielski A, Wawrzynski J and Wolnikowski J 1985. *Acta Physica Pol.* **67A**. 3 621
[10] Wolnikowski J, Wawrzynski J, Bielski A and Szudy J 1987 *Physica Scripta* **35**. 153
[11] Bielski A, Dokurno W, Szudy J and Wolnikowski J 1980 *J. Physica* B **60.** 101C. 113
[12] Dygdala RS 1988 *J. Phys.* B:At. Mol. Opt. Phys. **21** 2039
[13] Dygdala RS, Bobkowski R and Lisicki E J 1989 *J. Phys.* B:At. Mol. Phys. **22** 1563
[14] Czuchaj E, Rebentrost F and Stoll H, Preuss H 1997 *Chem.Phys.* **214** 277
[15] Okunishi M, Nakazawa H, Yamanouchi K and Suchiya S 1990 Revisited. *J.Chem.Phys.* **93**. 7526
[16] Kurosawa T, Ohmori K, Chiba H, Okunishi M, Ueda K and Sato Y 1998 *J. Chem. Phys.* **108** 8101
[17] Czuchaj E. and Stoll H 1999 *Chem.Phys.* **248** 1
[18] Czajkowski M, Bobkowski R and Krause L 1991 *Phys Rev* A. **44** 9 5730
[19] Czajkowski M and Koperski J 1999 *Spectrochimica Acta* A **55** 2221
[20] Inguar Wallace, Jarral Ryter and Breckenridge W H 1992 *J. Phys. Chem.* **96** 882
[21] Koperski J and Czajkowski M 2000 *Phys. Rev.* A. **62** 012505.

Journal of Physics: Conference Series **397** (2012) 012042 doi:10.1088/1742-6596/397/1/012042

Krypton influence on the spectral line shape of Cd 326.1 nm

G D Roston and Z F Ghatass*

Physics Department, Faculty of Science, Alexandria University P.O. Box 21511 Alexandria, EGYPT
* Department of Environmental Studies, Institute of Graduate Studies and Research, Alexandria University 163 Horrya Avenue, P.O. 832 Shatby, Alexandria, EGYPT

E-mail: dr.gamal_daniel@yahoo.com

Abstract. The line center of the Cd intercombination spectral line 326.1 nm (5^1S_o-5^3P_1) perturbed by Kr has been investigated using a high - resolution scanning Fabry-Perot interferometer. The van der Waals and Lennard -Jones potentials for Cd-Kr system has been calculated using the Coulomb approximation. The values of the pressure broadening (β) and shift (δ) coefficients for the studied line at temperature 468 K , density of cadmium N = 4.02 10^{12} cm^{-3} and gas Kr pressure ranged from 3 to 95 Torr has been obtained and compared with theoretical and experimental published values.

1. Introduction

Spectral line shape analysis is a powerful diagnostic tool for many media in extreme conditions of temperature and pressure. Measurements of the pressure broadening and shift of atomic spectral lines of van der Waals molecules are an important source of information about the interaction potentials between radiating and perturbing atoms. In the molecular state model the results are sensitive to the interaction potential, particularly in the region of long range interaction.

During the past several decades, there have been numerous investigations of van der Waals molecules especially for pure cadmium and cadmium inert gas systems [1-20]. There are some experimental and theoretical papers [10-20] that have been performed for obtaining the interatomic potentials for Cd + Kr using Cd intercombination spectral line at 326.1 nm (5 $^3P_1 \rightarrow 5\ ^1S_o$).

In the present paper the measurements of the width and shift of the natural Cd intercombination spectral line at 326.1 nm (5^1S_o - 5^3P_1) perturbed by Kr at low pressures are carried out. The broadening and shift coefficients β and δ for Cd line have been obtained experimentally and theoretical using the Coulomb Approximation (CA) for van der Waals and Lennard – Jones potentials . Also, in this work an attempt was undertaken to verify these potential curves obtained by different experimental and theoretical methods [10-20] using the pressure broadening and shift coefficients (β and δ) of the same spectral Line perturbed by Kr using the classical theory of Lindholm [21] and Foley [22].

2. Experimental part

2. 1. Experimental set-up

The measurements were made with the aid of a spectrometer with electrically tunable scanning Fabry-Perot etalon (FPI) (Spectra-Physics). The quartz cell with Cd and Kr gas is placed in the oven enabling the temperature stabilization up to 1K. The fluorescence spectrum of cadmium is excited by high frequency Cd lamp. For more detailed see the previous work [23].

The measurements of the pressure effect of the Cd line of 326.1 nm were carried out at temperature 468 K which corresponds to the number density of cadmium N = 4.02 10^{12} cm^{-3}. The Cd density was calculated using the formula proposed by Dichburn [24]. The perturbed gas Kr pressure ranged from 3 to 95 Torr at room temperature.

2.2. Spectral line profile
If the thermal motion of the emitting atom and its interactions with perturbing atoms are statistically independent, the resultant line shape may be expressed as a Voigt profile which is a convolution of the Gaussian and Lorentzian distributions

Due to the free thermal motion of the emitting and perturbing atoms, the interaction between the two atoms is treated as statistically independent and the resultant line shape may be expressed by Ballik's formula which is a convolution of Voigt and instrumental Airy profile for ideal FPI [25]:

$$I(\nu) = \frac{2}{\Omega}\left\{\frac{1}{2} + \sum_{n=1}^{\infty} R^n e^{-nL} e^{-n^2D^2/4} \cos\left[\frac{2n\pi}{\Omega}((\nu - \nu_o) - \Delta)\right]\right\} \tag{1}$$

where $L = \pi\gamma_L/\Omega$ and $D = \pi\gamma_D/\Omega\sqrt{\ln 2}$, ν_o is the unperturbed frequency of the line, Ω is the free spectral range , γ_L and γ_D are half-widths of the Lorentzian and Doppler components of the observed profile , and Δ is the pressure shift of the spectral line. Equation (1) is fitted to the measured profile using a least-squares algorithm given by Marquardt [26]. The fitting parameters are the Lorentzian (γ_L cm^{-1}) and Doppler (γ_D cm^{-1}) half-widths and the line shift (Δ cm^{-1}).

2.3. Experimental results
Figure 1 shows an example of the experimental and the best fit profile of the 326.1 nm Cd line perturbed by Kr.

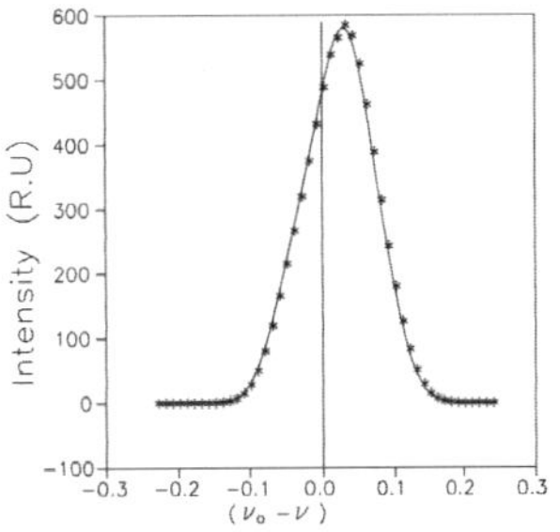

Figure 1. Experimental profile of Cd line 326.1 nm perturbed by Kr at pressure 60 Torr and temperature 468 K fit with a Ballik's formula. Detuning (ν_o- ν) from the line center is given in cm^{-1} .

Figure 2 show the relation between the Doppler half-width (γ_D) and the perturbing gas pressure (P) from 3 to 95 Torr. Figure 3 shows the plot of the Lorentzian half-width (γ_L) and the shift (Δ) of the Cd spectral line 326.1 nm perturbed by Kr as a function of the perturbing gas density (N) at temperature 468 K. The average experimental Doppler half-width (γ_D) amounts to (49.01±1.2) x 10^{-3} cm^{-1}.

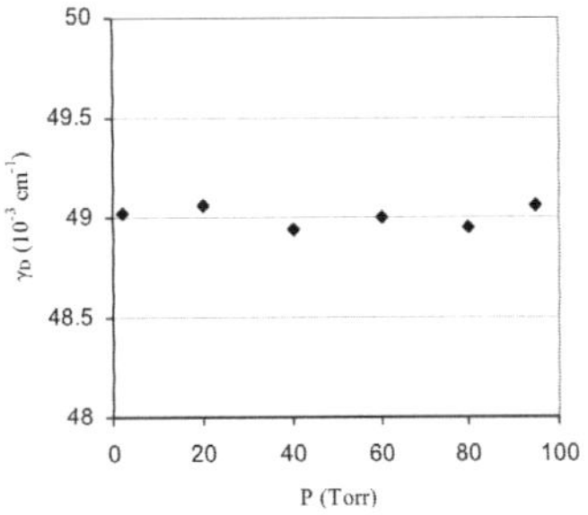

Figure 2. Doppler half-width (γ_D) of the Cd spectral line 326.1nm plotted against the perturbing gas pressure (in Torr) at temperature 468 K.

It is clear that the Lorentzian half-width (γ_L) and the shift (Δ) of the Cd spectral line are linearly dependent on the perturbing gas density N [9, 23]. These dependencies are described by the relations:

$$\gamma_L = \gamma_L^o + \beta N \quad \text{and} \quad \Delta = \Delta^o + \delta N \tag{2}$$

where γ_L^o and Δ^o are the residual half-width and shift of the spectral line. In this experiment, found that $\gamma_L^o = (2.28 \pm 0.4) \times 10^{-3}\ \text{cm}^{-1}$ and the residual value $\Delta^o = (1.38 \pm 0.3) \times 10^{-3}\ \text{cm}^{-1}$. From the slope of the two lines, the pressure broadening (β) and shift (δ) coefficients.

Figure 3. Lorentzian half-width (γ_L) and the pressure shift (Δ) of the Cd spectral line 326.1 nm perturbed by Kr at temperature 468 K plotted against the perturbing gas density.

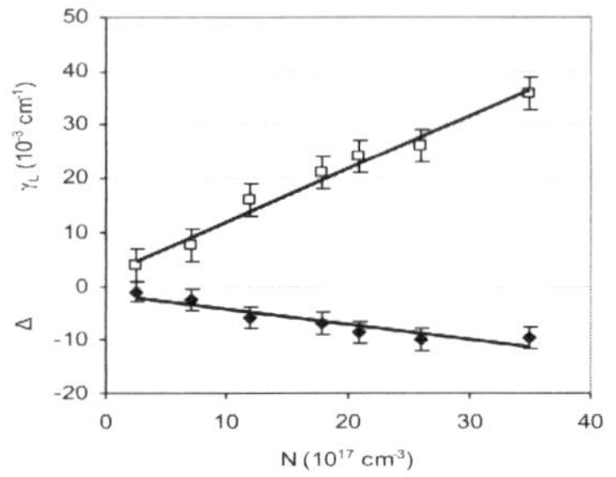

3. Theoretical calculation

3.1. Calculation of van der Waals and Lennard–Jones potentials

The experimental results were interpreted using the Lindholm-Foley version of the impact theory using van der Waals (VdW) and Lennard –Jones (LJ) potentials. For more details see [26, 27].

The van der Waals potenti al is given by:

$$V(R) = -C_6 / R^6 \tag{3}$$

where C_6 is the VdW force constant in the given state and R is the interatomic separation between the radiating and perturbing atoms. To calculate the VdW force constant C_6 the approximate formula given by Unsöld [28], has been used:

$$C_6 = \alpha e^2 \langle r^2 \rangle \tag{4}$$

where e is the electron charge, α is the dipole polarizability of the perturbing and $\langle r^2 \rangle$ is the quantum mechanical mean value of r^2 in a given state of the radiating atom. The values of $\langle r^2 \rangle$ have been calculated for both the ground and exited states using the Coulomb approximation:

$$\langle r^2 \rangle = (1/2) a_o^2 n_{eff}^2 [5 n_{eff}^2 + 1 - 3\ell(\ell + 1)] \tag{5}$$

where a_o is the Bohr radius , ℓ and n_{eff} are the azimuthal and effective quantum numbers. $\ell = 0, 1$ for the ground ($^1S_0$) and excited ($^3P_1$) states respectively.

The LJ potential is given by:

$$V(R) = \frac{C_{12}}{R^{12}} - \frac{C_6}{R^6} \tag{6}$$

where C_6 and C_{12} are the attractive and repulsive potential parameters and R is the interatomic separation between the radiating and perturbing atoms. For calculating the c_{12} constant, the Hindmarsh [29] formula has been used as:

$$C_{12} = q(R_A + R_B)^{12} \tag{7}$$

where $q = (0.9 \pm 0.3)\ 10^{-16}$ erg and R_A or R_B denotes the so –called Hindmarsh radius which is defined as the distance from the nucleus of atom A or B at which the unperturbed radial charge density has the value 0.012 a.u. These distances are calculated using the Froese-Fischer program [30] for the Hartree - Fock wave functions [27].

As the Cd intercombination spectral line at 326.1 nm is due to the transition (5^3P_1—5^1S_o) and the excited state $^3P_1$ have two molecular states $^3o^+$ and ^{31}, it is very interested to see which transition is more effected in the broadening and shift of the spectral line, for that we calculate the VdW ($\Delta C_6^o, \Delta C_6^1$ and ΔC_6) and L J ($\Delta C_{12}^o, \Delta C_{12}^1$ and ΔC_{12}) potentials parameters difference for different transition see Table 1, ($\Delta C_6, \Delta C_{12}$) for the ($5^3P_1$—$5^1S_o$) are calculated using the formulas

$$\Delta C_6 = (2/3)\Delta C_6^1 + (1/3)\Delta C_6^0 \quad \text{and} \quad \Delta C_{12} = (2/3)\Delta C_{12}^1 + (1/3)\Delta C_{12}^0$$

The calculated values of the potential parameters differences between the ground and excited states of Cd perturbed by Kr inert gas for the VdW (ΔC_6) and LJ (ΔC_{12}) are presented in Table 2.

Table 1. The used symbols for the VdW and LJ potential parameters differences with the corresponding broadening and shift coefficients for the different transitions

Transition	VdW parameters	LJ parameters	Broadening Coefficients β	Shift Coefficients δ
$A\,^3 0^+ - X\,^1 0^+$	ΔC_6^o	ΔC_{12}^o	β^0	δ^0
$B\,^3 1 - X\,^1 0^+$	ΔC_6^1	ΔC_{12}^1	β^1	δ^1
$5\,^3P_1 - 5\,^1S_o$	ΔC_6	ΔC_{12}	B	Δ

Table 2. The van der Waals (ΔC_6 x 10^6 cm^{-1} $Å^6$) and Lennard-Jones (ΔC_{12} x10^9 cm^{-1} $Å^{12}$) potential parameters differences between the ground and excited states of Cd perturbed by Kr inert gas

References	VdW parameters			L J parameters		
	ΔC_6^o	ΔC_6^1	ΔC_6	ΔC_{12}^o	ΔC_{12}^1	ΔC_{12}
This work (VdW-CA)	--	--	0.3357	--	--	12.535
Roston [10]	1.0326	1.1092	1.0837	0.0299	13.71	0.924
Roston [11]	0.1901	0.2943	0.2596	--	--	--
vdW-Grycuk *et* al. [12]	0.2998	0.4683	0.4122	--	--	--
Czajkowski [14]	0.0290	0.1433	0.1052	6.0274	9.471	4.305
Koperski [16, 17]	0.0232	0.6660	0.4517	4.1702	10.24	5.438
Czuchaj [18]	0.0171	1.3227	0.8875	5.862	17.57	9.762
Czuchaj 19]	0.0032	--	--	0.3817	--	--

3.2. Calculation of the broadening β and shift δ coefficients

The collision broadening and shift coefficients β and δ (in angular frequency units) for the Cd intercombination line 326.1using both van der Waals and Lennard-Jones potential parameters have been calculated as follows:

For the Van der Waal's potential, the broadening β and shift δ coefficients are calculated using the following formulas [31, 32]:

$$\beta = 8.16\bar{V}^{3/5}(\Delta C_6)^{2/5} \quad \text{and} \quad \delta = -0.357\beta \tag{8}$$

where $\bar{V}$ is the mean relative velocity. In the case of the Lennard-Jones potential, ΔC_6 and ΔC_{12} are the potential parameters difference between the ground and excited states, then β and δ are calculated using Hindmarsh et al. [29]:

$$\beta = 8\pi\,(3\pi/8)^{2/5}\bar{V}^{3/5}\,(\Delta\,C_6)^{2/5}\,B(\alpha) \tag{9}$$

$$\delta = 2\pi\,(3\pi/8)^{2/5}\bar{V}^{3/5}\,(\Delta\,C_6)^{2/5}\,S(\alpha) \tag{10}$$

where the broadening B(α) and shift S(α) functions are given by:

$$B(\alpha) = \int_0^\infty x\,\sin^2 0.5(\alpha\,x^{-11} - x^{-5})\,dx \tag{11}$$

$$S(\alpha) = \int_0^\infty x\,\sin(\alpha\,x^{-11} - x^{-5})\,dx \tag{12}$$

and

$$\alpha = 0.536\;\bar{V}^{6/5}\,(\Delta C_6)^{-11/5}\,\Delta C_{12} \tag{13}$$

The two functions *B*(α) and *S*(α) for different values of α are tabulated in Ref. [29]. The experimental, calculated and the published values of β and δ are presented in Table 3. To follow the labeled of the broadening β and shift δ coefficients see table I.

It is seen from Table III, the obtained experimental results for the broadening and shift coefficients β and δ are agreement with both the experimental result obtained by Brym and Domysfawska [9] and with theoretical values calculated using the Coulomb approximation and VdW potential. Concerning the comparison between the obtained results and those calculated using different published data can be concluded as:

1- The VdW potential gives an agreement results for the broadening and shift coefficients (β and δ) with the experimental results, specially using the potential parameters obtained by [10-12, 16-17], but it gives a poor agreement with the potential obtained from [14,18-19]. For that the long range interaction is more effected in the broadening and shift of the spectral lines.

2- The L J potential gives a good agreement results, using the potential parameters obtained by [10] and for the shift parameter using [16-17], but it gives a poor agreement with the Coulomb approximation and with that results calculated using the potential obtained by [14,18-19].

3- From the analysis of the obtained results, it was found that the more shallow state $^{3}1$ is more effected than the deepest state $^{3}0^{+}$ in this interaction.

Table 3. The experimental and calculated values of the broadening and shift coefficients β and δ (in units of 10^{-20} cm^{-1}/atom cm^{-3}) of the Cd line 326.1 nm perturbed by Kr inert gas and the corresponding experimental and theoretical published values

		References	Broadening Coefficient			Shift Coefficient		
			β^{o}	β^{1}	β	δ^{o}	δ^{1}	Δ
Experimental		This work	0.974 ± 0.05			-0.267 ± 0.04		
		Brym and Domysfawska [9]	1.00 ± 0.03			-0.27 ± 0.03		
Theoretical	Van der Waals	This work (VdW-CA)	—	—	0.906	—	—	-0.323
		Roston [10]	1.1420	1.1462	1.145	-0.31	-0.32	-0.317
		Roston [11]	0.7222	0.8599	0.818	-0.26	-0.31	-0.291
		vdW-Grycuk *et* al. [12]	0.8664	1.0356	0.984	-0.31	-0.37	-0.351
		Czajkowski [14]	0.3405	0.6449	0.570	-0.12	-0.23	-0.203
		Koperski [16, 17]	0.3108	1.1922	1.021	-0.11	-0.43	-0.364
		Czuchaj [18]	0.2754	1.568	1.337	0.098	-0.56	-0.477
		Czuchaj [19]	0.1412	—	—	-0.05	—	—
	Lenard Jouns	This work (LJ-CA)	—	—	0.601	—	—	-0.315
		Roston [10]	1.148	1.139	1.131	-0.27	-0.34	-0.252
		Czajkowski [14]	0.5167	0.523	0.399	0.04	-0.07	-0.089
		Koperski [16, 17]	0.482	1.160	0.997	0.044	-0.53	-0.499
		Czuchaj [18]	0.438	1.429	1.364	0.078	-0.58	-0.441
		Czuchaj [19]	0.3247	—	—	0.038	—	—

References

[1] Helmi M S, Grycuk T and Roston G D 1996 *Acta* B. **51** 633
[2] Helmi M S, Grycuk T and Roston G D 1996 *Chem. Phys.* **209** 53
[3] Roston G D, Grycuk T and Helmi M S. 1996 *Chem. Phys.* **213** 365
[4] Roston G D and Helmi M S 2000 *Chem. Phys.* 258 55
[5] Roston G D 2005 *Physica Scripta*. **72** 31
[6] Roston G D 2004 *J. Quant. Spect. Radiat. Transfer* **87** 83 94
[7] Roston G D and Helmi M S 2006 *J. Quant. Spect. Radiat. Transfer* **101** 201
[8] Roston G D, Ghatass Z F 2006 *J. Quant. Spect. Radiat. Transfer* **101** 205
[9] Brym S and Domysfawska 1995 *Physica Scripta* **52** 511
[10] Roston G D, Ghatass Z F 2008 *J. Quant. Spect. Rad. Transfer* **109** 2427
[11] Roston G D and Helmi M S 2009 *Chem. Phys.* **358** 30
[12] Grycuk T, Findeisen M and Sniecinska A. *Spectral Line Shapes*, Edited by L. Frommhold and J. W. Keto, New York, American Institute of Physics **6** 174
[13] Kowalski A, Czajkowski M and Breckenridge WH 1985 *Chem. Phys. Lett.* **121** 217
[14] Czajkowski M, Bobkowski R and Krause L 1991 *Phys. Rev* A. **44** 9 5730
[15] Kvaran A, Funk D J, Kowalski A and Breckenridge W H 1988 *J. Chem. Phys.* **89** 6069
[16] Koperski J, Łukomski M and Czajkowski M 2002 *Spectrochimica Acta* Part A **58** 2709

[17] Koperski J 2002 *Physics Reports* 369 177 326
[18] Czuchaj E and Sienkiewicz J 1984 *J. Phys.* B: At. Mol. Phys. **17** 2251
[19] Czuchaj E and Stoll H 1999 *Chem. Phys.* **248** 1
[20] Czuchaj E and Krosnicki M 2000 *Chem. Phys. Lett.* **329** 495 .
[21] Lindholm E 1945. *Ark. Mat. Astrono Ocho Fys.* 1945 **52**A 17
[22] Foley H M 1946 *Phys Rev.* **69** 616
[23] Roston G D and Ghatass Z F 2009 *J. Quant. Spect. Rad. Transfer* **110** 2174
[24] Dichburn R W and Glimour J C 1941 *Rev. Mod. Phys.* **13** 310
[25] Ballik E A 1966 *Appl. Opt.* **5** 170
[26] Marquardt D W 1963 *J. Soc. Industr. Appl. Math.* **11** 431
[27] Bielski A, Wasilewski J 1980 *Z. Naturforsch* A **35** 1112
[28] Unsöld A 1955. *Physik der sternatmospharen, mit besonderer berucksichtigung der sonne.* Springer, Berlin
[29] Hindmarsh W R, Petford A D and Smith G 1967 *Proc. Roy. Soc.*A. **297** 296
[30] Froese-Fischer C 1972 *Comput. Phys. Commun.* **4** 107
[31] Behmenburg W and Schuller F 1974 *Phys Rep.* **12** C 273
[32] Dygdała R S 1988 *J. Phys.* B: At. Mol. Opt. Phys. **21** 2039

Broadening of the H_2 ($X \rightarrow B$) transition lines in mixtures with rare gases and CF_4

Vadim A Alekseev[1] **and Ralph Püttner**[2]

[1]Institute of Physics, St.Petersburg State University, Ul'janovskaja St.1, Peterhof, 198504 St.Petersburg, Russia
[2]Institut für Experimentalphysik, Freie Universität Berlin, Arnimallee 14, D-14195 Berlin, Germany

E-mail: alekseev@va3474.spb.edu.

Abstract. The H_2 ($X \rightarrow B$) transition is measured in mixtures with rare gases and CF_4. The rotational lines exhibit a strong dependence of the pressure broadening effect on the rotational quantum number J. At pressures of ~1 atm the collision-induced spectral width for the lines R(0) and R(1) is ~10 cm^{-1} (FWHM), while for the transitions from the rotational levels with J > 3 the broadening effect is hardly visible at the spectral resolution used in the present study.

1. Introduction

The $B\,^1\Sigma^+_u$ state of molecular hydrogen is the lowest one which is coupled with the ground state $X\,^1\Sigma^+_u$ by a dipole allowed transition. The H_2 ($X \rightarrow B$) transition known as Lyman bands has been subject to many investigations both in absorption and in emission; the studies performed prior to 1994 are summarized in Ref. [1]. Besides the fundamental scientific interest in the benchmark system H_2, detailed knowledge of this transition is of prime importance also from a more applied point of view since hydrogen is the most abundant molecule in stellar atmospheres and interstellar clouds. Consequently, the emission of the H_2 molecule in the VUV region is observed in the spectra of many stellar objects and is one of the major sources of astrophysical information.

The shapes of spectral lines perturbed by collisions provide important information about concentration, temperature and composition of the gas surrounding the light-emitting atoms or molecules. To our knowledge, except for a brief study of the $0 \rightarrow 0$ and $0 \rightarrow 1$ bands in mixture with Ar at liquid nitrogen temperature [2], collisional effects in spectrum of the H_2 ($X \rightarrow B$) transition have not been studied previously. The present contribution reports on a spectroscopic study of this transition in mixtures with the rare gases He, Ne, Ar and the CF_4 molecule.

2. Experiment

Studies of absorption and luminescence excitation spectra of H_2 in mixtures with rare gases and CF_4 were conducted at the synchrotron radiation facility BESSY using a 10-m normal incidence monochromator [3]. The experiments were performed using a gas cell with a length of 8 cm that was separated from the ultra-high vacuum of the beamline by a LiF window. The light transmitted through the gas cell was detected by a GaAs photodiode or by a photomultiplier tube (PMT) installed behind a quartz window that was covered with a sodium salicylate luminophore. The luminescence subsequent to the excitation by synchrotron radiation was detected via two laterally located window ports. One window was made of MgF_2 to record luminescence in the VUV range by using a solar blind PMT. Emission in the UV and visible spectral ranges was recorded via the second on-side window with the use of conventional PMT.

3. Results

Due to the large rotational constants, the $X \rightarrow B$ absorption spectrum consists of well separated rotational lines. Figure 1 shows a part of the high-resolution spectra of pure H_2 and H_2-Ar mixtures which includes the lines R(0) and R(1) of the $0 \rightarrow 0$ band and the line P(5) of the $0 \rightarrow 1$ band. The differences in the widths of the various lines demonstrate that the broadening effect strongly depends on the individual rotational excitation. For the R(0) line the broadening coefficient is estimated to be ~10 cm^{-1}/atm.

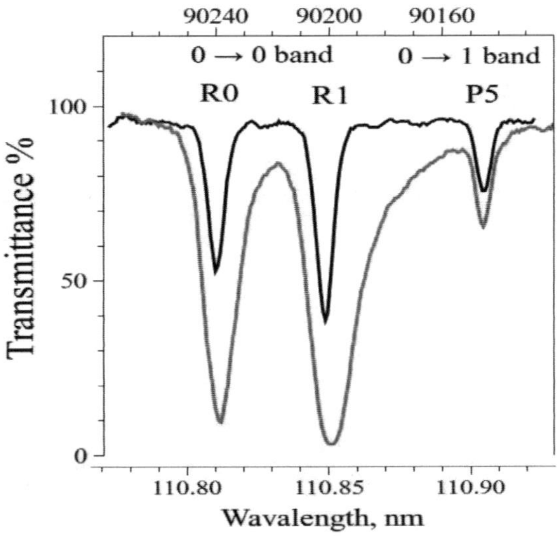

Figure 1. Absorption spectra: H_2 10mbar (black) and H_2 10 mbar + Ar 550 mbar (red).

The spectra displayed in Figure 1 were recorded using a 2400-groove/mm grating. Because of technical reasons most of spectra in the present study were recorded with a 300 grooves/mm grating. These spectra exhibit a substantially lower resolution so that the J-dependent collisional effects manifest itself in a stronger enhancement of low J transitions. Figure 2a shows an overview of the $0 \rightarrow 0$ band for pure H_2 and in a mixture with He. As can be seen in the spectra, the collision-induced enhancement of the absorption is a factor of 5 for the low-J lines R(0) and R(1) and about factor of 2 for the overlapping high-J lines P(4) and R(5). The effect also manifests itself in the luminescence excitation spectra. As is seen from Figure 2b, the relative intensities of the lines P(4) and R(5) in the normalized luminescence excitation spectra of pure H_2 and the H_2+ He mixture differ by a factor of 2. Effects on a comparable scale were observed in the mixtures with the other Rg gases and CF_4. Note that the shapes of the strongly broadened lines R(0) and R(1) depend on the rare gas used in the mixture. A detailed analysis of the line shapes requires results of high-resolution spectroscopic studies.

The energy separation between the lines P(J) and R(J+1) of the 0 → 0 band is small and for the P(4) and R(5) lines amounts only ~13 cm^{-1}. Larger separations between these lines in the spectra of 0 → $v' > 0$ bands opens the opportunity to study collisional effects for individual lines. For example, the spectra of 0 → 2 band in Figure 2c nicely illustrate that the lines P(J) are stronger enhanced by the interaction with helium (red curve) than the lines R(J+1). The 0 → $v' > 0$ bands can be found in the energy region with $\lambda < 110$ nm. Unfortunately, the optical absorption of Ar and Kr is very strong in this energy region so that complementary studies of mixtures with these rare gases are hampered.

In the present studies we also obtained first results on the broadening effect in pure H_2. The green curve in Figure 2a shows the spectrum of the 0 → 0 band at a pressure of 400 mbar. For reasons of comparison, this spectrum is scaled to the intensity of the overlapping lines P(4)/R5 in the low-pressure spectrum (black curve). A comparison of these two spectra clearly shows the J-dependence of the broadening effect.

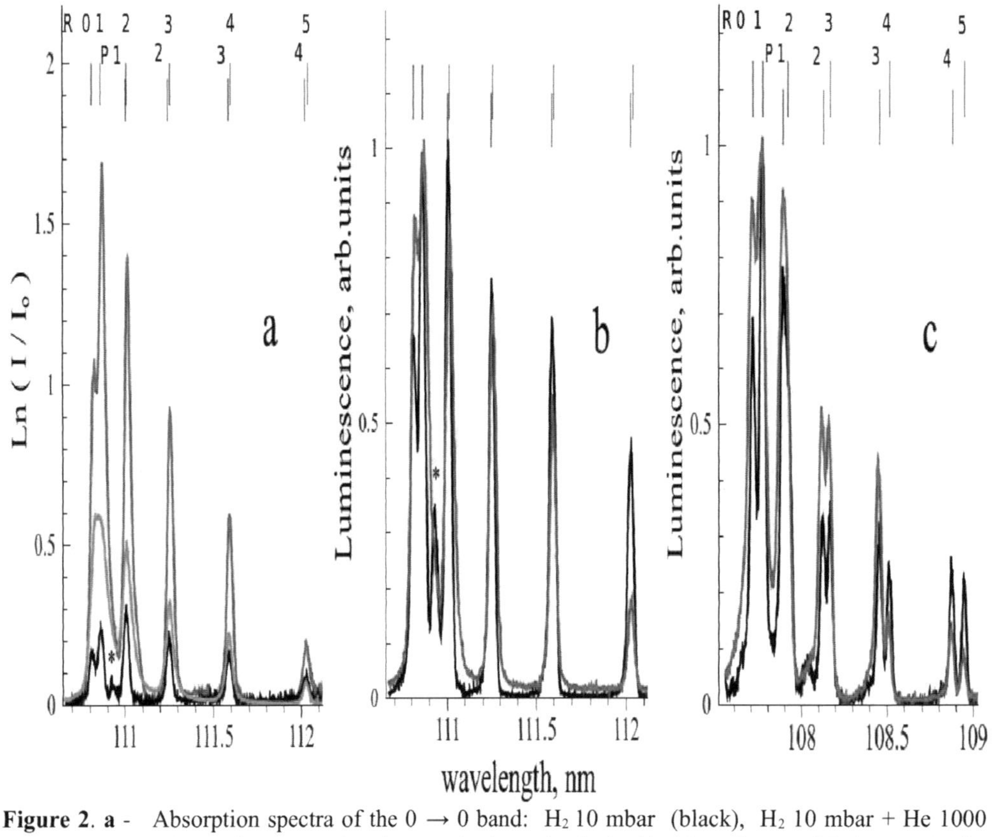

Figure 2. **a** - Absorption spectra of the 0 → 0 band: H_2 10 mbar (black), H_2 10 mbar + He 1000 mbar mixture (red), and H_2 400 mbar (green, the intensity is devided by a factor of 7. (see text); the overlapping lines R(0) and R(1) are saturated). The feature labelled ' * ' is the line P(5) of the 0 → 1 band. **b** - Luminescence excitation spectra of the 0 → 0 band: H_2 10 mbar (black), and H_2 10 mbar + He 1000 mbar mixture (red). **c** - Luminescence excitation spectra of the 0 → 2 band: pure H_2 5 mbar (black) and H_2 5 mbar + Ne 1000 mbar mixture (red). For the spectra shown in panels **b** and **c** the undispersed VUV luminescence was recorded using a solar blind PMT.

4. Discussion

The present study shows that the pressure broadening effect for the rotational lines of the H_2 ($X \rightarrow B$) transition exhibits a strong dependence on the rotational quantum number J. At pressures of ~1 atm the collision-induced spectral width for the lines R(0) and R(1) is ~10 cm^{-1} (FWHM), while for the transitions from the rotational levels with J > 3 the broadening effect is hardly visible at the spectral resolution used in the present study.

The pressure broadening effect has been extensively studied for rotational lines in the Raman spectra of the molecular hydrogen. In these studies it has been shown that the J-dependence of the self-broadening coefficients for the Q-branch transitions with J = 1-5 is not a monotone function [4]. At room temperature the main contribution to the broadening is due to rotational-rotational (R-R) inelastic process which has the maximum probability for the J = 3 level. The self-broadening coefficient for the line Q(3) is by a factor of two larger than those for the lines Q(0) and Q(5) which are approximately equal. At high temperatures the maximum for the line Q(3) is less distinct. In mixture with He, where the resonant R-R process is absent, the broadening coefficient shows the tendency to decrease with J, except for the J = 1 level which has the smallest coefficient [5]. The observed dependence is, however, not pronounced. For all studied temperatures, the difference in the broadening coefficients for Q(0) and Q(5) is ~30 %.

The pressure broadening coefficients for the lines in the Raman spectra of hydrogen and its mixtures with rare gases amount to ~10^{-3} cm^{-1}/atm which is a typical value for Raman spectra of other molecules as well. Consequently, a pressure of ~100 atm is required to obtain a collision-induced spectral width of ~10 cm^{-1}. The broadening effect for the rotational lines of electronic transitions is, however, significantly larger. For the O_2 ($X\ ^3\Sigma^-_g \rightarrow b\ ^1\Sigma^+_g$) magnetic dipole transition the self-broadening coefficient for the line J = 1 amount ~$6 \cdot 10^{-2}$ cm^{-1}/atm and it smoothly decreases by approximately a factor of 2 up to J = 30 [6]. For the CO ($X\ ^1\Sigma^+ \rightarrow B\ ^1\Sigma^+$) transition the self-broadening coefficient is one order of magnitude larger [7] and it shows, in contrast to the O_2 ($X \rightarrow b$) transition, no pronounced J-dependence of the broadening. This implies that the R-R inelastic collisions responsible for the J-dependence in the Raman spectra play only a minor role in the broadening mechanism for the CO ($X \rightarrow B$) transition [7]. No pronounced J-dependence of the rotational broadening coefficient was also observed for the NO ($X\ ^1\Sigma^+ \rightarrow A\ ^1\Sigma^+$) [8] and the OH ($X\ ^1\Sigma^+ \rightarrow A\ ^1\Sigma^+$) [9] transitions.

Comparing the broadening effect for the H_2 ($X \rightarrow B$) transition with the results from literature summarized above, two differences can be found. (i) The broadening coefficients for the H_2 ($X \rightarrow B$) lines are substantially larger than for the electronic transitions studied up to now. (ii) The J-dependence for the H_2 ($X \rightarrow B$) lines is very pronounced, with a decrease of the broadening coefficient by an order of magnitude from J = 0 to 5. The latter observation implies that the main contribution to broadening is due to R-R inelastic collisions. In this respect, the question on J-dependence of rotational relaxation rate is of significant interest. To our knowledge, rotational relaxation rates for the J-levels of the *B* state of molecular hydrogen are not known to date.

Experiments conducted in the course of the present study showed that the $H_2(B)$ luminescence is quenched in collisions with $H_2(X)$ and Ar. In particular, in mixture with 1000 mbar of Ar luminescence intencity is a factor of 10 weaker than in pure H_2 at pressure of 10 mbar. Interestingly, studies of the self-quenching process for the state $EF\ ^1\Sigma^+_g$ in H_2, which lies ~1 eV above the *B* state, have shown that the quenching-rate constant depends on J [10]. The J-dependence of the $H_2(B)$ quenching processes can not be excluded as well. This question requires further studies.

5. Conclusion

The present contribution mainly aims to provide first experimental results on the pressure broadening effect in the spectra of H_2 + Rg mixtures in the VUV region. A systematic investigation at high spectral resolution and for an extended pressure range is required to characterize the collisional induced broadening effects in more details. A more precise data analysis, and in particular a line shape modeling, would require the knowledge of the interaction potentials for the lower and the upper state of the collision complex. The ground states of Rg-H_2 van-der-Waals complexes have been the subject of many experimental and theoretical studies ([11] and references therein) and are known with high accuracy. First *ab initio* calculations of Rg-$H_2(B)$ interaction potentials have been performed complementary to the present experimental studies [12]. At present, the results are preliminary and are not reported here. The experimental studies of the pressure broadening effect and the *ab initio* studies of the relevant interaction potentials will be continued and the results shall be reported elsewhere.

Acknowledgements
We thank Prof. Dr. N. Schwentner for support of the present studies and many useful discussions. VAA acknowledges financial support from the Russian-German beamline project at BESSY.

References

[1] Roncin J-Y and Launay F 1994 *J. Phys. Chem. Ref. Data, Monograph* **4**
[2] Takezawa S and Iida Y 1995 *Annual Reports Gunma University* **16** 61
[3] Reichardt G, Bahrdt J, Schmidt J-S, Gudat W, Ehresmann A, Muller-Albrecht R, Molter H, Schmoranzer H, Martins M, Schwentner N and Sasaki S 2001 *Nucl. Instrum. Meth.* A **467–468** 462
[4] Rahn L A, Farrow R L and Rosasco G J 1991 *Phys. Rev.* A **43** 6075
[5] Forsman J W, Bonamy J, Robert D, Berger J Ph, Saint-Loup R and Berger H 1995 *Phys. Rev.* A **52** 2652
[6] Gordon I E, Rothman L S and Toon G C 2011 *J. Quant. Spectrosc. Radiat. Transfer.* **112** 2310
[7] Rosa M D, Roger L and Farrow R L *J. 2001 Quant. Spectrosc. Radiat. Transfer.* **68** 363
[8] Vyrodov A O, Heinze J and Meier U E, J. 1995 *Quant. Spectrosc. Radiat. Transfer.* **53** 277
[9] Kessler W J, Allen M G and Davis S 1993 *Quant. Spectrosc. Radiat. Transfer.* **49** 107
[10] Niki K, Fujiwara M, Motoshima Y, Kawauchi T, Fukutani K, 2011 *Chem. Phys. Lett.* **504** 136
[11] Barletta P 2009 *Eur. Phys. J.* D **53** 33
[12] Alekseev V A 2012 *21-International Conference on Spectral Line Shapes*

Nonlinear properties and collisional spectra in hydrogen-(heavy) noble-gas-atom mixtures

Waldemar Głaz[1], Tadeusz Bancewicz[1], George Maroulis[2], Anastasios Haskopoulos[2] and Jean-Luc Godet[3]

[1] Nonlinear Optics Division, Faculty of Physics, Adam Mickiewicz University, 61-614 Poznań, Poland

[2] Department of Chemistry, University of Patras, GR-26500 Patras, Greece

[3] Laboratorie de Photonique d'Angers, EA 4464, Université d'Angers, 2 boulevard Lavoisier, F49045 Angers Cedex 01, France

E-mail: glaz@amu.edu.pl

Abstract. In a series of previously published works the phenomenon of collision induced nonlinear hyper-Rayleigh scattering of light (CIHRS) was studied on the basis of the numerical quantum chemistry (QC) methods confronted with a theoretical approach. A class of super-molecular systems composed of hydrogen (H_2) and lighter inert gas atoms (Rg) was considered. In this report a development of this research work is presented with more massive and highly polarizable, Kr and Xe, perturbers involved. The collision-induced hyperpolarizability tensorial values, $\Delta\boldsymbol{\beta}$ obtained by means of the QC *ab initio* methods are applied in order to produce the CIHRS spectral distributions; the influence of $\Delta\boldsymbol{\beta}$ features on the line shapes is compared with the earlier theoretical predictions. Namely, the validity of the multipole-induced-multipole (MIM) mechanism is assessed. In particular, its relevance to reproduce long-range functional behavior of the so-called symmetry adapted (SA) components of the hyperpolarizability tensor $\Delta\boldsymbol{\beta}(R)$ is discussed. A thorough analysis of the translational CIHRS spectra is then performed to identify the role of the hyperpolarizability spatial distribution in forming particular sections of the line shapes. An extension of the dipole-induced-quadrupole analytical model is suggested and tested with regard to its ability to reproduce more accurate profiles.

1. Introduction

In view of the continuous development of experimental and numerical methods of investigation of still more subtle effects related to light scattered by molecular media, a growing demand for data sets of relevant molecular properties and for analysis of validity of their values is observed. Nonlinear hyper-Rayleigh scattering occurrences, particularly those in which a significant role is played by intermolecular interactions and correlations, no doubt belong to this category of phenomena.

The hyper-Rayleigh (HR) process can be observed, when there is a non-negligible probability of two absorbed photons of laser frequency ω to initiate a nonlinear response of the medium resulting in emission of a photon at a frequency of about 2ω. The most pronounced spectral properties of the HR scattered signal are mainly determined by the first hyperpolarizability tensor, $\boldsymbol{\beta}$, of a given microsystem. It may be expected, however, that there are also certain non-negligible contributions to the HR effect induced by interactions between the atoms and/or molecules during their *fly-by* encounters.This is the so-called collision-induced hyper-Rayleigh scattering (CIHRS) that can be especially conspicuous in molecular media that consist of microscopic entities whose intrinsic symmetry forbids the occurrence of the single-molecule HR effect. The hyperpolarizability property of the short-lived compounds can be evaluated both by means of numerical and theoretical procedures [1, 2], each having its specific domain of validity attributed, for instance, to characteristic frequency regions of the scattered light. Analysis of applicability of these methods is the main aim of the research presented.

2. Objectives

The above stated general idea of the study determines more detailed objectives to be achieved:

- to extend the set of the CIHRS-active compounds previously considered [3, 4, 5, 6, 7] to systems including the heavier noble gas atoms of Kr and Xe; as found earlier, the more polarizable the perturbers the greater is the intensity of the scattered signal, which offers a better chance for foreseeable experimental methods to detect it;
- to determine *ab initio* values of the supermolecules' collisional hyperpolarizabilities and their counterparts obtained analytically;
- to assess and compare the ability of the data sets obtained by means of the procedures mentioned to render particular frequency area of the CIHRS spectral lines.

3. Outline of the method

3.1. Spectral distributions

The intensity of the hyper-Rayleigh scattered light related to a collisional hyperpolarizability SA component $\Delta\boldsymbol{\beta}^{K}_{\Lambda L}(R)$, defined later on in Eq. 4, can be expressed on the basis of the quantum mechanical approach (QM) by the formula:

$$\left(\frac{\partial^2 I^{2\omega}}{\partial\Omega\partial\omega}\right)/I_0^2 \sim \sum_{\Lambda,L,j,j'} \Pi^2_{jj'}P_j \begin{pmatrix} j & \Lambda & j' \\ 0 & 0 & 0 \end{pmatrix}^2 \left|\int_0^\infty \hat{\Psi}^*(R;E+\hbar\omega,j')\Delta\beta^{(K)}_{\Lambda L}(R)\hat{\Psi}(R;E,j)dR\right|^2, \quad (1)$$

where the wave functions Ψ are obtained by means of solving numerically the radial Schrödinger equation [8, 9].

3.2. Hyperpolarizability

According to the symmetry properties of the H_2Rg systems, there are only two spherical tensor components of $\Delta\boldsymbol{\beta}$: the vector- and septor-like of rank 1 and 3, respectively. The *ab initio* (QC) evaluation of the SA components is performed on the following assumptions: all the calculations are accomplished at the coupled-cluster theory with singles and doubles level; the *Gaussian 03* computing procedure is used with the basis sets: [6s4p3d] for H_2, [8s7p6d1f] for Kr and [9s8p7d1f] for Xe. For the details of the method see [1, 10].

Within the MIM model some of the SA hyperpolarizability components can be expressed in terms of analytical formulae of the shape akin to the one given below [2]:

$$\Delta\beta^{(3)}_{03}(R) \quad = \quad -3\left({}^{(H_2)}\tilde{B}_{00}[(11)22]\,{}^{(He)}\tilde{\alpha}_{00}[11] - {}^{(He)}\tilde{B}_{00}[(11)22]\,{}^{(H_2)}\tilde{\alpha}_{00}[11]\right)R^{-4}; \quad (2)$$

$B_j[(11)\,22]$ is the dipole-dipole-quadrupole hyperpolarizability, $\alpha_{00}[1\,1]$ denotes the dipole-dipole polarizability tensor. After having substituted proper values of the molecular and atom quantities one can derive, for example, for H_2Xe:

$$\Delta\beta^{(3)}_{23}(R) \quad = \quad -2959.4R^{-4}. \quad (3)$$

This individual contribution of the lowest order effect, induced by the dipole-induced-quadrupole (DIQ) interaction, can be subsequently refined by extending the collisional series expansion into terms representing other kinds of intermolecular interactions, firstly—the higher order multipolar mechanisms (EMIM) and secondly—the overlap, dispersion and exchange forces. The details of the method are given in [2].

3.3. Spatial dependence of SA components

The SA adapted components of hyperpolarizability, defined in the formula below (in bold):

$$\Delta\beta_{\mu}^{(K)}(R) \sim \sum_{\lambda L} (2L+1)^{1/2}\, \boldsymbol{\Delta\beta_{\lambda L}^{(K)}(R)}\, Y_{\lambda\mu}(\Omega)\, C_{\lambda\mu\, L0}^{K\mu}, \quad (4)$$

are obtained by solving appropriate sets of equations resulting from this equation [2]. To this end, initial tensorial data sets of $\Delta\beta_{\mu}^{(K)}$ are calculated for three intra supermolecular orientations (0°, 45° and 90°) by the *ab initio* (QC) tools as well as by the multipole-induced-multipole analytical means. A simple model in which a kind of 'exchange interactions' influence is expressed in terms of a fitted exponential function combined with the EMIM model:

$$\Delta\beta(R) = (a_1R^2 + a_2R + a_3)\exp(a_4R + a_5) + EMIM, \quad (5)$$

is additionally taken into account in the considerations hereafter.

The approaches mentioned above are assumed to render a satisfactory description in different regions of colliding distances. Thus, their applicability of shaping particular parts of the CIHRS lines is worth determination, which is done by benchmarking the translational CIHRS spectra against model profile lines [6] calculated by using a kind of a 'best fit' (BF) $\Delta\beta(R)$ dependence. The form of the BF functions is established assuming that the relative root mean square deviation (RRMSD) of their shape from the points representing the *ab initio* data sets should not exceed a level of 0.5%. This requirement is fulfilled in general for the R distances below 10 a_0, whereas above this limit a relatively poorer precision of the QC methods used results in a more scattered *ab initio* point pattern. The RRMSD values in this case reach a level of up to several percent (the 123 and 323 components). On the other hand, the distance R equal 10 a_0 makes the lower bounds of the MIM model applicability region.

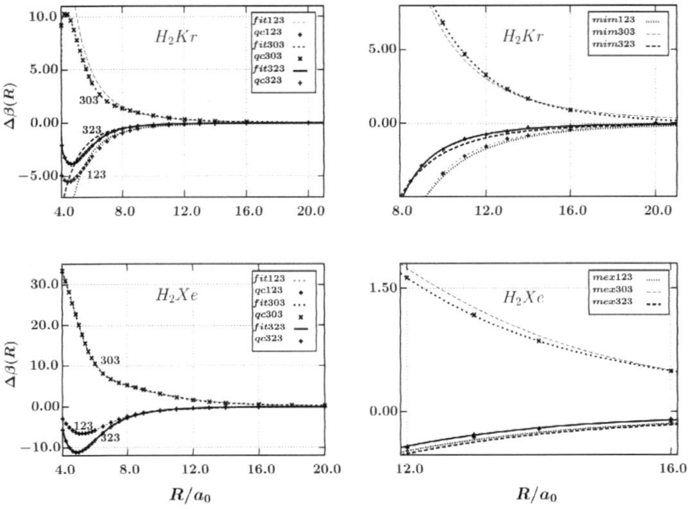

Figure 1. $\Delta\beta^{K}_{\lambda L}(R)$ SA component dependence on H_2-Rg distance for three possible sets of K, Λ, L: 123, 303 and 323; dots represent the *ab initio* quantum chemistry results, *fitKΛL* labels mark the *best quality* fitting functions, dashed profiles illustrate the pure EMIM (upper panel) and the combined EMIM + 'exchange' (lower panel) models.

In Fig. 1 the *ab initio* values are compared with the functional dependence $\boldsymbol{\Delta\beta}(R)$ resulting from the analytical models. In fact, the EMIM lines are in an acceptable agreement with the

Journal of Physics: Conference Series **397** (2012) 012044 doi:10.1088/1742-6596/397/1/012044

ab initio set for remote distances, beginning at $R \sim 9$ a$_0$; while the EMIM+'exchange' case, in almost every point considered, provides an accurately fitted plot. It must be noted, however, that for significantly larger *R*s the *ab initio* results happen not to follow a smooth functional pattern. Rather obviously, these constatations provide a rationale for the idea of blending the QC profiles with the MIM-obtained lines into overall spectral CIHR shapes.

4. Spectra

The complete roto-translational (R-T) spectra expressed via Eq. 1 are obtained as the convolution of the rotational (so called — 'stick') structure and superimposed translational envelope combined of contributions related to the SA translational components of $\Delta\boldsymbol{\beta}(R)$.

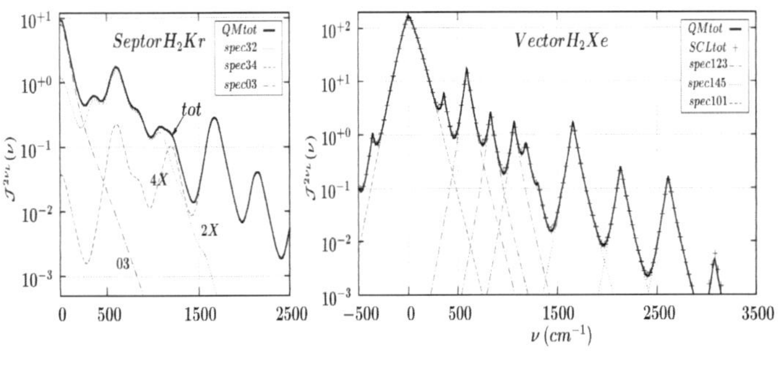

Figure 2. R-T total spectral lines. Translational profiles, ($specK\Lambda L$), are also included; $2X$ ($spec32$) and $4X$ ($spec34$) denote total Λ = 2, 4 contributions, whereas crosses mark semi-classical shape applied as a benchmarking device proving the computational accuracy [5].

Analysis of the diagrams in Fig. 2 shows that the section of the translational profiles significantly influencing the resulting R-T line shape spans within approximately 350 cm^{-1} of the frequency range to the sides of each rotational stick line. Therefore, only such a frequency

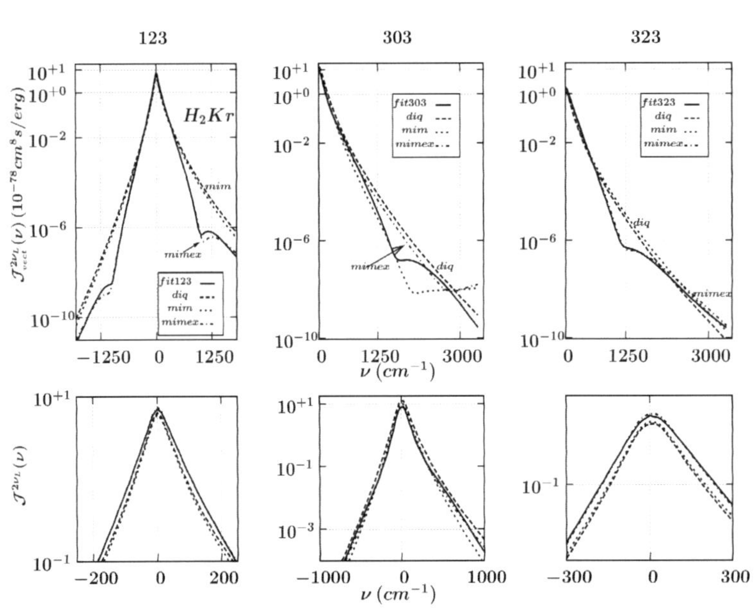

Figure 3. Translational spectral contributions obtained on the grounds of: *ab initio*/'best fit' approach (*fit3ΛL*), dipole-induced-quadrupole model (Eq. 3; *diq*), 'extended' EMIM method (*mim*), and EMIM/exchange approach (*mimex*).

displacement should be taken into consideration, when analyzing the role of the particular hyperpolarizability models on the individual translational spectral contributions.

The spectral patterns in Fig. 3 illustrate the rather widely accepted fact that the pure DIQ approximation cannot be used in order to reflect adequately enough the spectral CIHRS properties within the 350 cm^{-1} frequency limit. Indeed, the RRMS deviation from the 'best fit' shape in this case amounts to about 32%, 67% and 58% for $K\Lambda L = 123$, 303 and 323, respectively (all examples for H_2Kr). Moreover, it is also rather evident, that the extended multipolar series expansion, EMIM, does not seem to improve the picture (RRMSD $\sim$ 21%, 27% and 43%). Unfortunately, even though slightly better, these lines are not satisfactory enough in the line center as well (see the bottom enlargement panels, where RRMSD $\sim$ 10%, 11% and 12% for the EMIM shapes).

Instead, a tangible improvement can be noted, once the exponential 'exchange' term (Eq. 5) has been added to the model collisional mechanism (RRMSD $\sim$ 1.5%, 4% and 5%). Usually, this contribution is associated with the short-range intermolecular correlations (and thus— related to the frequency tails of the spectra), yet the examples in Fig. 3 suggest that its influence seems crucial even for the line cores so that to have them yielded properly.

5. Final remarks

To sum up, the analysis presented leaves one with the conclusion that none of the numerical and theoretical tools considered, if applied single-handedly, is capable of producing CIHRS R-T lines satisfactory enough from the experimental point of view, even within a specific area of applicability. On the one hand, the *ab initio* values of $\Delta\beta(R)$, though extremely precise for the short- and mid-range R values [6], should be supplemented by the data sets provided by the MIM long-distance relevant models. The latter procedure, on the other hand, need to be supported with correcting terms stemming from the other types of interactions. Therefore it turns out to be apparently inevitable to combine all the approaches mentioned so that a reliable overall functional dependence of $\Delta\beta$ on R could be determined. The procedures suggested in our previous work seem to pave the way towards this goal [6, 7].

Acknowledgments

This project has been supported by the Ministry of Science and Higher Education, Poland; No. NN 2020699/B/H03/2010/39

References

[1] Maroulis G 2000 *J. Phys. Chem.* **104** 4772
[2] Bancewicz T 2011 *J. Chem. Phys.* **134** 104309
[3] Głaz W, Bancewicz T, Godet J-L, Maroulis G and Haskopoulos A 2006 *Phys. Rev.* **A 73** 042708
[4] Bancewicz T, Głaz W, Godet J-L and Maroulis G 2008 *J. Chem. Phys.* **129** 124306
[5] Godet J-L, Bancewicz T, Głaz W, Maroulis G and Haskopoulos A 2009 *J. Chem. Phys.* **131** 204305
[6] Głaz W, Godet J-L, Haskopoulos A, Bancewicz T and Maroulis G 2011 *Phys. Rev.* **A 84** 012503
[7] http://zon8.physd.amu.edu.pl/~glaz/HYRAY/PUBLIC/BIBHYP.html
[8] Głaz W, Yang J, Poll J D and Gray C G 1994 *Chem. Phys. Lett.* **218** 183
[9] Głaz W and Tabisz G C 2001 *Can. J. Phys.* **79** 801
[10] Frish M J, Trucks G W, Schlegel H B *et al* 2004 *Gaussian 03* Revision D.01 (Wallingford, CT: Gaussian Inc.)

Journal of Physics: Conference Series **397** (2012) 012045 doi:10.1088/1742-6596/397/1/012045

Transition frequencies of oxygen B-band lines measured with optical frequency comb assisted cavity ring-down spectroscopy

J Domysławska[1], S Wójtewicz[1], D Lisak[1*], A Cygan[1], F Ozimek[2], K Stec[1], K Bielska[1], P Masłowski[1], C Radzewicz[2], R S Trawiński[1] and R Ciuryło[1]

[1] Instytut Fizyki, Uniwersytet Mikołaja Kopernika, ul. Grudziadzka 5/7, 87-100 Toruń, Poland
[2] Wydział Fizyki, Uniwersytet Warszawski, ul. Hoża 69, 00-681 Warszawa, Poland

E-mail: *`dlisak@fizyka.umk.pl`

Abstract. Transition frequencies of weak B-band lines of O_2 were measured with low uncertainties. These results were compared to data available in the literature. In order to achieve reported accuracy we took into account in the data analysis several subtle line-shape effects, such as speed dependence of collisional broadening and Dicke narrowing. We found that the speed-dependent Nelkin-Ghatak profile can properly describe experimental line shapes in the investigated pressure range of up to 22 Torr. Our experimental setup consists of the Pound-Drever-Hall locked frequency-stabilized cavity ring-down spectrometer. This spectrometer provides high-resolution and very high signal-to-noise ratio of measured spectra. It was recently linked to the optical frequency comb working in the visible region of spectrum. This upgrade allows for accurate determination of the absolute frequency axis of the CRD spectrum by direct measurement of the probe laser frequency.

1. Introduction

Recent progress in the optical frequency combs (OFC) technology [1] has greatly simplified ultra precise optical frequency measurement in a broad spectral region. OFC combined with a high-quality spectrometer [2, 3] can provide experimental data on molecular transition frequencies and subtle line-shape effects needed for improving molecular constants and spectroscopic databases, such as HITRAN [4]. Precise data on transition frequencies, intensities, and collisional effects are needed in such applications as spectroscopic temperature and pressure measurements, trace gases detection, modeling of atmosphere and weather forecast, climate changes and noninvasive medical diagnostics. Remote systems for global monitoring of atmosphere composition and pollution with high spatial resolution require precise spectroscopic data on used molecular transitions.

Continuous improvements in the frequency-stabilized cavity ring-down spectroscopy (FS-CRDS) [5] led to construction of precise spectrometer [6, 7, 8, 9] operating in the visible region of the spectrum, providing data with extremely high signal-to-noise ratio (SNR) [10] and having frequency axis directly linked to the frequency-doubled OFC [11]. Such configuration allows us to investigate weak molecular oxygen B-band ($b^1\Sigma_g^+ \leftarrow X^3\Sigma_g^-$, $(\nu' = 1) \leftarrow (\nu'' = 0)$) transitions and measure their line-shape properties as well as line intensities and transition frequencies.

The oxygen B-band is particularly interesting with respect to its potential applications to remote sensing of the atmosphere because these weak transitions do not saturate at long optical paths in the atmosphere and therefore gas properties can be determined from measured line profiles. In contrast to a relatively well characterized A-band ($b^1\Sigma_g^+ \leftarrow X^3\Sigma_g^-$, $(\nu' = 0) \leftarrow (\nu'' = 0)$), the weaker B-band located near 690 nm was measured by few authors, because of too low line intensities for traditional absorption techniques. Available O_2 *B*-band experimental data were measured by Fourier-transform spectroscopy [12], photo-acoustic spectroscopy [13], and CRDS [14, 15]. Also several interesting applications of the B-band spectra were proposed and realized including cloud parameters measurement [16, 17], stratospheric wind determination [18], measurement of pressure and temperature profiles of atmosphere [19], and chlorophyll fluorescence detection [20, 21].

2. Experimental setup

Our experimental setup consists of a Pound-Drever-Hall locked frequency-stabilized cavity ring-down spectrometer (PDH-locked FS-CRDS) linked to the frequency-doubled Er:fiber OFC used to measure absolute frequency of the probe laser at each point of a spectrum. The experimental setup and measurement procedure was improved comparing to this described in Ref. [11], where the absolute frequency of the probe laser was determined for the first and the last point of the spectrum.

In Fig. 1 a scheme of the experimental setup is presented. The ring-down cavity is actively

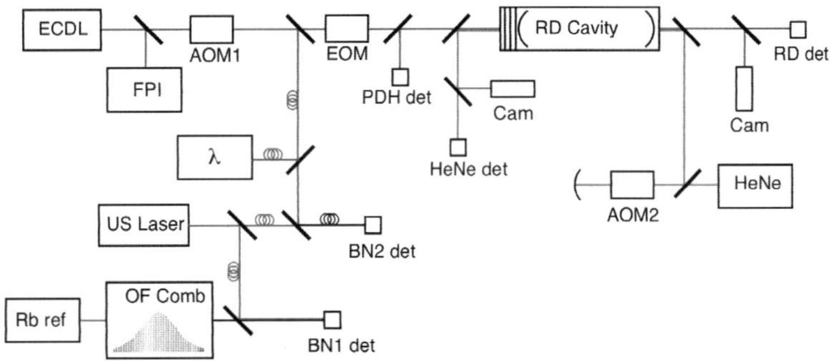

Figure 1. Experimental setup of the PDH-locked FS-CRDS linked to the OFC. ECDL - external cavity diode laser, AOM1, AOM2 - acousto-optic modulators, FPI - Fabry-Perot interferometer, λ - wavemeter, US laser - ultrastable laser, OF Comb - optical frequency comb, BN1 det, BN2 det - optical beat note detectors, Rb ref - Rubidium frequency reference, RD Cavity - ring-down cavity, HeNe - HeNe laser, HeNe det - detector for HeNe laser, Cam - cameras for transverse mode monitoring, RD det - detector for ring-down signals, EOM - electro-optic modulator, PDH det - detector for PDH lock.

locked to the stabilized HeNe laser with use of the FM technique. The cavity mirrors are designed to provide moderate reflectivity of about 98% for the HeNe laser wavelength which enables to use the FM locking scheme, and the feedback loop controls the cavity length by the PZT actuator [22, 23]. The probe laser (ECDL) is locked to the high-finesse ($F^* = 12600$) ring-down cavity with the PDH technique. 20 MHz modulation of the optical frequency is realized by an electro-optic modulator (EOM) and the PDH error signal is generated using the silicon detector (PDH det). The PDH-lock feedback loop controls the ECDL diode current and diffraction grating angle. The ring-down event is started by switching off the acousto-optic modulator, triggered by exceeding a given threshold of the transmission signal registered by the silicon detector (RD det). The PDH lock is repetitively interrupted for the ring-down event and relocked again, see Refs. [7, 8] for details. The scanning Fabry-Perot interferometer (FPI) serves for monitoring a single-mode operation of the ECDL laser. To measure a spectrum the ECDL is tuned and relocked to consecutive TEM_{00} modes of the RD cavity and typically 1000 ring-down events are acquired per each spectrum point.

In order to measure absolute frequency of the probe laser at each point of the spectrum it is compared to the OFC equipped with a frequency doubler to reach the probe wavelength of about 687 nm. The approximate frequency of the ECDL is measured by commercially available wavemeter (λ) with uncertainty of 65 MHz which is smaller than the repetition rate of the OFC (f_r = 250 MHz). Together with the beat note frequency between ECDL and the intermediate ultra stable laser (US laser) [24], measured by the detector (BN2 det), the wavemeter reading is used to determine absolute mode number n of the OFC closest to the optical frequency of the intermediate US laser. The US laser is an extremely narrow bandwidth (< 10 Hz) laser with a very slow frequency drift of about 300 Hz/s. Therefore the beat note frequency between OFC and US laser is almost constant during a spectrum measurement. Such approach simplifies automation of the absolute frequency measurements because there is no need to change the repetition rate f_r and the offset frequency f_o of the OFC during the spectrum measurement. The high-bandwidth BN2 detector allows to measure the frequency difference between the US and the ECDL lasers in a $\pm$20 GHz range. The wavemeter readings together with beat notes from BN1 and BN2 detectors and OFC f_r and f_o frequencies gives the ECDL frequency at each spectrum point. This procedure enables fast determination of the absolute frequency of the probe laser with an uncertainty smaller than 0.5 MHz.

3. Results and discussion

Our experimental transition frequencies ν_0 were determined as extrapolation to zero pressure of the line center determined from fits of the speed-dependent Nelkin-Ghatak profile (SDNGP) [25, 26] to the measured line profiles at different pressures, up to 22 Torr of pure O_2. Example spectrum with residuals from fits of the Voigt profile (VP) and SDNGP is presented in Fig. 2. The "W" shape of residuals from the VP fits is characteristic for profiles in which Dicke narrowing

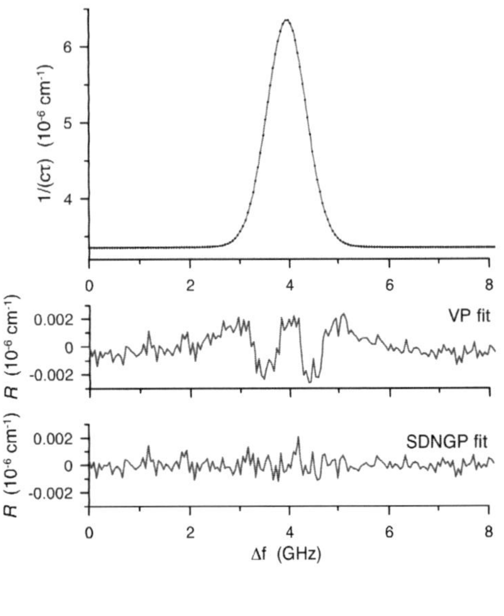

Figure 2. Experimental data of the self-broadened oxygen B-band R5 Q6 line at pressure 6 Torr and fitted SDNGP. Below are residuals (R) from the fits of VP and SDNGP.

and/or the speed-dependence of the collisional broadening play an important role. Both these effects are taken into account in the SDNGP, which can properly describe the observed line shapes of the O_2 B-band transitions [10]. It should be noted however that the difference of transition frequencies obtained from fits of the VP and SDNGP are below 150 kHz, which is less than standard uncertainties of ν_0.

In Fig. 3 our absolute transition frequencies ν_0 of the oxygen B-band transitions are compared to available literature data from the HITRAN database [4], Cheah et al. [12] and Gordon et al. [27]. Comparing to results presented in Ref. [11] this graph has been extended by adding three lowest rotational energy J'' transitions investigated using the new experimental setup. Results

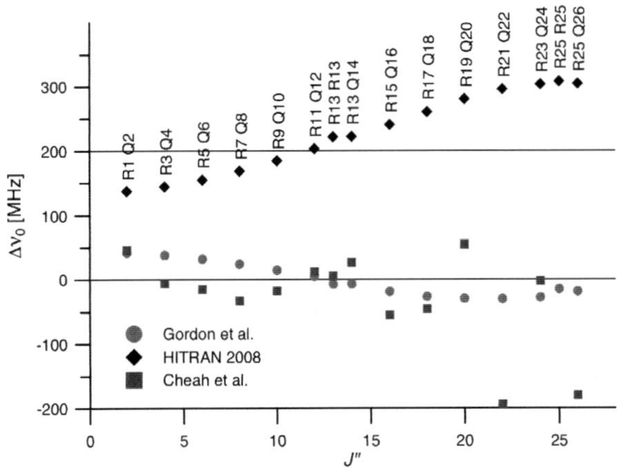

Figure 3. Differences between the absolute line positions ν_0 obtained in this experiment and values published elsewhere as a function of the total angular momentum quantum number for lower state J''.

of Gordon et al. [27] are the closest ones to our line positions, however observed differences of up to 40 MHz are much bigger than reported in [27] (0.3 - 3) MHz uncertainties and standard uncertainties of our ν_0 which are below 0.5 MHz for $J'' \leq 6$ and below 1.2 MHz in most other cases. Results from HITRAN [4] are systematically higher than ours by (140 - 300) MHz however this difference is still within the range of uncertainties given in [4]. Transition frequencies of Cheah et al. [12] have relatively big scatter for higher J'', but for lower J'' (≤ 18) they are different from our results by up to 50 MHz. It's worth noting that differences between our results and those from Ref. [12] are not correlated with differences between our results and data from Ref. [27]. Clearly further investigation and measurement of more transitions with high a signal-to-noise ratio of spectra, such as presented in this paper, is required to solve observed discrepancy between data from different laboratories.

Acknowledgements

The research is part of the program of the National Laboratory FAMO in Toruń, Poland and is supported by the Polish National Science Centre, Project No. DEC-2011/01/B/ST2/00491. The research was also supported by the Foundation for Polish Science TEAM Project co-financed by the EU European Regional Development Fund.

References

[1] Udem T, Reichert J, Holzwarth R and Hansch T W 1999 *Phys. Rev. Lett.* **82** 3568
[2] Mazzotti D, Cancio P, Castrillo A, Galli I, Giusfredi G and De Natale P 2006 *J. Opt.* A: *Pure Appl. Opt.* **8** S490
[3] Gatti D, Gambetta A, Castrillo A, Galzerano G, Laporta P, Gianfrani L and Marangoni M 2011 *Opt. Express* **19** 17520
[4] Rothman L S *et al* 2009 *J. Quant. Spectrosc. Radiat. Transf.* **110** 533

[5] Long D A, Cygan A, van Zee R D, Okumura M, Miller C E, Lisak D and Hodges J T 2012 *Chem. Phys. Lett.* **536** 1

[6] Lisak D, Masłowski P, Cygan A, Bielska K, Wójtewicz S, Piwiński M, Hodges J T, Trawiński R S and Ciuryło R 2010 *Phys. Rev.* A **81** 042504

[7] Cygan A, Lisak D, Masłowski P, Bielska K, Wójtewicz S, Domysławska J, Trawiński R S, Ciuryło R, Abe H and Hodges J T 2011 *Rev. Sci. Instrum.* **82** 063107

[8] Cygan A, Lisak D, Wójtewicz S, Domysławska J, Trawiński R S and Ciuryło R 2011 *Meas. Sci. Technol.* **22** 115303

[9] Wójtewicz S, Lisak D, Cygan A, Domysławska J, Trawiński R S and Ciuryo R 2011 *Phys. Rev. A* **84** 032511

[10] Cygan A, Lisak D, Wójtewicz S, Domysławska J, Hodges J T, Trawiński R S and Ciuryło R 2012 *Phys. Rev.* A. **85** 022508

[11] Domysławska J, Wójtewicz S, Lisak D, Cygan A, Ozimek F, Stec K, Radzewicz Cz, Trawiński R S and Ciuryło R 2012 *J. Chem. Phys.* **136** 024201

[12] Cheah S-L, Lee Y-P and Ogilvie J F 2000 *J. Quant. Spectr. Radiat. Transf.* **64** 467

[13] Barnes J E and Hays P B 2002 *J. Molec. Spectr.* **216** 98

[14] Scherer J J, Paul J B, Jiao H and O'Keefe A 2001 *Appl. Opt.* **40** 6725

[15] Motto-Ros V, Morville J and Rairoux P 2007 *Appl. Phys.* B **87** 531

[16] Kuze A and Chance K V 1994 *J. Geophys. Res.* **99** 14481

[17] Daniel J S, Solomon S, Miller H L, Langford A O, Portmann R W and Eubank C S 2003 *J. Geophys. Res.* **108** 4515

[18] Ortland D A, Skinner W R, Hays P B, Burrage M D, Lieberman R S, Marshall A R and Gell D A 1996 *J. Geophys. Res.* **101** 10351

[19] Nowlan C R, McElroy C T and Drummond J R 2007 *J. Quant. Spectrosc. Radiat. Transf.* **108** 371

[20] Meroni M, Busetto L, Colombo R, Guanter L, Moreno J and Verhoef W 2010 *Remote Sens. Environ.* **114** 363

[21] Guanter L, Alonso L, Gomez-Chova L, Meroni M, Preusker R, Fischer J and Moreno J 2010 *J. Geophys. Res.* **115** D19303

[22] Hodges J T, Layer H P, Miller W W and Scace G E 2004 *Rev. Sci. Instrum.* **75** 849

[23] Hodges J T and Ciuryło R 2005 *Rev. Sci. Instrum.* **76** 023112

[24] Lisak D, Cygan A, Bielska K, Piwiński M, Ozimek F, Ido T, Trawiński R S and Ciuryło R 2012 *Acta Phys. Pol.* A **121** 614

[25] Rautian S G and Sobelman I I 1966 *Usp. Fiz. Nauk.* **90** 209 [Rautian S G and Sobelman I I 1967 *Sov. Phys. Usp.* **9** 701]

[26] Lance B, Blanquet G, Walrand J and Bouanich J P 1997 *J. Mol. Spectrosc.* **185** 262

[27] Gordon I E, Rothman L S and Toon G C 2011 *J. Quant. Spectrosc. Radiat. Transf.* **112** 2310

Demonstration of the extremely high signal-to-noise ratio and advanced O_2 B-band line shape analysis in the PDH-locked FS-CRDS experiment

A Cygan[1]*, D Lisak[1], S Wójtewicz[1], J Domysławska[1], J T Hodges[2], R S Trawiński[1] and R Ciuryło[1]

[1]Instytut Fizyki, Uniwersytet Mikołaja Kopernika, ul. Grudziadzka 5/7, 87-100 Toruń, Poland
[2]National Institute of Standards and Technology, 100 Bureau Drive, Gaithersburg, Maryland 20899, USA

E-mail: *agata@fizyka.umk.pl

Abstract. We demonstrate an extremely high signal-to-noise ratio of 220000 : 1 obtained by the long-term spectra co-adding and averaging in the Pound-Drever-Hall-locked frequency-stabilized cavity ring-down experiment. The connection of this laser spectroscopy technique with the careful line-shape analysis revealed very subtle line-shape asymmetry of the R7 Q8 $^{16}O_2$ rovibronic B-band transition probed at pressure 933 Pa. This asymmetry could be ascribed to the speed dependence of the collisional shift, as will be shown. Reduction of the minimum detectable absorption coefficient of 2×10^{-10} cm^{-1} by an order of magnitude, according to the $1/\sqrt{N}$ dependence, shifted the bottom limit of measured line intensities toward values as low as 1.3×10^{-30} $cm^{-1}/(molecule\ cm^{-2})$. The demonstrated measurement precision enabled the quantification of systematic line-shape deviations, which were approximately 1 part in 80000 of the peak absorption. The influence of slowly drifting etaloning effects on the accuracy of the line-shape analysis is discussed.

1. Introduction

The combination of the high-resolution and high-signal-to-noise ratio absorption spectroscopy with an advanced line-shape analysis is crucial for understanding and solving many of physical problems investigated today. Study of molecular and atomic collisions [1, 2, 3], verification of interaction potentials [4, 5, 6], optical determination of physical constants [7, 8, 9, 10, 11] and testing their stability [12, 13], detection of isotopic composition [14], and deconvolution of hyperfine spectra [15] are only some of tasks considered by modern spectroscopy. Application of the difference frequency generation laser spectroscopy for absorption spectra measurements in the near- and mid-infrared regions [16, 17] provided for decades signal-to-noise ratios (SNR) high enough to detect many subtle line-shape effects in molecular spectra. The development of tunable diode lasers gave similar results in the visible and near-infrared regions [18, 19, 20, 21]. The precision of the Fourier-transform spectroscopy (FTS), usually used for complex molecular spectra study, was significantly improved in the past few years. In 2008 Brown *et al.* [22] obtained SNR of 200000 : 1 in the microwave region, while three years later Gerecht *et al.* [23] reported SNR of 100000 : 1 in the terahertz region. In both cases the FTS spectroscopy with the chirped-pulse technique was used. The cavity-enhanced spectroscopy techniques are well-known tools for probing very weak absorption spectra. That is why their range of applications concerns mainly an atmospheric study. High-accurate line parameter data designated by cavity-enhanced experiments and careful line-shape analysis allow for remote sensing of trace gases concentration with the subpercent precision demanded in satellite-based measurements. The highest previously reported SNR of 28000 : 1 was obtained by Long *et al.* [24] in 2011 for a single CO_2 spectrum measured at 1.6 μm by the frequency-stabilized cavity ring-down spectroscopy (FS-CRDS) technique. Review describing a development and application of the FS-CRDS technique can be found in [25]. This technique was also recently combined with the optical frequency comb in order to provide absolute frequency axis of recorded absorption spectra [26].

Here we demonstrate the extremely high SNR of 220000 : 1 we recently obtained by probing the R7 Q8 $^{16}O_2$ B-band transition near 687 nm by the Pound-Drever-Hall-locked frequency-stabilized spectrometer (PDH-locked FS-CRDS). To our knowledge, this is the highest SNR ever reported in the optical absorption spectroscopy experiment. The ultra-high precision of our spectrometer allowed us to observe very subtle asymmetry in the shape of the self-broadened R7 Q8 $^{16}O_2$ line recorded at low pressure 933 Pa. An improvement of the spectrometer detection limit, by spectra averaging technique and an investigation of an influence of etaloning effects on the line-shape analysis is presented in this work.

2. Experiment precision improvements

An early version of our CRDS spectrometer [27] was based on an active stabilization of cavity modes technique introduced by Hodges *et al.* in 2004 [28, 29] and provided spectra with SNR of about 2000 : 1. In 2011 the Pound-Drever-Hall (PDH) technique [30] of laser frequency stabilization was incorporated to the system [31]. Such solution allowed to inject more laser power to the cavity resonance and significant increase of ring-down events repetition rate made possible to average more (3000) decay signals for each spectrum point. Defined as a minimum detectable absorption loss (MDAL) the detection limit of the PDH-locked FS-CRDS spectrometer (2×10^{-10} cm^{-1}) was improved five times compared to the previous FS-CRDS version. Despite the fact that the typical value of the SNR of 6000 : 1 obtained in the PDH-locked FS-CRDS experiment was relatively high, we showed in [32] that it was not high enough to distinguish between fit qualities obtained from O_2 lines analysis with different theoretical line-shape models, more complex than the standard Voigt profile, at low pressure range. Further improvements of the PDH-locked FS-CRDS spectrometer were related to eliminate the PDH error signal offset, which interfered a stable work of the system, and some changes in the measurement procedure. The description of the control procedure and electronic setup for an active correction of the PDH error signal offset can be found in [33]. The more reliable work of the spectrometer together with the actively stabilized relative frequency axis gave an opportunity to improve spectrometer precision by spectra co-adding and averaging technique. Proceeding in this way the SNR of the order of 220000 : 1 has been achieved [34]. The history of our spectrometer precision improvements was shown in Fig. 1.

3. Analysis

In the line-shape analysis of the strongest in the $^{16}O_2$ B-band R7 Q8 transition, recorded at pressure 933 Pa, the Voigt profile and more sophisticated theoretical profiles were considered. Such line-shape effects as speed dependence of collisional broadening and shifting [35] and Dicke narrowing in case of soft [36] and hard [37, 38] collision model were incorporated. Mathematical expressions of these models were collected in [39]. In the cases of speed-dependent profiles we used a quadratic models for the speed dependence of collisional broadening and shifting as given by Priem *et al.* [40]. In this way formulas for reduced collisional width $B_W(x)$ and shift $B_S(x)$ functions are as follows:

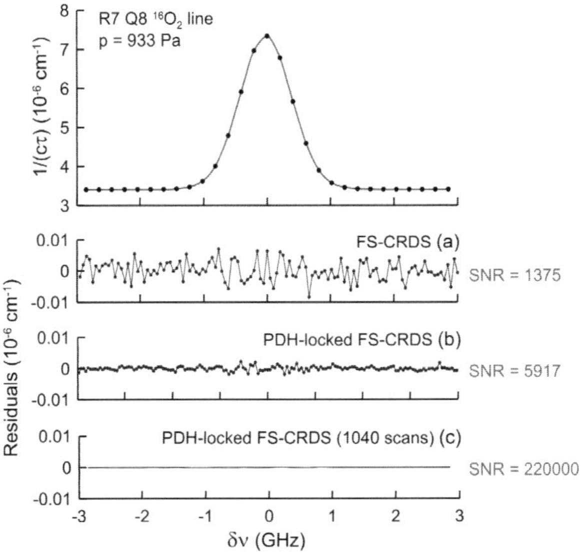

Figure 1. The self-broadened spectrum of the R7 Q8 $^{16}O_2$ B-band line recorded at pressure 933 Pa by subsequent versions of our CRDS spectrometer: (a) - FS-CRDS spectrometer from 2009 [27] (single spectrum), (b) - PDH-locked FS-CRDS spectrometer from 2011 [31] (single spectrum), (c) - PDH-locked FS-CRDS spectrometer with the PDH error signal offset correction procedure from 2012 [33, 34] (spectrum averaged from 1040 scans).

$$B_W(x) = 1 + a_W(x^2 - 3/2), \qquad B_S(x) = 1 + a_S(x^2 - 3/2), \tag{1}$$

where x denotes reduced absorber velocity and a_W, a_S are fitted parameters. We should note that in our previous investigations of the R7 Q8 $^{16}O_2$ transition, the speed dependence of the collisional shifting was ignored because of the small value of the pressure self-shifting determined for this line [27, 32]. Moreover, the value of SNR obtained in our previous experiments was to low too observe a possible subtle line-shape asymmetry caused by the speed dependence of the collisional shifting. However, in the present work, the enhanced spectrum SNR that we demonstrated for the R7 Q8 transition enabled us to carefully investigate the shape of this line. For this reason in the analysis with different line-shape models the parameter of the collisional shift was fitted and in case of speed-dependent profiles the a_S parameter was constrained to be equal to a_W.

In order to quantify and distinguish random effects and systematic distortions of the line profile we introduce two quantities: the signal-to-noise ratio (SNR) which specify statistical noise and the quality of the fit (QF) which is defined as the ratio of the peak absorption signal $\alpha_{\rm max} - \alpha_{\rm min}$ to the standard deviation $\tilde{S}_R$ of the fit residuals calculated from the whole spectrum:

$$\mathrm{QF} = (\alpha_{\rm max} - \alpha_{\rm min})/\tilde{\mathrm{S}}_{\mathrm{R}}. \tag{2}$$

Beyond the random noise, the QF parameter takes into account also systematic effects caused for example by etaloning effects or limitations in the line-shape model. Under conditions of low SNR, the value of the QF parameter is close to the value of SNR.

3.1. Asymmetry of the R7 Q8 O_2 B-band line

In Fig. 2. results of the line-shape analysis of the single and averaged (from 1040 scans) spectrum of the R7 Q8 $^{16}O_2$ transition with the use of the Voigt (VP) and speed-dependent Nelkin-Ghatak profiles (SDNGP) were presented. Unlike the VP profile, the SDNGP profile takes also into account speed-dependent effects and Dicke narrowing. Systematic limitations of the VP profile are clearly seen both in case of averaged as well as single spectrum of the investigated line [see

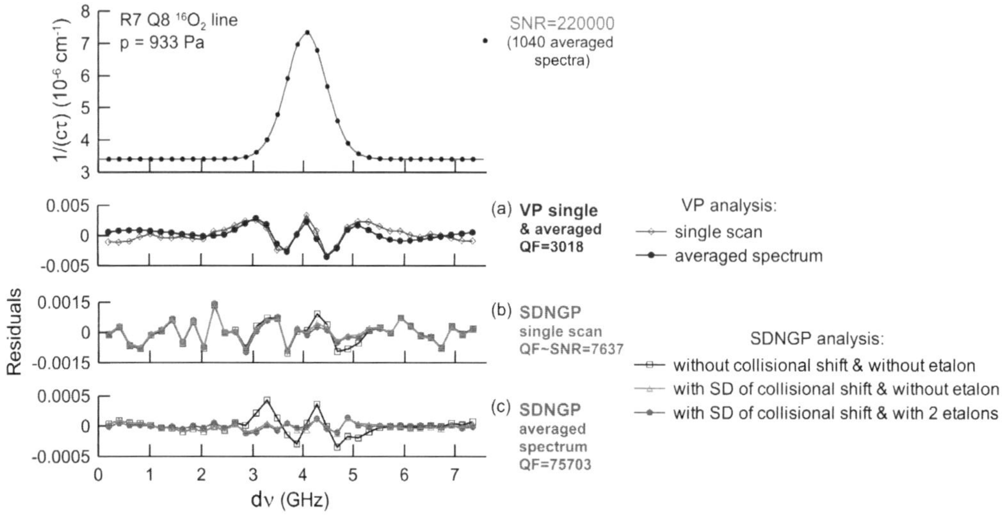

Figure 2. The self-broadened single and averaged from 1040 scans spectrum of the R7 Q8 $^{16}O_2$ B-band line recorded at pressure 933 Pa. Below are residuals from VP (a) and SDNGP (b), (c) fits. The abbreviation SD visible in the graph legend denotes *Speed Dependence*.

diamonds and circles in graph (a)]. The SDNGP analysis gives flat residuals with SNR equal to 7637 for single recorded scan of the R7 Q8 line, regardless of the consideration of the speed dependence of the collisional shift [compare squares and triangles in graph (b)]. However, in case of the averaged spectrum clear structure emerges on residuals [squares in graph (c)] when the speed dependence of the collisonal shift is not incorporated into analysis and it disappears when the speed-dependent effect of the collisional shifting is taken into account [triangles in graph (c)]. The asymmetry we found is the smallest one ever observed for the self-broadened molecular line recorded at such a low pressure. Recently De Vizia *et al.* [41] observed a small speed-dependent asymmetry for self-colliding H_2O molecules at pressure 500 Pa with SNR of 15000 : 1. We should note that under this precision we could not observed asymmetry for the studied R7 Q8 $^{16}O_2$ line. Simple comparison of ratios of the collisional shift to the total line width suggest that ten-fold smaller speed-dependent asymmetry should be expected in case of $O_2 - O_2$ system rather than in $H_2O - H_2O$. In this work we ascribed the asymmetry of the R7 Q8 line to the speed dependence of the collisonal shift based only on the observation of fit residuals. However, observed asymmetry can also be caused by several other physical effects, such as correlations between velocity- and phase-changing collisions, finite collision duration, or line mixing. In general, physical cause of this effect will be identified if advanced line-shape analysis will be realized in wider range of pressures.

3.2. Systematic effects in the line-shape analysis

Signal-to-noise ratio of 220000 : 1 that allowed us to observe the R7 Q8 line asymmetry is extremely high and previously was completely unattainable in optical absorption spectroscopy. However, the true information about the fit quality gives the QF parameter which is almost 3-fold smaller than SNR. Such big discrepancy between QF and SNR parameters could be explained by residual systematic effects introduced to line-shape analysis by imperfections of the used SDNGP model and etaloning effects observed as sine structures on the line profile background. In general, techniques of spectra co-adding and averaging should very well average and dissipate periodic

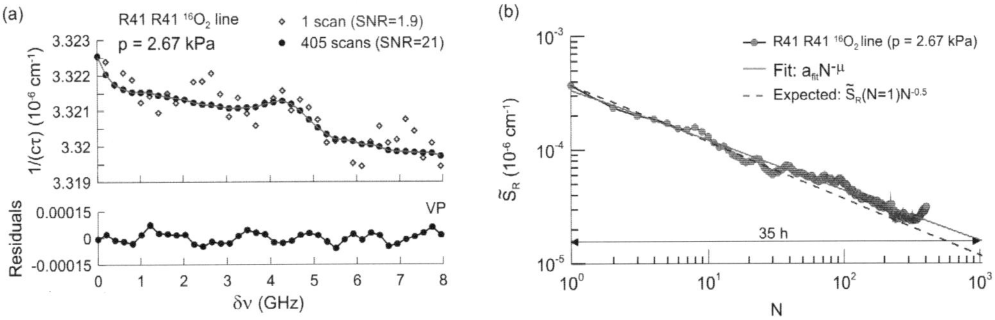

Figure 3. Detection limit improvements in the PDH-locked FS-CRDS experiment. (a) - the single and averaged from 405 scans spectrum of the R41 R41 $^{16}O_2$ B-band line recorded at pressure 2.67 kPa. Below residuals from VP fits were presented. (b) - the dependence of $\tilde{S}_R$ on the number of averaged scans N for R41 R41 $^{16}O_2$ line. The red line is the power-law fit. The blue dashed line corresponds to the expected $\tilde{S}_R(N=1)/\sqrt{N}$ dependence.

and temporally uncorrelated etaloning features. However, under conditions of slow etalonig drift and too short duration of spectra collections these etaloning averaging may be inefficient and its incorporation into the line-shape analysis may be necessary. In order to get high SNR and QF of the order of 80000 we had to include two etalons to the R7 Q8 line analysis. Although this improvement is not so clear in graph (c) (compare dots and triangles), we found that the etalons drift highly influences the evolution of QF parameter with the number of averaged scans N and the QF(N) relationship could be much slower than expected $\sqrt{N}$ dependence which takes place in case of statistical noise itself. The exhaustive discussion of the magnitude of etaloning effects in spectra averaging was presented in [34].

4. Detection limit improvements

Spectra averaging technique is the easiest way to improve data precision. In order to quantify the detection limit of our PDH-locked FS-CRDS spectrometer under conditions of spectra co-adding process we probed a very weak R41 R41 $^{16}O_2$ B-band transition having intensity at room temperature equal to 2.716×10^{-29} cm^{-1}/(molecule cm^{-2}). Line-shape analysis of the investigated transition was realized at pressure 2.67 kPa and demanded an incorporation of the R7 Q8 $^{16}O^{17}O$ line, 15 times stronger than the R41 R41 line and distant from it by $-5.182(9)$ GHz. Single and averaged from 405 scans spectrum of the R41 R41 is presented in Fig. 3. (a). Below are residuals from the VP fits to the averaged spectrum. As seen under conditions of SNR equal to 1.9 the examined line is completely non-observable. However SNR of 21 is high enough to do research. In Fig. 3 (b) the dependence of the standard deviation of fit residuals $\tilde{S}_R$ on the number of averaged scans N is shown. As would be expected in case of the weak absorption the $\tilde{S}_R(N)$ function is close to the $1/\sqrt{N}$ dependence. Determined a, $a_{\rm fit}$ and μ parameters of the power-law functions $aN^{-0.5}$ and $a_{\rm fit}N^{-\mu}$ fitted to experimental data are 0.00052×10^{-6} cm^{-1}, 0.00053×10^{-6} cm^{-1} and 0.45, respectively. The detection limit read from the graph is 2.4×10^{-11} cm^{-1} and is an order of magnitude lower than this obtained without spectra averaging. This detection limit corresponds to the line intensity of 1.3×10^{-30} cm^{-1}/(molecule cm^{-2}) measured with SNR of 1 : 1 at pressure 2.67 kPa.

5. Conclusions

In this work we discussed improvements of our CRDS spectrometer realized during last three years. Uncommon signal-to-noise ratio of 220000 : 1 we are able to get nowadays thanks to spectra averaging method allowed us to observe very subtle asymmetry in the shape of the R7 Q8 $^{16}O_2$ B-band line recorded at pressure 933 Pa. This is the first observation of such small asymmetry for self-colliding molecules investigated at such a low pressure. We ascribed this asymmetry to the speed dependence of the collisional shift, but we suggested further analysis of the R7 Q8 line for different pressures to find truly physical interpretation of this line-shape effect. During this study we also discussed influence of etaloning features and their drift on the quality of the line-shape analysis. We indicate the non-$\sqrt{N}$ dependence of the fit quality under presence of systematic effects. In case of low absorption we showed that improvement of the measurement precision follows the $1/\sqrt{N}$ relation. The detection limit of our spectrometer we have demonstrated is of the order of 2.4×10^{-11} cm^{-1} and allows us to measure absorption lines having intensities of the order of 10^{-30} cm^{-1}/(molecule cm^{-2}) or even weaker depending on the absorber pressure.

Acknowledgements

The research is part of the program of the National Laboratory FAMO in Toruń, Poland, and is supported by the Polish National Science Centre, Project No. DEC-2011/01/B/ST2/00491. The research was also supported by the Foundation for Polish Science TEAM Project co-financed by the EU European Regional Development Fund. A.Cygan is supported by the Polish NCN, Project No. N N202 2392 40.

References

[1] Szudy J and Baylis W E 1996 Phys. Rep. 266 p 127
[2] Hartmann J-M, Boulet C and Robert D 2008 *Collisional Effects on Molecular Spectra: Laboratory Experiments and Model, Consequences for Applications* (Amsterdam: Elsevier)
[3] De Vizia M D, Castrillo A, Fasci E, Moretti L, Rohart F and Gianfrani L 2012 Phys. Rev. A 85 p 062512
[4] Pine A S 1994 J. Chem. Phys. 101 p 3444
[5] Green S and Hutson J 1994 J. Chem. Phys. 100 p 891
[6] Wehr R, Victu A, Thibault F, Drummond J R and May A D 2006 J. Mole. Spec. 235 p 69
[7] Daussy C, Guinet M, Amy-Klein A, Djerroud K, Hermier Y, Briaudeau S, Bordé Ch J and Chardonnet C 2007 Phys. Rev. Lett. 98 p 250801
[8] Casa G, Castrillo A, Galzerano G, Wehr R, Merlone A, Di Serafino D, Laporta P and Gianfrani L 2008 Phys. Rev. Lett. 100 p 200801
[9] Yamada K M T, Onae A, Hong F-L, Inaba H and Shimizu T 2009 C. R. Phys. 10 p 907
[10] Lemarchand C, Djerroud K, Darquié B, Lopez O, Amy-Klein A, Chardonnet C, Bordé Ch J, Briaudeau S and Daussy C 2010 Int. J. Thermophys 31 p 1347
[11] Cygan A, Lisak D, Trawiński R S and Ciuryło R 2010 Phys. Rev. A 82 p 032515
[12] Uzan J P 2003 Rev. Mod. Phys. 75 p 403
[13] Karshenboim S G 2005 Can. J. Phys. 83 p 767
[14] Kerstel E and Gianfrani L 2008 Appl. Phys. B 92 p 439
[15] Long D A, Havey D K, Okumura M, Miller C E and Hodges J T 2010 Phys. Rev. A. 81 p 064502
[16] Pine A S 1974 J. Opt. Soc. Am. 64 p 1683
[17] Pine A S 1976 J. Opt. Soc. Am. 66 p 97
[18] Eng R S, Calawa A R, Harman T C, Kelley P L and Javan A 1972 Appl. Phys. Lett. 21 p 303
[19] Henry A, Hurtmans D, Margottin-Maclou M amd Valentin A 1996 J. Quant. Spectrosc. Radiat. Transfer 56 p 647
[20] Thibault F, Mantz A W, Claveau C, Henry A, Valentin A and Hurtmans D 2007 J. Molec. Spectrosc. 246 p 118
[21] Hurtmans D, Henry A, Valentin A and Boulet C 2009 J. Molec. Spectrosc. 254 p 126
[22] Brown G G, Dian B C, Douglass K O, Geyer S M, Shipman S T and Pate B H 2008 Rev. Sci. Instrum. 79 p 053103
[23] E. Gerecht, K. O. Douglass, D. F. Plusquellic, Opt. Expr. **19**, 8974 (2011).
[24] Long D A, Okumura M, Miller C E and Hodges J T 2011 Appl. Phys. B 105 p 471
[25] Long D A, Cygan A, van Zee R D, Okumura M, Miller C E, Lisak D and Hodges J T 2012 Chem. Phys. Lett. 536 p 1

[26] Domysławska J, Wójtewicz S, Lisak D, Cygan A, Ozimek F, Stec K, Radzewicz Cz, Trawiński R S and Ciuryło R 2012 J. Chem. Phys. 136 p 024201
[27] Lisak D, Masłowski P, Cygan A, Bielska K, Wójtewicz S, Piwiński M, Hodges J T, Trawiński R S and Ciuryło R 2010 Phys. Rev. A 81 p 042504
[28] Hodges J T, Layer H P, Miller W M and Scace G E 2004 Rev. Sci. Instr. 75 p 849
[29] Hodges J T and Ciuryo R 2005 Rev. Sci. Instrum. 76 p 023112
[30] Drever R W P, Hall J L, Kowalski F V, Hough J, Ford G M, Munley A J and Ward H 1983 Appl. Phys. B 31 p 97
[31] Cygan A, Lisak D, Masłowski P, Bielska K, Wójtewicz S, Domysławska J, Trawiński R S, Ciuryło R, Abe H and Hodges J T 2011 Rev. Sci. Instrum. 82 p 063107
[32] Wójtewicz S, Lisak D, Cygan A, Domysławska J, Trawiński R S and Ciuryło R 2011 Phys. Rev. A 84 p 032511
[33] Cygan A, Lisak D, Wójtewicz S, Domysławska J, Trawiński R S and Ciuryło R 2011 Meas. Sci. Technol. 22 p 115303
[34] Cygan A, Lisak D, Wójtewicz S, Domysławska J, Hodges J T, Trawiński R S and Ciuryło R 2012 Phys. Rev. A 85 p 022508
[35] Berman P R 1972 J. Quant. Spectrosc. Radiat. Transfer 12 p 1331
[36] Galatry L 1961 Phys. Rev. 122 p 1218
[37] Rautian S G and Sobelman I I 1966 Usp. Fiz. Nauk 90 p 209 [1967 Sov. Phys. Usp. 9 p 701]
[38] Nelkin M and Ghatak A 1964 Phys. Rev. 135 p A4
[39] Ciuryło R 1998 Phys. Rev. A 58 p 1029
[40] Priem D, Rohart F, Colmont J-M, Wlodarczak G and Bouanich J-P 2000 J. Mol. Struct. 517 p 435
[41] De Vizia M D, Rohart F, Castrillo A, Fasci E, Moretti L and Gianfrani L 2011 Phys. Rev. A 83 p 052506

Peculiarities of the C_2 $d^3\Pi \rightarrow a^3\Pi$ band system intensities in gas discharges through CO-contained mixtures

G.M.Grigorian

Senior Researcher, St.Petersburg State University, St.Petersburg, Russia

E-mail: galina_grigorian@yahoo.com

Abstract. The paper discusses the experimental results pointing to the efficient channel of the CO vibrational to the C_2 electronic energy-transfer. The radiation spectra of the $d^3\Pi_g$ electronic state of C_2 molecule are investigated and the relation of their kinetics to a vibrational excitation of CO molecules in the He-CO-O_2 plasma is discussed. The changes of CO vibrational energy distribution (VED) were imposed by an application of a laser resonator to the discharge tube under investigation. It was found that the modulation of laser radiation (and VED) led to a similar changes of the spontaneous radiation from $d^3\Pi_g$ state of C_2 molecules

1. Introduction

The growing interest observed for the kinetics of the C_2 molecules is related to the carbon nanoparticles production, studies of flame and carbon-containing-plasma properties as well as to the astrophysical investigations. Therefore, an information related to spectral investigations of C_2 molecules is already widely available in literature, see e.g.[1-4]. Even the processes leading to the C_2 production as well as the C_2 excitation-channels in a discharge plasma are not well known. However, the knowledge of kinetic processes with participation of these molecules and their rates is rather scarce.

A population inversion in v = 6 has been observed in a lot of experiments. There is general agreement to attribute this inversion to production of C_2 molecules in an electronic state crossing $d^3\Pi_g$ in the vicinity of v' = 6. In [5] a theoretically predicted metastable $^5\Pi_g$ state of C_2 molecules has been proposes as a candidate for the precursor state.

It was shown in our previous works [6-8], that changes in the CO vibrational population in CO containing plasmas influence intensities of Swan system (C_2: $d^3\Pi_g \rightarrow a^3\Pi_u$ transition), the vibrational distribution of $d^3\Pi_g$ state and the rotational temperature of v = 6 $d^3\Pi_g$. In conditions with high concentration of CO(w) molecules a population inversion in v = 6 of C_2 $d^3\Pi_g$ disappeared. One of the reasons of this phenomenon may be the changing of the excitation process of v = 6 $d^3\Pi_g$. Figure 1 shows that the electronic state of C_2 corresponds energetically to some vibrationally excited levels of CO($X^1\Sigma$, w) molecule, with the vibrational quantum numbers w > 10. This hints to a possible energy transfer from the vibrational to electronic states between these molecules (V-E processes). Up to date it is unclear which ones of the vibrational levels of CO($X^1\Sigma$, w>10) take part in processes followed the disappearing of the selective excitation of v = 6 $d^3\Pi_g$ C_2.

In the present work we study the influence of the changes of populations of different vibrational levels of CO(w) molecules on the concentration of v = 6 $d^3\Pi_g$ state of the C_2.

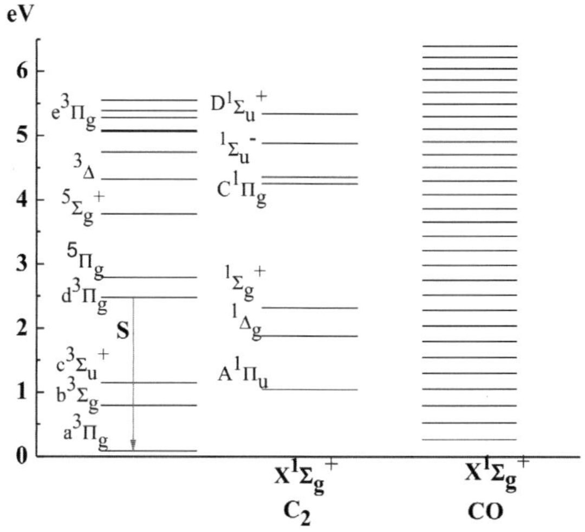

Fig.1 Scheme of the energetically-low levels of the electronic excited states of C_2 molecule and the vibrational states of CO molecule. Arrow shows the transition related to so called Swan band system.

2. Experimental set-up and methodology

The experimental set-up has been described in detail elsewhere [9]. The discharge was operated in DC mode. The experimental conditions were the typical for the active medium of electrical discharge CO-laser and the discharge tube with an appropriate resonator may work as a CO-laser. The molybdenum-glass discharge-tubes 15-20 mm in radius and 50 cm long, water- or nitrogen-cooled were used in our experiment. The hollow cylindrical electrodes from tantalum were installed in the side branches of the discharge tube at a distance of 4 cm from its axis. A discharge was excited using the DC currents of 20 - 100 mA. The relative CO concentrations in the applied He-CO-O_2 mixtures were varied in the range 3 – 15 %. The electric field in the discharge was determined by measuring the voltage drops across the discharge gap and the electrode sheaths. The electrode sheath voltages were found from the voltage measurements in experiments with different discharge lengths. A typical electrode sheath voltage was 350–400 V. The estimated value of the reduced electric field E/N was (1.5–2.5) 10^{-16} V cm^2

The spectra in the range 200 - 3000 nm were registered using the Czerny-Turner monochromator with a diffraction grating 300 and 1200 lines/mm. The recorded spectra allowed determination of concentrations of electronic excited species, gas temperature (through the analyses of the vibrational-rotational spectra) as well as vibrational populations of CO molecules (by analyzing the radiation in the first and second vibrational harmonics) - see also [10,11].

The changes of CO vibrational energy distribution were imposed by an application of a laser resonator to the discharge tube (in oscillation regime populations of vibrational levels of CO(v) molecules which take part in generation drastically decrease) – as can be seen in Fig. 2. It should be mentioned that the laser output was not very high, in order of few W. So, we have assumed that the plasma parameters, besides vibrational temperatures (i.e. electron and gas temperature, voltage, pressure, etc.) have not varied significantly during the modulation of the laser radiation. For example, we have found that the gas temperature decreased only by ~ 10 K, when laser was operating. At the same time, E/N was practically constant.

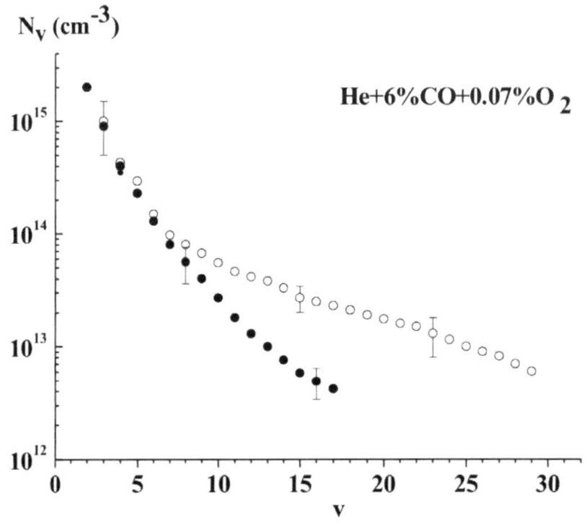

Fig.2 The distribution functions of the CO vibrational level under conditions of laser generation (full circles) and turn-off generation (open circles). P = 9 Torr, I = 30 mA

3. Results and discussion

Strong correlation has been found in experiments between concentrations of vibrationally excited CO molecules and the electronically excited state $d^3\Pi_g$ of C_2 molecule. In particularly, modulation of the intensity of lasing oscillation cased marked modulation of the intensity of the emission from $d^3\Pi_g$ (Swan bands). Fig.3 illustrates the spectrum of Swan bands of C_2 molecule under conditions of laser generation (a) and turn-off generation (b). Under conditions of laser generation the intensive emission from $C_2(d^3\Pi)$ was observed showing significant overpopulation of the sixth vibrational level. The population of the level v = 6 was 8 - 15 times larger than populations of lower levels. When lasing was absent the intensity of radiation from [$C_2(d^3\Pi$, v < 6)] increased 1.5-2 times. At the same time the transition from [$C_2(d^3\Pi$, v = 6)] decreased more than 1 order of magnitude.

It has been shown in [5] that overpopulation of the sixth vibrational level of the $d^3\Pi_g$ state of C_2 most likely caused processes with participation of the hypothetic quintet $^5\Pi_g$:

$$C_2O + C \rightarrow C_2(^5\Pi, v = 0) + CO \qquad (1)$$

$$C + C + M \rightarrow C_2(^5\Pi, v = 0) + M \qquad (2)$$

$$C_2(^5\Pi, v = 0) + M \rightarrow C_2(d^3\Pi, v = 6) + M \qquad (3)$$

C_2O molecules formed in the gas discharge plasma due to plasma chemical processes.

In conditions when concentration of vibrationally excited CO molecules is high we observe the change of the excitation mechanism [8]. Correlation between concentration of $C_2(d^3\Pi)$ and CO(v>10) may be cased by effective VE-process:

$$CO(v>10) + C_2 \rightarrow CO + C_2(d^3\Pi) \qquad (4)$$

The efficiency of these channels increases with the growth of populations of high enough CO vibrational states $\upsilon > 10$, e.g. when the laser radiation declines. The importance of these channels decreases with the fall of CO vibrational populations, when laser is turned-on. In addition, it can be assumed, that vibrationally excited CO molecules may quench metastable quintet $^5\Pi_g$.

To clarify the contribution of different vibrational states of CO molecule to the process (4), the consecutive changes of CO vibrational energy distribution were imposed by an application of a tunable laser resonator to the discharge tube.

It was found that the vibrational levels 12 < w < 20 of the CO molecules most effect on the concentration of the C_2 ($d^3\Pi_g$). In conditions with low concentrations of CO(w) molecules (in the lasing regime of the discharge tube) rotational temperature determined from (6,5) band of the C_2 Swan system - T_r^{C2} (v = 6) was ~ 150K, but T_r (v≠6) was equal to the gas temperature (~ 300-400 K) - Tg. Without lasing all rotational temperatures T_r^{C2} (v) was equal T_g and the inversion on the C_2 ($d^3\Pi_g$, v = 6) was absent.

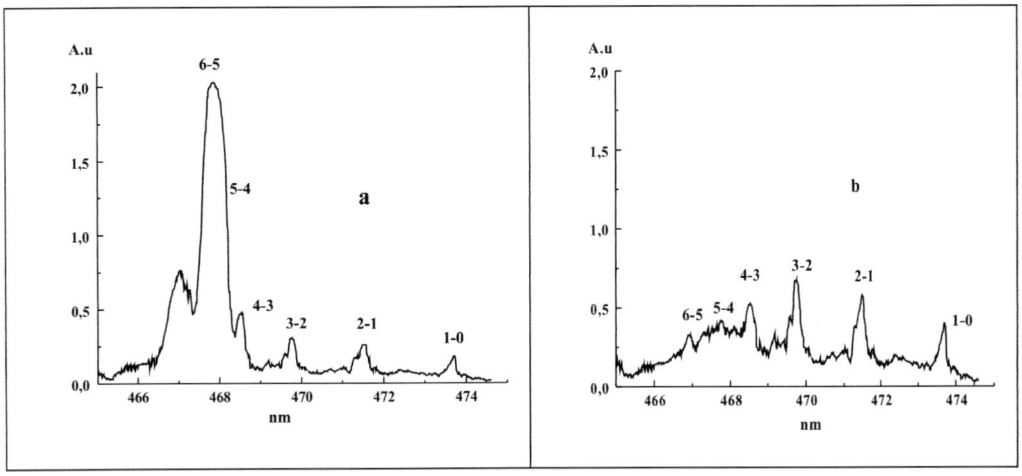

Fig.3 The spectrum of Swan bands (Δv=1) of C_2 molecule under conditions of laser generation (a) and turn-off generation (b). He+6%CO+ 0..2%O_2 , P = 9 Torr, I = 30 mA

This study was partially supported under grant № 07-02-01114a of Russian Foundation for Basic Research.

References

[1] Van Orden A, Saykally R J 1998 *Chem.Phys.* **98** 2313
[2] Weltner W, Van Zee R J 1989 *Chem.Phys.* **89** 1713
[3] Ballik E A, Ramsay D A 1963 *Astrophys.J.* **137** 84
[4] Ibragimova L B 1996 *Khim. Fizika [Chemical Physics]* **15** 10 (in Russian).
[5] Caubet P, Dorthe G 1994 *Chem.Phys.Lett,* **218** 529
[6] Grigorian G M, Dymshits B M, Ionikh Y Z 1989 *Sov J Quant Electron* **19** 889.
[7] Grigorian G M, Dymshits B M, Ionikh Y Z 1988 *Opt. Spectrosc. (USSR)* **65** 452
[8] Grigorian G M, Cenian A 2009 *Chem.Phys.Lett.*,
[9] Grigorian G M, Kochetov I V 2004 *Plasma Phys. Reports* **30** 788.
[10] Farrenq R, Rossetti C, Guelachvili G, Urban W 1985 *Chem. Phys.* **92** 389
[11] Farrenq R, Rossetti C 1985 *Chem.Phys.* **92** 401

Spectra of Rayleigh-Brillouin scattering in supersonic gas flows measured with a wide aperture spectrometer

Yu Anisimov, N Kosykh[1], I Mashek, S Smirnov
Department of Quantum Electronics, Saint Petersburg State University, Saint Petersburg, 198504, Russia

E-mail: ipnk5419@mail.ru

Abstract. We applied spontaneous Rayleigh-Brillouin scattering for measuring kinetic coefficients and velocity of flows. A kinetic model was used for spectral line shape calculation in case of wide aperture Fabri-Perot spectrometer. The calculation well corresponds to experimental spectra for stationary nitrogen and atmospheric air flow up to 3 Mach number.

1. Introduction

Laser Rayleigh-Brillouin scattering is useful tool for diagnostics of gases [1,2]. Scattered signal has spectral features reflecting the pressure, temperature and internal energy states of the gas. The method achieves high precision with using narrowband continuous wave lasers and thermostatic gas cell. Long collection of scattered radiation gives low level noise/signal ratio. Some kinetic models [3-5] predict spectral line shapes and define pressure macroscopic transport coefficients, such as the shear viscosity, the heat conductivity. The last S6 Tenti model [5] allows defining bulk viscosity, but also needs new approach for gas mixtures. In the work we used previous kinetic model [3] and applied it for interpreting of supersonic gas flows. Instability of real flows requires a short-time registration of spectrum and wide aperture.

2. Calculations based on theory of the scattering

Kinetic model describes spectral line shape for differential scattering cross section. The line shape depends on parameter y , which is the ratio of the scattering wavelength to the mean free path between collisions.

$$y = \frac{p}{kv_0\eta} = \frac{nk_BT}{kv_0\eta} \qquad (1)$$

where $k = |K_S - K_0|$ is the scattering wave vector, n is the number density T – temperature, v_0 is the thermal velocity, k_B is the Boltzman constant, η is shear viscosity. Value y depends on the scattering angle of the cross section, so it needs to take into account the angular dependence of the differential scattering cross-section. An elliptic line is the place in the plane of input diaphragm where y is the same. A parabolic line is place of the same Doppler shift.

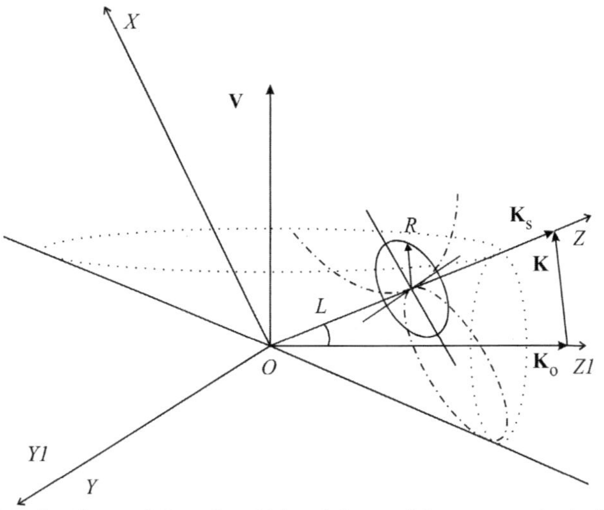

Figure 1. Forming lines of Doppler shift and the model parameter in the input plane.

Differential scattering cross section can be express:

$$DS(\vec{K},\omega) = [\frac{\sin\vartheta}{4\pi}(\frac{\partial\varepsilon}{\partial\rho})_T]^2 K_0{}^4 S(\vec{K},\omega) \tag{2}$$

where ϑ is the scattering polarization angle, ε is the dielectric constant, ρ is the gas density, $S(\vec{K},\omega)$ is the Van Have correlation function. Using wide aperture permits to get high intensity signal. The intensity is proportional square of the diaphragm; however the large area causes spectral line changing.

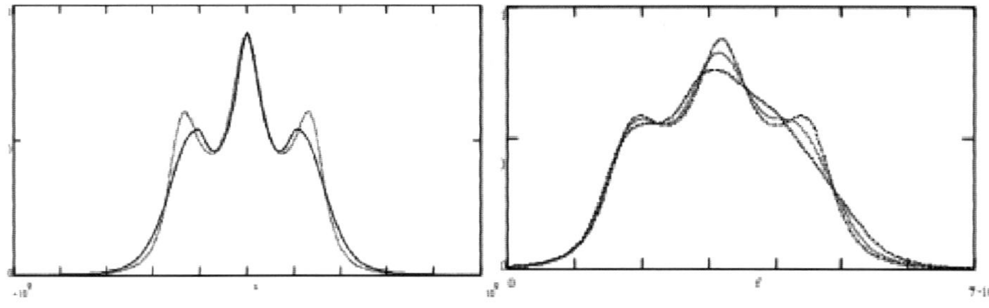

Figure 2. Calculation of line shape changing in stationary gas (left) and gas flow (right) by using round wide aperture.

Figure 2 shows line shape changing in the case of stationary gas (left). In the case of gas flow (right) the spectral line changes. Calculations were made by integrating scattering cross-sections in the input diaphragm plane.

3. Experimental result

We used experimental setup, consist of low power 50 mW Ar+ laser with wavelength λ = 0.488μm in a typical scheme of direct spectral analysis [6]. The frequency composition of the scattered light collected within the receiving aperture of a fast lens is analyzed with a high-resolution scanning spherical Fabry–Perot interferometer. The instrument function of the method coincides with the instrument function of the interferometer, whose width is 35 MHz at a free spectral interval of 1524 MHz. A photomultiplier tube operating in the photon-counting mode is used as a detector. The spectrum of the light scattered in moderate-density gases at small observation angles looks like a triplet of the central, or Rayleigh, peak and two Brillouin satellites.

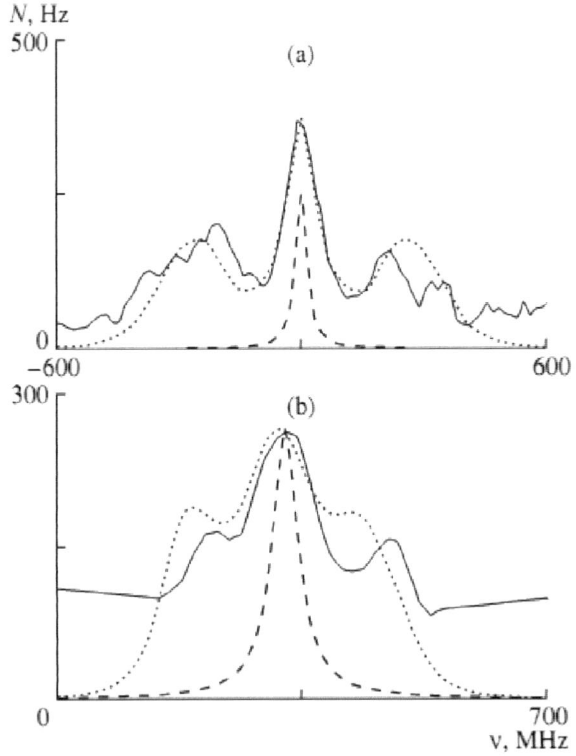

Figure 3. Experimental spontaneous scattering spectra (solid curve) (a) in steady-state nitrogen obtained under the normal conditions, with a scattering angle of 22°, a round aperture of 40 mm, at a distance of 150 mm and (b) in a supersonic (580 m/s) jet of rarefied (88 Torr) nitrogen at T = 140K, with a a scattering angle of 16°, a special diaphragm, at a distance of 150 mm from the scattering point as compared to model numerical calculations (dotted curve) and the spectrometer instrument function (dashed curve).

We will consider the experimental spectrum of the scattering at a subsonic nitrogen flow under the normal conditions (p =1 atm, T = 265 K), which is shown in Fig. 3(a). The scattered light was collected at the angle 22° to the incident radiation within the round aperture 35 mm in diameter, set at the distance L = 150 mm from the scattering point. The Brillouin components are separated by 260 ± 10 MHz from the fundamental harmonic. This corresponds to the local sonic speed 332 ± 15 m/s. It follows from Eq. (1) that the parameter y = 3.1 ± 0.1 corresponds to the shear viscosity η = (16.7 ± 0.5) $\times 10^{-6}$ Pa s, which is in agreement with the tabulated value of 16.65 $\times$ 10^{-6} Pa s for nitrogen at the atmospheric pressure. The mean free path found from the spectral data is equal to 6.6 $\times$ 10^{-8} m, which is also in agreement with the tabulated data (~10^{-7} m for most gases under the normal conditions). The spectral signal of the scattering at the supersonic nitrogen jet (Fig. 3 b) was obtained through the collection of the scattered light at the angle θ = 16° within the special diaphragm at the distance L = 150 mm. The scattering volume was V = 0.3 mm^3. The triplet as a whole is shifted with respect to the zero frequency by 325 ±5 MHz, which corresponds to the forward flow speed equal to 575 ± 10 m/s. The mean distance between the Brillouin peaks is 132 ± 10 MHz. Hence, the local sonic speed is 231 ± 20 m/s, and the flow temperature is equal to 130 ± 15 K. The parameter y is 1.2 ± 0.1, the static pressure is 88 Torr, and the viscosity is (9.8 ±0.8) $\times$ 10^{-6} Pa s as compared to the reference value of9.45 × 10^{-6} Pa s at 140 K (the reference data correspond to the pressure no lower than the atmospheric pressure).The mean free path is equal to 2.3 × 10^{-7} m.

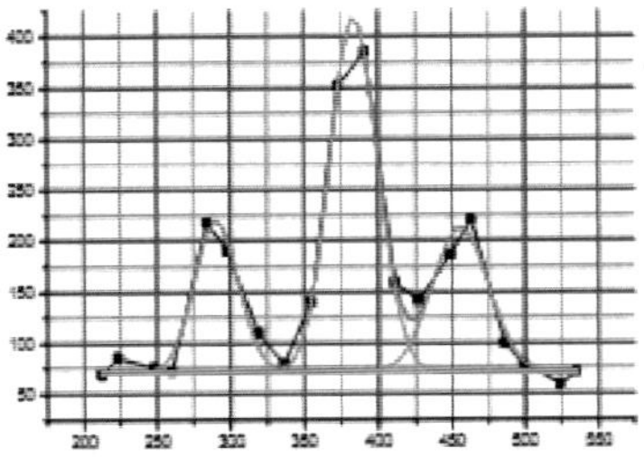

Figure 4. Spectrum of scattering in supersonic air flow (short-time registration within 10 sec).

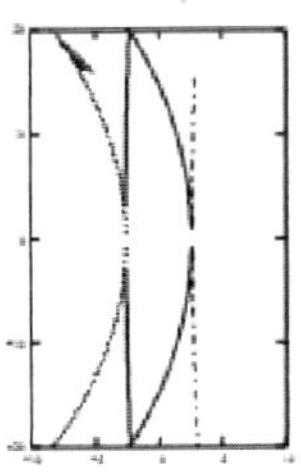

Figure 5. Optimal form of diaphragm is the cross section of parabolic and elliptic strips.

4. Conclusion

Laser-illuminated Rayleigh-Brillouin scattering is rapidly becoming an important diagnostic tool for study of statistical fluctuations in gases [7]. However small scattering signal demands special design of experimental setup, based on kinetic model calculations.

5. References

[1] Greytak T and Benedek G 1966 *Phys. Rev. Lett.* **17** 179
[2] Sandoval R and Armstrong R 1976 *Phys.Rev.*A **13** 752
[3] Sugawara A and Yip S 1976 *Phys. Fluids* **10** 1911
[4] Boley C, Desai R, Tenti G1972 *Can. J. Phys.* **50** 2158
[5] Boley C, Desai R, Tenti G1972 1974 *Can. J. Phys.* **52** 285
[6] Anisimov Yu, Kosykh N, Mashek I 2007 *Optics and Spectroscopy* **102** 467
[7] Pan X, Shneider M, Miles R 2004 *Phys. Rev.* A **69** 033814

Journal of Physics: Conference Series **397** (2012) 012049 doi:10.1088/1742-6596/397/1/012049

Hydrogen Atom Collision Processes in Cool Stellar Atmospheres: Effects on Spectral Line Strengths and Measured Chemical Abundances in Old Stars

Paul S. Barklem

Department of Physics and Astronomy, Uppsala University, Box 515 S75120, Uppsala, Sweden

paul.barklem@physics.uu.se

Abstract. The precise measurement of the chemical composition of stars is a fundamental problem relevant to many areas of astrophysics. State-of-the-art approaches attempt to unite accurate descriptions of microphysics, non-local thermodynamic equilibrium (non-LTE) line formation and 3D hydrodynamical model atmospheres. In this paper I review progress in understanding inelastic collisions of hydrogen atoms with other species and their influence on spectral line formation and derived abundances in stellar atmospheres. These collisions are a major source of uncertainty in non-LTE modelling of spectral lines and abundance determinations, especially for old, metal-poor stars, which are unique tracers of the early evolution of our galaxy.

Full quantum scattering calculations of direct excitation processes $X(nl) + H \leftrightarrow X(n'l') + H$ and charge transfer processes $X(nl) + H \leftrightarrow X^{+} + H^{-}$ have been done for Li, Na and Mg [1,2,3] based on detailed quantum chemical data, e.g. [4]. Rate coefficients have been calculated and applied to non-LTE modelling of spectral lines in stellar atmospheres [5,6,7,8,9]. In all cases we find that charge transfer processes from the first excited S-state are very important, and the processes affect measured abundances for Li, Na and Mg in some stars by as much as 60%. Effects vary with stellar parameters (e.g. temperature, luminosity, metal content) and so these processes are important not only for accurate absolute abundances, but also for relative abundances among dissimilar stars.

1. Introduction

Precise relative and absolute stellar chemical abundances from spectroscopy are needed to address many fundamental problems in modern astrophysics, e.g. the chemical and dynamical evolution of the Galaxy, the solar system and its place in the Galaxy, and understanding stellar processes such as mixing and gravitational settling. In particular, old metal-poor stars play a key role in understanding the evolution of the early universe and the formation of the first structures. While distant quasars and galaxies probe the early universe directly, the information they provide is rather general. On the other hand, the compositions and kinematics of old long-lived stars in the Milky Way also probe the early universe in an indirect but very detailed manner (so called "galactic archeology" [10]). Old stars are generally charactised by a lack of heavy elements, often referred to in astrophysics as "metals", and thus such stars are "metal-poor" (since the elements heavier than Li have been produced during the chemical evolution of the universe, and were not present immediately after the Big Bang). Since the very first stars may no longer be observable, we must look for the chemical signatures of the first generations of stars in these old stars, which presumably were formed in subsequent generations.

We routinely observe spectral line strengths with precision of order 0.01 dex[1] (~2%). Systematic errors dominate and *relative* abundances have estimated precision of typically an order of magnitude larger >0.1 dex (>20%), even ~1 dex (a factor of 10). This considerably limits our ability to accurately interpret stellar chemical abundances. An example of this is shown in an abundance-abundance diagram in figure 1, taken from [11]. Here we see that chemical evolution models predict structure in the relative abundances of metal-poor stars on a scale of 0.01 dex – the thin filamentary structures indicate sequences of stars enriched by only one supernova, i.e. true second generation stars. Such structures would tell us a great deal about the first generations of stars and the early chemical evolution of the Galaxy. Comparison with observations is difficult since the observations have errors of approximately one order of magnitude larger.

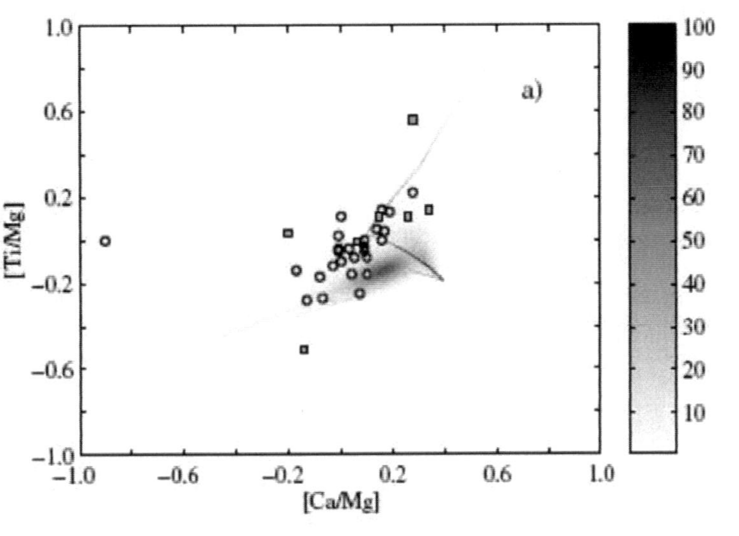

Figure 1. Abundance-abundance diagram showing ratios of Ti/Mg against Ca/Mg.[2] The points are observed data in metal-poor stars and the shading indicates the probability density function from chemical evolution modelling. Taken from [11] .

The current state-of-the-art in stellar abundance analysis attempts advances on three main fronts: 1) accurate descriptions of microphysical processes, 2) three-dimensional hydrodynamical models of the stellar atmosphere, and 3) non-local thermodynamic equilibrium (non-LTE) descriptions of the radiative transfer, and connections between these three fields. In this review I focus on the particular problem of microphysical data needed for non-LTE modelling, especially inelastic collisional processes involving hydrogen atoms. Non-LTE modelling requires accurate descriptions of all microphysical processes on atoms, i.e. all radiative and collisional processes on the atom of interest in the stellar atmosphere. Although basic hand-waving arguments can be made that collisions with electrons are the most important collisional processes in stellar atmospheres (based on the relative speeds of perturbers and the efficiency of collisions expected from the Massey criterion, e.g. [12,13]), the importance of low-energy collisions with hydrogen atoms, usually at least four orders of magnitude more abundant than electrons in line forming regions of cool star atmospheres, has been a

[1] "dex" indicates a logarithmic scale with base 10.

[2] The square bracket notation [X/Y] indicates a logarithmic dex scale where the solar ratio is at zero.

major source of uncertainty, e.g. [14]. This uncertainty becomes even more acute in old metal-poor stars where the abundance of hydrogen atoms may reach as much six orders of magnitude larger than that of electrons due to the scarcity of metals to provide electrons.

Thus, the question has been: does the greater abundance overcome the expected lower efficiency of hydrogen collisions with respect to electrons? About a decade ago there was quite little reliable information in this regard. There was one experimental result at intermediate energies [15], and one low-energy calculation [16] both for the Na(3s)+H → Na(3p)+H process, which demonstrated the importance of the avoided crossings with the ionic state. In astrophysics, the "Drawin formula" [14], an extension of a classical Thomson model, is still routinely employed for allowed transitions despite giving results *four orders of magnitude* too large for the Na(3s)+H → Na(3p)+H process. *Ad hoc* estimates are often used for forbidden transitions. In this paper I briefly review work to improve this situation by providing accurate data for inelastic hydrogen collisions for astrophysically important atoms and the impact the data have had on non-LTE stellar abundance analyses.

2. Physics of inelastic hydrogen collisions

The basic theory of atomic collisions has been reasonably well understood for some time [17]. However, the detailed calculation of accurate quantum-chemical data for quasi-molecules required as input to the relevant coupled-channel scattering equations, including for excited states, has been a major hurdle (see, for example, paper by Guitou et al. in these proceedings). Further, while the solution of the scattering equations in the standard adiabatic approach appears straightforward, it encounters difficulties such as the electron translation problem (see paper by Belyaev in these proceedings). Earlier studies such as [15,16] demonstrated, through comparison with the Landau-Zener model and detailed quantum scattering calculations, that the experimental data could be reasonably explained by nonadiabatic transitions associated with avoided crossings, also called pseudocrossings, in the NaH molecular potentials. At pseudocrossings, adiabatic potentials approach each other and the transition probabilities may become large even for slow collisions, despite that the Massey criterion based on atomic energy level spacings predicts the collisions to be nearly adiabatic.

Taking the particular case of low-lying states of NaH as an example, these pseudocrossings arise from interaction of the covalent Na+H and ionic $Na^{+}+H^{-}$ configurations, and are thus called avoided ionic crossings. The adiabatic potentials showing these pseudocrossings are shown in figure 2, where we see clearly the influence of the ionic configuration, which strongly perturbs the potentials from their behaviour at large internuclear distance. Similar effects are seen in other quasi-molecules such as LiH (e.g. [18]) and MgH [4] (also Guitou et al. in these proceedings). Quantum scattering calculations for Li+H [1], Na+H [2] and Mg+H [3] including all or most states below the ionic state asymptotic limit have found this to be an important mechanism for nonadiabatic transitions.

The basic physical interpretation of this mechanism is that, in a given X+H quasi-molecule, starting from a covalent state with appropriate symmetry where the electrons are associated to their respective atoms, at avoided ionic crossings the valence electron associated with the atom X has a probability to tunnel to the H atom resulting in a predominantly ionic charge distribution $X^{+}+H^{-}$. Later during the collision at a different avoided crossing, there is a probability that the electron may tunnel back to a different covalent molecular state leading finally to a different final state of the atom A, and excitation or deexcitation $X(nl)+H \rightarrow X(n'l')+H$. The electron may also stay with the H atom leading to ion-pair production $X+H \rightarrow X^{+}+H^{-}$. Calculations for Li+H, Na+H and Mg+H mentioned above have generally shown that large-valued cross sections for excitation are typically small compared with those for ion-pair production from certain states. This can be easily understood: a large-valued cross section for an excitation process results from passing two avoided ionic crossings, one of them (higher-lying) passed with a small transition probability, while for an ion-pair production process this region is passed with a large nonadiabatic transition probability due to small adiabatic energy splittings at highly excited crossings.

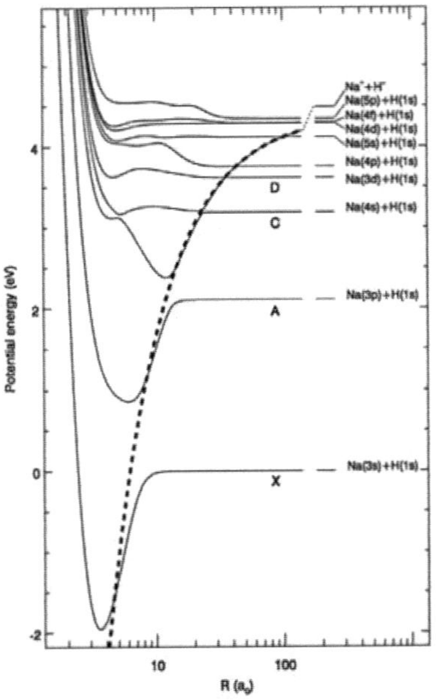

Figure 2. Adiabatic potential energies as a function of internuclear distance *R* for the lowest ten $^1\Sigma^+$ states of the NaH quasi-molecule from pseudo-potential calculations [19]. The atomic states and energies at dissociation are shown at the right-hand side of the figure. The thick, dashed (red) line shows the $1/R$ interaction corresponding to the interaction for the pure ionic configuration Na^++H^- at long range.

3. Comparison of results

Space does not permit a full comparison of all available cross section and rate coefficient results; a few key examples will be presented here, and the reader is referred to [20] for more thorough examination of the results for Li+H and Na+H and to [7] for Mg+H. In particular we note that figure 3 of [20] compares cross sections at low energy for Na(3s) + H $\rightarrow$ Na(3p) + H, the single case where experimental data is available. The quantum scattering calculations and the Landau-Zener model results agree quite well with experiment. Near the threshold, where there is no experimental data, the quantum scattering calculations show substantial differences depending on which quantum-chemical data are used. This highlights the sensitivity of the near-threshold cross sections to the uncertainties in the quantum-chemistry data, and gives an estimate of the uncertainties in the calculated cross sections. In this case the uncertainties seem to be around one or two orders of magnitude. However, it should be noted that the uncertainties vary strongly from transition to transition, and the transitions with the largest cross sections have the smallest uncertainties, as low as a factor of 2. Fortunately, these are the most important from an astrophysical perspective (see [6]).

The rate coefficients for Mg+H collisions are plotted against transition energy, ΔE, in figure 3. These plots are similar to those presented for Li+H and Na+H in [20], and include rate coefficients from quantum scattering calculations (upper panel) and those predicted by the commonly used Drawin formula (lower panel). Rate coefficients for charge transfer processes (ion-pair production) are shown in addition to those for excitation. A fit to the data for excitation processes is shown (dotted line), and the fits for Li+H and Na+H from [20] are shown for comparison (dashed lines). We emphasise that the actual data have a large scatter around these fits, and their purpose is to indicate the general behaviour. The most notable aspect of the Drawin formula results is that of the 21 excitation transitions considered, only 5 are optically allowed and can be calculated with the Drawin formula. This is due to the fact that Mg has two valence electrons leading to the presence of singlet and triplet spin terms, between which optical transitions are forbidden. As seen in Li and Na, there is a tendency for the Drawin results to be generally larger than those from quantum scattering calculations by a few orders of magnitude with a significant variation, here ranging from zero to four orders of magnitude larger.

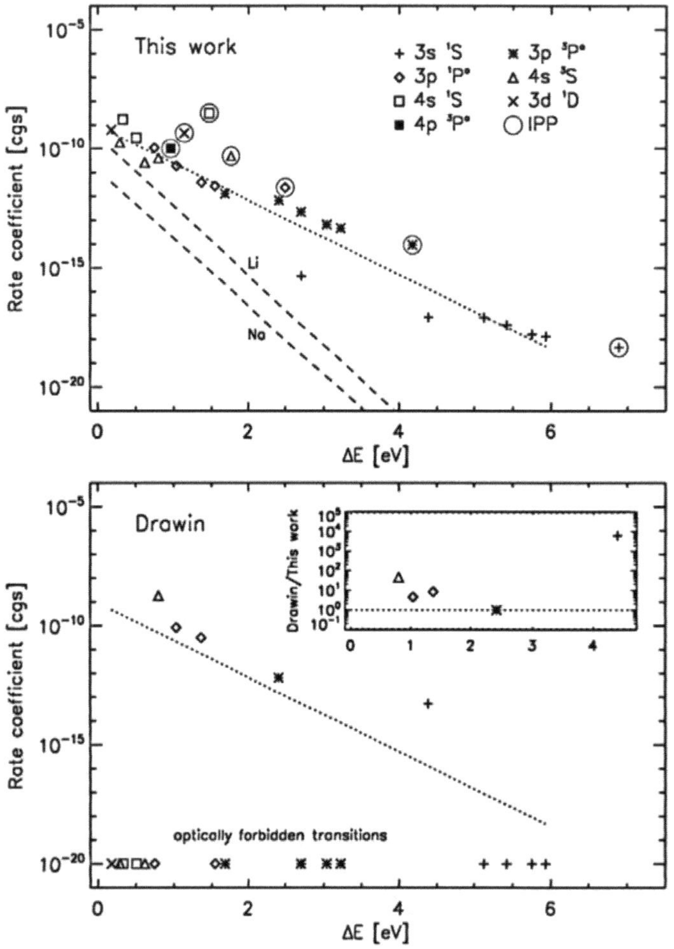

Figure 3. Rate coefficients at 6000 K for excitation and ion-pair production processes in Mg+H collisions, plotted against the transition energy ΔE. Different symbols are used to denote the initial state of the transition following the key given in the upper panel. The upper panel shows the results from this work based on detailed quantum scattering calculations [3]. Symbols inside circles refer to ion-pair production processes (IPP); i.e. the final state is the ionic state Mg^{+}+H^{-}. The dotted line shows a linear fit to the data for excitation processes (i.e. excluding ion-pair production), which is repeated in the lower panel to aid comparison. The dashed lines show the fits to the corresponding data for Li and Na from [20]. The lower panel shows the results of the Drawin formula, and the inset shows the ratio with the quantum scattering calculations. Optically forbidden transitions, where the Drawin formula is not applicable, are shown in the bottom of the panel.

4. Effects on line formation and stellar abundances

The effects on non-LTE line formation calculations and thus on measured stellar abundances has been examined for Li and Na in [8] and [9]. Work on Mg is presently underway. The most important result is that in both Li and Na it has been found that direct excitation and deexcitation processes, X(*nl*)+H→X(*n′l′*)+H where X is the atom of interest, are of little consequence for the statistical equilibrium. However, charge transfer processes, X(*nl*)+H→X^++H^-, particularly where *nl* corresponds to the first excited *S*-state, are important. The resulting effects on abundance corrections in solar-type stars are typically small, of order a few 0.01 dex (2%). In giants the influence is larger, of order 0.05 dex (~12%), and even larger still in metal-poor stars reaching of order 0.1–0.2 dex (25–60%). Thus, these processes are important for accurate absolute abundances, and also for accurate relative abundances of different elements and among dissimilar stars. An example of the evolution of the corrections for Li is shown in figure 4. This shows the increasing importance of hydrogen collision processes as one goes to the lowest metallicities, i.e. the oldest stars which trace the earliest times of the universe.

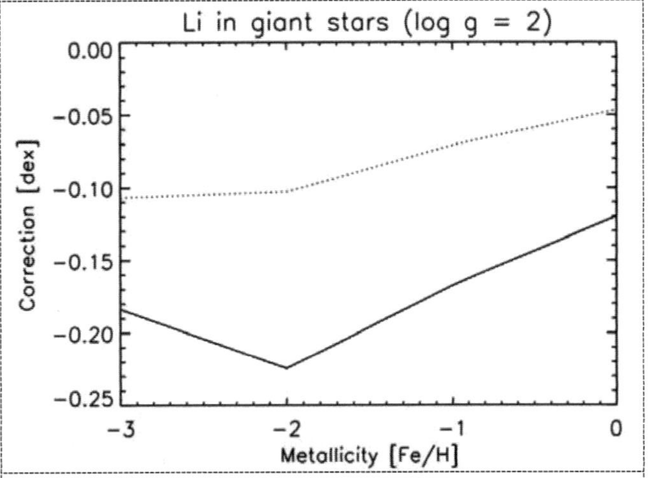

Figure 4. Effects on abundance corrections for Li in giant stars with metallicity. The dotted line shows the mean correction found (i.e. considering all possible stellar temperatures) and the full line shows the maximum correction found. Data taken from [8].

5. Conclusion and future

Inelastic processes involving collisions with hydrogen atoms, including charge transfer processes, have been found to be important in cool stellar atmospheres and have significant effects on measured abundances, especially in old metal-poor stars. Thus, accurate data for these processes is very important for understanding the nature of the early universe via galactic archeology. Accurate calculations have now been performed for Li, Na and Mg, and calculations for other astrophysically important elements such as Ca, O and Fe-group elements are planned. Tests of the data directly via measurements of the mutual neutralisation cross sections are planned at the DESIREE experiment [21]. The data and the line formation models can also be tested indirectly via spatially resolved observations on the sun, e.g. [22], and observations for Li and Mg lines were recently obtained at the Swedish Solar Telescope.

Acknowledgments
I acknowledge the important contributions of my collaborators to the work and results presented in this review, namely Andrey Belyaev, Alan Dickinson, Nicole Feautrier, Xavier Gadéa, Marie Guitou, and Annie Spielfiedel. I am a Royal Swedish Academy of Sciences Research Fellow supported by a grant from the Knut and Alice Wallenberg Foundation. My research is also supported by the Royal Swedish Academy of Sciences, the Wenner-Gren Foundation, Göran Gustafssons Stiftelse and the Swedish Research Council.

References

[1] Belyaev, A. K., Barklem, P. S., *Phys. Rev. A*, **68**, 062703 (2003)
[2] Belyaev, A. K., Barklem, P. S., Dickinson, A. S., Gadéa, F. X., *Phys. Rev. A*, 81, 032706 (2010)
[3] Belyaev, A. K., Barklem, P. S., Spielfiedel, A., M.Guitou, N. Feautrier, D. S. Rodionov, D. V. Vlasov, *Phys. Rev. A*, 85, 032704 (2012)
[4] Guitou, M., Spielfiedel, A., & Feautrier, N., *Chem. Phys. Lett.*, 488, 145 (2010)
[5] Barklem, P. S., Belyaev, A. K., Asplund, M., *A&A*, 409, L1 (2003)
[6] Barklem, P. S., Belyaev, A. K., Dickinson, A. S., Gadéa, F. X., *A&A*, 519, A20 (2010)
[7] Barklem, P. S., Belyaev, A. K., Spielfiedel, A., M.Guitou, N. Feautrier, *A&A* 541, A80 (2012)
[8] Lind, K., Asplund, M., Barklem, P. S., *A&A*, 503, 541 (2009)
[9] Lind, K., Asplund, M., Barklem, P. S., Belyaev, A. K., *A&A*, 528, A103 (2011)
[10] Freeman, K., Bland-Hawthorn, J., *ARA&A*, 40, 487, (2002)
[11] Karlsson, T., Gustafsson, B., *A&A*, 436, 879 (2005)
[12] Plaskett, H. H., *MNRAS*, 115, 256 (1955)
[13] Lambert, D., *Physica Scripta* T, 47, 186 (1993)
[14] Steenbock, W., & Holweger, H., *A&A*, 130, 319 (1984)
[15] Fleck, I., Grosser, J., Schnecke, A., Steen, W., & Voigt, H., *J. Phys. B At. Mol. Phys.*, 24, 4017 (1991)
[16] Belyaev, A. K., Grosser, J., Hahne, J., & Menzel, T., *Phys. Rev. A*, 60 2151 (1999)
[17] Mott, M. F., & Massey, H. S. W., *The theory of atomic collisions* - Oxford:Clarendon (1949)
[18] Croft, H., Dickinson, A. S., & Gadéa, F. X., *J. Phys. B At. Mol. Phys.*, 32, 81 (1999)
[19] Dickinson, A. S., Poteau, R., & Gadéa, F. X., *J. Phys. B At. Mol. Phys.*, 32, 5451 (1999)
[20] Barklem, P. S., Belyaev, A. K., Guitou, M., N. Feautrier, Gadéa, F. X., Spielfiedel, A., *A&A*, 530, A94 (2011)
[21] Thomas, R. D., Schmidt, H. T., Gurell, J., et al. *Journal of Physics: Conference Series*, 300, 1, 012011 (2011)
[22] Pereira, T. M. D., Kiselman, D., Asplund, M., *A&A*, 508, 1403 (2009)

Broad emission lines: A tool for studying nuclei of active galaxies

Dragana Ilić[1], Luka Č. Popović[2], Alla I. Shapovalova[3], Alexander N. Burenkov[3], Wolfram Kollatschny[4], Andjelka Kovačević[1], Vahram Chavushyan[5], Giovanni La Mura[6] and Piero Rafanelli[6]

[1] Department of Astronomy, Faculty of Mathematics, University of Belgrade, Serbia
[2] Astronomical Observatory, Volgina 7, 11060 Belgrade, Serbia
[3] Special Astrophysical Observatory of the Russian AS, Nizhnij Arkhyz, Karachaevo-Cherkesia 369167, Russia
[4] Institut fuer Astrophysik, Georg-August-Universitaet Goettingen, Germany
[5] Instituto Nacional de Astrofisica, Optica y Electronica, Apartado Postal 51-216, 72000 Puebla, Mexico
[6] Dipartimento di Fisica e Astronomia, Universita di Padova, Vicolo dell'Osservatorio, I-35122 Padova Italy

E-mail: `dilic@matf.bg.ac.rs`

Abstract. Active galactic nuclei (AGNs) are objects hosting in their center a super-massive black hole (SMBH) with an accretion disk surrounded by gas and dust. The mass of an SMBH can be derived from the dynamics of the gas gravitationally bounded to the SMBH. This is the case for the broad line region (BLR), i.e. a photoionized gas in the vicinity of an SMBH that emits broad emission lines (BELs), which properties can be used to estimate the mass of the SMBH. In spite of a number of papers devoted to the BLR research, its true nature is not well known. Therefore, it is still important to investigate the BLR structure (size, geometry, physics, etc.), where one of the aims is to better constrain the mass of the SMBH in the center of AGNs. The BELs are the only signatures of the BLR physics and geometry. They can be clearly identified in AGN spectra and they often show complex profiles. Their fluxes, profiles and ratios can provide much information about the BLR geometry and physics. Moreover, the BELs and continuum flux are very often varying in AGNs. Therefore, an investigation of the BEL flux and profile variability during a long period is another useful tool for mapping the geometrical and dynamical structure of the BLR. In this review we present and discuss some tools and techniques for studying the structure of the BLR using broad emission line properties.

1. Introduction

One of the important problem of active galactic nucleus (AGN) investigations is the most accurate estimation of the mass of the super-massive black hole (SMBH) in the center of an AGN. A direct determination of an SMBH mass is possible from dynamical methods (virial theorem) based on observations of the kinematics of stars orbiting around the SMBH. In the case of an AGN, the SMBH mass can be derived from the dynamics of the gas gravitationally bounded to the SMBH. The intensive broad emission lines (BELs), visible in the spectrum of Seyfert 1 galaxies and quasars (commonly called type 1 AGN), originate relatively close to the central power source, i.e. in the broad line region (BLR). The BLR is in the strong interaction with the radiation field produced by the central engine and is under the influence of the gravitational field of the SMBH, thus can be used to determine the SMBH mass using the virial theorem (see e.g. Gaskell 1988). Therefore, investigations of the BLR structure (size, geometry, physics, etc.) are very important, and for this the analysis of the BELs properties (flux, profiles, widths, etc.) is needed.

A lot of work has been devoted to understand the relationship between the BLR dynamics and the corresponding broad emission line profiles (see e.g. Sulentic et al. 2000, Gaskell 2009, and reference therein). The BLR is probably very complex, often with evidence for multiple components (Popović et al. 2004, Ilić et al. 2006, Bon et al. 2006, 2009), thus the BELs are the result of a combination of effects involving the gas motion and the radiation transfer across an environment which is only approximately understood. Moreover, in some AGNs (especially with a high variability in the line profiles) there may be an additional component (e.g. outflowing material or disk wind) present in the BLR that strongly affects the BEL profiles (as e.g. in NGC 4151, see Shapovalova et al. 2008, 2010a). As an example see Fig. 1, where the typical multi-Gaussian line fitting (see e.g. Popović et al. 2004, Sulentic et al. 2000, and references therein) is given for the case of the Hβ emission line of Mrk 817 (Ilić et al. 2006). The broad Hβ is fitted with three Gaussians (solid lines at the bottom of Fig. 1) indicating that there may be a disk-like emission component that contributes to the line wings, and an additional component contributing to the line core (Ilić et al. 2006).

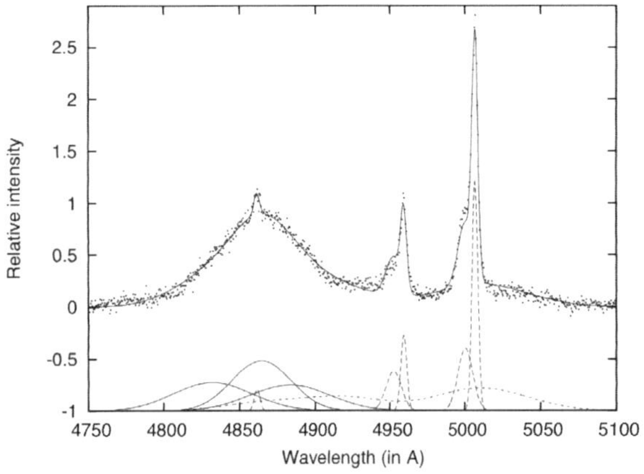

Figure 1. An example of the multi-Gaussian fitting of the Hβ emission line of Mrk 817 (Ilić et al. 2006). The dots represent the observation and solid line is the best fit. The Gaussian components are shown at the bottom. The dashed lines represent the Fe II template, [OIII] and Hβ narrow lines.

On the other hand, the physical properties (e.g. gas temperature, density, ionization parameter) of the emitting plasma in the BLR are still not well constrained. The BELs always come from the permitted line transitions (or sometimes semi-forbidden) which is obviously implying that the densities are high enough to collisionally suppress the forbidden emission lines. Thus, the standard nebular diagnostics developed for HII regions or planetary nebulae cannot be applied (Osterbrock & Ferland 2006). There are several numerical codes (as e.g. CLOUDY, Ferland et al. 1998), which use the photoionization model to explain the observed line ratios in AGNs and obtain the physical parameters of the BLR (see e.g. Korista & Goad 2004, Marziani et al. 2010). Moreover, the BEL fluxes could be used to probe the physical conditions of the BLR plasma using well-known laboratory plasma diagnostic tools, as e.g. the Boltzmann-plot (Griem 1997) that has been applied to the Balmer line series (see e.g. Popović 2003, 2006ab, Popović et al. 2008, Ilić et al. 2009, 2010a, 2012). It is believed that the photoionization is probably the main heating mechanism, but the line production can be influenced by other atomic processes, especially those caused by high density effects. For example, in the case of NGC 4151, a well-studied Seyfert 1.5 galaxy (see e.g. Peterson et al. 2004, Bentz et al. 2006, Metzroth et al. 2006, Shapovalova et al. 2008, etc.), during one observed period, the line fluxes are not correlating with the continuum flux and are saturating at higher continuum fluxes, indicating a presence of a non-radiatevly heated region in the BLR, or more ionizing to optical flux ratio than expected for a typical AGN (Shapovalova et al. 2008).

Moreover, the BELs of AGNs often exhibit variability, which is assumed to be caused by variation in the ionizing continuum strength and by dynamic evolution of the BLR gas on long timescales. A long-term optical spectral monitoring of the nucleus of some AGNs has revealed a time lag in the response of the broad emission lines relative to flux changes in the continuum (see e.g. Wanders and Peterson 1996, Kollatschny & Dietrich 1997, Peterson 2008), that is used as a tool for determination of the size of the BLR. Nowadays this method is called the reverberation mapping technique (Blandford & McKee 1982), and has motivated and initiated numerous monitoring campaigns (see e.g. the international "AGN Watch" campaign, Peterson 1999).

In this paper we will present some methods to study the physics and kinematics of the BLR using the emission lines and their properties (fluxes, profiles, variability), especially we will present some results based on our long-term optical spectral monitoring campaign of a sample of type 1 AGN (Shapovalova et al. 2004, 2009, 2010b, 2012).

2. Line fluxes: Plasma diagnostics of the BLR

The physical properties of the BLR are difficult to estimate directly from the BELs, that are basically the only signature of the BLR. The plasma conditions in the BLR are more closer to the conditions in stellar atmospheres than in gaseous nebulae (Osterbrock & Ferland 2006), that we deduce from the presence or absence of some emission lines. The temperatures ought to be less than 35 000 K since at higher temperatures the observed Fe II (that is sometimes even remarkably strong) would be effectively suppressed by the collisional ionization to Fe III. On the other hand, the electron density should be in the range $10^8\text{cm}^{-3} < n_e < 10^{14}\text{cm}^{-3}$, in order to suppress the emission of broad forbidden lines, while still allowing for the presence of permitted and semi-forbidden ones.

There are couple of methods how to determine the BLR physical properties using only observations of the BELs (Laor 2006, Marziani et al. 2001, Popović 2003, 2006). Marziani et al. (2001) found, using the CLOUDY photoionization computation, that the ratio between UV emission lines Si III] $\lambda1892$ and C III] $\lambda1909$ is a good density diagnostic in the density range $9.5 < \log n_e < 12$. Also, Laor (2006) considered electron scattering influence on the line shapes and determined the physical parameters (the electron density and optical depth) of the BLR by fitting the emission lines that have exponential wings, in the case of the low luminosity AGN NGC 4395. However, the problem of these direct methods for BLR diagnostics is that they are observationally constrained for either needing the UV observations or detecting BELs with strong exponential line wings.

Popović (2003, 2006) suggested that the Boltzmann-plot (BP) method, already well known to laboratory diagnostics of a high density plasma (Griem 1997), might be exploited to probe the BLR of some AGNs (see also Ilić et al. 2006, La Mura et al. 2007, Popović et al. 2008). The BP method uses the fluxes from one line transition series (as e.g. hydrogen Balmer line series) and the atomic parameters of the corresponding line series, to estimate the excitation temperature of the BLR plasma from the slope A of the line fluxes (normalized to the corresponding line atomic parameters) versus the energy of the upper level (Griem 1997, Popović 2003, 2006): $A = \log_{10}(e)/kT_{\text{exc}} \approx 5040/T_{\text{exc}}$. For example, in case of NGC 5548 the BLR temperature was estimated using the BP applied to the Balmer lines observed from 1996 to 2004 (Popović et al. 2008), and a high correlation between the variation of the optical AGN continuum and the BP temperature is found. The advantage of the BP method compared to the other BLR plasma diagnostics methods that use BELs, is that it requires only the measured Balmer line fluxes to estimate the excitation temperature. But one has to consider some possible drawbacks, concerning the use of complex BELs to infer the BLR physical properties, which in general appear in all methods (see Ilić et al. 2006 for discussion).

Recently, Ilić et al. (2012) investigated for what physical conditions the BP can be applied to the BLR plasma diagnostics, using the hydrogen Balmer line series obtained in three different ways: with the CLOUDY spectral synthesis code (Ferland et al. 1998), from the recombination theory (Storey & Hummer 1995), and from the Sloan Digital Sky Survey (SDSS) database (La Mura et al. 2007). It was shown that the plasma temperatures can be estimated using the BP, applied on the hydrogen Balmer line series, in the case of a high density BLR, i.e. the electron density is $n_e \sim 10^{14}\mathrm{cm}^{-3}$ (Ilić et al. 2012). Moreover, for these cases the electron temperature can be estimated from the BP excitation temperature. Fig. 2 presents the BP temperature as a function of the electron temperature, for cases when the BP fitting error is less than 10%. Clearly, there is a relation between these two temperatures, especially for the electron density of $10^{14}\mathrm{cm}^{-3}$ (open circles in Fig. 2).

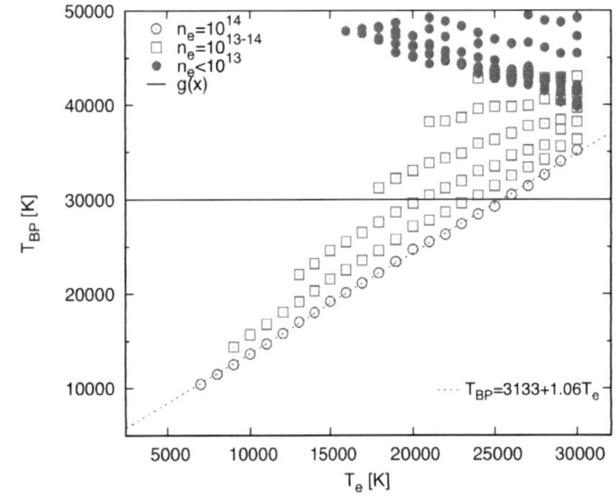

Figure 2. BP temperature plotted against the electron temperature for the case when the BP fitting-error is $\leq 10\%$. Open circles are for the electron density of $10^{14}\mathrm{cm}^{-3}$, open squares for $10^{13} \leq n_e < 10^{14}\mathrm{cm}^{-3}$, and full circles for $n_e < 10^{13}\mathrm{cm}^{-3}$. The linear best-fitting of the cases with $n_e = 10^{14}\mathrm{cm}^{-3}$ is given with the dashed line. The solid line marks the $T_{\mathrm{BP}} = 30000$ K (Ilić et al. 2012).

3. Line profiles: BLR geometry

The analysis of the BEL profiles (widths, shapes, asymmetries, and bumps) is a powerful tool to study the kinematics and geometry of the BLR (see e.g. Sulentic et al. 2000, Popović et al. 2004, Gaskell 2009). As noticed in the Introduction, the BEL profiles are usually very complex, showing a non-Gaussian shape with some characteristic features (e.g. asymmetries or bumps), implying that the BLR is most likely a complex region, that cannot be explained in terms of a simple single-region model. So far there are numerous geometrical models proposed to explain the kinematics and geometry of the BLR (e.g. biconical ejection, disk wind, combination of the disk-like and spherical component etc.), and we still have no self-consistent model that would explain the kinematics of the BLR in all AGNs (Peterson 2006). For example, there are indications that the BLR kinematics may be related to outflows and jets originating from a close vicinity of the supermassive black hole and accretion disk (e.g. Arshakian et al. 2010). Thus, one of the proposed models assumes that the BELs originate in an accelerating outflow affected by the gravitational field of the SMBH, and this model has been tested for several AGNs, as NGC 4151 or Mrk 668 (see Ilić et al. 2008, 2010b).

On the other hand, the Keplerian accretion disk model of the BLR has been often proposed to explain the BEL double-peaked profiles (see e.g. Chen et al. 1989, Chen & Halpern 1989,

Eracleous & Halpern 1994, 2003, Popović 2006b, Popović et al. 2002, 2004, Kollatschny & Bischoff 2002, Bon et al. 2006, 2009 etc.). The fraction of AGNs that clearly shows double-peaked profiles, and thus the evidence of the disk-like geometry, is small and statistically insignificant (e.g. Strateva et al. 2003). But, the presence of double-peaked lines is not required as a necessary condition for the existence of a disk-like geometry in the BLR. In the case of single-peaked lines, it could be that either the parameters of the disk (e.g. the inclination) are such that one observes single-peaked lines (e.g. Popović et al. 2004, Ilić et al. 2006) or the emission of the disk is masked by the emission of another emission line region (see e.g. Popović et al. 2002, 2004, Bon et al. 2006, 2009, Ilić et al. 2006, and references therein). The last proposed case is the so-called two-component model (see Popović 2006b, for a review). One possibility of the two-component model is to have a disk region where the line-wings are originating, and an additional spherical component, which is producing the line core and whose kinematics can be described with an isotropic velocity distribution (Popović et al. 2004, Bon et al. 2009). Another important model is a disk-like region with the disk wind that can also produce single-peaked BELs observed in most of AGNs (see e.g. Murray & Chang 1997, Flohic et al. 2012, and references therein).

Finally, one should consider BEL profile fittings with cautions as the same line profile can be modeled with more than one model. For an example, in Fig. 3 we present the BEL profiles of AGN Mrk 817 fitted with two different models (Ilić et al. 2006). The BLR of Mrk 817 might be composed from either disc-like+spherical regions (left panel, Fig. 3) or from two spherical regions, having different velocity distributions (right panel, Fig. 3).

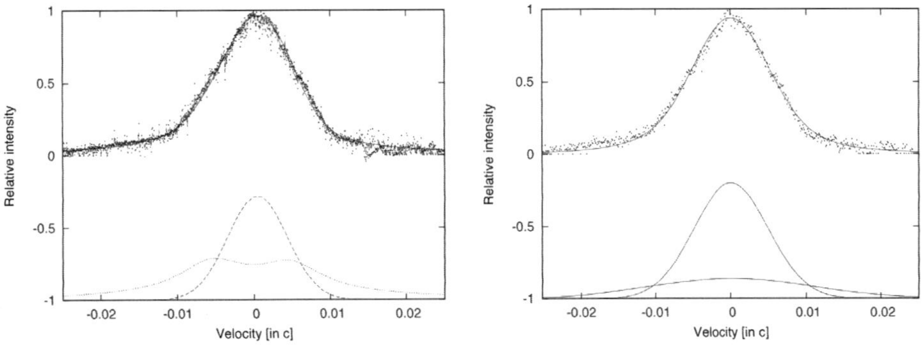

Figure 3. The averaged broad profile of the Hα and Hβ emission lines of Mrk 817 fitted with the two-component model: disk plus additional component (left). The broad profile of the Hβ emission line fitted with only two Gaussian components (right). For details about the models see Ilić et al (2006).

4. Line variability: Long-term optical monitoring campaign

The long-term optical monitoring campaign consisted from coordinated international observations of a sample of type 1 AGN, constantly carried out at four telescopes of Russia and Mexico during the period 1996–2012. The main aim of the campaign is to monitor the variability of BELs in the optical domain (spectral ranges near Hα and Hβ) and continuum, and investigate the structure (physics and kinematics) of the BLR in type 1 AGN.

The used instruments are: the 6-m and 1-m telescopes of SAO RAS, the 2.1-m telescope of INAOE (Cananea, México), and the 2.1 m telescope of OAN-SPM (Baja California, México).

High quality spectra (S/N > 50) around Hα and Hβ wavelength region (spectral range 4000-8000$\mathring{A}$, the resolution 3-15$\mathring{A}$) were obtained with long-slit spectrographs equipped with a CCD. For details on data acquisitions, data reduction and calibration see Sahpovalova at al. (2008, 2010b). The sample of observed AGNs contains well-known and studied objects: NGC 5548, NGC 4151, 3C390.3, Ark 564, and Arp 102B. So far, the results of this campaign have been presented and published in a number of papers (for more details see e.g. Shapovalova et al. 2001, 2004, 2008, 2009, 2010ab, 2011, 2012, Popović et al. 2008, 2011, Ilić 2007, Jovanović et al. 2010). Here we present in more details results obtained for 3C390.3 and Ark 564.

4.1. The case of 3C390.3

One of the monitored AGN is a radio-galaxy 3C390.3 showing double-peaked broad emission lines. The observations of 3C 390.3 were performed from 1995 to 2007. The results of these investigations were published in Shapovalova et al. (2001, 2010b), Popović et al. 2011 and Jovanović et al. (2010). The double-peaked Hα and Hβ lines clearly indicate a disk-like geometry of the BLR. But, the variations in line profiles show that there is an additional emission component contributing to the line core, that is originating in the region different then the disk (see the mean and rms Hα spectra in Fig. 4, and for more details see Fig. 12 in Shapovalova et al. 2010b). This additional component could be also explained as an extra emission coming from the perturbation in an accretion disk (Jovanović et al. 2010).

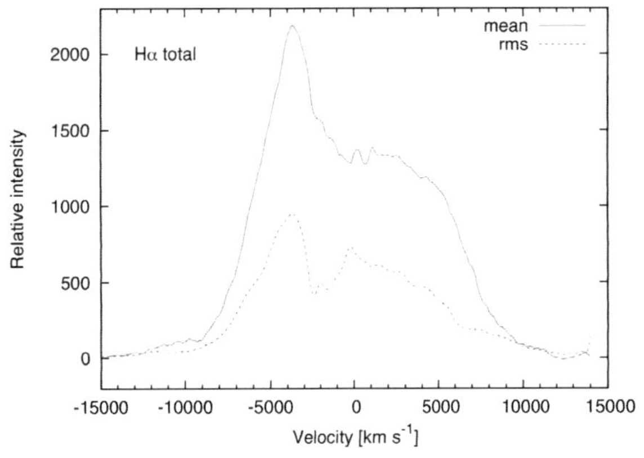

Figure 4. Mean and rms spectra of the Hα line of 3C390.3 from the whole monitoring period (1995–2007) (Shapovalova et al. 2010b).

Fig. 5 gives the light curves of the continuum flux (top) and the flux of the broad component of the Hβ line (bottom). Both light curves have similar behavior, but the cross-correlation analysis gives that the Hβ flux is delayed for $\sim$ 95 days (Shapovalova et al. 2010b). We can basically divide the monitoring period into two: period I - until 2002, and period II - after 2002, when there is a clear outburst seen in both light curves (red vertical line in Fig. 5).

One of the interesting results comes from the analysis of the line emission profiles from different epochs, i.e. the line fluxes of different line segments. Line profiles were divided along the velocity scale into nine line segments (see Table 2 in Popović et al. 2011), where e.g. line segments -4 and -3 represent together the blue wing flux and line segments 4 and 3 the red wing. We focus on the analysis of the line wings, since for them the emission of the accretion disk is contributing the most (Fig. 6). Using the disk model for the line profiles, we have shown that the change in the line profile wings in the period I (full circles), can be explained if the location of the disk emitting the broad lines is shifted along the accretion disk (blue line with

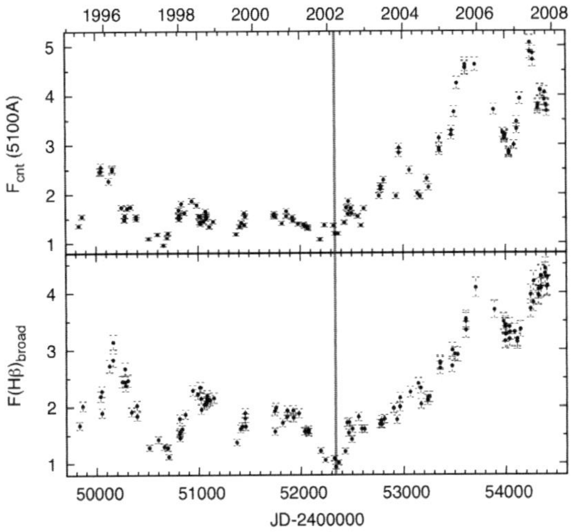

Figure 5. Light curves of the continuum (top) and broad Hβ (bottom). The continuum flux is in 10^{-15}erg cm^{-2} s^{-1} Å^{-1} and line flux in 10^{-13}erg cm^{-2} s^{-1}. The red line divides the monitoring period into two periods (Shapovalova et al. 2010b).

open circles), from $R_{\rm inn} = 250R_{\rm g}$ to $R_{\rm inn} = 550R_{\rm g}$ (Fig. 6). In period II (open squares), the position of the emitting part of the disk is fixed (see for more details Popović et al. 2011).

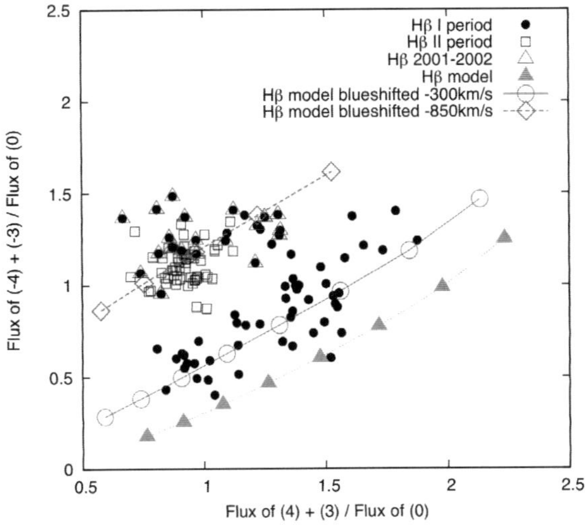

Figure 6. The blue-wing flux (two blue line segments, -4 and -3) vs. the red-wing (two red line segments, +4 and +3), both normalized to the central line segment flux. Observations in period I are denoted with full circles, in period II with open squares, while observations from 2001–2002 are also marked with open triangles. The notation of each curve is given in the upper right corner (Popović et al. 2011).

4.2. The case of Ark 564

In the following section, we will present in more details the latest research and results obtained for the narrow-line Seyfert 1 galaxy Ark 564. The observations of this AGN were performed from 1999 to 2010 (for details see Shapovalova et al. 2012). One of the reasons why narrow-line Seyfert 1 (NLS1) galaxies are interesting is the prominent emission of Fe II in the optical band around the Hβ line (Fig. 7), that is clearly visible due to the relatively small width of the emission lines in this type of AGNs. From the multi-Gaussian fitting of the wavelength range of 4000–5600 Å, and the analysis of the Gaussian line widths and shifts, we have shown that the

Journal of Physics: Conference Series **397** (2012) 012050 doi:10.1088/1742-6596/397/1/012050

optical Fe II lines of this AGN are coming from the intermediate line region (Shapovalova et al. 2012), as suggested before (see Kovačević et al. 2010, and references therein).

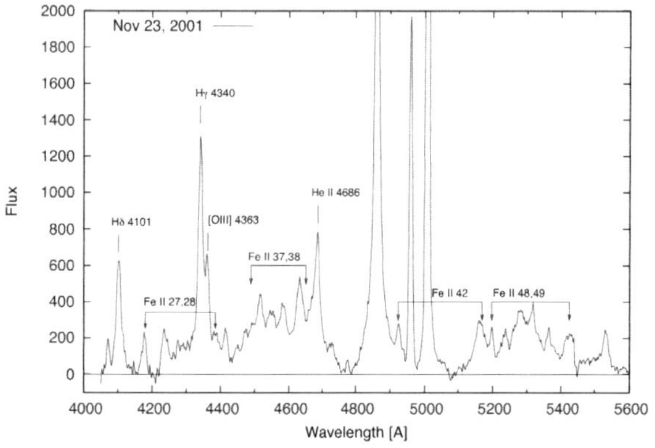

Figure 7. The Fe II emission around the Hβ line for the NLS1 galaxy Ark 564. Strong emission lines are labeled, as well as the Fe II multiplets (Shapovalova et al. 2012).

During the monitoring period, the mean continuum and lines fluxes are weakly varying, as the decrease of only ~20%-30% (see light curves in Fig. 5 in Shapovalova et al. 2012) from the beginning (1999) to the end of the monitoring campaign (2010) was detected. Also, in the light curves, five flare-like events (two prominent and three possible) lasting ~1-3 days were detected.

One obtained result is particularly intriguing. It was observed, that there is almost no correlation between the Hα and Hβ line fluxes (Fig. 8). Such behavior is unexpected if photoionization by the central continuum source would be the main mechanism responsible for the production of both lines. The lack of correlation could indicate very complex physical processes in the line forming region, i.e. beside the photoionization some additional physical processes may be present.

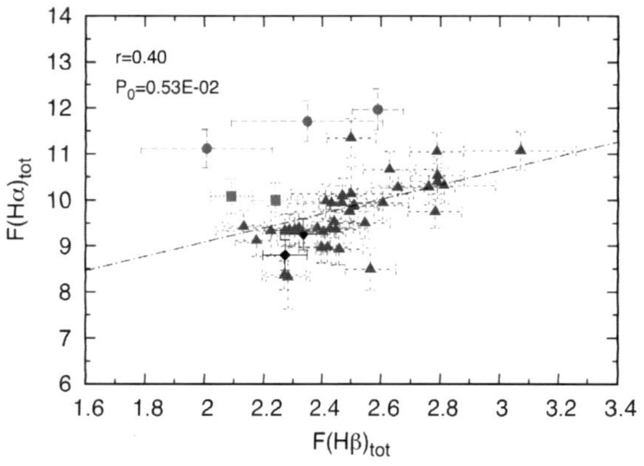

Figure 8. The Hα as a function of the Hβ line flux. The correlation coefficient and the corresponding P-value are given in the upper left corner. Data obtained with different telescopes are marked with different symbols: diamonds - 6m in Russia, circles - 1m Zeiss in Russia, triangles - 2.1m GHO in Mexico, squares - 2.1m SPM in Mexico (Shapovalova et al. 2012).

5. Conclusions

In this paper we present some methods and tools how to investigate the plasma properties of the BLR in AGNs, using the broad emission line parameters (e.g. line fluxes, ratios, widths, profiles, variability). Especially, we present the possibilities to use the long-term optical monitoring campaigns of variable AGNs in constraining the physics and structure of the BLR, and give some of the findings in two special objects. Here we outline our important conclusions:

i) We show that the Boltzmann-plot method can be exploited for the BLR plasma diagnostics, as it gives constraints to both the temperature and density of the BLR.

ii) The kinematics and geometry of the BLR is complex and most often there are contribution of more sub-regions with different geometry/motions to the total line profile. One possibility is that the BEL profiles can be described with the two-component model, where the disk-like region emission is contributing to the line wings, and an additional spherical component is producing the line core.

iii) The BLR physics is also complex, and there are most likely other ionization mechanism responsible for the line production, apart from the photoionization.

All of these findings should be taken into account when using the parameters of the BLR for the estimates of the mass of the supermassive black hole in the center of an AGN, and thus the investigations of the BLR properties are still in the top of the astrophysical research.

Acknowledgments
This work was supported by the Ministry of Education and Science of the Republic of Serbia through the project "Astrophysical Spectroscopy of Extragalactic Objects" (#176001). L. Č. P, W. K., and D. I. are grateful to the Alexander von Humboldt foundation for support in the frame of program "Research Group Linkage".

References

Arshakian T G, León-Tavares J, Lobanov A P *et al* 2010 *MNRAS* **401** 1231
Bentz M C, Denney K D, Cackett E M *et al* 2006 *ApJ* **651** 775
Blandford R. D. and McKee C. F. 1982 *ApJ* **255** 419
Bon E, Popović L Č, Gavrilović N, La Mura G and Mediavilla, E 2009 *MNRAS* **400** 924
Bon E, Popović L Č, Ilić D and Mediavilla E 2006 *New Astronomy Review* **50** 716
Chen K, Halpern J P and Filippenko, A.V. 1989 *ApJ* **339** 742
Chen K and Halpern J P 1989 *ApJ* **344** 115
Eracleous M. and Halpern J P 1994 *ApJS* **90** 1
Eracleous M. and Halpern J P 2003 *ApJ* **599** 886
Ferland G J, Korista K T, Verner D A, Ferguson J W, Kingdon J B and Verner E M 1998 *PASP* **110** 761
Flohic H M L G., Eracleous M and Bogdanović T 2012 *ApJ* **753** 133
Gaskell C M 1988 *ApJ* **325** 114
Gaskell C M 2009 *New Astronomy Review* **53** 140
Griem H R 1997 *Principles of plasma spectroscopy* (Cambridge: Cambridge University Press)
Ilić D 2007 *Ser. Astron. J.* **175** 15
Ilić D, Kovačević J and Popović L Č 2009 *New Astronomy Review* **53** 149
Ilić D, Popović L Č, Bon E, Mediavilla E and Chavushyan V H 2006 *MNRAS* **371** 1610
Ilić D, Popović L Č, La Mura G, Ciroi S and Rafanelli P 2010a *Journal of Physics Conference Series* **257** 012034
Ilić D, Popović L Č, La Mura G, Ciroi S and Rafanelli P 2012 *A&A* **543** 142
Ilić D, Popović L Č, León-Tavares J *et al* 2008 *Memorie della Societa Astronomica Italiana* **79** 1105
Ilić D, Popović L Č, Shapovalova A I *et al* 2010b *Memorie della Societa Astronomica Italiana Supplementi* **15** 166
Jovanović P, Popović L Č, Stalevski M and Shapovalova A I 2010 *ApJ* **718** 168
Kollatschny W and Bischoff K 2002 *A&A* **386** L19
Kollatschny W and Dietrich M 1997 *A&A* **323** 5
Korista K T and Goad M R 2004 *ApJ* **606** 749
Kovačević J, Popović L Č and Dimitrijević M S 2010 *ApJS* **189** 15
La Mura G, Popović L Č, Ciroi S, Rafanelli P and Ilić D 2007 *ApJ* **671** 104
Laor A 2006 *ApJ* **643** 112
Murray N and Chiang J 1997 *ApJ* **474** 91
Marziani P, Sulentic J W, Negrete C A, Dultzin D, Zamfir S and Bachev R 2010 *MNRAS* **409** 1033
Metzroth K G, Onken C A and Peterson B M 2006 *ApJ* **647** 901
Osterbrock D E and Ferland G J 2006 *Astrophysics of Gaseous Nebulae and Active Galactic Nuclei (2nd ed.)* (Sausalito, California: University Science Books)

Peterson M B 1999 *Structure and Kinematics of Quasar Broad Line Regions* (ASP Conference Series) **175** 49
Peterson M B 2008 *New Astron. Rev.* **52** 240
Peterson B M, Ferrarese L, Gilbert K M *et al* 2004 *ApJ* **613** 682
Popović L Č 2003 *ApJ* **599** 140
Popović L Č 2006a *ApJ* **650** 1217 (an Erratum)
Popović L Č 2006b *SerAJ* **173** 1
Popović L Č, Mediavilla E G, Bon E and Ilić D 2004 *A&A* **423** 909
Popović L Č, Mediavlilla E G, Kubičela A and Jovanović P 2002 *A&A* **390** 473
Popović L Č, Shapovalova A I, Chavushyan V H, Ilić D, Burenkov A N, Mercado A, Ciroi S and Bochkarev N G 2008 *PASJ* **61** 1
Popović L Č, Shapovalova A I, Ilić D *et al* 2011 *A&A* **528** 130
Shapovalova A I, Burenkov A N, Carrasco L *et al* 2001 *A&A* **376** 775
Shapovalova A I, Doroshenko V T, Bochkarev N G et al. 2004 *A&A* **422** 925
Shapovalova A I, Popović L Č, Collin S *et al* 2008 *A&A* **486** 99S
Shapovalova A I, Popović L Č, Bochkarev N G *et al* 2009 *New Astronomy Review* **53** 191
Shapovalova A I, Popović L Č, Burenkov A N *et al* 2010a *A&A* **509** 106
Shapovalova A I, Popović L Č, Burenkov A N *et al* 2010b *A&A* **517** 42
Shapovalova A I, Popovic L Č, Burenkov A N *et al* 2012 *ApJS* **202** 10
Shapovalova A I, Popović L Č, Ilić D *et al* 2011 *Baltic Astronomy* **20** 476
Storey P J and Hummer D G 1995 *MNRAS* **272** 41
Strateva I V, Strauss M A, Hao L *et al* 2003 *ApJ* **126** 1720
Sulentic J W, Marziani P and Dultzin-Hacyan D 2000 *ARA&A* **38** 521
Wanders I and Peterson B M 1996 *ApJ* **466** 174

Quasar Spectroscopy and Cosmology

D.A. Varshalovich[1,2], A.V. Ivanchik[1,2], S.A. Balashev[1,2]

[1]Ioffe Physical-Technical Institute of the Russian Academy of Sciences,
[2]St.-Petersburg State Polytechnical University

E-mail: varsh@astro.ioffe.ru

Abstract. Absorption spectra of distant quasars are the main source of our knowledge of physical conditions, structures, and composition of matter in the Early Universe. The detailed analysis of such spectra gave us a possibility to discover HD/H_2 molecular clouds existed 12−13 Gyr, investigate them and derive some important cosmological conclusions.

1. Introduction

Quasars are the most powerful sources of radiation and can be observed from distances up to 10−13 Gyr light years. In other words, the spectra of quasars measured at the present time were formed 10−13 Gyr ago. The emission of quasars themselves is used to probe remote clouds of interstellar and intergalactic gas along the line of sight. These clouds "imprint" their absorption lines with corresponding redshifts into the quasar spectrum. Most of the absorption lines are Lyman alpha lines at various redhshifts which are less then quasar redshift. These lines are associated with the higly ionized, high temperature intergalactic gas. Absorption systems formed in intergalactic medium can be used to probe large scale structure of the Universe. By means of correlation in Lyα forest absorptions in the spectra of distant quasars it is possible to determine Baryonic Acoustic Oscillation [1]. Some studies related to absorption lines of heavy element such as Mg II, C IV, which associated with the clouds in the halo of intervening galaxies (see e.g. [2]). Other studies are focused on Damped Ly-α systems which are thought to arise when the light from quasar pass through the intervening galaxies, and therefore provide an estimate of neutral gas density in the Universe and its evolution with time (e.g. [3]). There are also several cosmological problems that can be solved using quasar absorption system analysis: the determination of the baryon density in the Universe, the estimation of the temperature of the cosmic microwave background radiation at high redshifts, constrains on the possible variations of the fundamental physical constants and others. The first of these problems is closely related to Big Bang nucleosynthesis model.

As it is well known, at the early Universe there was a short period of synthesis of light nuclides from protons and neutrons. At that time, the Universe was filled with almost homogenous fully ionized plasma with the temperature T$\sim 10^9 - 10^8$ K. However, at that stage only the lightest nuclides D, ^{3}He, ^{4}He, ^{6}Li, ^{7}Li and ^{7}Be could be formed in sufficient amounts. As well as the rates of relevant reactions at certain energy are well known, the relative abundance of these nuclides formed during the BBN can be reliably calculated. Left panel of Fig. 1 shows the evolution of the abundance of light nuclides with time during the BBN nucleosynthesis. The only free parameter of such calculations is $\eta = (n_B/n_\gamma)$ - the relative baryon number density n_B to the

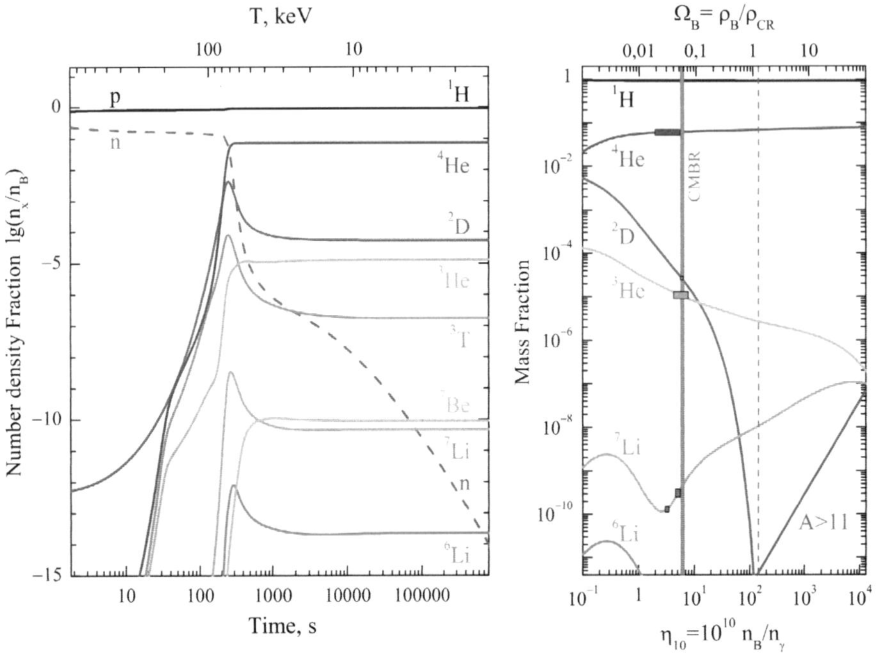

Figure 1. The results of primordial nucleosynthesis. Left panel: the evolution of abundance of light nuclides with time t and, accordingly, with the temperature T. Right panel: the relative abundance of the relic nuclides eventually formed as a function of the baryon-to-photon number density η_{10} or baryon density Ω_B. The solid lines are the theoretical calculations. The vertical strip shows the range of Ω_B obtained from different astronomical observations of primordial nuclide abundance. The dashed line corresponds to $\rho_B = \rho_{cr}$.

photon number density n_γ. This ratio has to be constant during the adiabatic expansion of the Universe. Therefore this parameter can be determined from comparison of observed abundance of primordial nuclides with the calculation results. On the right panel of Fig. 1 the dependencies of the relative abundance of light nuclides as a function of the $\eta_{10} = 10^{10}\eta$ (or the $\Omega_B = \rho_B/\rho_{cr}$ – baryon density ρ_B in units of critical density ρ_{cr}) are presented. It is clearly seen that the relative abundance of primordial deuterium $(D/H)_p$ is the most sensitive to the baryon number density. For example, if the density of baryon matter were equal to the critical one $\rho_B = \rho_{cr}$, we would not observe primordial deuterium at all. The isotopic abundance ratio (D/H) could only decrease during the subsequent evolution of the Universe because deuterium rapidly burns out into helium in stellar interiors. Therefore to infer baryon number density from (D/H) ratio, it is necessary to determine isotopic composition of the interstellar or intergalactic medium at as early cosmological epoch as possible, when the composition was close to the primordial one.

Until very recently, the relative abundance of (D/H) at high redshifts was measured using only atomic HI and DI by means of Lyman series lines detected in quasar absorption spectra [4]. However, such measurement encounters with severe obstacles. The spectra of HI and DI are virtually the same, with only all wavelengths are shifted at about 81 km/s. This value is closed to the typical velocity offset between multiply components seen in quasar absorption

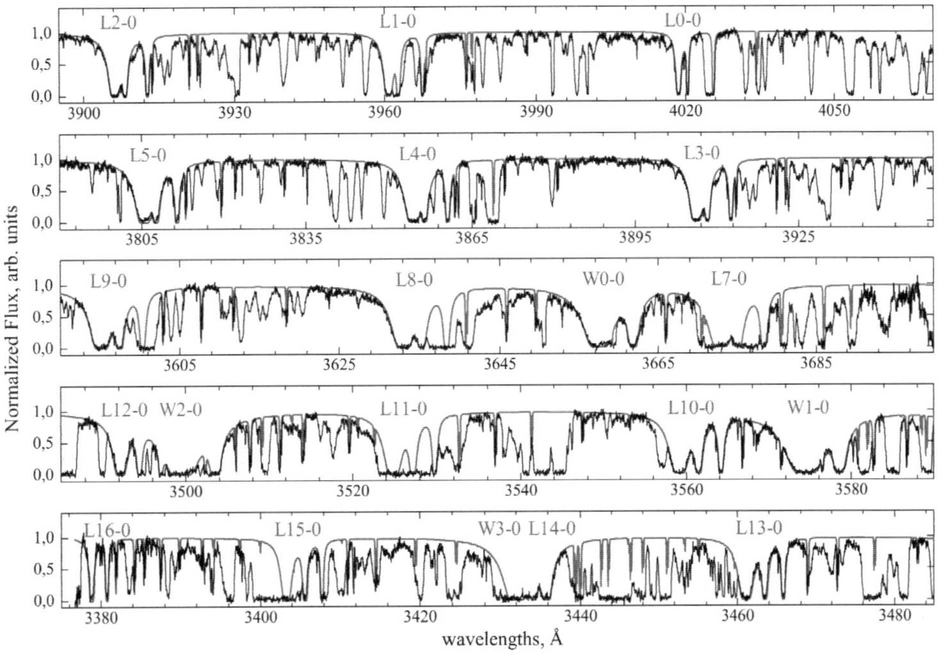

Figure 2. The high resolution optical spectrum of Q 0812+320 obtained at KECK telescope. Synthetic H_2 spectrum of the absorption system at z = 2.62649 fitted into the observed spectrum shown by red line. The bands of Lyman and Werner series of H_2 are marked.

spectra, which is believed to be several clouds on the line of sight located in gravitationally bounded systems like galaxy or protogalaxy. Also number density of DI is lower by 4-5 order of magnitude relative to HI, that makes line profile analysis very complicated. These facts can possibly explain the significant dispersion of values of the (D/H) ratio measured by atomic lines.

Difficulties with line identification did not arise in measurement of the relative abundance of HD and H_2 molecules, since their spectra are significantly differs and most of absoprtion lines are much narrow then atomic one. This methods applied only recently with first detection of HD molecules at high redshifts [5].

2. HD/H_2 absorption systems

One kind of absorption systems identified in quasar spectra are the molecular hydrogen absorption systems, which are very rare detected in quasar spectra. Only in $< 1\%$ of quasar spectra it is possible to detect H_2 absorption system. Additionally, H_2 absorption system can be identify only in high resolution spectra (R~50000), which is available only on the biggest optical telescope like Very Large Telescope and KECK. Molecular hydrogen absorption systems are supposed to be a diffuse interstellar clouds located in the intervening galaxy. These remote galaxies are detected in quasars spectra as Damped Lyman alpha (DLA) systems - absorption systems with damped Lyman alpha line, with column density of neutral hydrogen $\log(N) > 10^{20}$cm^{-2}. This amount of neutral hydrogen is sufficient to shield medium from diffuse cosmic UV background. Analysis of H_2 absorption systems allows us to study the physical conditions of diffuse clouds in remote galaxies. Numerous absorption lines of molecules H_2, in

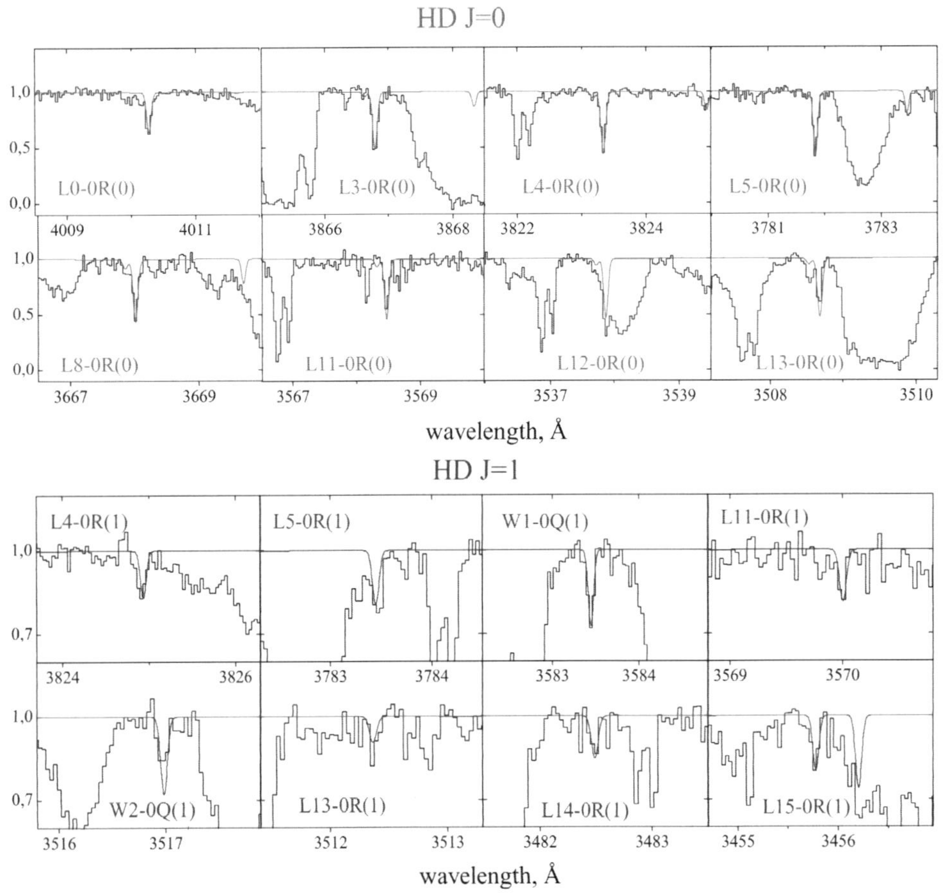

Figure 3. HD absorption lines, associated with H_2 absorption system at z=2.62649 in spectrum of Q 0812+320. Top and bottom panels show absorption lines for J=0 and J=1 level, respectively.

some cases HD and CO, and metals in different ionization states show that this cloud have typical temperatures $\sim 10 \div 100$ K and number densities in range $\sim 1 - 100\,cm^{-3}$. There are indications that size of these clouds are about parsec and subparsec, that is well agreed with what observed in nearby galaxies.

To date there are only 24 H_2 absorptions systems detected at high redshifts. Only in 6 of them have HD absorption lines. The detected HD/H_2 absorption systems are shown in Table 1. Fig. 2 shows spectrum of Q 0812+320 (obtained at 10 meter optical KECK telescope with high resolution spectrograph HIRES), in which one of the H_2 absorption system was detected [6] and it is shown by red profiles. Numerous (up to 80) absorption lines of H_2 from different rotational levels were detected in this systems. This allows us to obtain H_2 column density of this absorption system with great accuracy. Later, HD molecules were detected in this system [7], [8]. Some of the detected HD lines are shown on Fig. 3. The column densities of HD molecules are presented in Table 1. Additionally HD absorption lines corresponding J=1 rotational levels (see lower panels on Fig. 3) were detected in this system for the first time at high redshift.

Journal of Physics: Conference Series **397** (2012) 012051 doi:10.1088/1742-6596/397/1/012051

Table 1. List of known HD/H_2 absorption systems at high redshifts detected in quasars spectra. R_{mag} is the magnitude of QSO. $z_{\rm abs}$ and $z_{\rm em}$ - are the redshifts of absorption system and quasar, respectively. N(H I), N(H_2) and N(HD) - are log of the column densities (in cm^{-2}) of neutral hydrogen, H_2 and HD molecules of absorption systems, respectively.

Quasar	R_{mag}	$z_{\rm abs}$	$z_{\rm em}$	N(H I)	N(H_2)	N(HD)	reference
Q 0812+320	17.88	2.63	2.70	21.4	19.9	15.4	[7], [8]
Q 1232+082	18.40	2.34	2.57	20.9	19.7	15.5	[9], [10]
Q 1331+170	16.26	1.78	2.08	21.2	19.7	14.8	[11], [7]
Q 1439+113	18.07	2.42	2.58	20.1	19.4	14.9	[12]
J 1237+064	18.21	2.69	2.78	20.0	19.2	14.5	[13]
J 2123−005	15.83	2.06	2.26	19.2	17.6	13.8	[14]

H_2 and HD molecules are easily destroyed by the ultraviolet radiation. After absorbing UV photon in Lyman and Werner bands molecular hydrogen gets to the excited electronic states. Then it quickly returns in the electronic ground state. However about 13% of molecules fall in the continuum of ground electronic state, i.e. dissociated. When UV radiation penetrates into the cloud, Lyman and Werner lines becomes saturated, and rate of excitation decreases, therefore the rate of photodissociation decreases too. Calculation shows that at column density $N > 10^{15}$ cm^{-2} HD (as well as H_2) molecules are effectively shield itself from destructive UV radiation. Models of HD/H_2 clouds [15] demonstrate that in case of self-shielding all D will be in form of HD and all H will be in form of H_2. Therefore, for the self-shielded clouds the ratio of column densities N(HD)/2N(H_2) is well agreed with isotopic D/H ratio.

From 6 known HD/H_2 absorption systems we selected two systems corresponding the condition of self-shielding: z=2.33771 in spectrum of Q 1232+082 and z=2.62649 towards Q 0812+320. The abundance of heavy elements (obtained from measurement of associated absorption lines) was found to be about an order magnitude less then the solar one. It may indicate that these identified clouds have composition close to the primordial one. We have estimated D/H isotopic ratio for these clouds by using N(HD)/2N(H_2) ratio. It is consistent with the values of D/H determined from analysis of atomic D I and H I lines and the value obtained from analysis of the Cosmic Microwave Background fluctuations.

3. Conclusions

We have found for the first time [5] and investigated high-redshift molecular clouds containing HD as well as H_2 molecules, which existed at early cosmological epoch, 12-13 Gyr ago. The composition of these HD/H_2 clouds and the proper physical conditions were determined. Obtained low abundance of element heavier then H and He indicates that their composition were almost the primordial one. The typical value of abundance of heavy elements is one order lower then the solar abundance.

D/H isotopic ratio that existed in such clouds 12-13 Gyr ago has been estimated by new method – from abundance of HD to H_2 molecules. This ratio can be derived only for absorption systems with high HD column density, which corresponds condition that in the cloud all D is in form of HD and all H in form of H_2. From two HD/H_2 systems with higher HD molecules column density at z=2.33771 at spectrum of Q 1232+082 and z=2.62649 at Q 0812+320 spectrum we

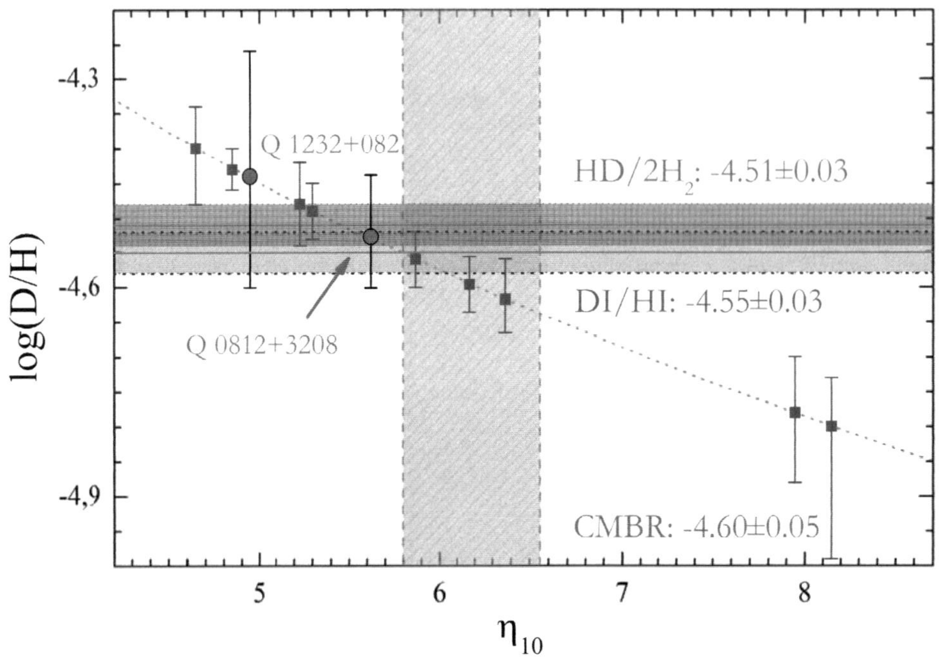

Figure 4. The results of the measurements of the isotopic D/H ratio at high redshifts. The blue points correspond to the atomic D I to H I absorption lines observed in quasar spectra. The two red points are the measurements of D/H isotopic values derived from the analysis of HD/H_2 absorption systems identified in Q 1232+082 and Q 0812+320. The blue and red horizontal stripe is mean values for blue and red points, respectively. The dotted line shows the dependence of D/H ratio as a function of η_{10} inferred from BBN. The green vertical strip is the value obtained from the analysis of the Cosmic Microwave Background fluctuations.

derived

$$D/H = N(\mathrm{HD})/2N(\mathrm{H_2}) = 2.97^{+0.52}_{-0.50} \times 10^{-5}$$

The baryon fraction in the Universe has been estimated from D/H isotopic ratio obtained on the base of the standard model of Big-Bang Nucleosynthesis, as we believed that all deuterium was formed just through this process

$$\Omega_B(\mathrm{BBN}) = \rho_b/\rho_{cr} = 4.1 \pm 0.3\%$$

This value of Ω_B (within errors) is consisted with value obtained from Cosmic Microwave Background fluctuations [16]

$$\Omega_B(\mathrm{CMB}) = \rho_b/\rho_{cr} = 4.6 \pm 0.3\%,$$

although the corresponding epochs are quite different.

Thus high resolution spectroscopy of high redshifted astrophysical objects may be a high-way to the fundamental cosmological parameters.

Acknowledgments
This work was supported by Ministry of Education and Science of Russian Federation (contract **No. 11.G34.31.0001** and Agreement **No. 8409**), by the State Program "Leading Scientific Schools of Russian Federation" (grant **NSh 4035.2012.2**), and by RFBR (grants **No. 11-02-01018-a** and **No. 12-02-31463**). SAB thanks Dynasty Foundation and Science and High-School Committee of Saint-Petersburg Administartion.

References

[1] Font-Ribera A, Miralda-Escudé J, Arnau E, Carithers B, Lee K G, Noterdaeme P, Pâris I, Petitjean P, Rich J, Rollinde E, Ross N P, Schneider D P, White M and York D G 2012 *ArXiv e-prints* (*Preprint* 1209.4596)
[2] Schulze S, Fynbo J P U, Milvang-Jensen B, Rossi A, Jakobsson P, Ledoux C, De Cia A, Krühler T, Mehner A, Björnsson G, Chen H W, Vreeswijk P M, Perley D A, Hjorth J, Levan A J, Tanvir N R, Ellison S, Møller P, Worseck G, Chapman R, Dall'Aglio A and Letawe G 2012 *AAP* **546** A20 (*Preprint* 1207.6088)
[3] Noterdaeme P, Petitjean P, Carithers W C, Pâris I, Font-Ribera A, Bailey S, Aubourg E, Bizyaev D, Ebelke G, Finley H, Ge J, Malanushenko E, Malanushenko V, Miralda-Escudé J, Myers A D, Oravetz D, Pan K, Pieri M M, Ross N P, Schneider D P, Simmons A and York D G 2012 *ArXiv e-prints* (*Preprint* 1210.1213)
[4] Pettini M, Zych B J, Murphy M T, Lewis A and Steidel C C 2008 *Mon. Not. of Royal Astron. Soc.* **391** 1499–1510 (*Preprint* 0805.0594)
[5] Varshalovich D A, Ivanchik A V, Petitjean P, Srianand R and Ledoux C 2001 *Astronomy Letters* **27** 683–685 (*Preprint* arXiv:astro-ph/0107310)
[6] Jorgenson R A, Wolfe A M, Prochaska J X and Carswell R F 2009 *Astrophys. Journal* **704** 247–254 (*Preprint* 0908.3485)
[7] Balashev S A, Ivanchik A V and Varshalovich D A 2010 *Astronomy Letters* **36** 761–772 (*Preprint* 1009.4186)
[8] Tumlinson J, Malec A L, Carswell R F, Murphy M T, Buning R, Milutinovic N, Ellison S L, Prochaska J X, Jorgenson R A, Ubachs W and Wolfe A M 2010 *Astrophys. Journal Letters* **718** L156–L160 (*Preprint* 1007.0030)
[9] Ivanchik A V, Petitjean P, Balashev S A, Srianand R, Varshalovich D A, Ledoux C and Noterdaeme P 2010 *Mon. Not. of Royal Astron. Soc.* **404** 1583–1590 (*Preprint* 1002.2107)
[10] Balashev S A, Petitjean P, Ivanchik A V, Ledoux C, Srianand R, Noterdaeme P and Varshalovich D A 2011 *Mon. Not. of Royal Astron. Soc.* **418** 357–369 (*Preprint* 1107.4808)
[11] Cui J, Bechtold J, Ge J and Meyer D M 2005 *Astrophys. Journal* **633** 649–663 (*Preprint* arXiv:astro-ph/0506766)
[12] Noterdaeme P, Petitjean P, Ledoux C, Srianand R and Ivanchik A 2008 *Astron. & Astrophys.* **491** 397–400 (*Preprint* 0809.3785)
[13] Noterdaeme P, Petitjean P, Ledoux C, López S, Srianand R and Vergani S D 2010 *Astron. & Astrophys.* **523** A80+ (*Preprint* 1008.0637)
[14] Malec A L, Buning R, Murphy M T, Milutinovic N, Ellison S L, Prochaska J X, Kaper L, Tumlinson J, Carswell R F and Ubachs W 2010 *Mon. Not. of Royal Astron. Soc.* **403** 1541–1555 (*Preprint* 1001.4078)
[15] Le Petit F, Roueff E and Le Bourlot J 2002 *Astron. & Astrophys.* **390** 369–381
[16] Komatsu E, Smith K M, Dunkley J, Bennett C L, Gold B, Hinshaw G, Jarosik N, Larson D, Nolta M R, Page L, Spergel D N, Halpern M, Hill R S, Kogut A, Limon M, Meyer S S, Odegard N, Tucker G S, Weiland J L, Wollack E and Wright E L 2011 *Astrophys. Journal Suppl. Ser.* **192** 18–+ (*Preprint* 1001.4538)

Journal of Physics: Conference Series **397** (2012) 012052 doi:10.1088/1742-6596/397/1/012052

Inelastic Collision Processes for Formation of Spectral Line Shapes in Stellar Atmospheres. Reprojection method.

Andrey K. Belyaev

Department of Theoretical Physics, Herzen University, St. Petersburg 191186, Russia

E-mail: belyaev@herzen.spb.ru

Abstract. Modeling of formation of spectral lines in stellar atmospheres requires accurate data on inelastic collision processes. The only source of this information is theoretical calculations mainly performed within the standard adiabatic Born-Oppenheimer approach. It is shown that the conventional application of this approach encounters severe difficulties, in particular, due to nonzero couplings in the asymptotic region. This results in the fact that the conventional applications do not provide convergences, which are necessary for calculations of inelastic cross sections, giving infinite cross sections. The proposed remedy is the quantum reprojection method within the standard adiabatic Born-Oppenheimer approach. The method takes nonvanishing asymptotic nonadiabatic couplings into account and distinguishes asymptotic currents in molecular-state channels and in atomic-state channels. The method is demonstrated for the example of low-energy inelastic Li + Na collisions for which the conventional application of the Born-Oppenheimer approach fails.

1. Introduction

Formation of spectral lines in stellar atmospheres is determined by many physical processes [1]. In addition to radiative processes directly providing spectral lines, different inelastic collision processes, mainly due to collisions of different atoms and ions with electrons as well as with hydrogen atoms and ions, play the important role, e.g., as pointed out in the talks at ICSLS 21 [2, 3]. The need for accurate data on low-energy inelastic atomic collisions involving hydrogen atoms for non-LTE formation calculations is well known, see, e.g., [1, 2, 3] and references therein. In principle, this kind of data could be obtained from experimental measurements and from theoretical calculations. However, in practice, experimental data on low-energy atomic collisions are not available and calculations remain the only source of information needed. That is why it is essential to be sure that theoretical approaches used for numerical calculations are reliable and provide accurate data on inelastic collision processes.

The majority of theoretical treatments of inelastic collisions involving atoms, ions, molecules, clusters, surfaces, and so on is performed within the standard adiabatic Born-Oppenheimer (BO) approach (or simply the BO approach), which is described, for example, in [4, 5]. The approach is based on the separation of electronic and nuclear motion. First the electronic fixed-nuclei Hamiltonian is treated and the electronic molecular states are determined, and then the nuclear dynamics is studied using an expansion of the total wavefunctions in terms of electronic molecular-state wavefunctions. The BO approach gives a clear physical picture of the scattering process and allows one to use the well-developed quantum-chemical methods and computer programs. In addition, the BO approach has as its limitation the Born-Oppenheimer approximation which is widely used in quantum chemistry and molecular physics.

Although the BO approach looks straightforward, it encounters severe difficulties. The point is that in order to calculate inelastic cross sections within the BO approach one has to reach three kinds of convergences: (i) a convergence with respect to a number of molecular states treated (which should be infinite in general, but finite in practice); (ii) a convergence with respect to an upper integration limit for a nuclear dynamical treatment (an internuclear distance R must go to infinity); (iii) a convergence with respect of a number of partial waves calculated (should be infinite). The first kind of convergence should be studied in each particular case individually, while two others are of general features. It has recently shown [6] that conventional applications of the BO approach do not lead to convergences of the second and the third kinds in many cases providing unphysical results, and hence a special care should be taken in order to obtain reliable cross sections for collision processes.

2. The standard adiabatic Born-Oppenheimer approach

The standard adiabatic BO approach is described in details in many papers, e.g., in Ref. [4, 5], see also Ref. [6, 7, 8, 9] and references therein. In order to use the BO approach, one needs to treat fixed-nuclei molecular states, so one of the coordinates should be the internuclear distance R. In order to reach the simplest and standard form of the coupled channel equations for nuclear

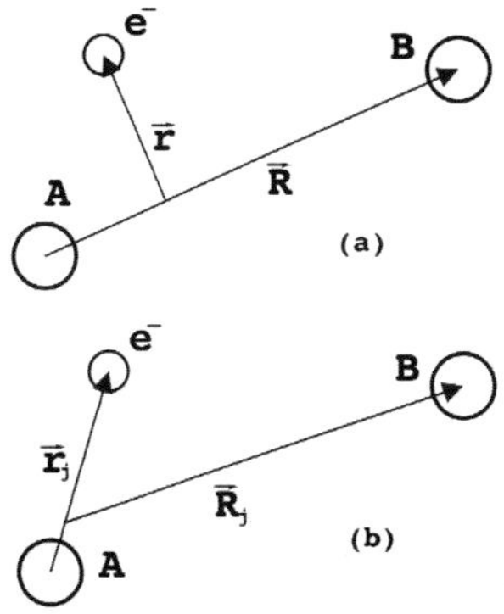

Figure 1. The two sets of the Jacobi coordinates for a three-body system (two nuclei A, B and an electron). The upper figure (a) shows the set used for a molecular representation, while the low figure (b) depicts the set used for a scattering representation in the asymptotic region.

dynamical treatment, the set of the Jacobi coordinates depicted in Fig. 1(a) should be used, that is, one of the Jacobi coordinates is the internuclear distance and the electronic coordinate should be measured from the Centre of Nuclear Mass (CNM). Using this set of the Jacobi coordinates, one calculates adiabatic potential energies $U_j(R)$ and nonadiabatic couplings, e.g., the radial couplings $\langle j|\frac{\partial}{\partial R}|k\rangle$, for selected molecular (fixed-nuclear) states. Solutions of the coupled channel equations result in nonadiabatic transition probabilities P_{jk}, and inelastic cross sections σ_{jk} are calculated via P_{jk} as follows

$$\sigma_{jk}(E) = \frac{\pi\hbar^2 p_j^{stat}}{2ME} \sum_{J=0}^{J_{max}} P_{jk}(J,E)(2J+1)\,, \tag{1}$$

M being the reduced nuclear mass, E being the collision energy, J being the total angular momentum quantum number, and p_j^{stat} being a statistic probability for population of the initial channel j. According to the conventional interpretation of the BO approach, nonadiabatic transition probabilities P_{jk} are determined as probabilities between molecular states, and the transitions take place in regions where nonadiabatic couplings are nonzero.

The first problem within the BO approach consists in the fact that some radial nonadiabatic couplings remain nonzero in the asymptotic region providing nonadiabatic transitions between molecular states up to an infinite internuclear distance. It has been shown analytically and confirmed numerically (see, e.g., [6, 7, 8, 9]) that the asymptotic radial couplings read

$$\langle j|\frac{\partial}{\partial R}|k\rangle_\infty = \gamma_k \frac{m}{\hbar^2}\left[U_j(\infty) - U_k(\infty)\right]\langle j|z^{at}|k\rangle\,, \tag{2}$$

where m is the mass of an electron, $\langle j|z^{at}|k\rangle$ is an atomic transition dipole moment, and γ_k is a scalar factor:

$$\gamma_k = \begin{cases} \gamma_A = -\frac{M_B}{M_A+M_B}\,, & \text{when } e^- \text{ is bound with } A \\ \gamma_B = +\frac{M_A}{M_A+M_B}\,, & \text{when } e^- \text{ is bound with } B \end{cases}\,. \tag{3}$$

Eq. (2) shows that asymptotic couplings between optically allowed transition states do not vanish at infinity. Choosing another origin for the electronic coordinate does not remove nonzero asymptotic coupling terms in the coupled channel equations [9]. Nonvanishing coupling terms in the coupled channel equations are the fundamental feature of the BO approach.

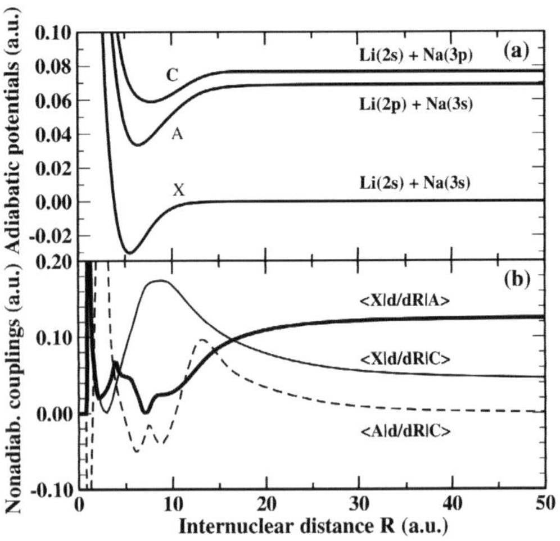

Figure 2. The adiabatic potentials (a) for the three lowest LiNa($^1\Sigma^+$) states (X, A, C) and the radial nonadiabatic couplings (b) between these states.

Fig. 2 shows an example of adiabatic potentials and nonadiabatic radial couplings for the three lowest $^1\Sigma^+$ states of LiNa. It is clearly seen that two asymptotic values of nonadiabatic couplings are nonzero in agreement with Eq. (2). Moreover, the nonvanishing asymptotic couplings have substantial values and cannot be neglected in the nuclear dynamics.

As the result of nonvanishing couplings, the nonadiabatic transitions between molecular states remain nonzero at an arbitrary large internuclear distance which is shown in Fig. 3 (left panels), where the nonadiabatic transition probabilities are plotted as a function of the upper integration limit R_0. It is clearly seen that the conventional application of the BO approach does not provide a convergence with respect to increasing of the upper integration limit.

Since the transitions between molecular states take place at an arbitrary large distance, nonadiabatic transition probabilities remain nonzero at any angular momentum quantum numbers J, see Fig. 4 (left panels). It results in the fact that the inelastic cross sections do not reach a convergence with respect to increasing of the upper summation limit J_{max} (which should go to infinity, in general), see Eq. (1). This is clearly seen in Fig. 5, where the inelastic cross sections are shown as a function of J_{max}. Note that the dashed lines plot the cross sections calculated by means of the conventional application of the BO approach.

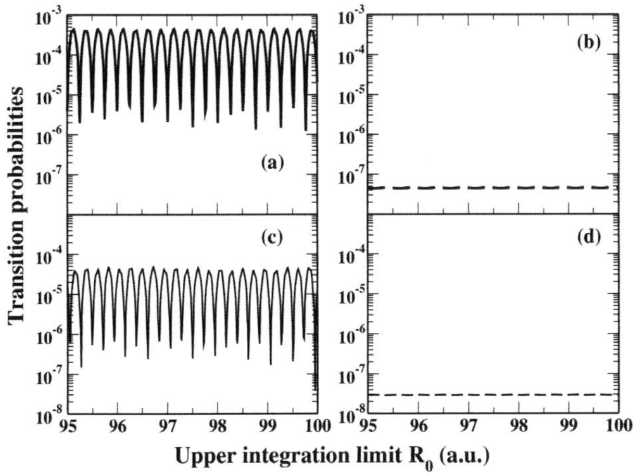

Figure 3. The transition probabilities for Li($2s \to 2p$) + Na($3s$) (upper panels (a) and (b)) and Li($2s$) + Na($3s \to 3p$) (lower panels (c) and (d)) excitation calculated for $E = 5$ eV and $J = 0$ by means of the conventional BO method (transitions between molecular states; solid lines) and by means of the reprojection method (transitions between atomic states; dashed lines) as a function of the upper integration limit R_0.

In fact, the nonvanishing asymptotic couplings are a consequence of a more fundamental shortcoming. The coordinates used to describe molecular states of the collision complex at small and intermediate distances are not suited for the description of the free atoms in the asymptotic region. Within the conventional applications of the BO approach, asymptotic wavefunctions read

$$^{BO}\Psi_j^{\pm}(\vec{r},\vec{R}) = Y_{JM_J}(\Theta,\Phi)\frac{\exp(\pm iK_jR)}{R}\phi_j(\vec{r},\vec{R}), \tag{4}$$

that is, the nuclear wavefunctions are written as functions of the internuclear distance R. The correct asymptotic incoming and outgoing wavefunctions [5, 6, 7, 8, 10]

$$\Psi_j^{\pm}(\vec{r}_j,\vec{R}_j) = Y_{JM_J}(\Theta,\Phi)\frac{\exp(\pm iK_jR_j)}{R_j}\phi_j(\vec{r}_j) \tag{5}$$

are written in another set of Jacobi coordinates [see Fig. 1(b)] and are different from the BO functions (4). The vector $\vec{R}_j$ connects the centers of mass of the atoms, in contrast to $\vec{R}$ which connects the nuclei, see Fig. 1. The conventional BO applications do not take these differences

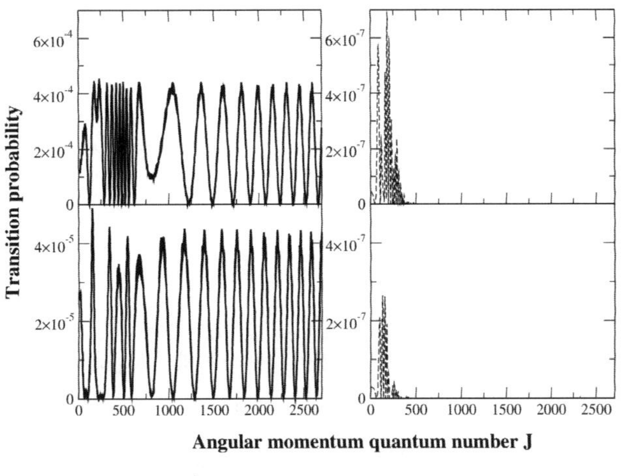

Figure 4. The transition probabilities for Li($2s \to 2p$) + Na($3s$) (upper panels) and Li($2s$) + Na($3s \to 3p$) (lower panels) excitation as a function of the total angular momentum quantum number J for $E = 5$ eV. The solid lines (left panels) show the probabilities obtained by means of the conventional BO method, while the dashed lines (right panels) depict the probabilities calculated by the reprojection method. Note different scales for different panels.

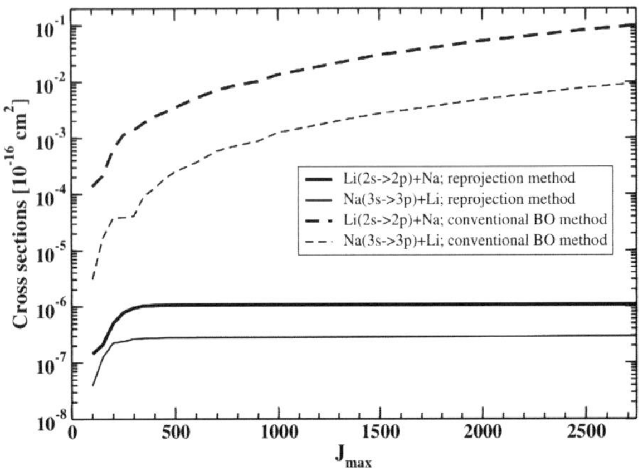

Figure 5. The Li($2s \to 2p$) + Na($3s$) [thick lines] and Li($2s$) + Na($3s \to 3p$) [thin lines] excitation cross sections calculated by means of the conventional BO method (dashed lines) and by means of the reprojection method (solid lines) as a function of the maximum value of the angular momentum quantum number J_{max} for $E = 5$ eV.

into account. The problem was first recognized in Ref. [10] and became known as the electron translation (ET) problem, although it should be called as the "molecular state (MS) problem."

3. The reprojection method

There are several remedies to deal with the MS problem. The first option is to go beyond the BO approach, for example, to use the hyperspherical coordinates or to solve the Faddeev equations. Within the BO approach, the proposed remedies are essentially based on (i) the inclusion of ET factors [10] or common translation factors [11] into the expansion of wavefunctions, or (ii) the use of state-specific reaction coordinates [12, 13], or (iii) the reprojection method [6, 7, 8, 14]. The first two remedies lead to modifications of basis functions, potentials, couplings, dynamical equations and consider all electrons explicitly, so they have been applied only to one-electron systems. The reprojection method is free from such limitations, in particular, Ref. [6] describes the multielectron reprojection method which does not require calculations of any additional quantum chemical data.

The reprojection method distinguishes molecular-state channels and atomic-state (scattering) channels, as well as probability currents in these channels. Within the reprojection method, the coupled channel equations are solved in the molecular-state representation from zero till relatively large internuclear distance R_0 keeping nonzero asymptotic couplings and using the correct asymptotic wavefunctions (5). Since the wavefunctions at the intermediate distances and in the asymptotic regions are written in different coordinates, they are reprojected into each other at R_0. In particular, the correct asymptotic wavefunctions in atomic-state (scattering) channels read

$$\Psi_j^{\pm}(\vec{r}, \vec{R}) = Y_{JM_J}(\Theta, \Phi) \sum_k \frac{\exp(\pm i K_j R)}{R} t_{kj}^{\pm} \phi_k(\vec{r}, \vec{R}) \,, \qquad (6)$$

where the reprojection coefficients are equal to

$$t_{kj}^{\pm} = \delta_{kj} \pm \frac{i K_j \hbar^2}{M\left(U_k(\infty) - U_j(\infty)\right)} \left\langle \phi_k \left| \frac{\partial}{\partial R} \right| \phi_j \right\rangle_{\infty} . \qquad (7)$$

These coefficients are used for accurate calculations of the scattering matrix from numerical solutions of the coupled channel equations, see Refs. [6, 7, 8, 14] for details and the formulae. It is seen from Eq. (7) that the nonzero asymptotic couplings are responsible for the construction of the correct asymptotic wavefunctions. It is also seen from Eq. (6) that an incoming or outgoing current in a *single atomic state* is distributed among *several molecular states* and vice versa.

Let us show how the reprojection method improves calculations of nonadiabatic transition probabilities and inelastic cross sections. The right panels of Fig. 3 show that the reprojection method leads to a convergence of the nonadiabatic transition probabilities with respect to an upper integration limit. Nonadiabatic transitions between molecular states still remain at an arbitrary large R, but they do not produce transitions between atomic states at large distances. The t-matrices correct the S-matrix and, finally, transition probabilities. Thus transitions between *atomic states* in the asymptotic region are unphysical, while transitions between *molecular states* in the same region are physical.

The right panels of Fig. 4 show that in contrast to the results of the conventional application of the BO approach, the transition probabilities obtained by means of the reprojection method are substantial only within a limited range of J, roughly up to $J \approx 500$ for $E = 5$ eV in Li + Na collisions. This leads to the convergence of the cross sections with increasing J_{max} (see Fig. 5), and finally, to the finite values of the inelastic cross sections.

It has thus been demonstrated that the conventional BO method applied to collision processes with nonzero asymptotic nonadiabatic couplings, which are fundamental features of the BO approach, leads to infinite inelastic cross sections even at low collision energies. The MS problem is severe for all collisions. The reprojection method takes into account nonzero asymptotic couplings and distinguishes asymptotic currents in molecular-state and atomic-state channels.

4. Concluding remarks

Let us compare a conventional application of the BO approach and the reprojection method. (i) Within a conventional application all asymptotic nonadiabatic couplings are wanted to be *zero* (though some are nonzero); the reprojection method keeps *nonzero* asymptotic nonadiabatic couplings (as they are obtained from quantum chemical calculations). (ii) Within a conventional application asymptotic wavefunctions have the BO form as a function of the *internuclear distance*; the reprojection method uses the correct asymptotic wavefunctions as a function of the *interatomic distance*. (iii) A conventional application assumes that currents in a *single atomic state* proceed into a *single molecular state* (and vice versa); within the reprojection method, currents in a *single atomic state* proceed into *multiple molecular states* (and vice versa). Finally, conventional applications do not provide proper convergences giving unreliable results, while the reprojection method yields necessary convergences leading to reliable results.

Acknowledgments

The author gratefully acknowledges the support of the Wenner-Gren Foundation (Sweden), the Max Planck Institute for Astrophysics (Germany), the CNRS and Paris Observatory (France).

References

[1] Asplund M 2005 *ARA&A* **43** 481
[2] Barklem P S 2012 *J. Phys.: Conf. Series* ICSLS 21 Proceedings (the present issue)
[3] Guitou M, Belyaev A K, Spielfiedel A, Feautrier N and Barklem P S 2012 *J. Phys.: Conf. Series* ICSLS 21 Proceedings (the present issue)
[4] Mott N F and Massey H S W 1949 *The Theory of Atomic Collisions* (Oxford: Clarendon Press)
[5] Maćias A and Riera A 1982 *Phys. Rep.* **90** 299
[6] Belyaev A K 2010 *Phys. Rev.* A **82** 060701(R)
[7] Grosser J, Menzel T and Belyaev A K 1999 *Phys. Rev.* A **59** 1309
[8] Belyaev A K, Egorova D, Grosser J and Menzel T 2001 *Phys. Rev.* A **64** 052701
[9] Belyaev A K, Dalgarno A and McCarroll R 2002 *J. Chem. Phys.* **116** 5395
[10] Bates D R and McCarroll R 1958 *Proc. R. Soc.* A **245** 175
[11] Errea L F, Mendez L and Riera A 1982 *J. Phys.* B **15** 101
[12] Mittleman M H 1969 *Phys. Rev.* **188** 221
[13] Thorson W R and Delos J B 1978 *Phys. Rev.* **18** 135
[14] Belyaev A K 2009 *Phys. Scr.* **80** 048113

Mg-H collision rates for non-LTE determination of stellar atmospheric parameters

M. Guitou[1], A. K. Belyaev[2], A. Spielfiedel[3], N. Feautrier[3], P. S. Barklem[4]

[1] Université Paris-Est, Laboratoire MSME UMR 8208 CNRS, 5 Bd Descartes, 77454, Marne-la-Vallée, France
[2] Department of Theoretical Physics, Herzen University, St. Petersburg 191186, Russia
[3] LERMA UMR 8112 CNRS, Observatoire de Paris, 92195 Meudon Cedex, France
[4] Department of Physics and Astronomy, Uppsala University, Box 515 S-75120 Uppsala, Sweden

E-mail: Marie.Guitou@univ-mlv.fr

Abstract. Non-LTE (Local Thermodynamical Equilibrium) modeling implies competition between radiative and collisional processes. The influence of inelastic hydrogen atoms collisions, dominant in cold atmospheres has been and remains to be a significant source of uncertainty for stellar abundance analysis. In the particular case of Mg atoms, a large number of electronic states of the MgH molecule as well as the associated couplings that mix the states during the collision were calculated by high level quantum chemical methods and then used in full quantum scattering calculations. This allows to treat the excitation processes between the seven lowest atomic states of magnesium in collision with hydrogen atoms, as well as the ion-pair production and the mutual neutralization processes. The detailed mechanisms involved during the collision process have been analysed in detail. Our calculations show that the usual approximate Drawin formula leads to errors by factors up to 10^5. As was already found in Li+H and Na+H collisions, excitation processes were found to be less important than charge transfer processes. However, unlike Li and Na, Mg has different spin terms, singlet and triplet, leading both to doublet molecular MgH electronic states. Collisional rates between spin-allowed and optically spin-forbidden atomic states are found to be of the same order of magnitude although optically spin-forbidden states are only collisionally coupled. Thus we may expect consequences on non-LTE calculations.

1. Context

Non-LTE modeling implies a competition between radiative and collisional processes. The radiative data are well known thanks to the Opacity and the Iron projects. The influence of inelastic hydrogen atom collisions dominant in cold atmospheres on non-LTE spectral line formation has been, and remains to be, a significant source of uncertainty for stellar abundance analyses, due to the difficulty in obtaining accurate data for such low-energy collisions, either experimentally or theoretically. For lack of a better alternative, the classical so-called Drawin formula is often used. Recent results on Li [1, 2] and Na [3, 4] in collisions with H atoms show that the Drawin formula may overestimate the cross sections by factors of one to five orders of magnitude. This implies new calculations of accurate cross sections and rates. We present

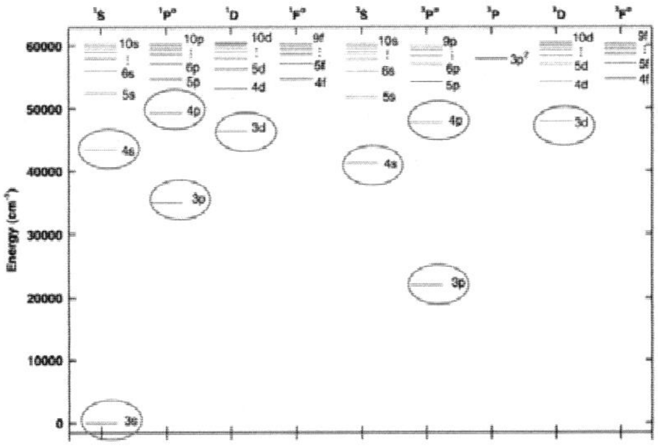

Figure 1. Diagram of Mg energy levels

in the following results for Mg-H collisions. Mg is of particular interest as giving some of the strongest absorption lines in stellar spectra which are easily detected.

2. MgH interaction potentials, related couplings and cross sections

In the standard adiabatic approach, the theoretical treatment of atomic collisions requires two steps: (i) calculations of fixed-nuclei potential energies and non adiabatic radial and rotational couplings, (ii) an appropriate treatment of the nuclear motion based on the previous calculated molecular data leading to the wave function for the nuclear motion. This leads in the Jacobi coordinates system to the usual close-coupled equations. But, most of the non-adiabatic couplings are nonzero when the internuclear distance goes to infinity. This corresponds to the fact that the Jacobi system is not appropriate for the description of the collisions partners long before and after the collision. To remove this difficulty, Belyaev *et al.* [5] proposed a way to connect the R-matrix calculated at some R_0 internuclear distance from close coupled equations in the Jacobi coordinates system to the asymptotic S-matrix allowing the calculation of cross sections.

The first step concerns quantum chemistry and the main challenge is to build large basis sets adapted to the study of highly excited states. All the electronic states arising from Mg+H for energies up to Mg $3s3d$ ^{1}D (see figure 1, levels identified in blue) and later to Mg $3s3d$ ^{3}D, Mg $3s4p$ ^{1}P and Mg $3s4p$ ^{3}P states (see figure 1, levels identified in red) were considered using large active spaces and basis sets [6, 9]. The energies and related couplings of 9 $^2\Sigma^+$, 5 $^2\Pi$, 2 $^2\Delta$, 2 $^4\Sigma^+$, 1 $^4\Pi$ were calculated using the version 2009.1 of the MOLPRO code [7]. The potential energy functions (PEFs) are represented on figure 2 as function of the internuclear distance. The more striking feature of these potentials is the presence of a series of avoided crossings among the $^2\Sigma^+$ states due to a strong mixing with the Mg^+-H^- ionic state. Those pseudo-crossings occur at larger and larger distances for the highest molecular states (see details in figure 3). This leads to possible ionisation of Mg atoms by collisions with H, ion pair production and to the reverse reaction, mutual neutralisation.

This perturbation due to the Mg^+-H^- ionic configuration leads to large non-adiabatic radial coupling terms among consecutive states (see figure 4). Strong radial couplings occur at short internuclear distances as well as rotational coupling terms [6]. It was shown in Ref. [8] that radial couplings between the lowest-lying MgH($^2\Sigma^+$) states dominate over rotational couplings involving states of other symmetries, in particular the $^2\Pi$ states. States of the same symmetry are coupled via non-adiabatic radial couplings. The ionic Mg^+-H^- configuration has $^2\Sigma^+$ symmetry

Journal of Physics: Conference Series **397** (2012) 012053 doi:10.1088/1742-6596/397/1/012053

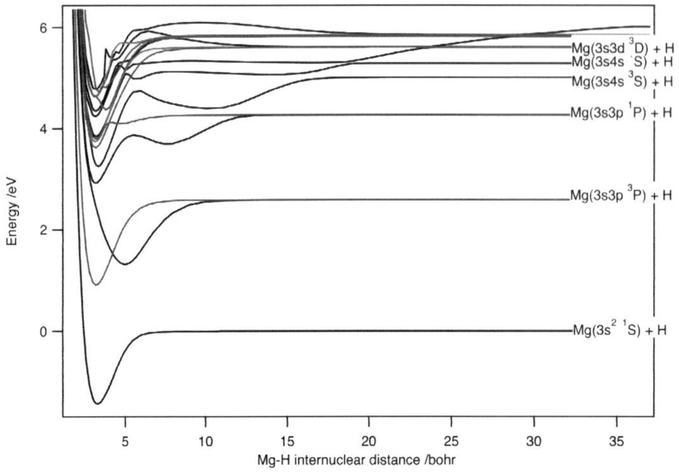

Figure 2. Potential energy functions of the lowest doublet electronic states of the MgH molecule. Black curves: $^2\Sigma^+$, blue curves: c and red curves: $^2\Delta$.

leading to large non-adiabatic radial couplings among consecutive states (see figure 4) which represent the most important mechanism for transitions among low-lying states. We present in the following calculations involving eight $^2\Sigma^+$ states up to and including the ionic channel.

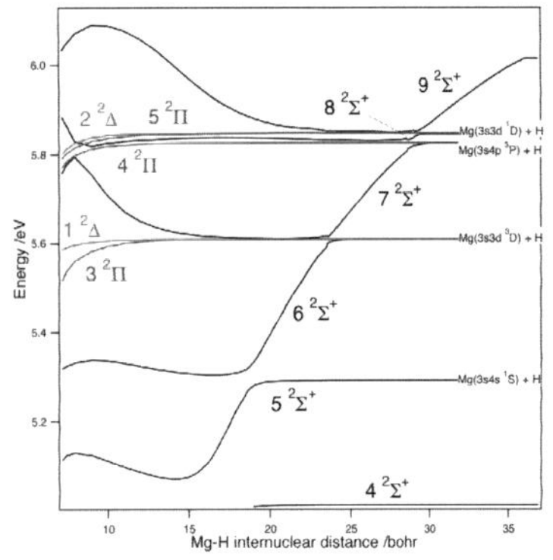

Figure 3. Zoom on the highest calculated potential energy functions

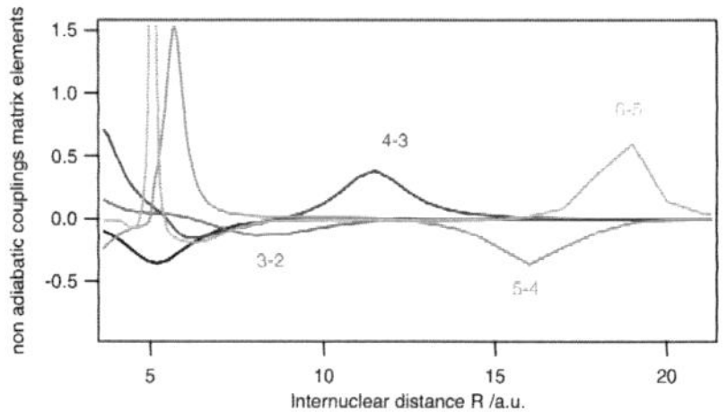

Figure 4. Non-adiabatic radial coupling matrix elements between consecutive (i,i+1) $^2\Sigma^+$ molecular states

The cross sections for the excitation and ion-pair production processes (see figure 5) show large variations in amplitude for different transitions, from as large as 120 to as small as 10^{-6} Å^2. We point out that cross sections for transitions between spin-allowed and spin-forbidden atomic states are of the same order of magnitude owing to relevant molecular mechanisms. The largest cross section is for ion-pair production from the Mg(3s4s ^1S) state. The physics for the inelastic processes in the treated collisions was carefully studied and several mechanisms were found [9]. It is shown that some of the mechanisms are determined by interactions between adjacent molecular states due to interactions between ionic and covalent configurations. However, mechanisms at

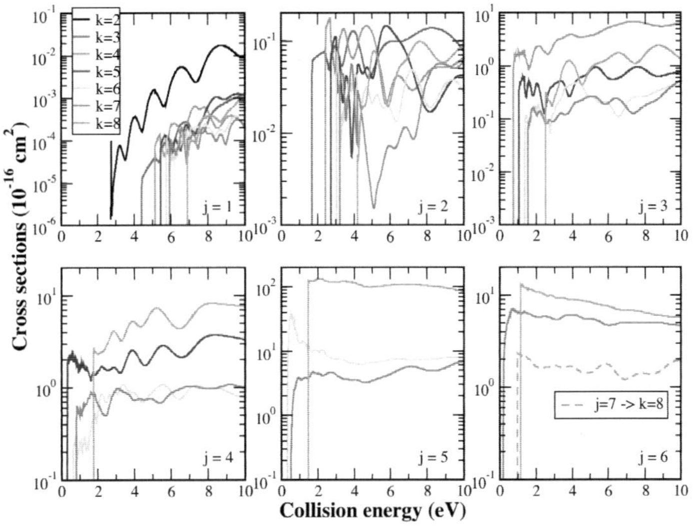

Figure 5. Inelastic cross sections $\sigma_{j,k}(E)$ for transitions $j \rightarrow k$ in low energy collisions[9] .

short distances, not restricted to adjacent states, are found to be important. These mechanisms explain the relatively large cross sections for excitation of highly excited states.

3. Mg+H rate coefficients

3.1. Full quantum results

From the cross sections, one can obtain the corresponding thermal rate coefficients at temperature T by an average over a Maxwellian velocity distribution. Rate coefficients at 4000 K for excitation and de-excitation processes :

$$\mathrm{Mg}(3\mathrm{s}\ nl\ ^{2S+1}L) + \mathrm{H}(1\mathrm{s}) \rightleftharpoons \mathrm{Mg}(3\mathrm{s}\ n'l'\ ^{2S'+1}L') + \mathrm{H}(1\mathrm{s})$$

and for the charge transfer processes, ion-pair production and mutual neutralisation :

$$\mathrm{Mg}(3\mathrm{s}\ nl\ ^{2S+1}L) + \mathrm{H}(1\mathrm{s}) \rightleftharpoons \mathrm{Mg}^{+}(3\mathrm{s}\ ^{2}S) + \mathrm{H}^{-}$$

are displayed in table 1. As expected, the rate coefficients follow the same trends as cross sections, i.e. large rate coefficients for ion-pair production/mutual neutralisation and large rates even for optically spin-forbidden transitions.

3.2. Comparison with Drawin approximate formula

The quantum rate coefficients for Mg+H are plotted (see figure 6) vs transition energy$\Delta\ E$ and data obtained from the Drawin formula are displayed in figure 7. The Drawin formula is an extension of the classical formula for ionisation of atoms by electron impact [11]. This formula has only two parameters,$\Delta\ E$ and the f-value of the atomic transition. The most remarkable aspect of the Drawin formula [4] results is that among the 21 excitation transitions considered, only 5 transitions are optically allowed and can be calculated with the Drawin formula. For the optically allowed transitions, the Drawin results are generally larger than the quantum results

Table 1. Mg+H rate coefficients at 4000 K.

Initial/Final states	3s ^{1}S	3p ^{3}P$_0$	3p ^{1}P$_0$	4s ^{3}S	4s ^{1}S	3d ^{1}D	ionic
3s ^{1}S	—	1.67 e-17	9.32 e-20	5.37 e-20	2.14 e-20	6.31 e-21	5.05 e-22
3p ^{3}P$_0$	4.87 e-15	—	2.76 e-13	7.95 e-14	2.07 e-14	4.35 e-15	1.47 e-16
3p ^{1}P$_0$	1.05 e-14	1.07 e-10	—	5.21 e-11	7.88 e-12	9.96 e-13	1.84 e-13
4s ^{3}S	5.26 e-14	2.67 e-10	4.52 e-10	—	1.38 e-10	1.18 e-11	9.14 e-12
4s ^{1}S	1.46 e-13	4.83 e-10	4.75 e-10	9.56 e-10	—	1.42 e-09	8.64 e-10
3d ^{1}D	2.23 e-14	5.28 e-11	3.12 e-11	4.28 e-11	7.41 e-10	—	1.73 e-10
ionic	2.42 e-13	2.42 e-10	7.84 e-10	4.48 e-09	6.10 e-08	2.35 e-09	—

by a few orders of magnitude [10]. The same trend was already found for Li [2] and Na [4] atoms in collision with H.

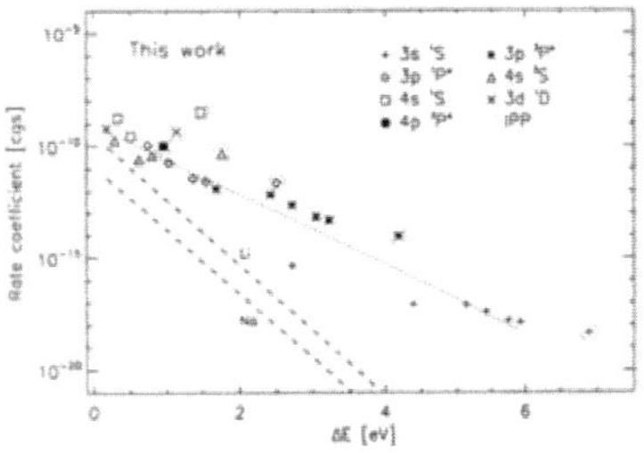

Figure 6. Rate coefficients at 6000 K in Mg+H collisions for excitation and ion-pair production processes plotted vs the transition energy ΔE: full quantum results.

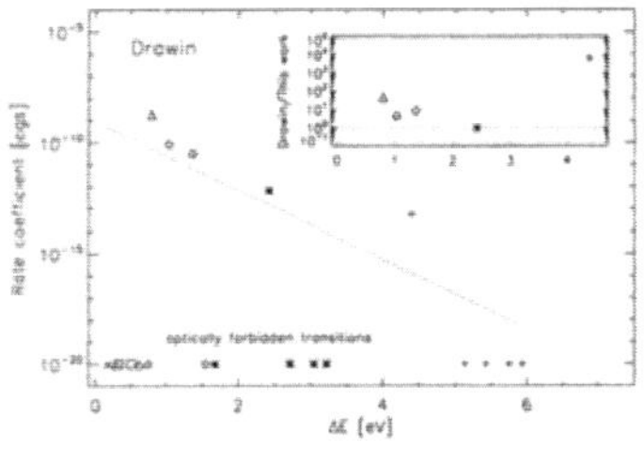

Figure 7. rate coefficients at 6000 K in Mg+H collisions for excitation and ion-pair production, processes plotted vs the transition energy ΔE: Results from Drawin formula.

4. Concluding remarks

As found previously in calculations for Li and Na, collisional excitation rate coefficients are smaller than rate coefficients for charge transfer. A comparison with the results found for Li and Na show that Mg-rate coefficients for excitation from the ground to the first excited states are roughly an order of magnitude larger [10]. Moreover, contrarily to Li and Na atoms, Mg has two spin symmetries and large collisional rates are found between singlet and triplet states which are only weakly radiatively coupled. This fact, together with the high rates lead one to expect that H-collisional processes could be important for non-LTE modeling. However, other factors, in particular the physical conditions of the stellar atmosphere (effective temperature, gravity, metallicity), contribute to the modeling. So, presently, any conclusion must await for detailed non-LTE calculations.

Acknowledgments

This research was supported by the Chemical Institute of the CNRS, the CNRS national program "Programme National de Physique Stellaire", Paris Observatory and the Royal

Swedish Academy of Sciences, Wenner Gren Foundation, Göran Gustafssons Stiftelse and the Swedish Research Council. The authors acknowledge the role of the SAM collaboration (http://www.anst.uu.se/ulhei450/GaiaSAM/) in stimulating this research through regular workshops.

References

[1] Belyaev A K, Barklem P S 2003 *Phys. Rev.* A **68** 062703
[2] Barklem P S, Belyaev A K and Asplund M 2003 *Astron. Astrophys.* **409** L1
[3] Belyaev A K, Barklem P S, Dickinson A S and Gadéa F X 2010 *Phys. Rev.* A **81** 032706
[4] Barklem P S, Belyaev A K, Dickinson A S and Gadéa F X 2010 *Astron. Astrophys.* **519** A20
[5] Belyaev A K, Egorova D, Grosser J and Menzel T 2001 *Phys. Rev. A* **64**, 052701
[6] Guitou M, Spielfiedel A and Feautrier N 2010 *Chem. Phys. Lett.***488** 145
[7] MOLPRO is a package of ab initio programs written by : H.-J. Werner and P. J. Knowles, with contributions from J. Almlöf, R. D. Amos, M. J. Deegan, S. T. Elbert,C. Hampel, W. Meyer, K. Peterson, R. Pitzer, A. J. Stone, P. R. Taylor, R. Lindh M. E. Mura, and T. Thorsteisson : http://www.molpro.net
[8] Guitou M, Belyaev A K, Barklem P S, Spielfiedel A and Feautrier N 2011*J. Phys.* B **448** 035202
[9] Belyaev A K, Barklem P S, Spielfiedel A, Guitou M, Feautrier N, Rodionov D S and Vlasov D V 2012 *Phys. Rev.* A **85** 032704
[10] Barklem P S, Belyaev A K, Spielfiedel A, Guitou M and Feautrier N 2012 *Astron. Astrophys.* **541** A80
[11] Drawin H W 1969 *Z. Phys.* **225** 483

The quasi-molecular absorption bands in UV region caused by the non-symmetric ion–atom radiative processes in the solar photosphere

A A Mihajlov[1], V A Srećković, LJ M Ignjatović[1], M S Dimitrijević[2] and A Metropoulos[3]

[1] University of Belgrade, Institute of Physics, P.O.Box 57, Pregrevica 118, Belgrade, Serbia
[2]Astronomical Observatory, Volgina 7, 11060 Belgrade, Serbia
[3]Theoretical and Physical Chemistry Institute, National Hellenic Research Foundation, Athens, Greece

E-mail: mihajlov@ipb.ac.rs

Abstract. The aim of this research is to show that the radiative processes in strongly non- symmetric ion- atom collisions significantly influence on the opacity of the solar photosphere in UV region. Within this work only the $He+H^+$ and $H+A^+$ ion-atom systems, where A is the atom of one of the metal (Mg, Si and Al), are taken in to account. It is caused by the fact that the needed characteristics of the corresponding molecular ions, i.e. molecular potential curves and dipole matrix elements, have been determined by now. Here the non-symmetric radiative processes are considered under the conditions characterizing the non-LTE standard model of the solar atmosphere (Vernazza J, Avrett E and Loser R 1981 *ApJS* **45** 635), which gives the possibility to perform all needed calculations and determined the corresponding spectral absorption coefficients. It is shown that the examined processes generate rather wide quasi-molecular absorption bands in the UV and VUV regions, whose intensity is comparable and sometimes even larger than the intensity of known one's caused by the $H+H^+$ radiative collision processes, which are included now in the solar atmosphere models. Consequently, the presented results suggest that the non-symmetric ion-atom absorption processes have to be also included in standard models of the solar atmosphere.

1. Introduction

The significant influence of some ion-atom radiative collision processes on the optical characteristics of the stellar atmospheres was already established. Here we keep in mind such symmetric radiative processes as photo absorption/emission and radiative charge exchange, which can be described by

$$\varepsilon_\lambda + A_2^+ \leftrightarrow A + A^+ \tag{1}$$

$$\varepsilon_\lambda + A^+ + A \leftrightarrow A + A^+ \tag{2}$$

where A = H(1s) or He(1s^2), A^+ and A_2^+ are the corresponding positive, single charged atomic and molecular ions in the ground electronic states, and ε_λ - the energy of the photon. It was shown that these processes can influence to the opacity of atmospheres of Sun (see e.g. [2] and [3]). However, the mentioned papers have leaved opened the questions of the significance of the non-symmetric radiative processes.

The main aim of this work is to draw attention to the possible significance of the non-symmetric radiative processes as factors which influence to the opacity of stellar atmosphere in UV and VUV region. Here we will consider the processes of absorption charge exchange and photo-dissociation of the type

$$\varepsilon_{\lambda} + AB^{+} \leftrightarrow A^{+} + B \quad (3)$$

$$\varepsilon_{\lambda} + A + B^{+} \leftrightarrow A^{+} + B \quad (4)$$

$$\varepsilon_{\lambda} + A + B^{+} \leftrightarrow (AB^{+})^{*} \quad (5)$$

where B is the ground state atom with the ionization potential I_B which is less than the ionization potential I_A of the atom A, AB^+ - the corresponding molecular ion in one of the electronic states which are asymptotically correlated with the state of the system $A + B^+$.

Let us note that the processes (3) and (4) represent the analogues of the processes (1) and (2), while the process (5) has not it's symmetric analog. In this work the non-symmetric radiative processes are considered under the conditions characterizing the models of the solar atmosphere presented in [1].

Namely, all data needed for the calculations of the spectral absorption coefficients are provided in tabular form only for the chosen models. In accordance with this model is possible that A = He($1s^2$) and B = H(1s), and A = H(1s) and B = Mg, Si and Al. Let us note that for the ions HeH^+ and $(HeH^+)^*$ the corresponding potential curves and the dipole matrix elements (as the functions of the internuclear distances) are taken from Green et al. [4] and [5], while for all other considered molecular ions the mentioned quantities are calculated within this work.
The corresponding spectral absorption coefficients, as the function of λ, the local temperature T and the relevant particle densities are determined in the region $40\text{nm} \leq \lambda \leq 230\text{nm}$.

2. Results and discussion

The non-symmetric processes (3) – (5) are schematically shown in figure 1, where: bf-, ff-, and fb denote the bound-free, free-free and free-bound radiative transitions; $\Delta I = I_A - I_B$; $E = E_{imp}$ and E'_{impa} are the impact energies of the corresponding ion-atom systems; U_{in} (R) and $U_{fin}(R)$ - are the adiabatic potential curves of the initial and final molecular electronic states ; v , v' , J, and J' - the quantum numbers of the corresponding bound (ro-vibration) and free states.

The contribution of the considered non-symmetric ion-atom absorption processes (3) - (5) to the opacity of the solar atmosphere is described here by the spectral absorption coefficient $\kappa_{ia;nsim}(\lambda; T)$. The behavior of $\kappa_{ia;nsim}(\lambda; T)$ for several values of λ is illustrated by figure 2, where h is the distance of considered layer from the referent one (h=0) in accordance with Vernazza et al. (1981) [1].

The behavior of the quantity $G = \kappa_{ia;nsim}(\lambda; T)/ \kappa_{ia;tot}(\lambda; T)$, where $\kappa_{ia;tot}(\lambda; T)$ characterize the total contribution of all ion-atom absorption processes, i.e. (1) - (5), is shown in figure 3. Than, in figure 4 is presented the behavior of the quantities $F_{sim} = \kappa_{ia;sim}(\lambda; T)/ \kappa_{ea}(\lambda; T)$ and $F_{tot} = \kappa_{ia;tot}(\lambda; T)/\kappa_{ea}(\lambda; T)$, dash and full lines respectively. Here $\kappa_{ia;sim}(\lambda; T)$ characterize the contribution of the symmetric ion-atom absorption processes (1) and (2), and $\kappa_{ea}(\lambda; T)$ - the contribution of the concurrent electron-atom processes (H^- continuum), which were treated until recently as the absolutely dominant.

Even the results presented in these figures shows that the neglecting of the contribution of the non-symmetric processes (3) - (5) to the opacity of the solar atmosphere, in respect to the contribution of symmetric processes (1) and (2) would caused significant errors.

From the presented material it follows that the considered non-symmetric ion-atom absorption processes can not be treated only as one of the channel among many equal channels of the influence on the opacity of the stellar atmospheres and should be included together with the processes studied in our previous papers [6-10].

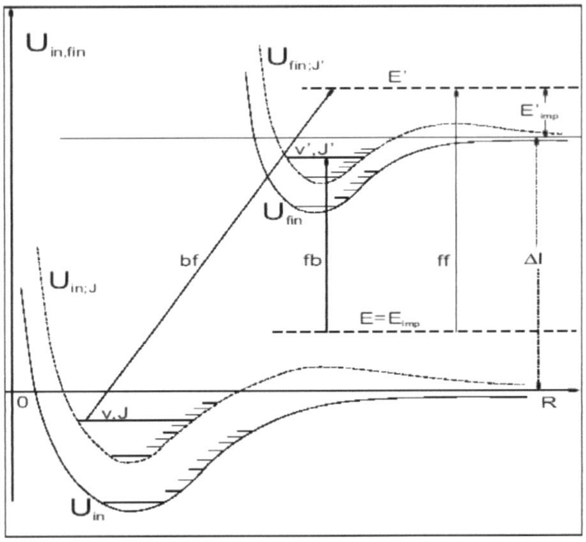

Figure 1. Schematic presentation of the non-symmetric processes (3-5) caused by the bound-free, free-free and free-bound radiative transitions.

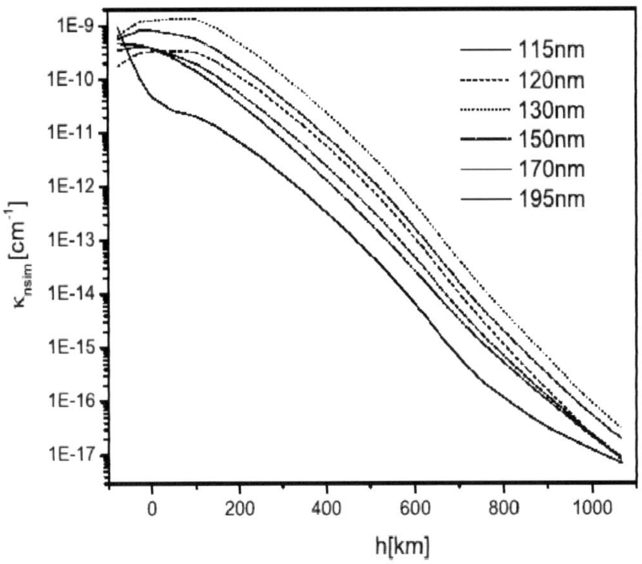

Figure 2. Quiet Sun. Spectral absorption coefficient κia;nsim(λ; T) for the spectral region 115nm < λ < 195nm.

From here it follows that the non-symmetric absorption processes (3)-(5) should be *ab initio* included in the same models.

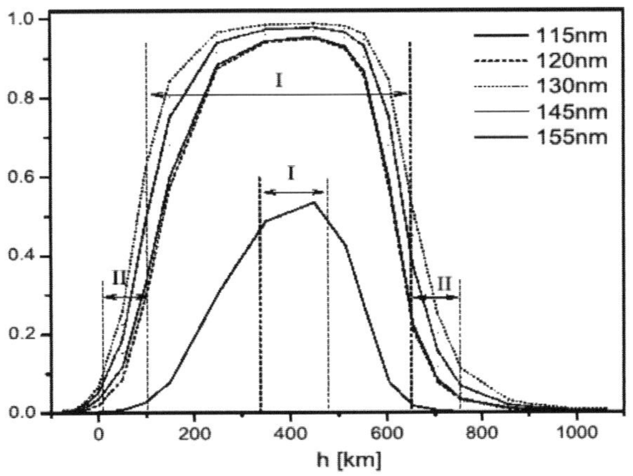

Figure 3. The behaviour of the quantity G = $\kappa_{ia;nsim}(\lambda;\, T)/\, \kappa_{ia;tot}(\lambda;\, T)$.

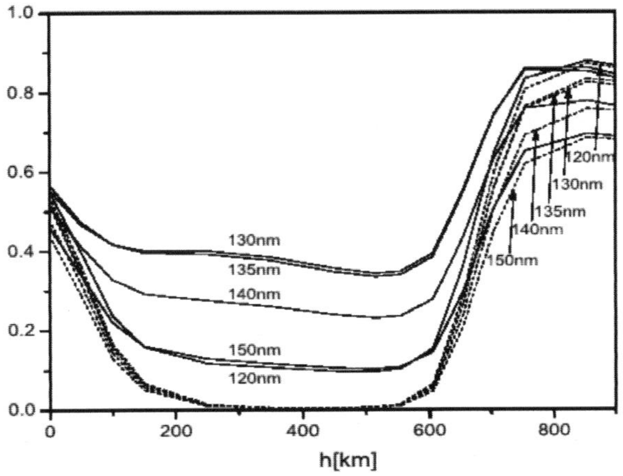

Figure 4. The behaviour of the quantities $F_{sim} = \kappa_{ia;sim}(\lambda;\, T)/\, \kappa_{ea}(\lambda;\, T)$ and $F_{tot} = \kappa_{ia;tot}(\lambda;\, T)/\kappa_{ea}(\lambda;T)$.

Acknowledgements
This work was supported by the Ministry of Education and Science of the Republic of Serbia as a part of the projects 176002, III4402.

References

[1] Vernazza J, Avrett E and Loser R 1981 *ApJS* **45** 635
[2] Mihajlov A, Dimitrijevic M and Ignjatovic Lj 1993 *A&A* **276** 187
[3] Mihajlov A, Ignjatović Lj , Sakan N and Dimitrijević M 2007 *A&A* **437** 1023
[4] Green T, Browne J, Michels H and Madsen M 1974 *J. Chem. Phys.* **61** 5198
[5] Green T, Michels H, Browne J and Madsen M 1974 *J. Chem. Phys.* **61** 5186
[6] Mihajlov A A, Sakan N M, Sreckovic V A and Vitel Y 2011 *J. Phys.*A **44** 9
[7] Mihajlov A A, Sreckovic V A, Ignjatovic Lj and Klyucharev A N 2012 *J. Cluster Sci.* **23** 47-75
[8] Mihajlov A A, Ignjatovic Lj, Sreckovic V A and Dimitrijevic M 2011 *Baltic Astronomy* **20** 566
[9] Mihajlov A A, Sakan N M, Sreckovic V A and Vitel Y 2011 *Baltic Astronomy* **20** 604
[10] Mihajlov A A, Ignjatovic Lj M, Sreckovic V A and Dimitrijevic M S 2011 *ApJS* **193**:2 2

Electron impact excitation for Ar VI

Haykel Elabidi[1,2]
[1] Deanship of the Foundation Year, Department of Physics, Umm Al-Qura University, P.O. Box 715, Makkah, Saudi Arabia
[2] Groupe de Recherche en Physique Atomique et Astrophysique, Faculté des Sciences de Bizerte, Université de Carthage, Tunisia

E-mail: haykel.elabidi@fsb.rnu.tn

Abstract. Radiative atomic and electron impact excitation data for Ar VI ion have been calculated using the AUTOSTRUCTURE code. The two-body non-fine structure operators of the Breit-Pauli Hamiltonian have been incorporated in AUTOSTRUCTURE, and for the collisional problem, we use the distorted wave approximation. We compare our radiative atomic results with the few available ones. To our best knowledge, there are no distorted wave collision strengths for Ar VI. Some discrepancies have been reported between our oscillator strengths and the only calculated ones using the MCHF method. We perform the same calculations with the UCL codes SST/DW/JAJOM and we compare with the present AUTOSTRUCTURE results.

1. Introduction

Atomic and collisional data are of great importance to accurate plasma modelling in astrophysics and in tokamaks. Many radiative processes, and many electron collision strengths are required for calculating level populations and spectral line intensities. These parameters can provide diagnostics of temperature and density of an emitting plasma, and of the abundance of elements in the plasma. Argon ions are important in astrophysics and in fusion [1, 2]. The Ar VI transition at $\lambda = 1303.87$ Å has been identified in [3] as an isolated line in the STIS spectrum of LSV+46°21, and it was noted in [3] that is the first time that Ar VI has been detected in the photosphere of any star.

The first results about the spectrum of Ar VI in the vacuum ultraviolet were published in [4]. A list of many theoretical and experimental works dedicated to the Ar VI before 1992 was reported in [5]. Recently, a new analysis of the Ar VI spectrum in the vacuum ultraviolet region was reported in [6], where adjusted and new energy levels and new classified lines were obtained. The only calculations that provide oscillator strengths or transitions probabilities for Ar VI are presented in [7] where the authors used the multiconfiguration Hartree-Fock (MCHF) and the multiconfiguration Dirac-Hartree-Fock (MCDHF) methods. Electron impact excitation calculations have been done for the argon isoelectronic sequence in [8] using the first-order many-body perturbation theory and in [2] using the Breit-Pauli R-matrix method. The existing results are presented as effective collision strengths. To our best knowledge, there no distorted wave fine structure collision strengths for Ar VI. This is the main motivation of the present work.

We present in this work radiative atomic data and electron impact collision strengths for the five times ionized argon (Ar VI). Comparisons of energy levels have been made with the values compiled by NIST [9]. Our oscillator strengths have been compared with the only existing theoretical results [7] where the MCHF method has been used.

2. Atomic structure and electron-ion scattering

The atomic structure is calculated using the AUTOSTRUCTURE (AS) code [10] by constructing target wavefunctions using radial wavefunctions calculated in a scaled Thomas-Fermi-Dirac-Amaldi statistical model potential using the Breit-Pauli intermediate coupling [11]. Besides the one-body and the two-body fine structure interactions, the two-body non-fine structure operators of the Breit-Pauli Hamiltonian, namely contact spin-spin, two-body Darwin and orbit-orbit are incorporated. It is shown in [10] how these interactions have been incorporated.

Recently, the Breit-Pauli Distorted Wave (BPDW) approach for electron impact excitation of atomic ions has been implemented in the AUTOSTRUCTURE code [12], which is used in the present paper for the scattering problem. We note that the distorted wave approximation (DW) is adequate for moderately and highly charged ions and the agreement between the DW and more sophisticated methods (close coupling for example) is good. For large l values, a "top up" for dipole transitions makes use of the sum rule of Burgess [13]. For higher multipoles, a geometric series in energy in combination with the degenerate energy limit [14] is used. We perform also the same calculations using the UCL codes SST/DW/JAJOM [15, 16, 17], where only the one-body and the two-body fine structure interactions are introduced in SST. We show the effect of the neglecting of these interactions on the structure and the collisional problems.

3. Results and discussions

16 configurations ($3s^23p - 3s^25s$, $3s3p^2$, $3s3p3d - 3s3p5s$, $3p^3$ and $3p^23d$), yielding 155 fine structure levels, have been used to study the atomic structure and the collisional problem of the Ar VI ion. Results of the atomic structure are presented in tables 1 and 2.

Table 1. Some Ar VI energy levels calculated with AS and SST compared with the MCHF [7] and the NIST [9] values.

i	Conf.	level	E_{AS}	E_{SST}	E_{MCHF}	E_{NIST}
1	$3s^23p$	$^2P^o_{1/2}$	0.0	0.0	0.0	0.0
2	$3s^23p$	$^2P^o_{3/2}$	2213.64	1978	2047.39	2207.88
3	$3s3p^2$	$^4P_{1/2}$	98521.41	103170	98746.57	100 157.5
4	$3s3p^2$	$^4P_{3/2}$	99320.29	103879	99486.49	100 957.6
5	$3s3p^2$	$^4P_{5/2}$	100563.05	105001	100619.02	102 191.6
6	$3s3p^2$	$^2D_{3/2}$	132593.05	138268	131753.79	132 462.7
7	$3s3p^2$	$^2D_{5/2}$	132698.44	138351	131850.54	132 574.7
8	$3s3p^2$	$^2S_{1/2}$	182217.67	185889*	170176.45	169 803.9
9	$3s3p^2$	$^2P_{1/2}$	186863.48	183493*	184439.40	182 182.1
10	$3s3p^2$	$^2PP_{3/2}$	188027.59	186563	185742.72	183 577.3
11	$3s^23d$	$^2D_{3/2}$	224874.85	241553	220303.14	218 595.9
12	$3s^23d$	$^2D_{5/2}$	224998.12	241708	220425.89	218 655.8
13	$3p^3$	$^2D^o_{3/2}$	259822.93	261109	260240.43	260 068.7
14	$3p^3$	$^2D^o_{5/2}$	260028.71	261268	260438.99	260 272.9
15	$3p^3$	$^4S^o_{3/2}$	276336.59	271030	271519.72	270 511.8
16	$3s3p3d$	$^4F^o_{3/2}$	290035.74	300117*	289823.12	–
17	$3s3p3d$	$^4F^o_{5/2}$	290479.99	300508*	290245.49	–
18	$3s3p3d$	$^4F^o_{7/2}$	291113.84	301060*	290847.72	–
19	$3s3p3d$	$^4F^o_{9/2}$	291950.14	301780*	–	–
20	$3p^3$	$^2P^o_{3/2}$	301977.11	299330*	294030.80	294086.0
21	$3p^3$	$^2P^o_{1/2}$	302144.00	299410*	294908.42	294101.3

Table 2. Oscillator strengths and transitions probabilities for some Ar VI allowed transitions calculated with AS and SST and compared with MCHF [7] and NIST [9].

$i-j$	f_{AS}	f_{SST}	f_{MCHF}	f_{NIST}	A_{AS}	A_{SST}	A_{MCHF}	A_{NIST}
1-3	0.00004	0.00004	0.00004		2.573E+05	2.381E+05	2.894E+5	
1-4	0.00000	–	0.000002		4.716E+03	–	6.038E+3	
1-6	0.05603	0.09715	0.07124	0.480	3.285E+08	6.019E+08	4.125E+08	2.8E+09
1-8	0.21743	0.18970	0.11720		4.815E+09	4.182E+09	2.265E+09	
1-9	0.30949	0.35450	0.40550		7.208E+09	8.330E+09	9.201E+09	
1-10	0.22506	0.22410	0.22400		2.654E+09	2.666E+09	2.578E+09	
1-11	0.96340	0.91281	0.93670		1.625E+10	1.758E+10	1.516E+10	
2-3	0.00001	0.00001	0.00002		1.523E+05	1.514E+05	2.039E+05	
2-4	0.00001	–	0.00001		9.188E+04	–	8.414E+04	
2-5	0.00005	0.00005	0.00006		2.337E+05	2.263E+05	2.409E+05	
2-6	0.00388	0.00734	0.00523	0.048	4.399E+07	8.822E+07	5.872E+07	5.4E+08
2-7	0.04663	0.08223	0.05965	0.440	3.530E+08	6.601E+08	4.469E+08	3.3E+09
2-8	0.04079	0.06001	0.07146		1.763E+09	2.587E+09	2.695E+09	
2-9	0.15854	0.14983	0.12320		7.211E+09	6.889E+09	5.468E+09	
2-10	0.54280	0.57812	0.54360		1.250E+10	1.297E+10	1.224E+10	
2-11	0.10163	0.09713	0.09849		3.361E+09	3.563E+09	3.130E+09	
2-12	0.86857	0.84929	0.84500		1.917E+10	2.078E+10	1.792E+10	
3-13	0.00001	0.00002	–		1.291E+05	2.256E+05	–	
3-15	0.22926	0.25299	0.22390		2.418E+09	2.694E+09	2.229E+09	
3-16	0.00007	0.00007	0.00007		8.585E+05	1.029E+06	8.222E+05	

Energies of the first 21 excited levels are presented in table 1 and compared with the present SST results, with the MCHF [7] and with the NIST [9] ones. Our energies from (AS) and the MCHF ones are closer to the NIST values (1%) than those calculated by SST (4%). An inversion is found between the $3s3p3d\ ^4\mathrm{F^o}$ levels (which are not provided by NIST) and the $3p^3\ ^2\mathrm{P^o}$ ones for SST and AS results (levels with asterisks in table 1). New and adjusted energy levels are reported in a revision study of Ar VI [6], but they are not yet taken into account by NIST. These new values are in agreement with the present results. The only calculated oscillator strengths and transition probabilities are those presented in [7] using the MCHF method. Within the 104 lines provided by NIST, there are only 3 lines with oscillator strength and transition probability values that are in disagreement with all the other results (AS, SST and MCHF). Sometimes there is a factor of 10 between them. Table 2 shows that the averaged difference between SST/AS and MCHF oscillator strengths is about 16 %, but the major disagreement is found to be for transitions involving the four levels: 6, 7, 8 and 9 (see table 1 for explanation of levels) where the difference can reach 40 %. The line at 1303.87 Å has been defined in [18] (and precedent papers) and has been used in [3] as $3s3p^2\ ^2\mathrm{P}_{3/2} - 3p^3\ ^2\mathrm{D}^o_{5/2}$ but in the revision study [6], it was defined as $3s3p(^3\mathrm{P})3d\ ^2\mathrm{P}^o_{3/2} - 3p^2(^3\mathrm{P})3d\ ^2\mathrm{P}_{3/2}$. In our calculations with AS, the $3s3p^2\ ^2\mathrm{P}_{3/2} - 3p^3\ ^2\mathrm{D}^o_{5/2}$ line is at 1388.87 Å and the $3s3p(^3\mathrm{P})3d\ ^2\mathrm{P}^o_{3/2} - 3p^2(^3\mathrm{P})3d\ ^2\mathrm{P}_{3/2}$ one is at 1351.71 Å which is closer to the experimental wavelength 1303.87 Å. More calculations of oscillator strengths and radiative decay rates are required to decide about some of the above disagreements.

We present in table 3 fine structure collision strengths for transitions from the ground and the first excited level to the first 10 lowest levels. To our best knowledge, there are no distorted wave fine structure collision strengths in the literature to compare with. Some discrepancies have been found between the AS and the SST results: at high energy (160 Ry) almost all the

Table 3. Fine structure collision strengths for the Ar VI $(1-j)$ and $(2-j)$ transitions (j from 2 to 10) calculated with AS and UCL-DW. Levels j are defined in Table 1.

Transition		Energy					
		10 Ry		40 Ry		160 Ry	
i	j	AS	UCL	AS	UCL	AS	UCL
1	2	6.044E−01	6.123E−01	5.799E−01	5.448E−01	5.723E−01	3.830E−01
1	3	1.584E−02	1.885E−02	3.652E−03	3.481E−03	2.433E−03	1.638E−03
1	4	2.103E−02	2.561E−02	2.827E−03	2.940E−03	1.511E−04	2.790E−04
1	5	1.301E−02	1.564E−02	1.734E−03	1.770E−03	6.945E−05	1.470E−04
1	6	1.151E+00	2.042E+00	1.581E+00	2.734E+00	2.291E+00	2.526E+00
1	7	6.492E−02	6.816E−02	4.261E−02	3.284E−02	4.478E−02	2.758E−02
1	8	2.662E+00	2.199E+00	3.625E+00	3.184E+00	5.640E+00	3.252E+00
1	9	3.676E+00	5.284E+00	5.000E+00	7.649E+00	7.804E+00	7.807E+00
1	10	2.640E+00	3.025E+00	3.582E+00	4.361E+00	5.600E+00	4.476E+00
2	3	9.647E−03	1.139E−02	2.264E−03	2.103E−03	1.542E−03	9.940E−04
2	4	2.689E−02	3.251E−02	4.726E−03	4.786E−03	1.884E−03	1.473E−03
2	5	6.072E−02	7.306E−02	1.233E−02	1.234E−02	6.766E−03	4.827E−03
2	6	2.451E−01	3.989E−01	2.772E−01	4.567E−01	3.753E−01	4.048E−01
2	7	2.013E+00	3.548E+00	2.725E+00	4.670E+00	3.922E+00	4.309E+00
2	8	1.010E+00	2.152E−01	1.372E+00	3.037E−01	2.134E+00	8.164E+00
2	9	3.815E+00	5.544E+00	5.191E+00	8.009E+00	8.092E+00	3.088E−01
2	10	1.294E+01	1.534E+01	1.763E+01	2.214E+01	2.751E+01	2.261E+01

UCL-DW results are lower than the AS ones. It seems to be a consequence of the neglect in SST of the two-body non-fine structure interactions. Nevertheless, collision strengths calculated with other approaches should be very important to conclude about this last interpretation.

Acknowledgments

This work has been supported by the Tunisian research unit 05/UR/12-04 and "the Faculté des Sciences de Bizerte".

References

[1] Burgos J M Munoz, Ballance C P, Loch S D and Boivin R F 2009 *Astron. Astrophys.* **500** 1253
[2] Ludlow J A, Ballance C P, Loch S D and Pindzola F S 2010 *J. Phys. B: At. Mol. Opt. Phys.* **43** 074029
[3] Rauch T *et al.* 2007 *Astron. Astrophys.* **470** 317
[4] Phillips L W and Parker W L 1941 *Phys. Rev.* **60** 301
[5] Raineri M *et al.* 1992 *Phys. Scr.* **45** 584
[6] Raineri M, Gallardo M, Borges F O, Trigueiros A G and Reyna Almandos J 2009 *Phys. Scr.* **79** 025302
[7] Froese Fischer C, Tachiev G and Irimia A 2006 *J. Quant. Spectrosc. Radiat. Transfer* **92** 607
[8] Colgan J, Zhang H L and Fontes C J 2008 *Phys. Rev.* A **77** 062704
[9] Ralchenko Yu V *et al.* and NIST ASD Team (2012), http://physics.nist.gov
[10] Badnell N R 1997 *J. Phys. B: At. Mol. Opt. Phys.* **30** 1
[11] Bethe H A and Slapeter E E 1977 *Quantum Mechanics of One-and Two-Electron Atoms* (New York: Plenum)
[12] Badnell N R 2011 *Comput. Phys. Comm.* **132** 1528
[13] Burgess A 1974 *J. Phys. B: At. Mol. Phys.* **7** L364
[14] Burgess A, Hummer D G and Tully J A 1970 *Phil. Trans. R. Soc. Lond.* A **266** 225
[15] Eissner W, Jones M and Nussbaumer H 1974 *Comput. Phys. Comm.* **8** 270
[16] Eissner W 1998 *Comput. Phys. Comm.* **114** 295
[17] Saraph H 1978 *Comput. Phys. Comm.* **15** 247
[18] Saloman E B 2010 *J. Phys. Chem. Ref. Data* **39** 033101

Electron-Impact Broadening of C II Spectral Lines

Neila Larbi-Terzi[1], **Nébil Ben Nessib** [1,2], **Sylvie Sahal-Bréchot** [3], **Milan S. Dimitrijević** [3,4]

[1]Groupe de Recherche en Physique Atomique et Astrophysique, Institut National des Sciences Appliquées et de Technologie, University of Carthage, Centre Urbain Nord B. P. No. 676, 1080 Tunis Cedex, Tunisia

[2]Department of Physics and Astronomy, College of Science, King Saud University. PO Box 2455, Riyadh 11451, Saudi Arabia.

[3]Laboratoire d'Étude du Rayonnement et de la Matière en Astrophysique, Observatoire de Paris, UMR CNRS 8112, UPMC, Bâtiment Evry Schatzman, 5 Place Jules Janssen, F-92195 Meudon Cedex, France

[4]Astronomical Observatory, Volgina 7, 11060 Belgrade 38, Serbia

E-mail: `nebilbennessib@planet.tn`, `sylvie.sahal-brechot@obspm.fr`, `mdimitrijevic@aob.bg.ac.rs`

Abstract. Using semiclassical perturbation approach in the impact approximation, we have obtained Stark broadening parameters for 149 C II multiplets. Energy levels and oscillator strengths are taken from the TOPbase database. Results are obtained as a function of temperature, for a perturber densities of 10^{14}, 10^{17} and 10^{18} cm^{-3}. In addition to electron-impact full half widths and shifts, Stark broadening parameters due to singly ionized carbon-impacts have been calculated, in order to provide Stark broadening data for the important charged perturbers in the atmospheres of carbon white dwarfs. The complete results will be published elsewhere, as well as the comparison to the existing experimental and other theoretical data. As an example of obtained results, we present data for a part of multiplets with wavelengths in the visible part of the spectrum.

1. Introduction

Recently, Dufour et al. [1], [2] discovered a new type of white dwarfs with the surface composition mostly composed of carbon. In order to consider their origin and evolution, it is necessary to develop a new generation of accurate models. Broadening by electron impact and C II ion interactions (Stark broadening) is the dominant line broadening mechanism at the temperatures and pressures of interest (effective temperatures within 19,000-23,000 K, electron density within $10^{15}cm^{-3}$-$10^{18}cm^{-3}$), so that the inclusion of accurate Stark broadening parameters is essential for this type of white dwarf atmosphere modelling.

In order to provide data needed for analysis and investigations of such stars, we have undertaken to calculate a great number of widths and shifts of C II lines colliding with electrons and C II ions for plasma conditions of interest for these carbon white dwarfs. For this work, we have used the impact semi-classical perturbation (SCP) method [3], [4], [5] for electron and ion collisions. We have used the needed atomic data from the TOPbase, (R-matrix with innovative asymptotic techniques), [6] http://cdsweb.u-strasbg.fr/TOPbase/. It is suitable for CII and offers a great number of levels and line strengths for the transitions.

Table 1. Calculated C II full widths at half intensity maximum and shifts for wavelengths 3500 Å- 7500 Åcovering the visible part of the spectrum for the electron density $N_e = 10^{14}$ cm^{-3}. In the first column, we give multiplets, wavelengths of the multiplets in Å and C values which give an estimate for the maximum perturber density for which the line may be treated as isolated if it is divided by the corresponding full width. Non-empty cells preceded by an asterisk mean that the impact approximation reaches its limit of validity ($0.1 < NV \leq 0.5$). Empty cells which are denoted by an asterisk mean that the impact approximation is not valid. The energy levels (terms) and the wavelengths of the multiplets are those of TOPbase.

Multiplet	T [K]	W_e [Å]	d_e [Å]	W_{CII} [Å]	d_{CII} [Å]
2s^2 4s ^{2}S - 2s^2 3p ^{2}P^o	5000.	0.106E-02	0.571E-03	0.417E-04	0.454E-04
	10000.	0.758E-03	0.464E-03	0.587E-04	0.556E-04
3935.7 Å	20000.	0.678E-03	0.378E-03	0.730E-04	0.657E-04
C=0.82E+17	30000.	0.657E-03	0.327E-03	0.832E-04	0.739E-04
	50000.	0.643E-03	0.278E-03	0.948E-04	0.799E-04
	80000.	0.643E-03	0.232E-03	0.101E-03	0.835E-04
2s^2 6s ^{2}S - 2s^2 4p ^{2}P^o	5000.	0.838E-02	0.461E-02	0.614E-03	0.593E-03
	10000.	0.702E-02	0.431E-02	0.776E-03	0.718E-03
5362.0 Å	20000.	0.664E-02	0.358E-02	0.923E-03	0.803E-03
C=0.51E+17	30000.	0.664E-02	0.330E-02	0.932E-03	0.880E-03
	50000.	0.703E-02	0.280E-02	0.111E-02	0.934E-03
	80000.	0.725E-02	0.238E-02	0.118E-02	0.103E-02
2s^2 7s ^{2}S - 2s^2 4p ^{2}P^o	5000.	0.108E-01	0.644E-02	0.103E-02	0.991E-03
	10000.	0.940E-02	0.645E-02	0.125E-02	0.112E-02
4326.7 Å	20000.	0.875E-02	0.543E-02	0.152E-02	0.132E-02
C=0.19E+17	30000.	0.900E-02	0.477E-02	0.153E-02	0.135E-02
	50000.	0.933E-02	0.430E-02	0.164E-02	0.151E-02
	80000.	0.942E-02	0.358E-02	0.187E-02	0.149E-02
2s^2 7s ^{2}S - 2p^3 ^{2}P^o	5000.	0.236E-01	0.141E-01	0.222E-02	0.215E-02
	10000.	0.205E-01	0.143E-01	0.268E-02	0.246E-02
6368.7 Å	20000.	0.187E-01	0.120E-01	0.332E-02	0.288E-02
C=0.41E+17	30000.	0.191E-01	0.106E-01	0.333E-02	0.294E-02
	50000.	0.196E-01	0.965E-02	0.359E-02	0.327E-02
	80000.	0.196E-01	0.807E-02	0.401E-02	0.325E-02
2s^2 8s ^{2}S - 2s^2 4p ^{2}P^o	5000.	0.149E-01	0.945E-02	0.163E-02	0.160E-02
	10000.	0.135E-01	0.961E-02	0.218E-02	0.177E-02
3864.2 Å	20000.	0.128E-01	0.797E-02	0.224E-02	0.197E-02
C=0.98E+16	30000.	0.133E-01	0.709E-02	0.227E-02	0.211E-02
	50000.	0.138E-01	0.628E-02	0.284E-02	0.220E-02
	80000.	0.139E-01	0.519E-02	0.297E-02	0.267E-02
2s^2 8s ^{2}S - 2p^3 ^{2}P^o	5000.	0.293E-01	0.187E-01	0.321E-02	0.315E-02
	10000.	0.266E-01	0.190E-01	0.429E-02	0.347E-02
5414.9 Å	20000.	0.249E-01	0.159E-01	0.439E-02	0.389E-02
C=0.19E+17	30000.	0.258E-01	0.141E-01	0.448E-02	0.416E-02
	50000.	0.267E-01	0.126E-01	0.552E-02	0.435E-02
	80000.	0.267E-01	0.104E-01	0.587E-02	0.529E-02

Table 1. Continued

Multiplet	T [K]	W_e [Å]	d_e [Å]	W_{CII} [Å]	d_{CII} [Å]
2s2p(^{3}P^o) 3p ^{2}P - 2s^2 4p ^{2}P^o	5000.	0.203E-02	-0.676E-03	0.204E-03	-0.474E-04
	10000.	0.161E-02	-0.520E-03	0.232E-03	-0.601E-04
5089.1 Å	20000.	0.142E-02	-0.420E-03	0.255E-03	-0.729E-04
C=0.20E+17	30000.	0.136E-02	-0.373E-03	0.266E-03	-0.808E-04
	50000.	0.132E-02	-0.315E-03	0.270E-03	-0.913E-04
	80000.	0.130E-02	-0.272E-03	0.281E-03	-0.972E-04
2s^2 3p ^{2}P^o - 2s^2 3s ^{2}S	5000.	0.139E-02	-0.281E-04	0.605E-04	-0.648E-05
	10000.	0.104E-02	-0.334E-04	0.814E-04	-0.105E-04
6741.6 Å	20000.	0.837E-03	-0.287E-04	0.924E-04	-0.146E-04
C=0.63E+18	30000.	0.784E-03	-0.310E-04	0.989E-04	-0.167E-04
	50000.	0.750E-03	-0.348E-04	0.106E-03	-0.191E-04
	80000.	0.730E-03	-0.308E-04	0.109E-03	-0.214E-04
2s^2 3p ^{2}P^o - 2s 2p^2 ^{2}P	5000.	0.589E-03	0.172E-03	0.357E-04	0.744E-05
	10000.	0.459E-03	0.111E-03	0.482E-04	0.110E-04
5165.8 Å	20000.	0.377E-03	0.851E-04	0.549E-04	0.150E-04
C=0.37E+18	30000.	0.350E-03	0.768E-04	0.589E-04	0.167E-04
	50000.	0.329E-03	0.675E-04	0.628E-04	0.191E-04
	80000.	0.315E-03	0.586E-04	0.649E-04	0.214E-04
2s^2 4p ^{2}P^o - 2s^2 3d ^{2}D	5000.	0.251E-02	0.717E-03	0.276E-03	0.555E-04
	10000.	0.199E-02	0.523E-03	0.313E-03	0.718E-04
5921.7 Å	20000.	0.177E-02	0.397E-03	0.344E-03	0.865E-04
C=0.19E+18	30000.	0.174E-02	0.345E-03	0.355E-03	0.956E-04
	50000.	0.174E-02	0.290E-03	0.368E-03	0.107E-03
	80000.	0.174E-02	0.239E-03	0.369E-03	0.118E-03
2p^3 ^{2}P^o - 2s^2 3d ^{2}D	5000.	0.816E-03	-0.281E-04	0.237E-04	-0.124E-04
	10000.	0.644E-03	-0.297E-04	0.328E-04	-0.172E-04
4115.6 Å	20000.	0.541E-03	-0.247E-04	0.384E-04	-0.210E-04
C=0.31E+17	30000.	0.503E-03	-0.177E-04	0.414E-04	-0.232E-04
	50000.	0.471E-03	-0.168E-04	0.457E-04	-0.263E-04
	80000.	0.447E-03	-0.143E-04	0.498E-04	-0.296E-04
2s^2 5p ^{2}P^o - 2s^2 4s ^{2}S	5000.	0.384E-02	0.234E-03	0.661E-03	0.892E-04
	10000.	0.356E-02	0.195E-03	0.731E-03	0.109E-03
5390.0 Å	20000.	0.366E-02	0.132E-03	0.758E-03	0.130E-03
C=0.72E+17	30000.	0.381E-02	0.116E-03	0.784E-03	0.144E-03
	50000.	0.402E-02	0.104E-03	0.788E-03	0.156E-03
	80000.	0.413E-02	0.107E-03	0.794E-03	0.167E-03
2s2p(^{3}P^o) 3s ^{2}P^o- 2s^2 4s ^{2}S	5000.	0.450E-02	-0.135E-02	0.510E-03	-0.519E-03
	10000.	0.383E-02	-0.109E-02	0.643E-03	-0.615E-03
4669.4 Å	20000.	0.343E-02	-0.907E-03	0.759E-03	-0.708E-03
C=0.44E+16	30000.	0.323E-02	-0.788E-03	0.882E-03	-0.778E-03
	50000.	0.302E-02	-0.655E-03	0.864E-03	-0.830E-03
	80000.	0.283E-02	-0.551E-03	0.980E-03	-0.864E-03

Table 1. Continued

Multiplet	T [K]	W_e [Å]	d_e [Å]	W_{CII} [Å]	d_{CII} [Å]
$2s^2$ 6p $^2P^o$ - $2s^2$ 4s ^2S	5000.	0.514E-02	0.253E-02	0.944E-03	0.389E-03
	10000.	0.490E-02	0.215E-02	0.100E-02	0.462E-03
3887.0 Å	20000.	0.497E-02	0.167E-02	0.108E-02	0.535E-03
C=0.11E+17	30000.	0.523E-02	0.144E-02	0.112E-02	0.581E-03
	50000.	0.551E-02	0.120E-02	0.121E-02	0.656E-03
	80000.	0.566E-02	0.990E-03	0.110E-02	0.681E-03
$2s^2$ 6p $^2P^o$ - $2s^2$ 4d ^2D	5000.	0.173E-01	0.740E-02	0.286E-02	0.107E-02
	10000.	0.168E-01	0.639E-02	0.308E-02	0.126E-02
6739.8 Å	20000.	0.170E-01	0.515E-02	0.323E-02	0.145E-02
C=0.35E+17	30000.	0.177E-01	0.447E-02	0.340E-02	0.161E-02
	50000.	0.184E-01	0.376E-02	0.341E-02	0.169E-02
	80000.	0.186E-01	0.312E-02	0.322E-02	0.181E-02
$2s^2$ 7p $^2P^o$ - $2s^2$ 5s ^2S	5000.	0.356E-01	0.168E-01	0.688E-02	0.307E-02
	10000.	0.358E-01	0.140E-01	0.721E-02	0.352E-02
7473.6 Å	20000.	0.386E-01	0.107E-01	0.779E-02	0.396E-02
C=0.35E+17	30000.	0.410E-01	0.928E-02	0.763E-02	0.427E-02
	50000.	0.436E-01	0.762E-02	0.760E-02	0.455E-02
	80000.	0.451E-01	0.608E-02	0.799E-02	0.468E-02
$2s^2$ 7p $^2P^o$ - $2s^2$ 4d ^2D	5000.	0.197E-01	0.968E-02	0.362E-02	0.161E-02
	10000.	0.194E-01	0.849E-02	0.380E-02	0.183E-02
5388.4 Å	20000.	0.204E-01	0.671E-02	0.413E-02	0.206E-02
C=0.18E+17	30000.	0.213E-01	0.596E-02	0.403E-02	0.224E-02
	50000.	0.223E-01	0.489E-02	0.405E-02	0.244E-02
	80000.	0.228E-01	0.401E-02	0.421E-02	0.248E-02
$2s^2$ 8p $^2P^o$ - $2s^2$ 5s ^2S	5000.	0.467E-01	-0.165E-02	0.865E-02	-0.345E-02
	10000.	0.480E-01	-0.104E-02	0.939E-02	-0.402E-02
6335.0 Å	20000.	0.517E-01	-0.331E-03	0.980E-02	-0.447E-02
C=0.63E+16	30000.	0.544E-01	-0.130E-03	0.988E-02	-0.474E-02
	50000.	0.570E-01	0.170E-03	0.108E-01	-0.536E-02
	80000.	0.584E-01	-0.950E-04	0.105E-01	-0.607E-02
$2s^2$ 8p $^2P^o$ - $2s^2$ 4d ^2D	5000.	0.272E-01	0.202E-03	0.494E-02	-0.194E-02
	10000.	0.278E-01	0.448E-03	0.536E-02	-0.227E-02
4770.3 Å	20000.	0.295E-01	0.671E-03	0.554E-02	-0.252E-02
C=0.36E+16	30000.	0.308E-01	0.941E-03	0.560E-02	-0.266E-02
	50000.	0.320E-01	0.842E-03	0.612E-02	-0.296E-02
	80000.	0.326E-01	0.651E-03	0.604E-02	-0.346E-02
$2s^2$ 3d ^2D - $2s^2$ 3p $^2P^o$	5000.	0.173E-02	-0.845E-04	0.120E-03	-0.117E-04
	10000.	0.132E-02	-0.470E-04	0.147E-03	-0.179E-04
7226.4 Å	20000.	0.109E-02	-0.278E-04	0.166E-03	-0.246E-04
C=0.72E+18	30000.	0.102E-02	-0.436E-04	0.177E-03	-0.272E-04
	50000.	0.976E-03	-0.357E-04	0.185E-03	-0.311E-04
	80000.	0.951E-03	-0.307E-04	0.191E-03	-0.349E-04

Table 1. Continued

Multiplet	T [K]	W_e [Å]	d_e [Å]	W_{CII} [Å]	d_{CII} [Å]
$2s^2$ 4d ^2D - $2p^3$ $^2D^o$	5000.	0.348E-02	0.939E-03	0.367E-03	0.279E-03
	10000.	0.323E-02	0.833E-03	0.439E-03	0.334E-03
6022.4 Å	20000.	0.303E-02	0.678E-03	0.528E-03	0.395E-03
C=0.29E+17	30000.	0.293E-02	0.634E-03	0.564E-03	0.425E-03
	50000.	0.281E-02	0.553E-03	0.596E-03	0.466E-03
	80000.	0.270E-02	0.465E-03	0.648E-03	0.513E-03
$2s^2$ 5d ^2D - $2s^2$ 4p $^2P^o$	5000.	0.139E-01	0.368E-02	0.172E-02	0.142E-02
	10000.	0.129E-01	0.306E-02	0.213E-02	0.166E-02
6289.5 Å	20000.	0.123E-01	0.242E-02	0.223E-02	0.184E-02
C=0.79E+16	30000.	0.121E-01	0.215E-02	0.252E-02	0.203E-02
	50000.	0.117E-01	0.177E-02	0.297E-02	0.217E-02
	80000.	0.113E-01	0.144E-02	0.340E-02	0.245E-02
$2s^2$ 5d ^2D - $2p^3$ $^2D^o$	5000.	0.440E-02	0.152E-02	0.583E-03	0.500E-03
	10000.	0.412E-02	0.121E-02	0.736E-03	0.583E-03
3713.2 Å	20000.	0.392E-02	0.103E-02	0.771E-03	0.652E-03
C=0.28E+16	30000.	0.381E-02	0.924E-03	0.868E-03	0.711E-03
	50000.	0.366E-02	0.765E-03	0.102E-02	0.759E-03
	80000.	0.349E-02	0.629E-03	0.116E-02	0.853E-03
$2s^2$ 6d ^2D - $2s^2$ 4p $^2P^o$	5000.	0.128E-01	0.363E-02	0.179E-02	0.137E-02
	10000.	0.127E-01	0.257E-02	0.213E-02	0.161E-02
4658.2 Å	20000.	0.129E-01	0.224E-02	0.223E-02	0.175E-02
C=0.62E+16	30000.	0.129E-01	0.196E-02	0.253E-02	0.204E-02
	50000.	0.128E-01	0.149E-02	0.251E-02	0.229E-02
	80000.	0.125E-01	0.129E-02	0.258E-02	0.224E-02
$2s^2$ 6d ^2D - $2p^3$ $^2P^o$	5000.	0.297E-01	0.900E-02	0.413E-02	0.323E-02
	10000.	0.294E-01	0.677E-02	0.495E-02	0.379E-02
7114.0 Å	20000.	0.296E-01	0.595E-02	0.514E-02	0.411E-02
C=0.14E+17	30000.	0.297E-01	0.508E-02	0.598E-02	0.478E-02
	50000.	0.293E-01	0.397E-02	0.588E-02	0.538E-02
	80000.	0.283E-01	0.341E-02	0.600E-02	0.526E-02
$2s^2$ 7d ^2D - $2p^3$ $^2P^o$	5000.	0.305E-01	0.105E-01	0.475E-02	0.338E-02
	10000.	0.316E-01	0.804E-02	0.575E-02	0.390E-02
5761.9 Å	20000.	0.335E-01	0.692E-02	0.592E-02	0.432E-02
C=0.82E+16	30000.	0.342E-01	0.569E-02	0.594E-02	0.444E-02
	50000.	0.343E-01	0.459E-02	0.712E-02	0.478E-02
	80000.	0.336E-01	0.395E-02	0.751E-02	0.580E-02
$2s^2$ 7d ^2D - $2s^2$ 4p $^2P^o$	5000.	0.150E-01	0.496E-02	0.235E-02	0.165E-02
	10000.	0.156E-01	0.374E-02	0.283E-02	0.192E-02
4037.8 Å	20000.	0.165E-01	0.321E-02	0.293E-02	0.212E-02
C=0.41E+16	30000.	0.169E-01	0.262E-02	0.293E-02	0.217E-02
	50000.	0.171E-01	0.210E-02	0.354E-02	0.236E-02
	80000.	0.167E-01	0.181E-02	0.369E-02	0.285E-02

So, we have modified the SCP code of Sahal-Bréchot by coupling it to the energy level and oscillator strength data for C II data in the TOPbase ([6], http://cdsweb.u-strasbg.fr/TOPbase/), in order to calculate Stark broadening parameters for a large number (149) of C II multiplets.

2. Results and discussion

The Stark broadening parameters (Full Width at Half intensity Maximum - FWHM) have been obtained for 149 C II multiplets and the complete results are available on line in electronic form as Supporting information for Ref. [8], given in tables S1, S2 and S3. There is presented, as well, the comparison with existing experimental and other theoretical results and analysis of determined Stark broadening parameters. The results are presented for temperatures from 5000 up to 80000 K and electron densities 10^{14} (Table S1 in Supporting information for Ref. [8]), 10^{17} (Table S2, see Ref. [8]) and 10^{18} cm^{-3} (Table S3, see Ref. [8]). As an example of obtained results here are selected from the tables in electronic form given as additional information for Ref. [8], Strak broadening parameters for 27 C II multiplets for wavelengths 3500 Å- 7500 Åcovering the visible part of the spectrum, for the electron density $N_e = 10^{14}$ cm^{-3}. They are presented in Table 1. Values of the parameter C, given also in Table 1, divided by the corresponding full widths, give an estimate for the maximum perturber density for which the line may be treated as isolated (see Eq.(5) in Ref. [7]). For each value given in tables, the collision volume V multiplied by the perturber density N is much less than unity and the impact approximation is valid. Values for $NV > 0.5$ are not given but denoted by an asterisk and values for $0.1 < NV \leq 0.5$ are given but preceded by an asterisk.

The results will be also implemented in the STARK-B database, http://stark-b.obsmp.fr [9], [10], the result of a collaborative project between LERMA (Laboratoire d'Étude du Rayonnement et de la Matière en Astrophysique, Observatoire de Paris) and AOB (Astronomical Observatory, Belgrade). The STARK B database is currently developed at Paris Observatory. It is in free access, and is on line since the end of 2008. It is a part of the atomic and molecular databases of the Paris Observatory, and there is a link to the Serbian Virtual Observatory (SerVO, http://servo.aob.rs/ darko). A mirror site is planned at AOB. STARK-B is, as well, one of databases of VAMDC (Virtual Atomic and Molecular Data Centre - http://vamdc.eu and http://portal.vamdc.eu) [11], [12], the FP7 European project created in summer 2009 for 3.5 years. Its aim is to create and maintain an interoperable e-Infrastructure for exchange of atomic and molecular data, to serve as a forum for data producers and users, and to develop standards and softwares for exchange of atomic and molecular data. ore details on VAMDC and STARK-B are given in Ref. [10] in this issue.

References

[1] Dufour P, Liebert J, Fontaine G and Behara N 2007 *Nature Letters* **450** 522
[2] Dufour P, Fontaine G, Liebert J, Schmidt G D and Behara N 2008 *ApJ* **683** 978
[3] Sahal-Bréchot S 1969 *A&A* **1** 91
[4] Sahal-Bréchot S 1969 *A&A* **2** 322
[5] Sahal-Bréchot S 1974 *A&A* **35** 319
[6] Cunto W, Mendoza C, Ochsenbein F and Zeippen C J 1993 *A&A* **275** L5
[7] Dimitrijević M S, Sahal-Bréchot S and Bommier V 1991 *A&A* **89** 581
[8] Larbi-Terzi N, Sahal-Bréchot S, Ben Nessib N and Dimitrijević M S 2012 *MNRAS* **423** 766
[9] Sahal-Bréchot S, Dimitrijević M S and Moreau N 2012 STARK-B database, [online]. Available: http://stark-b.obspm.fr [Jul 15, 2012]. Observatory of Paris, LERMA and Astronomical Observatory of Belgrade
[10] Sahal-Bréchot S, Dimitrijević M S and Moreau N 2012 *J. Phys.: Conf. Series* this issue
[11] Dubernet ML, Boudon V et al. 2010 JQSRT 111 2151, http://www.vamdc.eu/
[12] Rixon G, Dubernet M L, et al. 2011 7th International Conference on Atomic and Molecular Data and their Applications -ICAMDATA-2010, AIP Conf. Proc. 1344 107

CARS at a «hard-to-realize conditions»: lineshape spectroscopy at high temperatures in a real flame

K A Vereschagin[1], A K Vereschagin[1], V V Smirnov[1], O M Stel'makh[1], V I Fabelinsky[1], W Clauss[2] and M Oschwald[2]

[1]General Physics Institute, Vavilov Str.38, 119991, Moscow, Russia;
[2]German Aerospace Center Space Propulsion Institute, Langer Grund, 74239, Hardthausen, Germany.

veresch@kapella.gpi.ru

Abstract. On the example of the study of the collisional broadening and shift of the hydrogen Q-branch lines due to collisions with water molecules in wide temperatures range [2000 K - 3000 K], we display the use of single-short CARS-spectroscopy for lineshape analisys at experimental conditions in which the object naturally does not exist, but can be created due to some physical and/or chemical processes for some time, small in comparison with the time necessary for stationary laboratory researches. Importance of light statistics as well as some specific features of CARS spectroscopy, which are the most actual from the point of view of use of CARS as a tool f or lineshape spectroscopy, are discussed.

1. Introduction

When CARS- spectroscopy is used for combustion studies [1-4] the precision of temperature determination critically depends on both correctness of theoretical model, used for CARS-spectra simulations, and on the reliability of the available experimental spectroscopic and kinetic data about lines positions and line broadening of Raman transitions of the «probe» molecules in a wide range of temperatures and pressures. From a more fundamental point of view, accurate measurement of Raman line shape parameters can provide valuable information on the intermolecular potentials.

Till recently available data concerning the broadening and shift for H_2-X (X: H_2, He, N_2) mixtures at temperatures up to 1200 K were obtained by means the stimulated Raman spectroscopy (SRS), providing very accurate determination of linewidths and lineshifts. SRS-study of broadening for H_2-H_2O mixtures in temperature range 600 – 1800 K has been done [5], but there are no data concerning the shift for H_2-H_2O were reported.

In our recent papers [6,7] we reported on measurements of collisional broadening and shift coefficients of Q_1, Q_3, Q_5, Q_7 and Q_9 lines of hydrogen molecules caused by collisions with water molecules at high temperatures and pressures under conditions taking place in a pulsed high-pressure H_2-O_2 combustion chamber

In the present work, on the example of the study of the collisional broadening and shift of the hydrogen Q-branch lines due to collisions with water molecules in wide temperatures range [2000 K - 3000 K], we display the use of single-short CARS-spectroscopy for lineshape analisys at «hard-to-realize conditions». Under a «hard-to-realize conditions»-we imply experimental conditions in which the object naturally does not exist, but can be created due to some physical and/or chemical processes for some time, small in comparison with the time necessary for stationary laboratory researches.

2. Some specific features of CARS, essential for lineshape spectroscopy

From 70th years of the last century CARS has found both spectroscopic [1] and diagnostic application [2]. The theory and techniques of CARS are described in various reviews elsewhere [1,3,4], however, rather more generally. Some specific features of CARS spectroscopy, which are the most actual from the point of view of use of CARS as a tool for lineshape spectroscopy, are discussed below.

CARS is nonlinear-optical process induced by the lowest-order nonlinear term in the induced molecular polarization of the medium. The polarization may be written in terms of the susceptibility tensors χ as:

$$P_i(t) = \chi^{(1)}_{i\alpha} \cdot E_\alpha(t) + \chi^{(2)}_{i\alpha\beta} \cdot E_\alpha(t)E_\beta(t) + \chi^{(3)}_{i\alpha\beta\gamma} \cdot E_\alpha(t)E_\beta(t)E_\gamma(t) + \ldots \quad (1)$$

Interaction of three pump waves with molecules of a medium results in appearance of polarization at combined frequencies, and among others, at Anti-Stokes frequency

$$\omega_{CARS} = \omega_1 - \omega_2 + \omega_3 \quad (2)$$

Amplitude of the anti-Stokes wave depends on amplitudes of pump electric fields as following:

$$E_{CARS}(\omega_{CARS}) \propto \chi^{(3)}(\omega_1, \omega_2) \cdot E_1(\omega_1) \cdot E_2^*(\omega_2) \cdot E_3(\omega_3) \quad (3)$$

Intensity of CARS signal

$$I_{CARS} \propto E_{CARS} \cdot E^*_{CARS} \quad (4)$$

and at given pump electric fields is

$$I_{CARS}(\omega_{CARS}) \propto I_1(\omega_1) \cdot I_2(\omega_2) \cdot I_3(\omega_3) \cdot \left|\chi^{(3)}(\omega_1, \omega_2)\right|^2 \quad (5)$$

Taken into account finite spectral widths of lasers used CARS intensity can be approximately written as following:

$$I_{CARS}(\omega) \propto \int_{-\infty}^{\infty} I_3(\omega - \xi) \cdot \left|\chi^{(3)}(\xi)\right|^2 \cdot \int_{-\infty}^{\infty} I_1(x - \xi) \cdot I_2(x) dx d\xi \quad (6)$$

For degenerated broadband CARS spectral width of pump wave radiation usually is about 10^{-3} cm^{-1} and of Stokes wave radiation is about 150-300 cm^{-1}. In such a case

$$I_{CARS}(\omega) \propto \left|\chi^{(3)}(\omega)\right|^2 \cdot I_{CARS\_nonresonant} \quad (7)$$

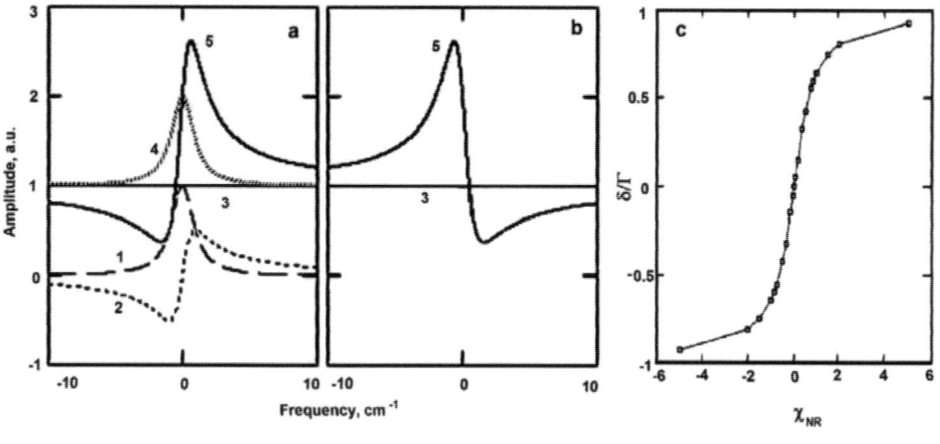

Figure 1. CARS line profile (**a,b**) for Lorentzian spectrum and reduced shift (**c**) of maximum of CARS-profile versus non-resonant background χ_{NR} (ratio of "non-resonant" to maximum "resonant" susceptibilities). 1- $Im(\chi^{(3)}_R)$, 2- $Re(\chi^{(3)}_R)$, 3 – $|\chi_{NR}|^2$, 4 – $\{[Im(\chi^{(3)}_R)]^2 + |\chi_{NR}|^2\}$, 5 – CARS spectral profile. **a** - $\chi_{NR} > 0$, **b** - $\chi_{NR} < 0$, Γ=1 cm^{-1}, χ_{NR} = 1. With the use of any device for spectra analysis reduced shifts become dependent on the slit function of the spectrometer

By dividing an experimental «resonant» CARS spectrum onto «nonresonant» one we can obtain spectrum of $\left|\chi^{(3)}(\omega)\right|^2$, containing all spectral and kinetic information about molecules under investigation. Spectral resolution in case of broadband CARS is determined completely by dispersive device used for CARS spectra analysis and photodetector in use.

While CARS itself can be used to deduce line-broadening and line-shift information, the CARS spectrum depends on both the absorptive (*imaginary*) and dispersive (*real*) parts of the non-linear susceptibilities, as well as a non-resonant background contribution:

$$\chi^{(3)}_{CARS} = \chi^{(3)'} + i\chi^{(3)''} + \chi^{(3)NR} \tag{8}$$

Whereas for collisional regime the lineshape of a single resonance is given by Lorentzian contour (like as for spontaneous Raman), in Doppler (collision-free) regime the lineshape of a single resonance is given by CARS-Doppler profile, distinguishing from Raman-Doppler profile, given by Gaussian contour [8].

Besides, interference from the non-resonant background produces asymmetry in the CARS line shape and additional line-shift, depending on magnitude and sign of non-resonant background. As an example, the CARS-profiles of a single homogeneously broadened Raman transition at presence of a non-resonant backgroung

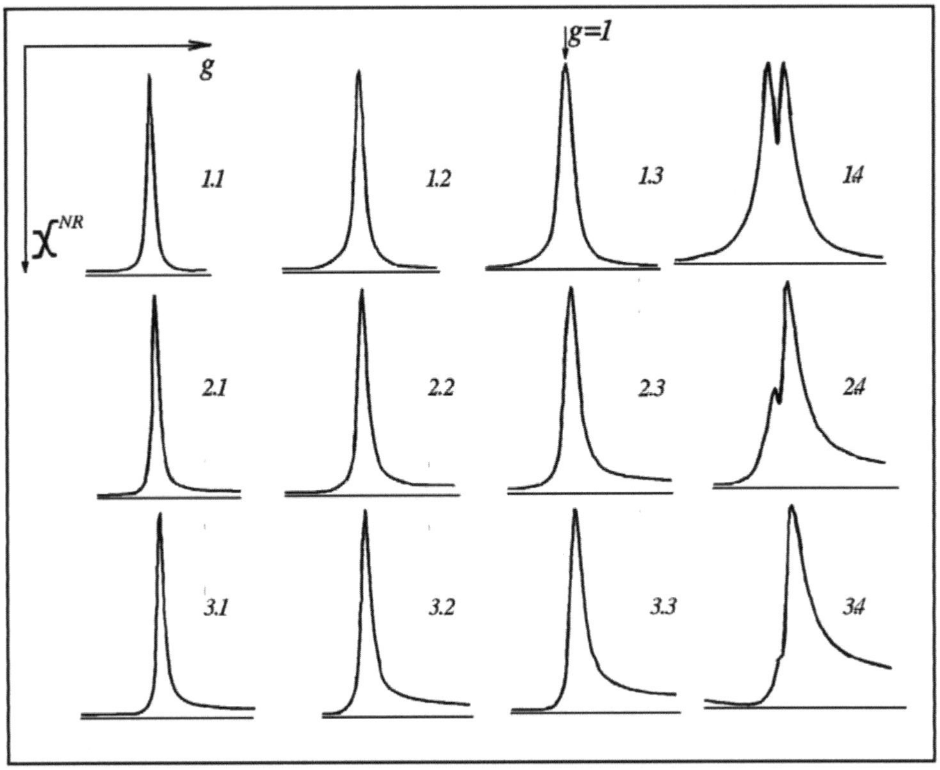

Figure 2. Evolution of the shape of a single line with increase in parameter of saturation g (rows) and magnitude of not-resonant background (columns).

$$\left|\chi_{CARS}^{(3)}\right|^2 = \frac{A^2}{\Delta\omega^2 + \Gamma^2} + \frac{2A\Delta\omega}{\Delta\omega^2 + \Gamma^2}\cdot\chi^{(3)NR} + \left(\chi^{(3)NR}\right)^2 \quad (9)$$

depending on detuning from the Raman resonance $\Delta\omega = \Omega_0-(\omega_1-\omega_2)$, are presented at Fig.1 (coefficient A contains all factors independent on frequency).

Strong biharmonic fields cause considerable redistribution of molecules over energy levels (saturation effect). Saturation is characterized by so-called "saturation parameter" g: $g=I_1I_2/I_s^2$; I_1,I_2 – intensities of laser beames, I_s –intensity of two-photon saturation [9].

In absence of saturation and in neglect of a not-resonant background the shape of the single homogeneously broadened line is described by a Lorentzian contour with width (HWHM) of Γ_L, amplitude of $A{\sim}1/\Gamma_L^2$ and frequency aligned on a resonance ($\Delta\omega$=0). At $0\leq g\leq 1$ shape of a contour becomes non-Lorentzian, remaining a bell-shaped curve with one maximum, the amplitude of a contour A decreases, and halfwidth Γ grows with increase of saturation parameter g. At $g>1$ contour of the line has two maxima with a characteristic hole in the line center; with increase of saturation parameter g the amplitudes of the maxima decreases, depth of the hole increases (Fig. 2, figs. 1.1-1.4).

Presence of non-resonant background results in asymmetry of a contour, shift of a maximum and some increase of its amplitude (see Fig. 2, fig.1.1-3.1). Peculiarity of saturation of a contour, distorted by presence of non-resonant background, is that in an initial (nonsaturated) contour the frequency of a maximum doesn't coincide with the resonance center whereas the the largest amplitude distortion, caused by saturation, takes place at the center of resonance. Distortions of a contour amplitude and halfwidth of a saturated line depend on size of non-resonant background χ_{NR}.

All mentioned above peculiarities of CARS make the analysis of the collisional processes responsible for the forming of the spectral line shape more complicated.

3. Single-shot lineshape spectroscopy and light statistics

The single-shot lineshape spectroscopy implies some special requirements to laser sources. Namely, complete contour of a line should be registered at a moment, but with the sufficient spectral resolution. In such a case experimenter is forced to use "broadband"-CARS technique (BB-CARS) [3] when "broadband" spectrum of Stokes laser should be wide enough to "cover" at once complete spectrum of researched transition; for spectral analysis of a single-shot line contours of spectra under investigation a dispersive device with appropriate spectral resolution should be used and contour of a spectral line can entirely be registered at a multichannel photodetector (OMA) "in one stage" (during one laser shot). Such approach can be applied to express-measurements of spectral lines profiles in non-stationary conditions and in «short-living» objects. A «price» for it is: not very high spectral resolution, being defined in this case by properties of the dispersive device used for the spectral analysis.

Usually a broadband laser radiation has not a continuous spectrum, but – mode's structure that means: radiation is present only on certain frequencies of a spectrum and is absent in intervals between

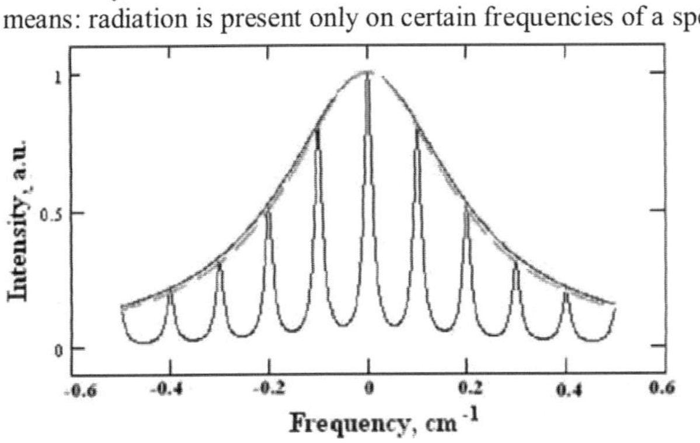

Figure 3. Spectral structure of line contour probed by multimode laser.

Dashed bell-shaped curve – original line contour.

Solid bell-shaped curve – convolution of original line contour with slit function (HWHM=0.01 cm^{-1}).

Multipeak curve – response of CCD-detector on probe by multimode laser.

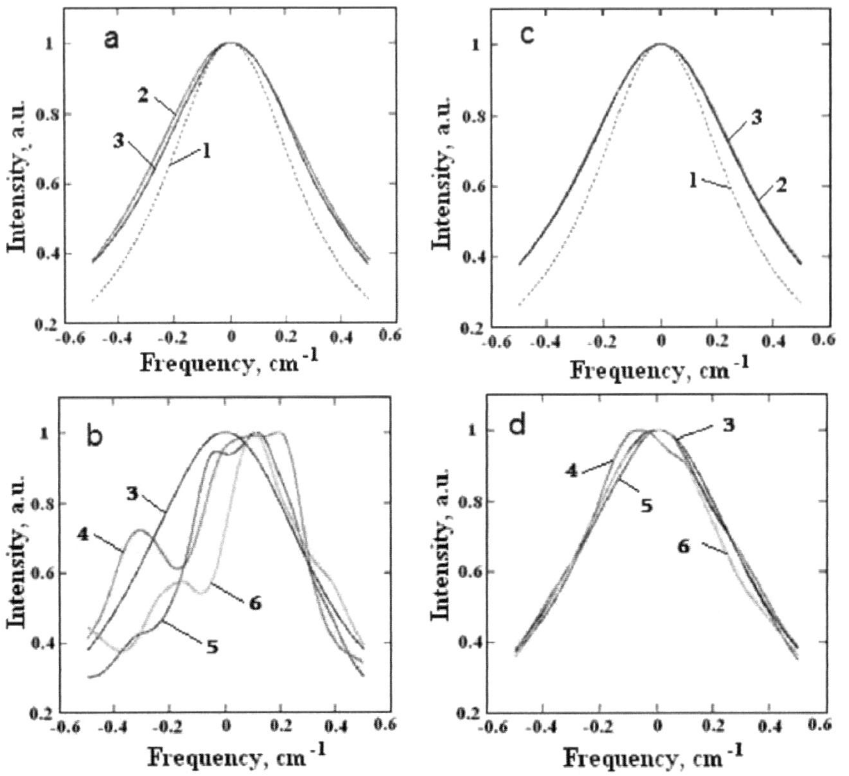

Figure 4. Lineshapes obtained for transition probed by multimode laser with different light statistics. **a,b** – modes statistics is exponential with mean value is equal to 1, and corresponding ***R***elative ***S***tandard ***D***eviation (***RSD***) is equal to 100%. **c,d**, – modes statistics is normal with mean value is equal to 1, and corresponding ***RSD*** = 20% .***1*** – original line contour. ***2*** – convolution of original line contour with slit function .***3*** – averaged over 50 single shot spectra.***4,5,6*** – consequent individual realizations of single shot spectra. Modes separation ~ 0.033 cm-1, slit function (HWHM)~0.09 cm^{-1}.

modes [10]. With use such a laser the contour of a spectral line will be defined in several (by number of modes) points and with the accuracy depending on fluctuations of mode's amplitudes and phases, being belonged to the spectral range occupied by a contour on frequencies scale.

Thus, the contour of a spectral line is being "painted" not continuously, but – discretely, only at mode's localization places over the spectrum of laser radiation; in some sense it is similar to «painting» of a contour "point-by-point" by frequency scanning of a narrow-band laser source with a step equal to the mode spacing. The response of the spectral device on each mode represents by itself a profile of slit function, and superposition of slit functions profiles from each of modes is forming a resulting signal shape on the detector (Fig.3). With equal amplitudes of modes, aggregate of peak values from each of slit function profile at frequencies of modes location will describe a total spectral line profile. At casual distribution of amplitudes and phases of modes of broadband radiation, the contour of a spectral line can be considerably deformed, up to full unrecognizability (Fig.4).

So, at «single-shot» lineshape spectroscopy, the basic role is played by mode's structure of laser sources radiation and by statistics of this radiation [11]. Even at possibility of a signal accumulation (work under "accumulation/averaging on N shots" mode of operation) the average value of the measured amplitudes of a signal in each of contour points can differ considerably, and arbitrarily, from it's mathematical expectations. In full measure this statement concerns the case of "scanned" variant of the lineshape spectroscopy with use of a narrow-band tunable laser under "accumulation/averaging on N shots" mode of operation.

We have found that when using a special configuration of the CARS-experiment (DBB-CARS [12]) there is a possibility to average a spectral noise over single-shot spectra of pump lasers in use [13]. The numerical analysis has showed [14], that, due to nonlinearity of CARS-process, the change of light statistics could take place – from «exponential» to «normal» - with decrease of variation of stochastic process depending on efficiency of DBB-adjustment. For correct realization of DBB-approach it is important to adjust properly combs of mode structures of pump lasers.

4. Investigation of Collisional Broadening and Shift of the Hydrogen Q-branch Transitions by Water at High Temperatures

To carry out the research of spectral lines shapes at high temperatures and pressures we have had to work with high pressure H_2/O_2 pulsed burner [15]. The flame of H_2/O_2 high pressure pulsed combustion chamber (HPPCC) is characterized by high gas temperatures (up to 3500 K) and pressures (up to 150 bar) with combustion products content: H_2, H_2O (major species) and atoms and radicals (minor species). Gas parameters can fluctuate from shot to shot, and single-shot mode for diagnostics is preferable.

We have developed a high-resolution (~0.1 cm^{-1}) spectroscopic technique that allows single-shot *CARS*-spectra of hydrogen *Q*-branch molecular transitions to be obtained in non-stationary media. The method is based on the application of a *F*abry-*P*erot interferometer to the spectral analysis of a broadband *CARS*-spectral line profile with detection of an interference pattern by a *CCD*-detector [13].

The three-channel measurement set-up and the pulsed chamber are described elsewhere [6, 13, 15]. Combustion process was repetitive with a rate of 1 Hz. After the H_2-O_2 mixture burned out, the outflow of the gas from the combustion chamber led to a pressure and temperature drop. This, in turn, allowed obtaining the data at various temperatures and pressures by varying the delay between CARS-probing and the ignition.

Pressure measurements were carried out by the high-speed piezoelectric pressure gauge Kistler 6061B, built in a wall of the combustion chamber (1-st channel).

Temperature measurement in the combustion chamber were done with the help of a broadband CARS-spectrometer, that allowed registering a significant fragment of a hydrogen *Q*-branch spectrum during one laser shot (2-nd channel). A Jobin-Ivon THR-1000 grating spectrograph with a slit function width of ~1.5 cm^{-1} performed spectral analysis of the CARS signal in this second channel. CARS-spectrum was detected by a multichannel CCD detector.

High resolution single shot spectra of a selected H_2 Q-branch line were measured simultaneously by another CARS-spectrometer also during one laser shot (3-rd channel). Spectral analysis of the CARS signal in this third channel was done by a LOMO IT-28-30 Fabry-Perot interferometer having a free spectral range (FSR) of 1.66 cm^{-1} and a slit function width of ~0.09 cm^{-1}.

Fabry-Perot interferometer interference rings were detected by a two-dimensional CCD array

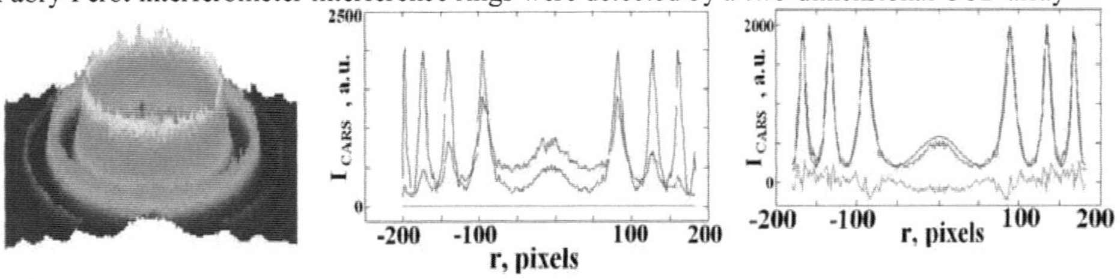

Figure 5. Fabry-Perot interferogram (left), result of the angular averaging with/without correction on transversal CARS-intensity distribution (middle) and fitting of the corrected interferogram by a convolution of Airy function with a Lorentzian contour (right).

(Fig.5, left). A specially developed software package made angular integration of a circular fringe structure (to minimize the influence of a noise), normalized interference patterns on the light intensity distribution over a cross-section of a CARS-beam (Fig.5, middle) and realizes a fitting procedure (Fig.5, right) of experimental interferogram by a convolution of Airy function (interferometer slit function) with Lorentzian contour (spectral line profile).

In the limits of experimental accuracy fitting of Fabry-Perot interferogram by a convolution of Airy function either with CARS-profile or with corresponding Lorentzian contour. results in the same value of the fitting parameter « linewidth », while values of the fitting parameter «central wavelength» were found slightly different. This circumstance allows us to avoid varying of non-resonant background and to realize the fitting procedure only with four fitting parameters (baseline, scaling factor, linewidth and central wavelength) providing line position Ω_0 and spectral line contour width Γ.

Additional CARS line shift arising due to specific CARS lineshape resulting from interference with non-resonant background (Fig. 1, c) was taken into account at the next step of data processing. This additional shift ("CARS-shift") depends on nonresonant background value, linewidth and slit function of the spectrometer. For calculation of the shift coefficients we have corrected every of obtained line position Ω_0 for the "CARS-shift". Magnitude and sign of nonresonant background were determined from BB-CARS-spectra processing (second channel), and transition linewidths (obtained together with line positions from "fitting" procedure), as well as Fabry-Perot interferometer slit function, were taken into account during "CARS-shift" calculations. Consideration of additional line shift due to the interference with non-resonant background allows allocating the contribution from J-dependent collisional shift in the total change of transition frequency. Stark shift was neglected for all lines since it was negligible compared to "CARS-shift" and collisional self-shift.

5. Summary

Experimental CARS-spectra should seriously be interpreted in sense of discussed above peculiar properties of CARS-lineshape for getting correct information concerning spectral lineshape of Raman transition.

Thanks to nonlinearity of CARS-process change of light statistics could take place – from «exponential» to «normal» with decrease of variation of stochastic process – that could improve quality of single shot CARS spectra and provide possibility of high resolution single-shot CARS lineshape spectroscopy.

CARS is not the best way for fine researches of spectral features of a line contour, however, probably, there hardly exist any alternatives if it is necessary to perform the measurements in a non-stationary object like the flame of High Pressure Pulsed Ccombustion Chamber.

References

[1] Nibler J W and Knighten G W 1979 *Raman Spectroscopy of Gases and Liquids* ed Weber A. (Springer-Verlag, Berlin-Heidelberg-New-York)
[2] Régnier P R and Taran J-P 1973 *Appl. Phys. Lett.* **23** 240
[3] Greenhalgh D A 1988 *Advances in Nonlinear Spectroscopy* ed R J H Clark and R E Hester (John Wiley & Sons, New York)
[4] Roy Sukesh, Gord James R and Patnaik Anil K 2010 *Progress in Energy and Combustion Science* **36** 280
[5] Shaussard F, Michaut X, Saint-Loup R, Berger H, Joubert P, Lance B, Bonamy J and Robert D 2000 *J.Chem.Phys.* **112** 158
[6] Vereschagin K A, Vereschagin A K, Smirnov V V, Stelmakh O M, FabelinskyV I, Clauss W and Oschwald M 2008 *J.Raman Spectrosc.* **39** 722
[7] Vereschagin K A, Vereschagin A K, Smirnov V V, Stel'makh O M, Fabelinsky V I, Clauss W and Oschwald M 2010 *J.Raman Spectrosc.* **41** 897
[8] Henesian M A and Byer R L 1978 *J.Opt.Soc.Am.* **68** no.5 648
[9] Zadkov V N and Koroteev N I 1985 *Optika i spektroskopiya* **59** 67

[10] O. Svelto. Principles of Lasers. Springer; 5th ed. edition (2009).
[11] Goodman J W 1985 *Statistical Optics* (Wiley-Interscience Publ., New York)
[12] Eckbreth A C and Anderson TJ 1985 *Appl.Opt.* .**24** 2731
[13] Vereschagin K A, Smirnov V V, Stelmakh O M, Fabelinsky V I, Clauss W, Klimenko D N, Oschwald M and Vereschagin A K 2005 *J. Raman Spectrosc.* **36** 134
[14] Vereshchagin K A and Vereshchagin A K 2007 *Proc. SPIE*. Vol. 6727, pp. 67271K
[15] Clauss W, Klimenko D N, Oschwald M, Vereschagin K A, Smirnov V V, Stelmakh O M and Fabelinsky V I 2002 *J.Raman Spectrosc.* **33** 906

Raman line shape studies of hydrogen cryosolutions

I Verzhbitskiy[1], W Herrebout[2], B van der Veken[2] and A Kouzov[1,*]

[1]Department of Physics, St. Petersburg State University, Ulyanovskaya str. 3, Peterhof, St. Petersburg 198504, Russia
[2]Department of Chemistry, University of Antwerp, Groenenborgerlaan 171, 2020 Antwerp, Belgium
*E-mail: alex@AK1197.spb.edu

Abstract. The pure rotational and rotation-vibration Raman transitions in H_2 dissolved in liquid argon (LAr, T=87 K) and nitrogen (LN_2, T=80 K) are studied at different hydrogen concentrations. Whereas the rotational $S_0(J)$ (J=0, 1) shapes of H_2-LAr can be accurately fitted by Lorentzian profiles, the wings of the same H_2-LN_2 lines demonstrate a distinct sub-Lorentzian behaviour. The difference between these patterns is attributed to the change of the time scale of anisotropic interactions which perturb the rotational motion of H_2. They should be slower for the H_2-LN_2 system in which the long-range quadrupole-quadrupole interactions develop. The latter dominate in the formation of the symmetric H_2-LN_2 $S_1(J)$ line shapes. By contrast, the same H_2-LAr lines are mainly monitored by the vibrational broadening and their shapes are distinctly asymmetric.

1. Introduction

Hydrogen molecule, H_2, and its isotopic substituents are unique molecular objects which pertain almost unhindered quantized rotation even in condensed phases. Their line shapes contain a wealth of state-resolved information on couplings of molecular degrees of freedom with the bath and can therefore shed light on the microdynamics of liquids. Intriguing line asymmetry in the roto-vibrational Raman spectra of hydrogen in liquid SF_6 has been observed [1,2], while the rotational lines remained symmetric [1,3]. In other reference cryosolvents (LAr, LN_2), the asymmetry remained undetectable whatsoever [2,3]. However, the latter result may be doubted because of the insufficient Signal-to-Noise (S/N) ratios attained in the earlier studies. Generally, the existing data on the line profiles in the considered spectra are poor, semi-quantitative and lacking systematic treatment. To clarify the issue, the Raman line shapes of H_2 dissolved in liquid argon and nitrogen are now revisited with the use of the state-of-the-art spectroscopic techniques.

2. Experimental

The conventional 90°-detection experimental setup exploiting the triple Princeton Instruments TriVista TR557 spectrometer was employed. The native perpendicular to the scattering plane polarization of the Ar^+ laser light source (514.5 nm) was not varied since the recorded transitions are known to be totally depolarized. The spectral resolution reached the value of 0.2 cm^{-1}, sufficient to avoid the slit distortions of the studied profiles. The S/N ratios were 1-2 decimal orders better than attained in older experiments [1-3]. The Raman setup was equipped with liquid-nitrogen cryostat that allows measuring the spectra at low temperatures with stability better than 1 K.

3. Results

The rotational $S_0(J)$ and roto-vibrational $S_1(J)$ lines (J=0, 1) of hydrogen were studied in LAr (T=87 K) and LN_2 (T=80 K). In order to trace the concentration effects on the line profiles, the hydrogen concentration x was varied in the range 0.02-0.08 mole fractions. However, we were unable to observe any effect in both cryosystems, in contrast to the earlier reported slight self-broadening of the $S_0(J)$ lines in LN_2 [3]. We emphasize that the $S_0(J)$ shapes of H_2-LAr at detunings z from the line centers accessible for the investigation ($|z|\leq 20$ cm^{-1}) are found to strictly follow the Lorentzian distribution corrected by the quantum asymmetry factor $Q(z)=2/[1+\exp(\hbar z/kT)]$ (see figure 1). The appearance of a Lorentzian shape is remarkable, for it is an indication of the fast varying molecule-bath interaction [4].

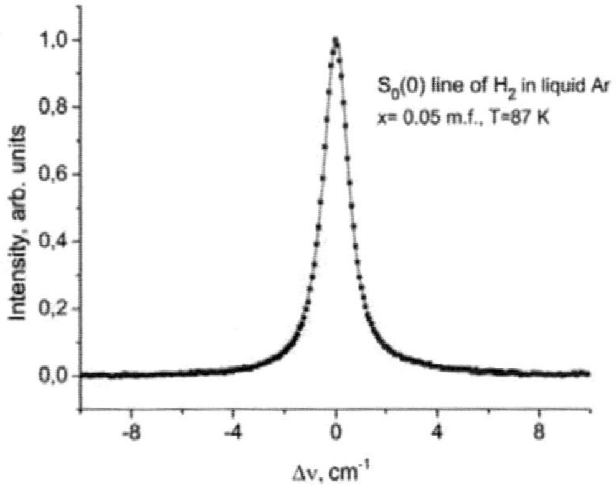

Figure 1. Measured rotational Raman line shape of H_2-LAr cryosystem (dots) fitted by Lorentzian function (solid line).

Generally, the memory function theory [4] predicts the isolated-line intensity to vary as $I(z)\sim \mathrm{Re}[iz+\Gamma(z)]^{-1}$, where $\Gamma(z)$ is the one-sided Fourier transform of the time autocorrelation function characterizing the molecule-bath interactions. At detunings satisfying $zt_c << 1$ (where t_c characterizes the decay rate of the interaction correlation function), one has $\Gamma(z)\approx\Gamma(0)$ and the line shape becomes Lorentzian. If, additionally, $\Gamma(0)\cdot t_c<<1$ (impact limit), the dispersion intensity distribution applies not only to the central part of the line but also to its near wings. At even larger detunings, $\Gamma(z)$ damps out and the intensity distribution in the wing becomes sub-Lorentzian, $I(z)\sim \mathrm{Re}\Gamma(z)/z^2$. In this range, $F(z)=I(z)z^2$ decreases with z, as is evident from figure 2 where the symmetrized functions $\widetilde{F}(z^*)=F(z)/Q(z)$ are plotted for the $S_0(1)$ line for both cryosolutions. To visualize the different dynamical properties of two cryosystems, the dimensionless frequencies $z^*=z/\Gamma$ were used, with the half widths $\Gamma=\mathrm{Re}\Gamma(0)$ found (see table 1) by the least-square fitting of the experimental profiles by the extended (quantum-corrected) Lorentzians. The different behavior of the line wings on figure 2 is striking. While no regular deviations from the extended Lorentzian distribution are observable for H_2-LAr, which confirms the impact character of the anisotropic H_2-Ar interactions, $\Gamma(z)$ for the nitrogen solution begins to decrease as the detuning exceeds three half widths. For the latter case, rough estimation gives the effective half width of the $\mathrm{Re}\Gamma(z)$ spectrum to be about 40 cm^{-1}. If one assumes that the quadrupole-quadrupole H_2-N_2 interactions dominate and that their spectrum is formed mainly by the hindered rotation of N_2, then the width of $\mathrm{Re}\Gamma(z)$ can be independently estimated from the collision-induced absorption by liquid nitrogen [5]. The resulting value based on such IR data is 40-50 cm^{-1}, in close agreement with our result, which can serve as an indication of the rotational coupling nature of the H_2-LN_2 solution. The H_2-LAr interactions are shorter-ranged than the H_2-LN_2 ones and, due to this, produce the $\Gamma(z)$ spectrum so broad that its decay can not be spectroscopically traced.

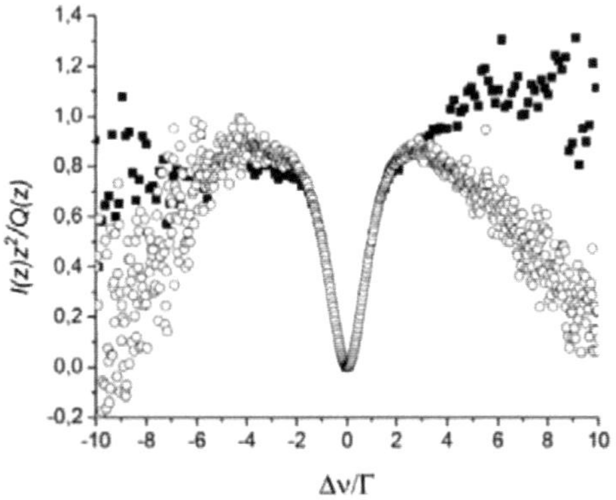

Figure 2. Example of experimental distributions for rotational $S_0(J)$ lines as functions of dimensionless detuning $z^*=z/\Gamma$ for H_2-LAr ($S_0(0)$ line, filled squares) and H_2-LN_2 ($S_0(1)$ line, open circles) solutions. Γ values were found by Lorentzian fits.

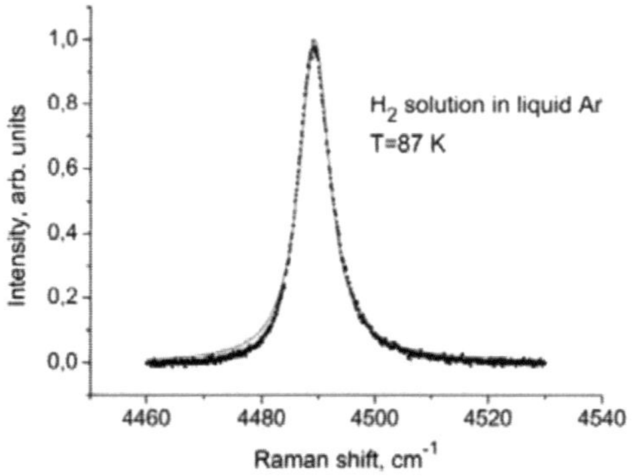

Figure 3. Lorentzian fit (solid line) of $S_1(0)$ line (H_2-LAr solution) showing its shape asymmetry.

Analysis of the rotation-vibration line shapes also led to interesting results. Strikingly, the $S_1(J)$ (J=0,1) shapes of H_2-LAr regularly deviate from the extended Lorentzian profiles at lower frequencies, but not at higher ones, as seen on figure 3. In contrast to this, similar deviations detected in the H_2-LN_2 $S_1(J)$ profiles are symmetric and resemble those observed in the rotational spectrum of this system.

Data from table 1 allow to conclude that the vibrational contributions dominate in the roto-vibrational line broadening in the H_2-LAr system. Besides, preliminary data on the $Q_1(J)$ line shapes show that, in contrast to the results on the pure rotational shapes of this cryosystem, the simulated Lorentzian $Q_1(J)$ wings are somewhat higher than the observed ones. Since the $Q_1(J)$ lines are solely broadened by the vibrational molecule-bath interactions, this can indicate on the slower evolution of the vibrational interactions in comparison to the anisotropic ones which relax in this system almost instantaneously.

Table 1. Measured shifts (δ) and half widths (Γ) of the $S_0(J)$ and $S_1(J)$ Raman lines (J=0, 1) of hydrogen in cryosolutions. Units are wavenumbers, cm^{-1}. Transition frequencies of free molecule were taken from Ref. [5].

Transition	H_2 in liquid argon, T=87 K		H_2 in liquid nitrogen, T=80 K	
	δ	Γ	δ	Γ
$S_0(0)$	-0.8	0.61	2.5	6.01
$S_0(1)$	-1.1	0.61	0.5	6.38
$S_1(0)$	-9.6	3.24	-7.3	8.30
$S_1(1)$	-8.5	3.42	-3.8	8.46

Also, table 1 shows that the vibrational contributions to the H_2-LN_2 roto-vibrational line widths are three times smaller than the rotational ones. The balance in this case is thus in favor of the rotational broadening.

Considering the aforementioned facts altogether, we may suggest that the asymmetry of the roto-vibrational lines in the H_2-LAr cryosolution is an effect of the pronounced non-impact vibrational relaxation accompanied by a fast decorrelation of anisotropic molecule-bath interactions.

4. Conclusion

As found by the present study, the shapes of the rotational and roto-vibrational Raman lines of hydrogen dissolved in liquid argon and nitrogen are nearly Lorentzian. To interpret the observed deviations from the dispersion profiles, the vibrational and rotational couplings of H_2 with the bath are to be characterized by finite time scales. Depending on the relative roles of these couplings as well as on their characteristic times, the deviations can show up in the roto-vibrational line asymmetry as was indeed detected for the H_2-LAr cryosolution.

References

[1] Le Duff Y and Holzer W 1974 *Chem. Phys. Lett.* **24** 212
[2] Orlova N D and Pozdniakova L A 1980 *Opt. Spektrosk.* **48** 1086 (translated as 1980 *Optics and Spectroscopy* **48** 594)
[3] Orlova N D and Khamitov R 1988 R *Opt. Spektrosk.* **65** 567 (translated as 1988 *Optics and Spectroscopy* **65** 338)
[4] Foster D 1975 *Hydrodynamic Fluctuations, Broken Symmetry and Correlation Functions* (Benjamin, Reading, MA).
[5] Birnbaum G 1980 *Vibrational Spectroscopy of Molecular Liquids and Solids*, edited by S. Bratos and R M Pick (NATO Advanced Study Institutes Series) (Plenum Press)
[6] Stoicheff B P 1957 *Canad. J. Phys.* **35** 730

Spectroscopy of Atomic Vapors in Nanometer Cells: Dicke Narrowing and Beyond

T A Vartanyan, V V Khromov

St. Petersburg National Research University of Information Technologies Mechanics and Optics, Kronverkskii pr. 49, St. Petersburg, 197101, Russian Federation.

E-mail: tigran@vartanyan.com

Abstract. Sub-Doppler spectroscopy of gaseous media confined in thin pillbox-shaped cells was pioneered by R.H. Dicke. In the past, this idea attracted much less attention compared to "Dicke narrowing" in buffer gas where the atoms or molecules perform a diffusive motion instead of being bounced back and forth between the walls of the cell in a completely predetermined nature. The situation is going to be changed as atomic spectroscopy becoming an essential part of mobile devices for civil and military applications that require tiny spectroscopic cells. In the pillbox shaped cells, the role of the fast moving atoms is diminished, while the slowly moving atoms contribute most to the absorption as well as to the fluorescence. The role of the slowly moving atoms and their transient polarization in selective reflection spectroscopy was highlighted by J.L. Cojan. By merging these two approaches we have developed a theoretical description of optical reflection from and transmission through the narrow slice of atomic vapours.

1. Two different Dicke narrowing mechanisms

More than fifty years ago R.H. Dicke proposed two methods of the Doppler width suppression. Both methods rely on the fact that the extent of the Doppler broadening depends on the phase difference between the atomic oscillator and the external field rather than on the instantaneous value of the atomic velocity. Hence, the Doppler width may be diminished in a room temperature vapor cell provided the spatial shifts of the atoms are restricted. In the first, most prominent method, this is accomplished by the gas phase collisions [1]. In the second method, which, in the past, attracted much less attention, the atoms are placed in a narrow pillbox-shaped cell [2]. Thus, the atom-wall collisions serve to restrict the atomic motion. Both methods, to be used in high-resolution spectroscopy, have to resolve the problem of collisional broadening that can surpass the Doppler width and compromise an attempt to enhance the resolution.

1.1. Dicke narrowing due to collisions with buffer gas atoms

In a cell with rarefied vapours the mean free path of atoms or molecules is much larger than the wavelength of electromagnetic radiation corresponding to the transition in question. In this case Doppler broadening dominates. In denser samples, the mean free path is shortened and the Doppler width is diminished. To avoid collisional broadening, which is very effective in the case of collisions between the same atoms, buffer gases may be employed. The method works perfectly if the internal state of the radiating particle and the phase of its polarization are preserved in the course of the collision event. There is a vast literature on this subject.

1.2. Dicke narrowing in a pillbox

The main idea of [2] is that in a pillbox shaped cell the slowly moving atoms contribute most to the absorption as well as to the fluorescence simply because they spend more time in their free flights between the walls. In this case preservation of the internal state of the radiating particle is less important, while the distance between the cell walls gains crucial importance. Dealing with the microwave region, R.H. Dicke employed a standing wave in a cell one half wavelength thick. This particular choice of the width guarantees that all atoms experience the field oscillations of the same phase.

2. Cojan sub-Doppler reflection spectroscopy

The role of the slowly moving atoms and their transient polarization in the presence of only one boundary was highlighted by J.L. Cojan [3]. His theory of the sub-Doppler spectral line shapes in reflection was further developed in [4]. An essential part of the theoretical description consists in the statement that the atoms that move out of the boundary between the transparent solid material and the vapor itself are not in the steady state. In optical domain the atomic polarization is completely destroyed after the atom-wall collision. Even more, in most cases the atoms that leave the surface are in their ground state [5]. As the polarization of these atoms is zero at the surface, there is no reason to expect that their contribution to the reflectivity of the boundary is equal to the contribution of the atoms with the steady state polarization. A somewhat surprising result is that the atoms that leave the surface with the velocity v make just the same contribution to the reflection spectrum as the atoms that are approaching the surface with the velocity $-v$. Hence, integration over the velocity distribution runs from minus infinity to zero rather than from minus infinity to plus infinity. Such integration is known to lead to the sharp Doppler-free features in the reflection spectra. More details may be found in [6-8].

Narrow sub-Doppler resonances in selective reflection were successfully employed to study collisional broadening of atomic spectral lines otherwise hidden under much larger Doppler width [9].

From the point of view of the electrodynamics of continuous media, sub-Doppler features in the selective reflection spectra are due to the spatial dispersion of a Doppler-broadened gas. As the Doppler broadening dominates, it means that the free path of the atoms exceeds the wavelength. Hence, the dielectric response of the gas is highly nonlocal. Macroscopic electrodynamics fails in this case and needs additional boundary conditions that may be supplied only by the microscopic theory. Nevertheless, when the microscopic theory is developed the results may be cast in the form of the admittance of the gas boundary. This value may be used in the ordinary Fresnel formulas of macroscopic electrodynamics as a substitute of the refractive index.

3. Nonlinear effects in selective reflection

The study of nonlinear effects in the selective reflection was started with an ordinary saturation effect in a two level system [10]. It was shown that the contribution of the departing atoms is not equal to the contribution of the arriving atoms in this case. The saturation intensities for both contributions differ as well. Nevertheless, the sub-Doppler features remain in the nonlinear reflection spectra. Nonlinear selective reflection from an atomic vapor at arbitrary incidence angle was studied in [11]. Nonlinear effects in selective reflection were employed to induce the self-diffraction of light on the boundary of resonance vapours [12]. Then, the nonlinear effects in selective reflection were studied in a pump-probe scheme experimentally [13] and theoretically in a two-level system [14] and in a three-level system [15].

4. Line-shape modification via optical coating of the interface

The shape of the spectral lines obtained in selective reflection at the boundary of a rarefied gaseous medium is usually dominated by the real part of its admittance. The situation may be changed by the optical coating of the transparent dielectric material that confines the gas. The task here is somewhat different from the ordinary antireflection coating. As the gas is rarefied, diminishing of the reflectivity of the bare dielectric material will lead to a very weak reflection proportional to the square of the gas density. If, on the contrary, one want to enhance reflectivity proportional to the imaginary part of the surface admittance the coating thickness and its refractive index should be chosen properly. The exact formulas for one and two layer coatings are given in [16].

An important consequence of this consideration is that the spatial inhomogenuity of gas phase in the close proximity of the surface may lead to the mixing of real and imaginary parts of the surface admittance of the gas and the corresponding distortion of the line shapes.

5. van der Waals interaction of excited atoms with solid surfaces

As the shapes of the lines recorded in selective reflection are narrow and sensitive to the surface conditions they were successfully employed to study the van der Waals interactions of the excited atoms with the solid surfaces [17]. Even more intriguing, selective reflection spectroscopy provides with the evidence of van der Waals repulsion of the excited state atoms when there is a resonance between the atomic transitions and the internal excitations of the solid surface [18].

6. Reflection from and transmission through the narrow slice of atomic vapors

By merging the ideas of Dicke and Cojan we have developed a theoretical description of optical reflection from and transmission through the narrow slice of atomic vapors [19]. Contrary to the microwave regime, the standing wave approximation was not used assuming an antireflection coating to be applied on the rear wall of the cell. A manifold of line shapes was obtained depending on the width of the cell. An unusual interference between transient polarizations with the period doubled compared to familiar Fabri-Perot resonances is shown to lead to four fold enhancement of the reflection amplitude as well as switching between the even and odd spectral line shapes.

7. Invention of the Extremely Thin Cell

It was not until the invention of an Extremely Thin Cell by D. Sarkisyan [20] that the observation of these effects becomes possible in the optical domain. Since that time the field is flourishing with a number of linear and nonlinear optical effects observed in the ETC in different laboratories [21,22].

An additional feature of ETC is a possibility to explore the atom-wall interaction. Although important information has been obtained from the position and distortion of the central part of the spectral line [23], far wings provide information on the interaction at still smaller distances from the solid surface.

8. Applications

The atomic resonances are indispensable in the role of the absolute frequency reference in mobile devices for civil and military applications. Hence, the need for the tiny spectroscopic cells mount on the chip will grow [24].

Acknowledgments

This work was supported by a grant No 11-02-01135 of Russian Foundation for Basic Research.

References

[1] Dicke R.H. 1953 The effect of collisions upon the Doppler width of spectral lines *Phys. Rev.* **89** 472-3

[2] Romer R. H. and Dicke R.H. 1955 New technique for high-resolution microwave spectroscopy *Phys. Rev.* **99** 532-6

[3] Cojan J.L. 1954 Contribution a l'etude de la reflection selective sur les vapeurs de mercure de la radiation de resonance du mercure *Ann. Phys.* (Paris) **9** 385-440

[4] Schuurmans M.F.H. 1976 Spectral narrowing of selective reflection *J. Phys.* (Paris) **37** 469-85

[5] Przhibelskii S.G. and Khromov V.V. 2000 Exchange between Translational and Electronic Energies of Atoms in Their Impacts with a Surface *Optics and Spectrosc.* **88** 17-21

[6] Burgmans A.L.J., Woerdman J.P. 1976 Selective reflection from sodium vapor at low densities

1976 *J. Phys.* (Paris) **37** 677-81
[7] Burgmans A.L.J., Schuurmans M.F.H. and Bölger B. 1977 Transient behavior of optically excited vapor atoms near a solid interface as observed in evanescent wave emission. *Phys.Rev.* **16** 2002-7
[8] Schuurmans M.F.H. 1980 The fluorescence of atoms near a glass surface *Contemp. Phys.* **21** 463-82
[9] Akul'shin A.M., Velichanskii V.L., Zibrov A.S., Nikitin V.V., Sautenkov V.V., Yurkin E.K. and Senkov N.V. 1982 Collisional broadening of intra-Doppler resonances of selective reflection on the D_2 line of cesium *JETP Letters* **36** 303-7
[10] Vartanyan T.A. 1985 Resonant reflection of intense optical radiation from a low-density gaseous medium *Sov. Phys. JETP* **61** 674-7
[11] Nienhuis G., Schuller F. and Ducloy M. 1988 Nonlinear selective reflection from an atomic vapor at arbitrary incidence angle, *Phys.Rev.* A **38** 5197-205
[12] Akulshin A.M., Velichanskii V.L., Vartanyan T.A., Gamidov R.G., Sautenkov V.A. and Filimonov S.I. 1989 Self-diffraction of resonance radiation on a selective gas mirror *Opt. Spectrosc.* (USSR) **66** 423-4
[13] Velichanskii L., Gamidov R.G. , Pak G.T. and Sautenkov V.A. 1990 Nonlinear selective reflection of bichromatic light from interface between transparent dielectric and resonant gas *JETP Letters* **52** 136-40
[14] Schuller F., Nienhuis G. and Ducloy M. 1991 Selective refection from an atomic vapor in a pump-probe scheme *Phys.Rev.* A **43** 443-54
[15] Schuller F., Gorceix O. and Ducloy M. 1993 Nonlinear selective refection in cascade three-level atomic systems *Phys.Rev.* A **47** 519-28
[16] Vartanyan T. A. and Träger F. 1994 Line shape of resonances recorded in selective reflection: influence of an antireflection coating *Opt. Commun.* **110** 315-20
[17] Oria M., Chevrollier M., Bloch D., Fichet M. and Ducloy M. 1991 Spectral observation of surface-induced van der Waals attraction on atomic vapor *Europhys. Lett.* **14** 527-32
[18] Fichet M., Schuller F., Bloch D. and Ducloy M. 1995 van der Waals interactions between excited-state atoms and dispersive dielectric surfaces *Phys. Rev.* A **51**, 1553-64
[19] Vartanyan T.A. and Lin D.L. 1995 Enhanced selective reflection from a thin layer of a dilute gaseous medium *Rhys. Rev.* A **51** 1959-64
[20] Sarkisyan D., Bloch D., Papoyan A., Ducloy M. 2001 *Opt. Comm.* **200** 201-8
[21] Cartaleva S., Saltiel S., Sargsyan A., Sarkisyan D, Slavov D., Todorov P. and Vaseva K. 2009 Sub-Doppler spectroscopy of cesium vapor layers with nanometric and micrometric thickness *JOSA* B **26** 1999-2006
[22] Naumov A.N., Podshivalov A.A., Drabovich K.N., Miles R.B. and Zheltikov A.M. 2001 Theory of Doppler-free spectroscopy with λ-thick vapor cells Phys. Lett. A **289**, 207
[23] Laliotis A. *et al.* 2007 Testing the distance-dependence of the van der Waals interaction between an atom and a surface through spectroscopy in a vapor nanocell *Proc. SPIE* **6604**, 660406-1 - 11
[24] Hasegawa M., Chutani R.K., Gorecki C., Boudot R., Dziuban P., Giordano V., Clatot S., Mauri L. 2011 Microfabrication of cesium vapor cells with buffer gas for MEMS atomic clocks *Sensors and Actuators* A **167** 594-601

Peculiarities of atomic lines in sonoluminescense spectra

M V Kazachek[1] **and T V Gordeychuk**[2]
[1,2]V.I.Il`ichev Pacific Oceanological Institute,
43, Baltiyskaya Street, Vladivostok, 690041, Russia

E-mail: mihail@poi.dvo.ru

Abstract. Alkali-metal lines in SL spectra are broadened, asymmetrically shifted toward the red spectral region, unshifted narrow parent peaks are observed. The shape is influenced a number of parameters. The effects have no explanation. We model a line shape, assuming that line broadening arises from a density and line asymmetry results from superposition of spectra generated at different densities of a perturbing medium. Simulation shows that broad-band emission occurs within the density range of 10-400 Amg for KCl, LiCl, NaCl aqueous solutions under Ar at the ultrasound frequency of 20 kHz. The lower limit of the range shifts to higher density in order of K<Li<Na following the increase of exited state energy. The effect of surfactant on the shape of Na D-line is also observed, which probably indicates the essential influence of nano-size layers on the dynamics of emitting process.

1. Introduction

Sonoluminescence (SL) was discovered about 80 years ago by Frenzel and Shultes as light emission from bubbles in liquids undergoing ultrasound irradiation [1]. Later it was called as multibubble SL (MBSL). It is considered that emission originates from bubbles violently collapsed by acoustic pressure - "hot-spot" hypothesis [2]. As result a bubble can focus sound energy by 12 orders of magnitude. About 20 years ago a great experimental result was gotten by Gaitan. He found regime for stable single bubble SL (SBSL) [3]. SBSL spectra are referred to blackbody, bremsstrahlung, eximer's emission. The shape of MBSL spectra differs from SBSL shape and usually consists of a continuum and separate lines. The lines are emitted by atoms and molecules of irradiated solution, and their fragments.

Peak conditions inside the bubbles are really extreme (the values for pressure of 500 atm and temperature of 15000 K are experimentally defined). A lot of intrabubble estimations were received using SL spectra. MBSL temperature was estimated from the intensity of metal lines [4]. SBSL temperature and the plasma density are estimated using the asymmetry of argon line [5]. Ionization degree for such electron density is about 3. Broadening and shift of metal lines are used for estimation of pressure and density of emission medium, for example, using high-volatile metal carbonyls dissolved in organic liquid with low vapor pressure [6]. In this work the pressure is evaluated as 300 bars and relative density as 20 Amg.

Alkali-metal atomic emission is one of unresolved problems of SL. Firstly the mechanism of non-volatile salt hitting into the bubble is unclear. Now most of experimental results support the injected-

[1] To whom any correspondence should be addressed.

droplet model [2]. The structure of the lines of emission from metal atoms is distorted. The lines exhibit broadening, asymmetry and a long-wavelength shift as compared to the flame fluorescence spectra. Narrow peaks of the parent emission doublet are observed on the background of these broadened lines. The first estimations of bubble-collapse parameters from Na line broadening and shift were made in classic work and were appeared as: density about 40 Amg, pressure about 300 bars and temperature about 3000 K [7].

The shape of atomic lines is influenced different parameters, but most of the effects haven't any explanation. The shape of K line is different in water and octanol [8]. However in [8] the shape is independent from dissolved gas, which fact is surprising. It is known that helium shifts the line toward violet side in contrast to argon. Recently it was shown that helium really shifts line toward violet side, and the line shape changes for different gases [9]. Also the noticeable ultrasound frequency effect is obtained for K line [10]. Line width grows when sound frequency falls. Authors do not explain the effect. It is interesting that frequency effect for Na is opposite than for K [11]. In [11] is also obtained that in case of water the wide component of spectrum is more pronounced as compared with the case of acid. Authors propose that something is superimposed on the Na line in water case. Results [11] show the shape of Na line is influenced by alcohol likewise for K line. Its width grows with increasing of ethanol content in water. Also the shape changes with ultrasound intensity. In spectra from [11], as like as in spectra presented in [9], one more peculiarity of SL line shape is obvious - the presence of unshifted narrow peaks of metal atoms in SL spectra. If one assumes the emission comes from very dense medium, it is strange. They should be spread over the spectrum under conditions near bubble collapse.

2. Results and discussion

In general, until recently, the width of a line was being considered, but asymmetry and presence of unshifted peaks were not been explained. Nevertheless, the line shape can keep the information about the conditions during the time when the emission occurs. Recently it was shown that metal emission occupies a rather long period of time during collapse phase [12]. So we assume that complex profile of metal line forms due to superposition of the spectra generated at different densities of a perturbing medium. The simple model based on the hypothesis of integral character of metal-line shape in SL successfully reproduces all features of experimental spectra (involving shift, broadening, asymmetry, and parent emission peaks) [13]. Simulation of the experimental spectra allows one to make a conclusion about the character of density variations during the emission, and to determine the interval of densities in which the emission from alkali-metal atoms occurs.

Figure 1 shows the high-resolution (0.26 nm) spectra in the vicinity of atom emission peaks (thick solid lines) and the results of simulation, heeding shift and the broadening by density during the emission (dotted lines). The spectra were measured from Ar-saturated 3M KCl, NaCl, LiCl aqueous solutions. Ultrasound frequency was 20 kHz. The procedure was described elsewhere [13]. The fluorescence spectra of a flame with salts additives, measured at the same resolution as that used for the SL spectra, are shown for the comparison (thin solid lines). The spectra are normalized at the first doublet component in respective flame spectrum.

The unshifted peaks are not prominent in Na spectrum in Fig. 1 whereas they are observed in some experimental spectra [9, 11]. From the viewpoint of our model the only origin of parent peaks is "low-density" emission with the value of about 1 Amg [13]. Recently [12], it was experimentally demonstrated that Na emission can occur in "low-density" phase far from the point of full collapse.

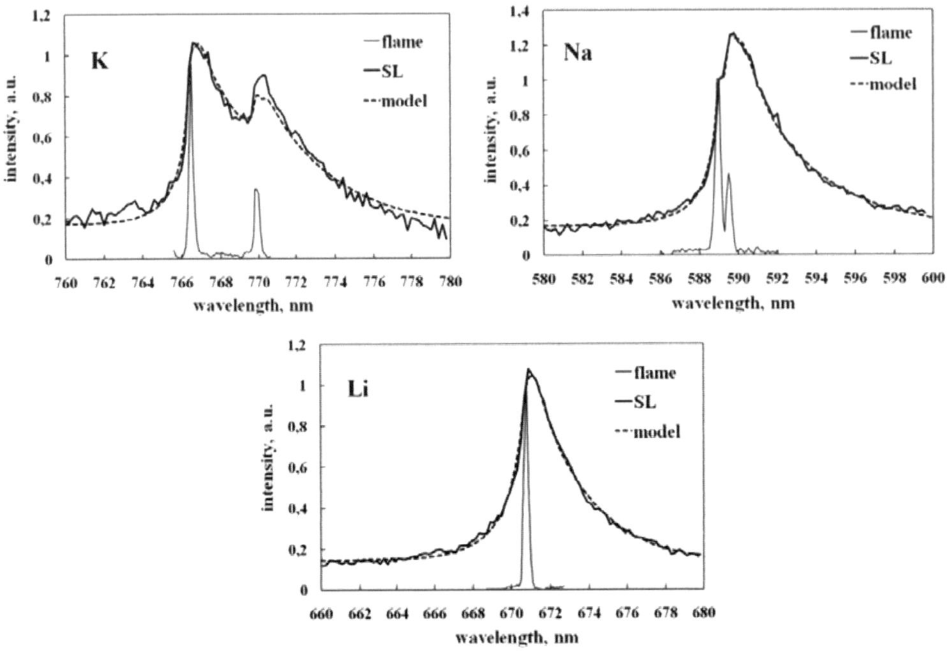

Figure 1. MBSL alkali-metal high-resolution (0.26 nm) experimental spectra and their models; alkali-metal fluorescence spectra in a flame (resolution 0.26 nm).

Using the model the lower and the upper values of density, when alkali-metal broad-band emission occurs, were estimated (Table 1). The model calculations show the increase of the lower density limit in order of K<Li<Na, which is probably caused by the growth of excited energy of atomic states in the same order. It is not unlikely that K, whose excited state energy is smaller than that for Na and Li, begins to emit at earlier stage of collapse and, hence, at lower density of perturbing medium. Thereupon Li starts to emit. Na emission begins at higher density.

Table 1. Density range of alkali-metal emission during SL.

Metal	Wavelength, nm	Excited Energy, eV	Minimum Density, Amg	Maximum Density, Amg
Na	590	2.11	41	415
Li	671	1.85	16	279
K	768	1.62	10	203

The upper limit of density range also shifts toward the higher values with the increase of the energy of alkali-metal excited states. The reason why the alkali-metals cease to emit at high density, when the emission of continuum occurs, is unclear. Possibly, the emitting layers inside the collapsing bubble become optically opaque or, which is more likely, the duration of "high-density" emission is too short for noticeable contribution to SL spectrum.

Figure 2 shows a new result demonstrating the influence of surfactant on the shape of Na line in SL spectra. Earlier it was reported that small amount of surfactant leads to the growth of intensity of Na line, as well as in a case of large concentration of salt [14]. It was also calculated that the thickness of a salt-solution layer of about 16 nm near bubble boundary is enough for explanation of Na line intensity [15]. In the case of surfactant addition a thinner layer is required. We propose that vaporization of the thin layer begins at less density. Possibly it is the reason why we obtain more narrow line for the case of addition of surfactant. The density range evaluated from the shape is shown in Table 2.

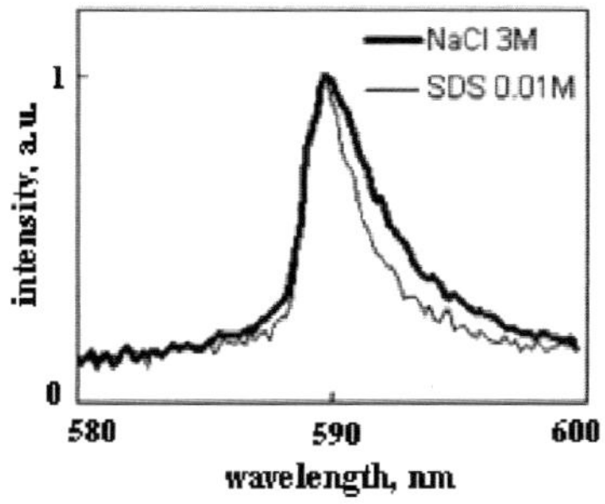

Figure 2. Na D-line shape in MBSL high-resolution (0.26 nm) spectra from 3M NaCl Ar-saturated aqueous solution and with addition of 0.01M of surfactant.
Ultrasound frequency is 20 kHz.

Table 2. Density range of Na emission from aqueous NaCl solution and with addition of surfactant.

	Minimum Density, Amg	Maximum Density, Amg
3M NaCl	38	364
0.01M SDS	28	285

We resume that the question about the peculiarities of atomic lines in SL spectra remains open and needs further investigations.

References

[1] Frenzel H and Schultes H 1934 *Z. Phys. Chem. Abt. B.* **27B** 421
[2] Xu H, Eddingsaas N and Suslick K 2009 *J. Am. Chem. Soc.* **131** 6060
[3] Gaitan D, Crum L, Churh C and Roy R 1992 *J. Acoust. Soc. Am.* **91** 3166
[4] Didenko Y, McNamara W and Suslick K 2000 *Phys.Rev.Lett.* **84** 777
[5] Flannigan D and Suslick K 2010 *Nature Physics Letters* **4** 598
[6] McNamara W, Didenko Y and Suslick K 2003 *J.Phys.Chem.B* **107** 7303
[7] Sehgal C, Steer R, Sutherland R and Verrall R 1979 *J.Chem.Phys.* **70** 2242
[8] Flint E and Suslick K 1991 *Science* **253** 1397
[9] Hayashi Y and Choi P-K 2010 *Proc.20thIntern.Congr.Acoust.ICA2010* (Sydney: Australia) p 248
[10] Abe Sh and Choi P-K 2008 *Nonlin. Acoust.: Fundamentals and Applications (ISNA 18)* (Stockholm, Sweden/AIP Conf. Proc.) **1022** p 201
[11] Hatanaka S, Hayashi S and Choi P-K 2010 *Japan. Journ. Appl. Phys.* **49** 07HE01
[12] Chu H, Vo S and Williams G 2009 *Phys. Rev. Lett.* **102** 204301
[13] Kazachek M and Gordeychuk T 2011 *Tech. Phys. Lett.* **37** 262
[14] Sunartio D, Yasui K, Tuziuti T, Kozuka T,Iida Y,Ashokkumar M and Griezer F 2007 *Chem.Phys.Chem.* **8** 2331
[15] Lepoint-Mullie F, Voglet N, Lepoint T and Avdi R 2001 *Ultrason. Sonochem.* **8** 151

Vibrational spectra of molecular fluids in nanopores

V G Arakcheev[1], **V B Morozov**[1,2]

[1] International Laser Center, M.V. Lomonosov Moscow State University, 119991 GSP-1, Moscow, Russia
[2] Faculty of Physics, M.V. Lomonosov Moscow State University, 119991 GSP-1, Moscow, Russia

E-mail: morozov@phys.msu.ru

Abstract. Coherent anti-Stokes Raman spectroscopy (CARS) is applied for quantitative analysis of carbon dioxide phase composition in pores of nanoporous glass samples at nearcritical temperatures. Measurements of the 1388 1/cm Q-branch were made in a wide pressure range corresponding to coexistence of gas (gas-like), adsorbed and condensed phases within pores. At temperatures several degrees below the critical value, CARS spectra behavior is easy to interpret in terms of thermodynamic model of surface adsorption and capillary condensation. It allows estimating mass fractions of different phase components. Moreover, spectra measured at near critical temperatures 30.5 and 33°C have pronounced inhomogeneous shapes and indicate the presence of condensed phase in the volume of pores. The effect obviously reflects the fluid behaviour near the critical point in nanopores. Pores with smaller radii are filled with condensed phase at lower pressures. The analysis of the CARS spectra is informative for quantitative evaluation of phase composition in nanopores.

1. Introduction

Phase behavior of molecular fluids in pores of natural and synthetic porous materials is of considerable practical and theoretical interest to such areas as filtration and separation of molecules and species, chemical catalysis, biosensing, research and development of nanocomposite materials. A fluid confined in pores of nanometer scale diameters can reveal phase transition properties essentially different from those in a bulk volume [1-7]. Referring to the development of supercritical fluid technologies [8, 9], where molecular fluids can be used as a chemical reagent or a transport media, it is important to have a tool for rapid non-perturbing local probing microstructure and phase behavior of the medium filling pores in the volume of nanoporous samples. Vibrational spectroscopy methods and Raman techniques in particular [10] are widely used for study and diagnostics of molecular fluids, including near and supercritical ones [11-16]. Quite naturally, Raman spectroscopy is proposed as an effective tool for probing fluids filling pores of transparent nanoporous host materials [2, 6].

Carbon dioxide, with the critical values of temperature and pressure T_C =31.1°C and p_C =72.8 atm correspondingly, is a most popular media for supercritical technologies. It has well studied structure of molecular ro-vibrational levels and spectral lines [17]. In the domain of Raman spectroscopy Fermi-dyad $\nu_1 / 2\nu_2$ (1285 cm^{-1} and 1388 cm^{-1}) is a "face-brand" of carbon dioxide since earliest period of Raman spectroscopy [18,19], its pre-laser [20] and laser periods [21]. Two intense peaks are used as spectral markers to probe carbon dioxide, at near-critical and supercritical conditions in particular [12, 15, 16, 22-24]. At the same time, some peculiarities of the dyad spectral transformations due to collisional dephasing [25-28], dimer aggregation and clustering [24, 29] are necessary to investigate as they clarify data concerning microstructure of the fluid and intermolecular interactions. In the present study we probed high-frequency peak 1388 cm^{-1}, with close to homogeneous spectral profile in pressure range from 40 atm and higher.

In pores of several nanometers in diameter, the amount of fluid adsorbed on walls of pores can be comparable with the amount of gaseous or condensed fluid in the pore volume. Each phase is characterized by definite values of spectral widths and shifts, therefore the structure of multi-component phase composition can be recognized by Raman spectra analysis.

Based on concept of adsorption and condensation inside pores, spectral behavior of molecular medium can be associated with thermodynamic parameters (pressure, density, temperature), as well as with the characteristics of porous material structure (the average pore radius distribution, void ratio, specific surface area etc.). Available transparent materials like nanoporous glasses, polymers, aerogels, zeolites provide a wide range of pore radiuses and morphology and allow realizing different confinement conditions. Recently we applied coherent anti-Stokes Raman spectroscopy (CARS) method [30] to study spectra of carbon dioxide filling pores of nanoporous glasses at temperatures near 20°C (several degrees lower than the critical temperature) and found that adsorption and condensation in pores are accompanied by a rather vividly interpreted spectral behavior [6, 31-35]. In the present paper, CARS spectra measurements for carbon dioxide in pores of nanoporous glasses were extended to nearcritical temperatures.

2. Experimental CARS spectra measurements

Two kinds of nanoporous glass samples, Vycor [36] and DV-1M [37], have mean pore radii 2 and 3.5 nm, respectively, and the same radius distribution widths of ~0.5 nm. The samples were placed inside a thermostabilized high-pressure cell made of stainless steel (inner volume is ~ 1 cm^3) between input and output quartz windows [6, 31].

CARS spectrometer was based on nanosecond passively Q-switched Nd:YAG oscillator with an amplifier and second harmonic unit and a broadband dye-laser [6, 16]. Equally linearly polarized beams of second harmonic radiation ω_1 (532 nm wavelength, 0.05 cm^{-1} spectral bandwidth, 15 ns pulse duration and 0.2 mJ pulse energy) and tuned dye-laser radiation ω_2 (571-577 nm central wavelength, 15 cm^{-1} spectral bandwidth, 12 ns pulse duration and 1 mJ pulse energy) focused into the cell by a lens of 17 cm focal length were used in broadband collinear CARS-scheme. The instrumental function of the registration system was ~0.5 cm^{-1}.

Examples of CARS spectra transformations with pressure at 20.5°C are shown in figure 1 for the Vycor sample. DV-1M sample demonstrated qualitatively similar behaviour [35]. At reasonably low pressures, the profiles of the measured spectra are Lorentzian and close to that in bulk gas. Nonresonant background is negligible. As the pressure is increased towards the saturation value, another red shifted peak appears; this peak corresponds to the contribution of the molecules adsorbed on the walls of pores. The peak gradually grows with respect to the gas peak as the pressure approaches the bulk saturation pressure in accordance with the increase of the mass fraction of molecular layer adsorbed on walls of pores and growing condensation inside pore channels. At the same time, the gas contribution still exists; it is mainly caused by the gaseous fluid situated in the gaps between the sample and the cell windows and also by remains the gas phase in large pores. Schematic representation of different stages of the process of pore filling is shown in figure 2. When the pressure reaches the bulk saturation value, the spectrum profile takes a Lorentzian shape again and coincides with that in bulk liquid. It indicates that the fluid turns completely into the condensed phase inside the cell and pore volume. Each series of measurements with a given sample at a given temperature was performed with a fixed position of the sample in the cell and with a fixed alignment of laser beams. In all experimental series with both nanoporous samples, the transformations of CARS spectra were qualitatively similar. Figure 3 illustrates normalized spectral contributions corresponding to different carbon dioxide phases in nanopores at 20.5°C, gas peak width and frequency position correspond to pressure 53 atm.

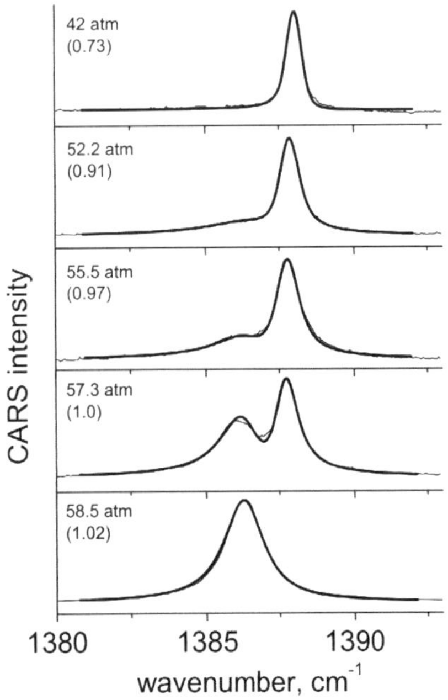

Figure 1. Experimental (thin line) CARS spectra of the 1388 cm^{-1} Q-branch of carbon dioxide measured with the Vycor glass sample at 20.5°C (p_{sat}=57.3 atm) at different pressures up to condensation and spectra calculated (thick line) according the model (1)-(6); normalized pressures are shown in brackets

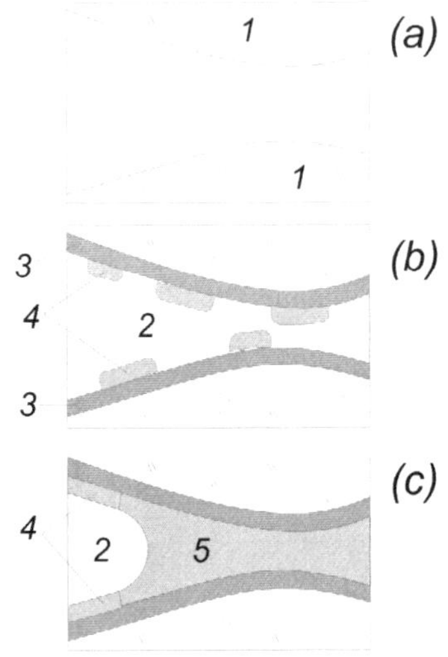

Figure 2. Schematic representation of different stages of the process of pore filling: an empty pore (a), adsorption on walls of pores (b), and capillary condensation (c); 1: glass, 2: gas phase within pores, 3: the first monolayer, 4: adsorbed layers, 5: liquid.

3. Theoretical modelling

3.1. CARS spectrum profile

In the scalar theory of CARS , the anti-Stokes wave intensity is expressed through the interference of resonant and nonresonant responses of the nonlinear media [30, 38]:

$$I_a(\omega_1-\omega_2)\propto\left|\lambda_R^{(3)}+\lambda_{NR}^{(3)}\right|^2 I_1^2 I_2, \tag{1}$$

where I_1, I_2, I_a are the intensities of the first (ω_1) and the second (ω_2) pumping and anti-Stokes waves respectively; $(\omega_1-\omega_2)$ is tuned to the Raman resonance frequency Ω; $\lambda_R^{(3)}$ and $\lambda_{NR}^{(3)}$ are the coefficients of resonant and nonresonant interaction in four-wave mixing process. Resonance signal is induced in Raman-active fluid inside and outside the nanoporous samples. At temperatures below the critical value, the nanoporous sample is surrounded by gas at pressures $p<p_0$ (p_0 is a saturation pressure) or liquid at $p>p_0$, according to the equation of state [39]. Resonance frequency in liquid is red-shifted respective to one in gas [15, 16, 24, 28]. Inside the nanopores, significant amount of the fluid can be adsorbed on the walls of pores. The molecular vibrations in the adsorbed layers are modified, therefore the corresponding spectral contributions generally differs from ones of gas and liquid. In the central part of pore channels, the fluid can be in gaseous or condensed states, depending on the thermodynamic conditions. Spectral responses of fluid inside pores at temperatures below the critical value have strongly inhomogeneous line shapes consisting of two or three components that corresponds to contributions of gas, adsorbed and condensed (liquid) phases [34, 35]. Widths and positions of gas and liquid contributions from pores volume are equal to that from the bulk; the liquid contribution is red-shifted with respect to the gas one. Spectral dependences for the bulk carbon dioxide were thoroughly investigated heretofore [28] and now are used as a reference data. Accurate analysis also shows that the spectral contribution of adsorbed layers lies between gas and liquid peaks, and its linewidth almost twice broader than one of liquid phase [6, 33-35], see figure 3. Thereby, the resonant part of the anti-Stokes signal can be calculated as follows:

$$\lambda_R^{(3)} = \sum_k \frac{A_k \bar{\lambda}_k^{(3)}}{-i-\Delta_k}, \tag{2}$$

where coefficients A_k describe qualitative proportions of different molecular fractions, $\bar{\lambda}_k^{(3)}$ is the relative spectral amplitude of a fraction k, and $\Delta_k = \omega_1 - \omega_2 - \Omega_k / \Gamma_k$ is the frequency detuning of the biharmonic pump from the Raman resonance frequency Ω_k normalized to the line width Γ_k.

3.2. *Thermodynamic approach*

In our previous works [34, 35], it was demonstrated that spectral behavior at temperatures reasonably below the critical value satisfactory agrees with phase fraction estimation based on thermodynamic description of the surface adsorption and capillary condensation in pores with a definite pore radius distribution.

For pore specification we use following dimensionless function expressing volume fraction of pores with radius less than r:

$$V(r) = 0.5\left\{1+\tanh\left(\frac{r^2-r_0^2}{r_0\delta r}\right)\right\}, \tag{3}$$

where r_0 is the mean pore radius, δr=0.58 nm, that corresponds to distribution width ~0.5 nm of pore radius distribution defined as a derivative of (3) $U(r) = \mathrm{d}V(r)/\mathrm{d}r$:

Adsorption on the nanopore walls is described on the basis of the BET (Branauer-Emmet-Teller) model [40] in the form:

$$\frac{n_m}{n} = \left(\frac{F^*}{F}-1\right)\left(\frac{1}{C}+\frac{C-1}{C}\frac{F}{F^*}\right). \tag{4}$$

There n is the total adsorption, n_m is the monolayer capacity, F is fugacity, which replace pressure in calculations in order to extend the model for the case of nonideal gases, F^*=73 atm and C=11 are parameters of the model.

Condensation conditions of a real gas are governed by the Kelvin equation [41], also including fugacity instead pressure [42]:

$$\frac{F}{F_{sat}} = \exp\left(-\frac{2\sigma V_m}{r_m RT}\right), \tag{5}$$

F_{sat} denotes the fugacity at saturation pressure, V_m is the molar volume of the liquid phase, R is the universal gas constant, r_m is the meniscus curvature radius, σ is the surface tension coefficient that is expressed as $\sigma(T) = b(1-T/T_C)^{1.253}$ [42], b=7.7·10^{-2} J/m^2. According to (4), at given values of F (or pressure) and T, all pores with radius less than $r = r_m + d$ (d denotes a thickness of the adsorbed layers) are filled with condensed phase, while broader channels contain gas phase in a central part of pore volume and adsorbed layers on the pore walls (see Fig.2,c).

4. Results and discussion

Calculations of spectral profiles are fulfilled on the basis of expressions (1)-(2). Mass fractions of each phase were calculated according (3)-(5). For each temperature, fitting parameters are extracted on the basis of the whole series of experimental spectra measured for a given sample at a fixed temperature. The line width and shift parameters for gas and condensed phases were taken from experimental results for carbon dioxide in a bulk volume [28]. Calculated curves for T =20.5°C are shown in figure 1 together with the experimental ones and demonstrate quite good correspondence. Detailed fitting analysis of several experimental series shows [34, 35] that molecules from the first monolayer nearest to the walls of pores do not contribute to the observed coherent signal. More precisely, the amount of molecules in the "silent" adsorbed layer corresponds to 1.3 monolayer. Obviously, molecules can become non-Raman active due to interaction with pore walls. Figure 4 displays an example of calculated dependences of pore volume and mass fractions of different coexisting phases of carbon dioxide in pores at T =20.5°C: gas, adsorbed layers on walls of pores and condensed liquid in pores).

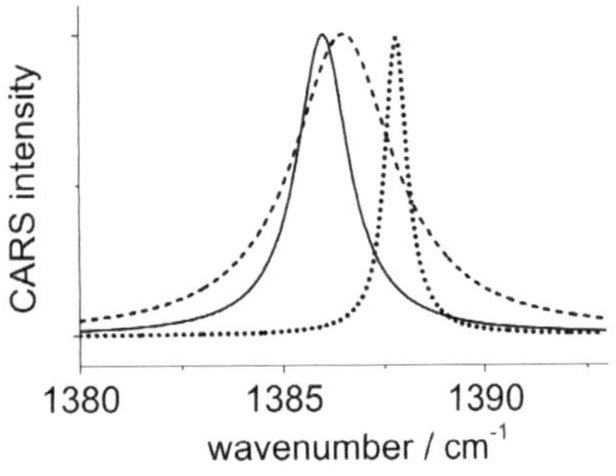

Figure 3. Normalized spectral contributions corresponding to different carbon dioxide phases in nanopores at 20.5°C and at 53 atm: gas (doted line), the adsorbed layer (dashed line), the condensed liquid (solid line).

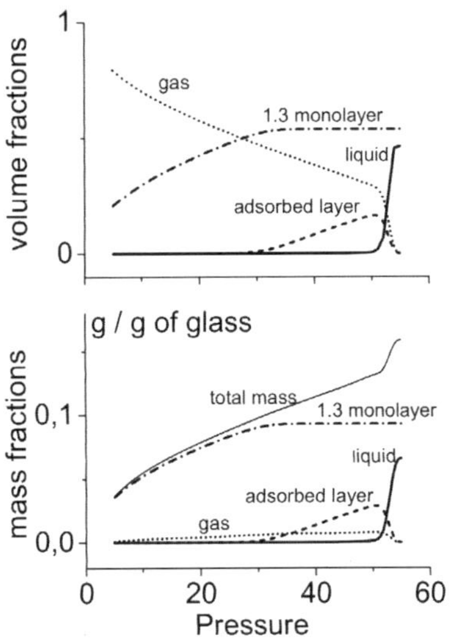

Figure 4. Calculated dependences of pore volume and mass fractions of different coexisting phases and total mass (thin solid line) of carbon dioxide in pores of Vycor at T =20.5°C: gas (doted line), adsorbed on pore walls layers (dashed line), first monolayer (dash-doted line), condensed in pores liquid (thick solid line).

Despite the significant volume fraction of gas phase inside pores, its mass fraction remains small in comparison with the total mass. It should be recalled that in view of the experimental geometry, in which the samples were placed within the cell charged with carbon dioxide, gas (or liquid) phase presents not only inside pores; also bulk fluid fills the gaps between nanoporous samples and quartz windows. The gaps are crossed by the periphery part of focal region of the laser beams but still make a significant addition to the anti-Stokes signal. The experimental geometry (the samples positions and laser beams focusing conditions) remained unchangeable during each series of experiments. The contribution of the fluid from the gaps to the total anti-Stokes signal was assumed to be proportional to second degree of concentration calculated using the equation of state [39]. Results of spectral analysis show that in almost the entire measured pressure range essential part of carbon dioxide molecules within nanopores are in adsorbed state. As was discussed above, molecules that fill the nearest to the pore walls, "silent" adsorbed layer (1.3 monolayers, d =0.5nm), do not contribute to the anti-Stokes signal, that is why their fractions are shown in the figure 4 separately. When the first adsorbed "silent" monolayer becomes filled, the adsorbed layer continues its filling with molecules that contribute into the signal. Total amount of adsorbed molecules permanently increases as long as condensation process does not prevail.

As the temperature rises and approaches the critical value, distinguishing the contributions of gas, adsorbed and condensed phases is becoming increasingly difficult. The reason is a decrease of density difference between coexisting phases of gas and liquid according to equation of state [39] and corresponding reduction in frequency difference [28]. However, the effective numerical analysis based on the discussed above model is still possible up to the temperatures several degrees lower than the critical value. Analyzing the spectra in this case one must pay attention to the behavior of their width and asymmetry. In the nearest vicinity of T_C, detailed interpretation of the phase behavior of near-critical and supercritical fluids in nanopores becomes ambiguous. We show that some useful quantitative information can be obtained by CARS.

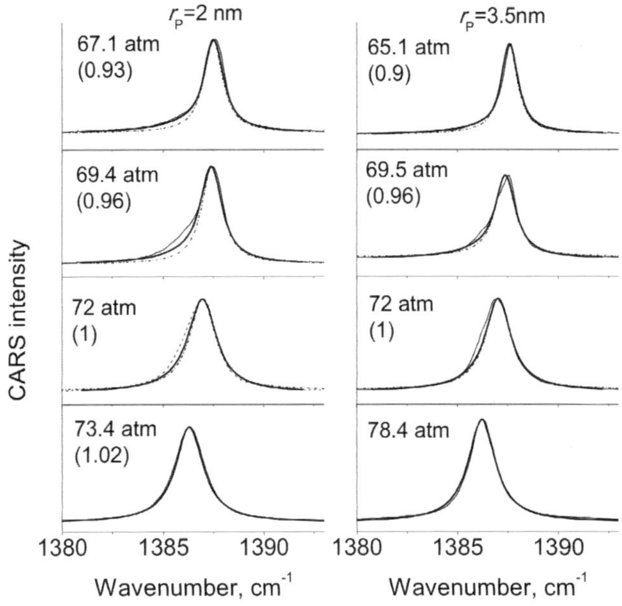

Figure 5. Experimental CARS-spectra (thin solid lines) of carbon dioxide at different pressures obtained with the Vycor (left) and DV-1M (right) samples at near-critical temperature 30.5°C, calculations (thick solid line) according (1)-(6) allowes to demonstrate contribution of surface adsorption into the low-frequency wing. Lorentzian profiles corresponding to bulk fluids at the same conditions are shown to display asymmetry of the spectra.

We studied spectral responses at T =30.5 and 33°C. Spectral shapes are very sensitive to small pressure variations. Surface tension value is extremely low at 30.5°C ($\sigma \sim 3 \cdot 10^{-9}$J/m^2) and absents at 33°C, then capillary condensation should not be taken into account, but adsorption of molecules on surface of pore walls still takes place. A significant increase in the low-frequency spectra wing indicates the presence of adsorbed and condensed phases inside pores. The analysis carried out for T =30.5°C by modeling the adsorption in pores for both samples shows that the increase in low-frequency wing can not be explained only by surface adsorption. Obviously, the condensed phase of carbon dioxide exists inside the pores at ~69.5 atm (see figure 5). At supercritical temperature 33°C, inhomogeneous broadening is similar to that at T =30.5°C. At sufficiently high pressures, pore volume is occupied by fluid with density of liquid; at the same time, the density in substantial part of pore volume is gas-like. Fitting the experimental spectra according to (1)-(2) allows us to calculate the mass and volume fractions of condensed and gas (gas-like) phases inside nanopores. Quantitative analysis shows that at the same pressure values, relation between fractions of gas-like and condensed phases strongly depends on temperature. Pores with smaller radii are filled with condensed phase at lower pressures. For example, at T =30.5°C and p =72 atm (see figure 5, the third graphs from the top) estimation for volume fraction of liquid in the Vycor sample is close to 100%, while in the DV-1M is only ~40%. The pattern described above obviously corresponds to the critical fluid behavior.

5. Conclusions

CARS-spectra transformations were observed experimentally for carbon dioxide at compressing at near-critical temperatures in two nanoporous samples with mean pore radii 2 and 3.5 nm. Measurements at 30.5 and 33°C obviously displayed inhomogeneous shapes of obtained spectra. An increase in the intensity of the low-frequency spectral wing can not be explained only by adsorption on the pore walls and, apparently, indicates the presence of the condensed phase in the volume of pores. Relation between condensed and gas (gas-like) phases strongly depends on temperature. Fitting the experimental spectra and analysis of spectral contributions allows us to calculate amount and volume fractions of condensed and gas phases inside nanopores. It was shown that pores with smaller radii are filled with condensed phase at lower pressures. The effects obviously reflect near critical fluid behavior in glass nanopores. Thereby, the CARS spectra analysis is applicable for quantitative evaluation of phase composition of near critical carbon dioxide at nanoporous confinement.

The work was supported by Russian Foundation for Basic Research (projects 11-02-01309-a, 11-02-12112-ofi-m-2011, 12-02-01067-a). The authors acknowledge partial support from M.V.Lomonosov Moscow State University Program of Development.

References

[1] Travalloni L, Castier M, Tavares F M, Sandler S I 2010 *J. Supercrit. Fluids* **55** 455
[2] Li D, Huang Y 2011 *J. Ram. Spectr.* **42** 1396
[3] Smirnov P, Yamaguchi T, Kittaka S, Takahara S, Kuroda Y 2000 *J. Phys. Chem. B* **104** 5498
[4] Wong A P Y, Kim S B, Golburg W I, Chan M H W 1993 *Phys. Rev. Lett.* **70** 954
[5] Hiejima Y, Yao M 2004 *J. Phys. : Condens. Matter* **16** 7903
[6] Arakcheev V G, Valeev A A, Morozov V B, Olenin A N 2008 *Las. Phys.* **18** 1451
[7] Vishnyakov A, Piotrovskaya E M, Brodskaya E N, Votyakov E V, Tovbin Y K 2001 *Langmuir* **17** 4451
[8] Darr J A, Poliakoff M 1999 *Chem. Rev.* **99** 495
[9] Ma X, Tomasko D L 1997 *Ind. Eng. Chem. Res.* **36** 1586
[10] Weber A 1979 *Raman Spectroscopy of Gases and Liquids* (Berlin: Springer-Verlag)
[11] Chesnoy J 1986 *Chem. Phys. Lett.* **125**, 267
[12] Strauss H L, Mukamel W A 1984 *J. Chem. Phys.* **80** 6328
[13] Howdle S M, Bagratashvili V N 1993 *Chem. Phys. Lett.* **214** 215
[14] Musso M, Matthai M, Keutel D, Oehme K-L 2002 *J. Chem. Phys.* **116** 8015
[15] Nakayama H, Saitow K, Sakashira M, Ishii K, Nishikawa K 2000 *Chem. Phys. Lett.* **320** 323
[16] Arakcheev V G, Bagratashvili V N, Valeev A A, Gordiyenko V M, Kireev V V, Morozov V B, Olenin A N, Popov V K, Tunkin V G and Yakovlev D V 2003 *J. Ram. Spectr.* **34** 952
[17] Herzberg G 1991 *Molecular Spectra and Molecular Structure II. Infrared and Raman Spectra of Polyatomic Molecules* 2nd ed (Malabar, Florida: Krieger)
[18] Rasetti F 1929 *Nature* **123** 205
Rasetti F 1929 *Nature* **123** 7577
[19] Fermi E 1931 *Zeitschrift fur Physik* **71** 250
[20] Welsh H L, Pashler P E, Stoicheff B P 1952 *Can. J. Phys.* **30** 99
[21] Wright R B, and Wang C H 1973 *J.Chem.Phys.* **58** 2893

[22] Garrabos Y, Tufeu R, Le Neindre B, Zalczer G, Beysens D 1980 *J. Chem. Phys.* **72** 4637
[23] Deacon C G, Clouter M J, Kiefte H 1985 *J.Chem.Phys.* **83** 446
[24] Cabaco M I, Longelin S, Danten Y, Besnard S 2007 *J. Phys. Chem. A* **111** 12966
[25] Lavorel B, Millot G, Saint-Loup R, Berger H, Bonamy L, Bonamy J, Robert D 1990 *J. Chem. Phys.* **93** 2176
[26] Arakcheev V G, Kireev V V, Morozov V B, Olenin A N, Tunkin V G, Valeev A A, Yakovlev D V 2007 *J.Ram.Spectr.* **38** 1046
[27] Arakcheev V G, Morozov V B 2009 *JETP Lett.* **90** 574
[28] Arakcheev V G, Bagratashvili V N, Valeev A A, Morozov V B, Popov V K 2010 *Russian Journ. Phys. Chem. B* **5** 75
[29] Cabaco M I, Besnard M, Longelin S, Danten Y 2010 *J. Mol. Liq.* **153** 15
[30] Shen Y R 1984 *The Principles of Nonlinear Optics* (New-York, Chicester, Brisbane, Toronto, Singapore: John Wiley&Sons)
[31] Arakcheev V G, Bagratashvili V N, Dubyanskiy S A, Morozov V B, Olenin A N, Popov V K, Tunkin V G, Valeev A A, Yakovlev D V 2008 *J. Ram. Spectr.* **39** 750
[32] Arakcheev V G, Bagratashvili V N, Valeev A A, Morozov V B, Olenin A N, Popov V K, Tunkin V G 2008 *Moscow University Physics Bulletin* **6** 388
[33] Arakcheev V G, Bagratashvili V N, Valeev A A , Morozov V B, Olenin A N, Popov V K, Yakovlev D V 2009 *Russian Journal of Physical Chemistry B* **3** 57
[34] Arakcheev V G, Valeev A A, Morozov V B, Farizanov I R 2011 *Moscow University Physics Bulletin* **66** 147
[35] Andreeva O V, Arakcheev VG, Bagratashvili V N, Morozov V B, Popov V K, Valeev A A 2011 *J.Ram.Srectr.* **42** 1747
[36] Haller W 1983 *Solid Phase Biochemistry* ed Scouten WH (New York: Wiley) ch 11 535
[37] Andreeva OV, Obyknovennaya I E, Gavrilyuk E R 2005 *Opt. Journ.* (in Russian) **72** 37
[38] Akhmanov S A, Koroteev N I 1981 *Methods of Nonlinear Optics in Light Scattering Spectroscopy* (in Russian) (Moscow: Nauka)
[39] Span R, Wagner W 1996 *J. Phys. Chem. Ref. Data.* **25** 1509
[40] Brunauer S, Emmett P H, Teller E 1938 *J. Am. Chem. Soc.* **60** 309
[41] Thomson W 1870 *Proc. R. Soc. Edinburgh* **7** 63
Thomson W 1871 *Proc. R. Soc. Edinburgh* **42** 448
[42] Cazorla-Amoros D, Alcaniz-Monge J de la Casa-Lillo M A, Linares-Solano A 1998 *Langmuir* **14** 4589
[43] Rathjen W, Straub J 1977 in *Heat Transfer in Boiling* (New York: Academic Press)

High resolution spectroscopy of Cs vapor confined in optical cells of few-micron thicknesses

S Cartaleva[a], A Krasteva[a], A Sargsyan[b], D Sarkisyan[b], D Slavov[a], T Vartanyan[c]

[a]Institute of Electronics, BAS, boul. Tzarigradsko shosse 72, 1784 Sofia, Bulgaria
[b]Institute for Physical Research, NAS of Armenia, Ashtarak-0203, Armenia
[c]St.Petersburg National Research University of Information Technologies, Mechanics and Optics, Kronverkskiy pr. 49, St.Petersburg 197101, Russian Federation

e-mail: stefka-c@ie.bas.bg

Abstract. We present here the new behavior of Electromagnetically Induced Transparency (EIT), Velocity Selective Optical Pumping (VSOP) and Velocity Selective Excitation (VSE) resonances observed in Cs vapor confined in unique cells with thicknesses $L = 1.5\lambda$ and $L = 6\lambda$. It is shown experimentally that in both cells, the EIT resonance is significantly narrower than would be expected from the ground state dephasing rate due to atomic collisions with the cell windows. The enhanced absorption (fluorescence) narrow VSOP resonance at the closed transition transforms into reduced absorption (fluorescence) one with small increase of atomic concentration or light intensity. A striking difference appears between the VSE resonance broadening due to excited atom thermalization, in $L = 6\lambda$ and conventional $L = 2.5$ cm cells.

1. Introduction

Laser spectroscopy of alkali vapor contained in optical cells is widely used in various laboratory experiments and photonic devices. The reduction of cell thickness L is of importance not only for optical photonic sensor miniaturization but it also results in observation of new phenomena with L approaching the wavelength λ of the irradiating light. Recently, significant effort has been devoted to the development of miniaturized atomic clocks and magnetometers based on Electromagnetically Induced Transparency (EIT) resonances. The EIT resonances have been prepared in Cs atoms confined in sub-millimetre-thickness cell with high pressure buffer gas added to prevent frequent collisions of alkali atoms with the cell walls, which destroy atomic coherence [1]. The EIT resonance contrast measured in such cells is similar in magnitude to that obtained in centimetre-sized cells, but substantially more laser intensity is needed when sufficient buffer-gas pressure is used. Buffer gas broadens optical transitions, which limits the pumping efficiency and compromises the performance of vapor cell clock and magnetometer. A promising strategy for increasing atom-light interaction time consists in the use of anti-relaxation-wall coated cells; however, high atomic-vapor densities are limited owing to coating degradation with temperature [2].

Our communication concerns the high resolution laser spectroscopy of Cs vapor confined in unique optical cells with thickness of few microns, further on called Micrometric Cell (MC). The dimensions of such cell differ significantly. The distance between the high-quality cell windows L varies from 1 μm to 6 μm. At the same time, the window diameter is (1÷2) cm. If the irradiating laser light is adjusted in orthogonal to the cell window direction, a strong spatial anisotropy will present for the time of interaction between the atoms and the laser radiation. Two kinds of atoms can be distinguished – "slow" and "fast". Slow atoms involve those flying predominantly in direction orthogonal to the laser beam (typically of a millimeter diameter) and reach the steady state in the interaction with the laser light. The fast atoms have significant velocity projection on the laser beam propagation direction, which results in a time of atomic flight between the MC windows shorter than the lifetime of the excited state of the atom. Due to transient effects, this anisotropy leads to: (i) observation of a significant difference between the fluorescence and transmission (absorption) spectra [3] and (ii) occurring of Velocity Selective Optical Pumping (VSOP) resonances centered at hyperfine optical transitions [4]. A series of VSOP resonances are observed, which are subject of intensive study, motivated by the possible applications for development of wavelength references as well as for investigation of atom-atom and atom-cell window collisions.

More specifically, here we report about the new behavior of the EIT and VSOP resonances observed in Cs vapor confined in $L = 1.5\lambda$ and $L = 6\lambda$ cells, where $\lambda = 852$ nm is the wavelength of the laser light resonant with Cs D_2 line. The $L = 1.5\lambda$ cell provides the possibility to study the EIT jointly with the Dicke-type coherent narrowing of optical transition, namely at such thickness of optical cell the effect of Dicke coherent narrowing first revival has been observed [4]. At the same time, for $L = 6\lambda$, the Dicke effect vanishes, thus allowing the two different cases to be compared.

In addition, we demonstrate that (i) in MC a reasonably narrow EIT resonance can be observed without buffer gas using and (ii) at the closed hyperfine transition, a high-contrast Velocity Selective Excitation (VSE) resonance occurs, which is not observed for conventional optical cells, under the same experimental conditions.

2. Experimental set up for sub-Doppler and sub-natural width resonance study

Both used in the experiment micrometric cells are filled with Cs vapor and irradiated by bi-chromatic laser light from two narrow-band (with spectral width of about 2 MHz) distributed feedback (DFB)

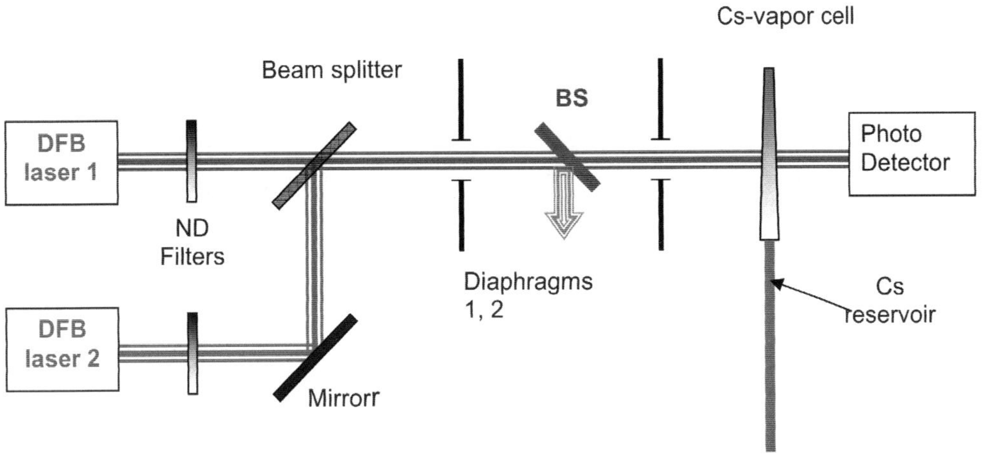

Fig.1 Experimental setup for VSOP, EIT and VSE resonance observation by pump-probe spectroscopy.

diode lasers – pump and probe one (Fig.1). The frequency of the pump laser is fixed at certain hyperfine optical transition while that of the probe one is scanned along a set of hyperfine transitions starting from a single ground-state level. The beams of the lasers are with orthogonal to each other polarizations. After careful overlapping, the two light beams propagate in direction orthogonal to MC windows. Transmission (absorption) and fluorescence (in an orthogonal to the laser beam direction) of Cs vapor were measured on the D_2 resonance line.

3. Sub-Doppler-width resonances of Cs vapor confined in optical cell with L = 1.5λ

We illustrate (Fig.2) the results related to the MC spectrum on the set of transition starting from the ground hyperfine level $F_g = 4$, obtained by scanning of the probe laser. In this case, the pump laser is switched off. If the probe laser is of low intensity, narrow Dicke resonances occur, centered at the $F_g = 4 \rightarrow F_e = 3,4,5$ transitions - curve (1). Of highest amplitude is the resonance at the closed $F_g = 4 \rightarrow F_e = 5$ transition, which does not suffer any atomic population loss due to hyperfine and Zeeman optical pumpings. To obtain an EIT resonance, a pump laser is involved and fixed at the $F_g = 3 \rightarrow F_e = 4$ transition. The EIT resonance, centered at $F_g = 4 \rightarrow F_e = 4$ transition - curve (2), is of natural width ($\Delta\nu = 5$ MHz), proving that it is based mainly on atoms flying parallel to cell windows. In fact, the dephasing rate of ground levels for the atoms moving orthogonal to the windows is ~ 70 MHz, as in the L = 1.5λ cell the atomic interaction time with a laser beam is highly anisotropic. Atoms moving nearly parallel to the cell windows have much longer interaction times compared to atoms moving in the direction perpendicular to the windows. The advantage of such approach is that mainly the slow atoms contribute to the EIT signal, which exhibits very good contrast.

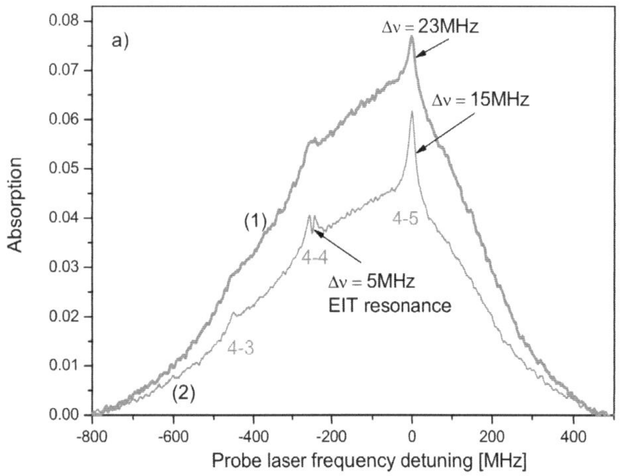

Fig.2. EIT and Dicke resonance at L = 1.5λ: (1) probe laser is scanning on the $F_g = 4$ set of transitions, no pump; (2) pump laser is fixed at 3 - 4 transition and probe scanned as in (1);
T = 105°C, i.e. atomic density ~ 2.10^{13} at/cm^3,
$W_{probe} = 0.3$ mW/cm^2, $W_{pump} = 35.4$ mW/cm^2.

While it is clear that the centered at the $F_g = 3 \rightarrow F_e = 4$ transition laser causes accumulation on the $F_g = 4$ level of Cs atoms with small velocity projection on the laser beam direction, it should be stressed that the Dicke resonance amplitude at the $F_g = 4 \rightarrow F_e = 5$ transition is strongly enhanced by the pump laser. Moreover, some narrowing of the Dicke resonance is observed. Hence, beside for optical cell miniaturization study, the proposed two-laser approach could be advantageous for further Dicke effect study in MC, because still it is not experimentally clarified what are the contributions of different-velocity-class atoms to the Dicke narrowing. Our experiment has shown that if the pump laser frequency goes out of exact resonance with the transition, the amplitude of the narrow resonance decreases very fast and it is getting much broader, i.e. the narrow Dicke resonance at the $F_g = 4 \rightarrow F_e = 5$ transition is mainly due to the "slow" atoms flying along the cell window surface.

4. Sub-Doppler and sub-natural width resonances observed in cell with L = 6λ - difference between open and closed optical transitions

In order to check if the reason for the large amplitude of the sub-Doppler-width resonance at the $F_g = 4 \rightarrow F_e = 5$ transition (Fig.2, curve (2)) is namely the Dicke phenomenon, the second cell with

$L = 6\lambda$ is used, where no Dicke effect is expected. Moreover, our aim is also to make an accurate comparison of the $L = 6\lambda$ cell behavior with that of a conventional cell with L = 2.5cm. For this, a beam splitter, BS, is inserted (denoted in blue in Fig.1), directing both precisely overlapped pump and probe beams to the centimeter cell as well.

To make the analysis simpler, we use very low powers of both lasers, thus working in a linear regime. From Fig.3a, it can be seen that even without Dicke effect a significant accumulation of "slow" atoms occurs, that are cycling at the $F_g = 4 \rightarrow F_e = 5$ transition. As for $L = 6\lambda$ cell this narrow resonance appears only with the pump laser switched on, it is determined as Velocity Selective Excitation, VSE, resonance. An EIT resonance is observed in the $F_g = 4 \rightarrow F_e = 4$ transition absorption, measured by the probe beam (Fig.3a). While the EIT resonance width stays constant (less than 2 MHz) with the atomic source temperature enhancement, the VSE resonance suffers a measurable broadening (Fig.3b). Further theoretical studies are in progress for clarifying the physical processes behind the VSE resonance broadening under different experimental conditions.

The comparison of the EIT/VSE resonance for $L = 6\lambda$ and L = 2.5 cm cells shows two unexpected results (Fig.3c): (i) despite the much higher rate of atomic collisions with the walls in the case of $L = 6\lambda$ cell, the width of the EIT resonance is similar for both cells (about 2 MHz, mainly determined by the laser widths); (ii) the thermalization of selectively excited atomic population at the $F_e = 4$ level

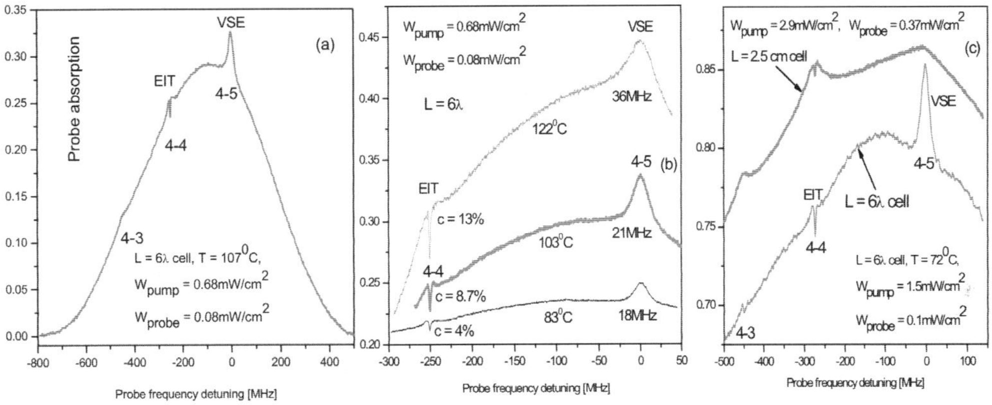

Fig.3 Probe absorption spectra for $L = 6\lambda$ and L = 2.5 cm cells: (a) EIT resonance at 4-4 transition and VSE resonance at 4-5 transition; (b) EIT and VSE resonance behavior with atomic source temperature; (c) illustration of the striking difference between the thermalizations of atomic population selectively excited by the narrow-band pump beam, for the $L = 6\lambda$ and L = 2.5 cm cells. Atomic density: $(2.3.10^{12} - 5.3.10^{13})$ at/cm^3.

by the pump laser is dramatically lower for the $L = 6\lambda$ cell than that observed in the L = 2.5 cm cell. Both results will be analyzed theoretically in further study. Here, in relation with the different thermalizations of the excited by the pump laser atoms, we assume that it could be due to the radiation trapping effect [5] which is much more pronounced in the L = 2.5 cm cell than in the MC. In fact, the extremely small thickness of $L = 6\lambda$ cell reduces the probability of the fluorescence re-absorption by Cs atoms, because of the extremely short time and distance of the fluorescence propagation within the MC.

In general, our experiments have shown that the MC is much less sensitive to the magnetic field gradients, laser beam overlapping and the mutual coherence of the two independent lasers.

5. VSOP resonance sign reversal

In addition to the first observation of narrow and well resolved fluorescent spectra in Extremely Thin Cell (ETC) (of L < 1 μm) [4], recently it has been demonstrated [6] that even narrower dips appear in the already narrowed fluorescence profiles, at ETC thickness L = λ. Those saturation dips have been reported only for the open hyperfine transitions of Cs D_2 line, which suffer atomic population loss due to hyperfine or/and Zeeman optical pumping. No dip has been observed in the fluorescence of completely closed 4-5 transition. For later transition and using a narrow band (few MHz) diode laser, even a tiny peak in the absorption has been demonstrated [7].

Here, results are presented related to dark (reduced absorption) and bright (enhanced absorption) VSOP resonances observed in MC with thickness L = 6λ. The two open $F_g = 4 \rightarrow F_e = 3, 4$ transitions show dark VSOP resonances in the transmission and fluorescence spectra, at all experimentally examined Cs source temperatures. However, the profile of the closed $F_g = 4 \rightarrow F_e = 5$ transition shows different sign VSOP resonance at different atomic source temperatures T. At T=60^{0}C, very well pronounced narrow bright VSOP in the absorption (i.e. enhanced absorption), as well as a sharp top of the fluorescence profile is observed (Fig.4a). This interesting absorption peak is observed only at low atomic concentration and low light intensity. However, rising the temperature to about 84^{0}C leads to VSOP resonance sign reversal, i.e. narrow dip occurs in the absorption (Fig.4b). Moreover, the fluorescence profile of the $F_g = 4 \rightarrow F_e = 5$ transition shows a well pronounced dip at T = 84^{0}C. For lower temperatures of Cs source, the increase of light intensity results in the same effect – transformation of the bright resonance to a dark one. Note that no VSOP resonance broadening is observed during the resonance sign transformation.

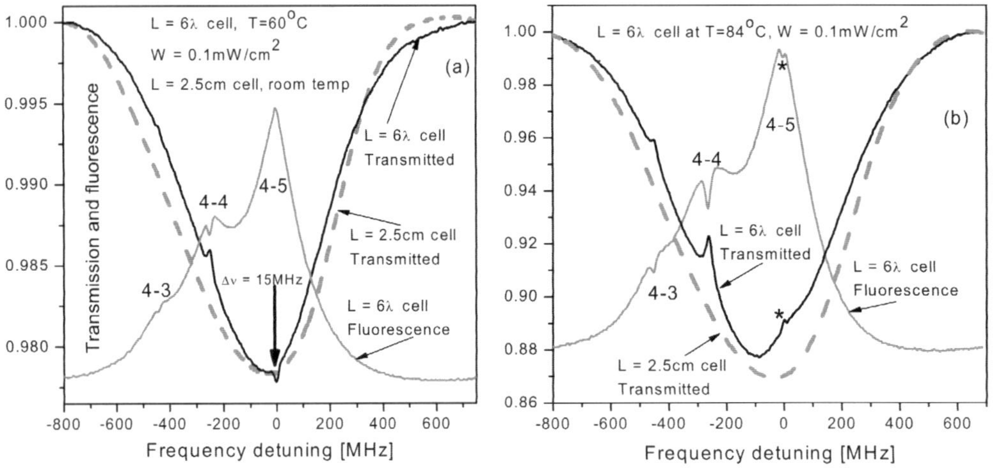

Fig.4 (a) Narrow VSOP resonances of different signs in the L = 6λ cell fluorescence and transmission spectra, for low atomic concentration (9.2.10^{11} at/cm^3) and very low light intensity – comparison with a conventional L = 2.5 cm cell profile, where no sub-Doppler-width feature occurs; (b) Illustration of the bright resonance (in the fluorescence and absorption of $F_g = 4 \rightarrow F_e = 5$ transition) transformation into a dark one (denoted by asterisks) with atomic concentration enhancement (5.2.10^{12} at/cm^3).

Physical processes behind the resonance sign reversal are attributed to the depolarization of the excited level, which transforms the completely closed system at low light intensity and atomic concentration to one with effective loss in the excitation process, for higher intensity and Cs vapor pressure. Two main processes can be responsible for population mixing of magnetic sublevels on the excited state, namely: (i) the elastic interaction between Cs atoms at enhanced atomic vapor pressures and (ii) the elastic interaction between the cell windows and the Cs-atoms travelling parallel to the windows surface [8].

6. Conclusions

VSOP, VSE and EIT resonances are experimentally observed in MC with few micron thickness confining Cs atoms, on the D_2 line of Cs. The resonance registration is performed by means of the pump-probe spectroscopy, using two narrow-band DFB lasers. In the L = 6λ cell, very high amplitude and narrow VSOP resonance occurs at the $F_g = 4 \rightarrow F_e = 5$ transition. In opposite, such narrow VSOP resonance is not observed for the L = 2.5 cm cell. This observation is attributed to the higher rate of radiation trapping in the last cell.

Very interesting result is that related to the EIT resonance width, which for the L = 6λ cell is about an order of magnitude less than estimated from the atomic collision rate with the cell windows. The comparison with the EIT resonance observed in the L = 2.5 cm cell under the same experimental conditions shows that in both cells EIT resonances of similar width are observed.

The obtained experimental results are of significant importance for the development of spectroscopy of alkali atoms confined in optical cells with micrometric thickness, for the purpose of photonics sensor miniaturization, as well as for new scientific effects study.

Acknowledgements: This work was supported by the Bulgarian Science Fund (grant No: DO 02-108), as well as by a Marie Curie International Research Staff Exchange Scheme Fellowship within the 7th European Community Framework Programme.

References

[1] Knappe S, Hollberg L and Kitching J 2004 *Opt. Lett.* **29** 388
[2] Budker D and Romalis M 2007 *Nat. Phys.* **3** 227
[3]. Sarkisyan D, Bloch D, Papoyan A and Ducloy M 2001 *Opt. Commun.* **200** 201
[4] Sarkisyan D, Varzhapetyan T, Sarkisyan A, Malakyan Yu, Papoyan A, Lezama A, Bloch D and Ducloy M 2004 *Phys. Rev.* A **69** 065802
[5] Allegrini M, Huennekens J, Namiotka R K, Sagle J and Jabbour Z J 1995 *Phys. Rev.* A **51** 4472
[6] Andreeva C, Cartaleva S, Petrov L, Saltiel S, Sarkisyan D, Varzhapetyan T, Bloch D and Ducloy M 2007 *Phys. Rev.* A **76** 013837
[7] Cartaleva S, Saltiel S, Sargsyan A, Sarkisyan D, Slavov D, Todorov P and Vaseva K 2009 *J. Opt. Soc. Am.* B, **26** 1999
[8] Krasteva A, Slavov D and Cartaleva S 2011 *Int. J. Opt.* **2011** ID 683415

Spectroscopy of the atom-wall interactions in a nanocell

V V Khromov[1], A E Logunov[1], A S Pazgalev[1,2], S G Przhibel'skii[1], D Sarkisyan[3] and T A Vartanyan[1]

[1] St. Petersburg National Research University of Information Technologies, Mechanics and Optics, Kronverkskiy pr. 49, St.Petersburg 197101, Russian Federation
[2] Ioffe Phys.-Technical Institute of the Russian Academy of Sciences, Politekhnicheskaya st. 26, St.Petersburg 194021, Russian Federation
[3] Institute for Physical Research, NAS of Armenia, Ashtarak-0203, Armenia

E-mail: Tigran.Vartanyan@mail.ru

Abstract. The optical spectroscopy of rubidium atoms confined to a nanocell was employed to study the quenching and energy transfer processes in the course of atom-wall collisions. The distance between the sapphire windows of the cell varies between 150 and 500 nm forming a vapour wedge. At moderate pressures, the frequency of the atom-wall collisions greatly exceeds the frequency of atom-atom collisions. Hence, the line shapes and intensities recorded in a nanocell provide valuable information about interactions of excited atoms with the surface of the window material.

1. Introduction

The atom-solid surface interactions play an important role in many applications and are of fundamental interest for atomic physics as well as for solid state physics. In this contribution we report on the results of our experimental investigations of the energy transfer processes and quenching of the excitation of rubidium atoms due to collisions with the monocrystalline sapphire surface.

An extremely thin sapphire cell with the diameter of 24 mm and the smoothly changing thickness in the range of 150-500 nm, thus, forming a vapour wedge, was used in order to increase the number of atomic collisions with the sapphire surface. Rubidium vapours were excited by means of two *cw* diode lasers with the powers of $I_P \sim 20$ mW and $I_D \sim 40$ mW. The wavelengths of the lasers were adjusted to $5S_{1/2}$–$5P_{3/2}$ and $5P_{3/2}$–$5D_{5/2}$ transitions, respectively. Figure 1 plots the rubidium levels and the transitions involved in the study.

The pumping rates of the atomic states $5P_{3/2}$ and $5D_{5/2}$ exceeded the relaxation rates of the corresponding levels, γ_P and γ_D. On the other hand, the pumping rates were smaller than the inverse time of the free flight between the cell walls as well as the Doppler width. Under these conditions, a very anisotropic distribution of the excited atoms over the cell volume is established. The atoms, slowly moving in the direction to the walls, interact with the resonant radiation more effectively than those atoms that move faster [1].

The populations of the atomic level, n_P and n_D, depend on the pumping intensities I_P and I_D in a specific way: $n_P \sim I_P^{1/2}$ and $I_D \sim n_P I_D^{1/2}$. This dependence is due to the gradual saturation of the atoms with different velocities that correspond to different detunings of the atomic transition relative to the pump [2].

Published under licence by IOP Publishing Ltd

In the experiment we have measured the fluorescence intensities of the rubidium vapour on the above mentioned transitions as well as on the wavelength of 420 nm that corresponds to the transition $6P_{3/2}$–$5S_{1/2}$. The fluorescent light was analyzed by a wide-aperture monochromator and registered by a photo-multiplier tube connected to a lock-in amplifier.

2. Methods and results

In an extremely thin cell that was used in our experiments the fluorescence intensity was substantially reduced due to the effective quenching of the excited atoms on the cell walls. Nevertheless, we were able to measure and to compare the fluorescence intensities on the transitions $5P_{3/2}$–$5S_{1/2}$ and $5D_{5/2}$–$5P_{3/2}$, although their intensities were several times smaller than the corresponding intensity of the pump scattered by the walls. This goal was achieved by a special modulation technique: instead of the pump light modulation, the population of the atomic levels was modulated. For instance, while measuring the fluorescence intensity on $5S_{1/2}$–$5P_{3/2}$ transition, only the intensity of the laser that is resonant to $5D_{5/2}$–$5P_{3/2}$ transition was modulated, and vice versa. A lock-in amplifier was employed to reject the scattered light and detect the weak fluorescence on the studied transitions. In all case dc component of the scattered light was not detected.

The detection part of the experimental set-up consists of a diffraction grating monochromator and a photo-multiplier tube. It was calibrated in a range of 400-800 nm with Xenon discharge lamp.

The main outcome of these measurements is that the fluorescence intensity ratio of the $5P_{3/2}$–$5S_{1/2}$, $5D_{5/2}$–$5P_{3/2}$ and $6P_{3/2}$–$5S_{1/2}$ transitions is appeared to be 500:2:1.

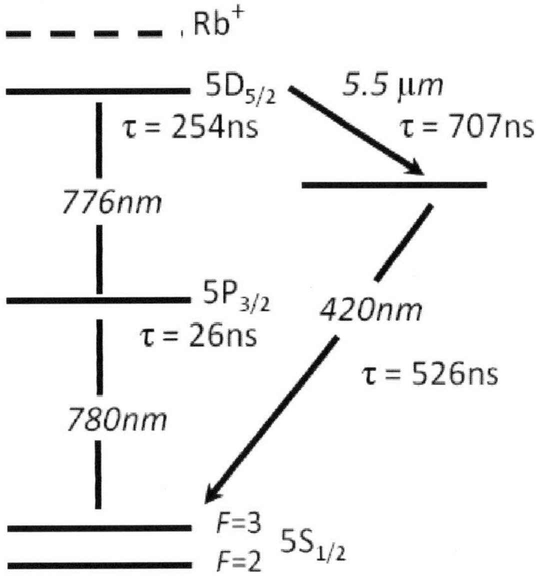

Figure 1. Rubidium levels and transitions involved in the study. Two diode lasers operate at the wavelengths of 780 nm and 776 nm.

3. Discussion

The first part of the ratio quoted above, 500:2, is in a fair agreement with the estimations of the corresponding levels populations based on the oscillation strengths, lifetimes and light intensities in use. Such agreement supposes that the quenching probabilities of $5P_{3/2}$ and $5D_{5/2}$ states are close to each other. Note, that even the order of magnitudes of these probabilities is not known. As far as we know the only one precise measurement of the excited alkali atom quenching rate under collisions with the sapphire surface was performed with cesium [3]. The probability of the cesium $6P_{3/2}$ excited state deactivation after collision with the wall was reported to be 0.98±0.02. If we suppose that the quenching efficiency is of the same order for rubidium atom, we could expect that quenching probability is close to 100% for both $5P_{3/2}$ and $5D_{5/2}$ states as well. Unexpectedly, the observed fluorescence intensities of $6P_{5/2}$–$5S_{1/2}$ and $5D_{5/2}$–$5P_{3/2}$ transitions differ only twice. This is difficult to explain, because spontaneous transition from $5D_{5/2}$ to $5P_{3/2}$ state should occur during the time of flight between cell walls. For our thin cell at room temperatures this time is about 1 ns, while spontaneous decay lifetime is around 700 ns. Then, without any other processes, which lead to population of $6P_{3/2}$ state, each atomic collision with the wall should deactivate both states with probability close to 100%. A as a result the total population of $6P_{3/2}$ state cannot exceed 1/700-th part of the population of $5D_{5/2}$ state. If we additionally take into account that the $5D_{5/2}$–$5P_{3/2}$ transition is approximately two times faster than the $6P_{3/2}$–$5S_{1/2}$ transition and that the $6P_{3/2}$ state has also a nonzero probability to decay to a lower lying $6S_{1/2}$ state, the expected fluorescent intensity of $6P_{3/2}$–$5S_{1/2}$ transition must be 2000 times weaker than that of $5D_{5/2}$–$5P_{3/2}$ transition. Contrary to that, the measured values differ only by a factor of two.

Two alternative explanations of the observed effect may be considered. First, extremely small quenching efficiency of $6P_{3/2}$ state may be the reason. But it is quite improbable, that $6P_{3/2}$ state is quenched in a substantially different way than $5P_{3/2}$ state. Another possibility is an "incomplete quenching" of $5D_{5/2}$ state, which implies that during the collisions with the wall the atoms are not quenched completely, but are transferred into $5P_{3/2}$ state with a high enough probability. In order to check both alternatives, additional measurements were performed.

We have studied the fluorescence intensity ratio of $5D_{5/2}$–$5P_{3/2}$ and $6P_{1/2}$–$5S_{1/2}$ transitions in its dependence on the thickness of the cell. It was observed that the fluorescence intensity ratio does not depend on the thickness. This excludes the first alternative and testifies an effective incomplete quenching of $5D_{5/2}$ state as a main process in atom-wall interaction. A plausible explanation of this large efficiency of the population transfer from $5D_{5/2}$ to $6P_{3/2}$ level is that the energy gap between these levels is small (it corresponds to an IR transition at the wavelength of 5.5 μm), and lies in the range of the phonon absorption spectrum of the sapphire substrate. Hence, high efficiency of this "incomplete quenching" may be due to matching between the sapphire phonon spectrum and the energy difference between the $5D_{5/2}$ and $6P_{3/2}$ states of rubidium [4].

One can expect that the "incomplete quenching" should manifest itself also in other spectroscopic features. In particular, a distant wing of the fluorescence spectrum of the excited atomic states may be suppressed or disappear provided the initial state has an opportunity to make a transition to a lower lying states with high probability.

Acknowledgments
This work was supported by a grant No 11-02-01135 of Russian Foundation for Basic Research.

References

[1] Sarkisyan D., Bloch D., Papoyan A., Ducloy M. 2001 *Opt. Comm.* **200** 201-8
[2] Vartanyan T A 2000 *Optics and Spectrosc.* **88** 564-7
[3] Przhibelskii S G and Khromov V V 2000 *Optics and Spectrosc.* **88** 17
[4] Failache H, Saltiel S., Fichet M., Bloch D. and Ducloy M. 2003 *Eur. Phys. J.* D **23** 237

Calculation of the ortho–para conversion of hydrogen in a p−type silicon lattice using a dwell time approach

R M Herman[1], A Suarez[1], J Sofo[1] and J C Lewis[2]

[1]104 Davey Laboratory, Department of Physics, Pennsylvania State University, University Park, PA 16802 USA

[2]Department of Physics and Physical Oceanography, Memorial University of Newfoundland, St. John's NL Canada A1B 3X7

E-mail: john.lewis@mun.ca

Abstract. Quantitative spectroscopic studies of hydrogen in a p−type silicon lattice at room temperature and at reduced temperature have led to rates for the ortho–para conversion process. The characteristic relaxation time at room temperature is about 8 hours. Explanations of this rate on the basis of the interaction between the interstitial H_2 and naturally occurring ^{29}Si using the Wigner rate expression encounter several difficulties, the principal being that the decay would involve multiexponential decay, in contradiction to observation.

In an earlier work we calculated the rate assuming that the ortho–para conversion was effected during scattering of holes from the hydrogen molecules. The result was smaller than observed by several orders of magnitude. In the present work it is assumed that sp_z holes diffuse randomly throughout the Si lattice, dwelling on effective areas associated with sp_z sites. The transition matrix elements are the same as for the scattering mechanism. The resultant characteristic time at room temperature we find to be 1000 hr. Considering the uncertainties in the calculation the discrepancy between our result and observation is not sufficient as to negate our physical picture.

1. Introduction

It is possible to obtain interstitial H_2 in crystalline Si by cooling the silicon from a melt in a hydrogen atmosphere. H_2 has been observed in Si by Raman [1] and by IR spectroscopy [2], the IR spectrum being ascribed to an interaction-induced dipole moment [3]. The ortho and para isomers were identified through Raman spectroscopy by Chen *et al.*[4]. Quantitative spectroscopic studies of hydrogen in a p−type silicon lattice at room temperature and at reduced temperature have led to determination of rates for the ortho–para conversion process [6, 7]. The characteristic relaxation time at room temperature is about 8 hours. An attempt to explain this rate on the basis of the interaction between the interstitial H_2 and naturally occurring ^{29}Si using the Wigner rate expression [7] was given by Hiller et al. [5]. Their explanation was found to be incompatible with experiment by Peng et al. [8], the principal difficulty being that the Wigner mechanism would necessitate a superposition of different decaying exponentials corresponding to the presence of ^{29}Si in nearest, next nearest, etc. neighbours, leading to a complicated multiexponential decay, as opposed to observation. A second difficulty is discussed below.

It is worth noting that the ortho–para conversion of a given ortho molecule in pure hydrogen has been quantitatively explained as due to the magnetic dipolar fields of neighbouring ortho molecules [9]. At very high densities a different mechanism becomes important, in which the emerging excitations are coupled to the converting molecules via electric quadrupole–quadrupole interactions [10].

All of the mechanisms given here involve the nuclear spins of the hydrogen molecule interacting with other nuclear spins. In this work and its predecessor [11] we will evaluate the effects of holes interacting with the proton spins of the hydrogen. While the holes are typically of very low density, say $10^{15}\,\mathrm{cm}^{-3}$, their magnetic moments are $\mathcal{O}\left(10^3\right)$ times nuclear spin magnetic moments, and they are highly mobile. This mechanism therefore merits consideration as a serious contender in OP conversion of hydrogen in silicon lattices. In this paper, therefore, the ortho–para conversion is assumed to be due to the unpaired electron spins accompanying holes, interacting with the nuclear spins of the interstitial H_2 primarily by the Fermi contact interaction, and the electron density at these spins is enhanced by exchange at the H_2 nuclear spins. The exchange interaction is readily included using a technique introduced by Herman[12].

We first consider the geometry of the Si lattice, and then derive the interaction between holes in a silicon lattice, and interstitial H_2.

2. H_2 tetrahedral Trapping Sites in Crystalline Silicon

Crystalline silicon is known to have the zincblende structure, with Si atoms occupying both the Zn and S sites in the structure. To visualise the situation, consider a face centred cubic structure with corners located on a cubic lattice having side length $a = 5.40\,\mathrm{a.u.}$, with sides oriented parallel to Cartesian axes, x, y, z. It is well known that any atom in the fcc structure could be placed at the origin of coordinates, $(x, y, z) = (0, 0, 0)$ without preference. To obtain the zincblende structure, a similar fcc structure is added, with each atom displaced by $(\frac{1}{4}, \frac{1}{4}, \frac{1}{4})$ from each of the original atoms in the fcc structure, with displacements understood to be in units of a. For purposes of discussion we will refer to these as the primary and secondary sublattices. Of course, the secondary sublattice could have been produced with any similar displacement from the primary but once established, the displacements will continue as being similar throughout the entire crystal. So let us discuss the situation as described above.

To understand the tetrahedral trapping sites, let us note that on the principal diagonal in the $(1, 1, 1)$ direction there are atoms located at $(0, 0, 0)$ and $(1, 1, 1)$ (primary lattice sites) and at $(\frac{1}{4}, \frac{1}{4}, \frac{1}{4})$ (secondary site within the original cubic volume described above). There are no atoms located at $(\frac{1}{2}, \frac{1}{2}, \frac{1}{2})$ nor at $(\frac{3}{4}, \frac{3}{4}, \frac{3}{4})$, and these can be regarded as typical locations for tetrahedral vacancies into which H_2 molecules can become trapped. In practice, H_2 is diffused into Si at high temperatures, then upon cooling the H_2 remain trapped. The site given by $(\frac{1}{2}, \frac{1}{2}, \frac{1}{2})$ is surrounded by four Si atoms at equivalent distances, belonging to the secondary lattice, located at $(\frac{1}{4}, \frac{1}{4}, \frac{1}{4})$, $(\frac{1}{4}, \frac{3}{4}, \frac{3}{4})$, $(\frac{3}{4}, \frac{1}{4}, \frac{3}{4})$ and $(\frac{3}{4}, \frac{3}{4}, \frac{1}{4})$. All Si tetrahedral atoms lie a distance $2.34\,\mathrm{a.u.}$ from the trapping site. The three latter sites mentioned, which can be viewed as a basal plane in the present coordinate system, lie further from the origin than the originally cited apex atom at $(\frac{1}{4}, \frac{1}{4}, \frac{1}{4})$. The H_2 trapping site located at $(\frac{3}{4}, \frac{3}{4}, \frac{3}{4})$ is similarly surrounded by Si atoms at the primary lattice locations $(1, 1, 1)$, $(1, \frac{1}{2}, \frac{1}{2})$, $(\frac{1}{2}, 1, \frac{1}{2})$ and $(\frac{1}{2}, \frac{1}{2}, 1)$ with all tetrahedral sites again residing a distance $2.34\,\mathrm{a.u.}$ from the tetrahedral vacancy. In this case, the basal plane is located nearer to the origin than the (primary) apex $(1, 1, 1)$. Of course, the Si atom located at $(0, 0, 0)$ would belong to a similar tetrahedron, with the other (basal plane) atoms at $(0, -\frac{1}{2}, -\frac{1}{2})$, $(-\frac{1}{2}, 0, -\frac{1}{2})$ and $(-\frac{1}{2}, -\frac{1}{2}, 0)$. Each Si atom is surrounded by four nearest–neighbours for which bonds can be supported. They also lie at a distance $\sqrt{3}a/4 = 2.34\,\mathrm{a.u.}$ Therefore, it appears to be the case that for every Si–Si bond, there will exist two tetrahedral vacancies lying on the bond axis, to either side of the bond. Since all Si atoms in the primary lattice have identical surroundings it follows that there are as many tetrahedral vacancies as there are Si atoms themselves. Of course the other apex atoms forming the tetrahedral sites will be similarly bonded to partner Si atoms in directions leading away from the tetrahedral centres. These directions could have served as the directions along which tetrahedral centres could have been described. Because there are four of these special directions within the silicon crystal, and because there are four apex atoms per tetrahedral site, we conclude that all of the sites have become locatable through the analysis above (with the $(1,1,1)$ direction being specially labelled), and that the sites so located are properly shared by all of the atoms in the sample.

It should be noted that the tetrahedral atoms surrounding the H_2 molecule do not bond to one another. The distances separating these atoms are given by $a/\sqrt{2} = 3.82\,\text{a.u.}$ The Si–Si bonds, other than the ones already noted as lying on the far side of the apex atoms will be occupying tetrahedral directions more of less forming an "umbrella" structure cradling the H_2 molecules. It is these four sets of three bonds each that physically constrain the H_2 to translational states centred about the tetrahedral symmetry points. It is these bonds that effect the H_2 trapping.

Now reassign the former $(1,1,1)$ direction as a new $z-$direction. Without loss of generality the rates of transition between the $(J,M) = (0,0)$ and $(J,M) = (1,0)$ states of H_2 can be analysed relative to these coordinates. Relaxation rates between the other (J,M) and $(0,0)$ states will be identical. In this way of looking at the situation, each primary lattice Si atom will be bonded to a partner Si atom located in the $z-$direction, spaced by a distance $\sqrt{3}a/4 = 2.34\,\text{a.u.}$, as well as three other Si partners located in the three other tetrahedral directions (as yet unspecified, though once chosen, to be the same for each of the primary Si lattice atoms, etc. for the secondary lattice). The conduction bands of Si are prolate ellipsoidal in nature with minimum energy being associated with relatively large $\boldsymbol{k}-$values, with directions parallel or antiparallel to the tetrahedral bonds present in the crystal. For now, consider the band centered about the large $\boldsymbol{k}-$values $\pm\hat{\epsilon}_3 k_0$. The corresponding hole states will be of sp_z type.

3. Hole – H_2 interaction

Ortho–para conversion occurs between molecular states such as $|JMm_{s_A}m_{s_B}\rangle = |1\,0 -\frac{1}{2} -\frac{1}{2}\rangle$, and $|J'M'm'_{s_A}m'_{s_B}\rangle = |0\,0\,\frac{1}{2} -\frac{1}{2}\rangle$, A and B referring to the two H_2 nuclei, together with $|0\,0 -\frac{1}{2}\,\frac{1}{2}\rangle$. The $|-\frac{1}{2} -\frac{1}{2}\rangle$ spin state is a triplet state accompanying a $J=1$ rotational state, while the singlet state is $(1/\sqrt{2})\left(|\frac{1}{2} -\frac{1}{2}\rangle - |-\frac{1}{2}\,\frac{1}{2}\rangle\right)$, accompanying (mostly) the $J=0$ rotational state. Any interaction that could cause a transition would have to satisfy three conditions:

- it must be a magnetic interaction with the proton spins in H_2;
- it must have a spatial dependence that can couple to the orientational angles of H_2; and
- there must be a thermal bath capable of absorbing the energy released (or supplying the energy needed in a reverse process) to make the transition $E_{J=1} \to E_{J=0}$.

The third condition above is stringent. For example, in the proposed spin exchange with ^{29}Si mentioned above this spin exchange would not have been very likely at the outset, simply because the energy exchanged would have to be released into very long wavelength phonons (corresponding to 170 K, compared with the Debye temperature 658K) or would have to be exchanged with translational energies of H_2 constrained to the Si lattice, which are much greater ($\sim$ 1070 K, by our calculation).

Thus, for this last reason alone, transitions based upon spin exchange with surrounding paramagnetic atoms/nuclei will be considerably suppressed.

The transition matrix elements are the same as for the scattering mechanism [11] with the same exchange enhancement. The contact interaction Hamiltonian responsible for the ortho–

para conversion is

$$\mathcal{H}' = -\left(\frac{16\pi}{3} g_n \mu \mu_n\right) \sum_j \boldsymbol{\sigma}_{\sigma_z j} \cdot \left[\boldsymbol{\sigma}_A\, \delta\left(\boldsymbol{x}_j - \boldsymbol{R}_A\right) + \boldsymbol{\sigma}_B\, \delta\left(\boldsymbol{x}_j - \boldsymbol{R}_B\right)\right] \tag{1}$$

where μ and μ_n are the Bohr and nuclear magnetons, $g_n = 5.59$ is the proton gyromagnetic ratio, and the $\boldsymbol{\sigma}$ are the Pauli spin matrices. $\boldsymbol{R}_A$ and $\boldsymbol{R}_B$ are the positions of the A and B protons respectively, while $\boldsymbol{x}_j$ is the position of the jth electron. In eq. (1) j is summed over all Si sites, though only four nearest neighbour sites are important. The R^{-6} terms from the proton magnetic dipoles make little contribution, principally because they are not exchange–enhanced.

For H_2, the lowest–lying configuration can be approximated with Heitler–London wave functions, with effective nuclear charge 1.166 as determined by Wang [13]. The small ionic contributions to the H_2 ground state can be shown to be unimportant, as can more highly excited added configurations. Accordingly, the H_2 ground state can be represented by

$$\varphi = \frac{1}{2\sqrt{1+S_{AB}^2}} \left[u_A\left(1\right) u_B\left(2\right) + u_A\left(2\right) u_B\left(1\right)\right] \left(\alpha_1\beta_2 - \beta_1\alpha_2\right) \tag{2}$$

where, as above, A and B represent the two hydrogen nuclei and 1 and 2 denote the electronic coordinates associated with electrons 1 and 2. S_{AB} represents the overlap integral $\langle u_A | u_B\rangle$. The functions u_A and u_B are used to represent the modified $1s$ states centered on the their respective nuclei with α and β indicating spin up or down states respectively.

Hole wavefunctions were calculated by Density Functional Theory as described previously [11].

After considerable further analysis it is found that the electron spin densities at the two nuclei satisfy the relationship

$$\text{Spin density at } A - \text{Spin density at} B = -0.0274 \cos\Theta \tag{3}$$

where Θ is the $A \to B$ displacement vector angle with the z–axis, for H_2 above the apex Si atom, with A being further removed from the Si site than B for $\Theta = 0$. Eq. (3) is not exact; but the next odd term in $\cos\Theta$ will not connect $J = 0$ to $J = 1$ rotational states. After integration over electronic coordinates the interaction Hamiltonian yields

$$V_{fi} = \left(\frac{4\pi g_n \mu^2}{3 \times 1836}\right) \times 3.08 \cos\Theta\ \boldsymbol{\sigma}_{sp_z} \left(\boldsymbol{\sigma}_A - \boldsymbol{\sigma}_B\right). \tag{4}$$

4. Hole motility

In the present work it is assumed that sp_z holes moving with group velocity v diffuse randomly throughout the Si lattice, dwelling on effective areas associated with sp_z sites. The dwell time is taken to be the separation between sp_z– containing planes divided by velocity v.

The dwell time we equate with the planar separation d divided by velocity. The ortho–para conversion rate will then be

$$\mathcal{R} = N_{\text{hole}}\, v\, \mathcal{A}\ (\text{conversion amplitude with dwell time } \tau = d/v)^2 \tag{5}$$

with effective area per sp_z site being $\mathcal{A} = \left(N_{sp_z} d\right)^{-1} = 60.2\, a_0^2$, where a_0 is the Bohr radius. The actual conversion amplitude is $\mathcal{H}_{fi}\tau/i\hbar$.

Overall we have

$$\mathcal{R} = N_{\text{hole}} N_{sp_z}^{-1} \left(d/v\right) \left(\frac{4\pi^2}{3}\right) \left(\frac{g_n}{1836}\right)^2 (0.0137)^2 \left(\frac{4\mu^2}{\hbar a_0^3}\right)^2 \tag{6}$$

The resultant characteristic time at room temperature we find to be about 1000 hr. Considering that the result is the product of several large and small terms each with considerable uncertainty the discrepancy between this and observation is not sufficient as to negate our physical picture.

References

[1] Pritchard R, Ashwin M, Tucker J, and Newman R 1998 *Phys. Rev.* B **57**, 15048
[2] Leitch A, Alex A, and Weber J 1998 *Phys. Rev. Lett.* **81** 421
[3] Zhou J A, Chen E, and Stavola M 1999 *Physica* B **200** 273–4
[4] Chen E E, Stavola M, Fowler W B, and Zhou J A 2002 *Phys. Rev. Lett.* **88**, 245503
[5] Hiller M, Lavrov E V, and Weber J 2007 *Phys. Rev. Lett.* **98** 055504
[6] Hiller M, Lavrov E V, and Weber J 2007 *Phys. Rev. Lett.* **99** 209901
[7] Wigner E P 1933 *Z. phys. Chem.* B **23** 28
[8] Peng C, Stavola M, Fowler W B, and Lockwood M 2009 *Phys. Rev.* B **80** 125207
[9] van Kranendonk J 1983 *Solid Hydrogen* (New York: Plenum Press)
[10] Strzhemechny M A and Hemley R J 2000 *Phys. Rev. Lett.* **85** 5595
[11] Herman R M, Suarez A, Sofo J and Lewis J C 2010 in *20th Int. Conf. Spectral Line Shapes* CP1290 AIP 284
[12] Herman R M 1965 *Phys. Rev.* **137** A1133–8
[13] Wang S C 1928 *Phys. Rev.* **31** 579

Physical interpretation of the blue shift of spectra obtained by corona discharge in liquid helium

N F Allard[1,2]
[1] Observatoire de Paris, GEPI, UMR8111, CNRS, 61, Avenue de l'Observatoire, F-75014, Paris, France
[2] Institut d'Astrophysique, UMR7095, CNRS, Université Paris VI, 98bis Boulevard Arago, F-75014, Paris, France

E-mail: `nicole.allard@obspm.fr`

Abstract. We discuss the major importance of temperature and perturber density in the broadening and shift of the He 706 nm triplet $3s$-$2p$ transition within a unified line shape semi-classical theory which contains the impact as well the quasistatic limits. In a previous paper we presented the first calculations of He-He collisional profiles for this transition using *ab-initio* molecular potentials. The excellent agreement between our theoretical determination of the near wing of the line profiles and reported experimental spectra of a corona discharge in gaseous He at 300 K established the validity of our interpretation of the spectra and the accuracy of the interaction potentials. In liquid He, the experimental line profiles are formed under conditions where two or more perturbers are likely to act simultaneously. Such phenomena determine the character of a spectral line at high perturber density. In this work we show that it is possible to determine the physical conditions of temperature and pressure from the experimental spectra of an electric discharge in liquid He. We show that at 4 K the theoretical line shift is negative in disagreement with an incorrect analysis reported in a recent paper and also presented at this conference.

1. Unified profile calculations

A unified theory of spectral line broadening yields the complete profile from the line center to the far wing, and includes the effect of the dependence of the transition probability on interatomic distance for all states that contribute to the line. Complete details and the derivation of the theory are given by [1]. In our theoretical work, the spectral line is computed from the Fourier transform (FT) of a dipole autocorrelation function. A pairwise additive assumption allows us to calculate the total profile $I(\Delta\omega)$, when all the perturbers interact as the FT of the Nth power of the autocorrelation function $\phi(s)$ of a unique atom-perturber pair. We obtain for a perturber density n_p

$$\Phi(s) = e^{-n_p g(s)}, \tag{1}$$

where decay of the autocorrelation function with time leads to atomic line broadening.

For a transition $\alpha = (i, f)$ from an initial state i to a final state f, we have

$$\begin{aligned} g_\alpha(s) &= \frac{1}{\sum_{e,e'}{}^{(\alpha)} |d_{ee'}|^2} \sum_{e,e'}{}^{(\alpha)} \int_0^{+\infty} 2\pi\rho d\rho \int_{-\infty}^{+\infty} dx \; \tilde{d}_{ee'}[\,R(0)\,] \\ &\quad [e^{\frac{i}{\hbar}\int_0^s dt\, V_{e'e}[R(t)]} \; \tilde{d}^*{}_{ee'}[R(s)] - \tilde{d}_{ee'}[R(0)]\,]. \end{aligned} \tag{2}$$

The e and e' label the energy surfaces on which the interacting atoms approach the initial and final atomic states of the transition as $R \to \infty$ (R denotes the internuclear distance between the radiator and the perturber). $\Delta V(R)$, the difference potential, is given by

$$\Delta V(R) \equiv V_{e'e}[R(t)] = V_{e'}[R(t)] - V_e[R(t)] \ , \tag{3}$$

and represents the difference between the electronic energies of the quasi-molecular transition. The interatomic interactions are the main physical quantities needed for a good understanding of collisional processes. The *ab initio* computation of the adiabatic potential energy curves of He_2 have been carried out using the MOLPRO 2009 package [1] (see figure 1). Complete details are given by Allard *et al.* 2009 [2]. In a previous study, we have checked that this approach leads, for the lowest quintet, results comparable to a full CI, within less than a few thousandths of a wavenumber. Moreover errors due to BSSE are also far below the wavenumber thanks to the huge basis set involved.

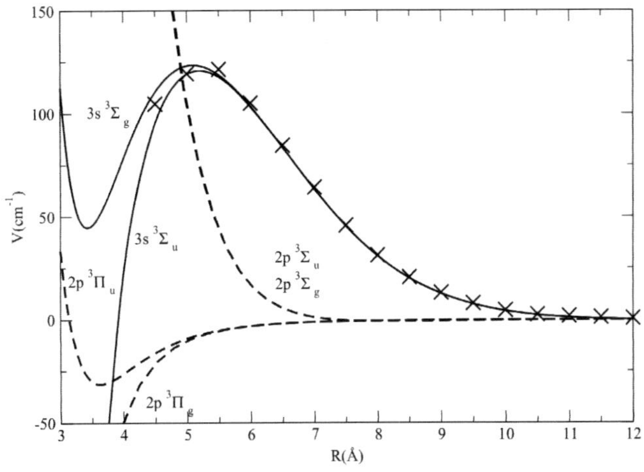

Figure 1. Potential energies for the triplet $3s$ (full line) and $2p$ (dashed line) states of the He_2 molecule. For comparison, a full CI potential calculation for the $3s\ ^3\Sigma_g$ state is overplotted (extracted from table 1 of [3]).

Emission measurements of the self-broadened lines of He due to 3s ^{1,3}S - 2p ^{1,3}P transitions show that beyond a symmmetrical Lorentzian center these lines exhibit a blue asymmetric behaviour [2, 4]. The observed asymmetries are a consequence of maxima in the $3s$-$2p$ potential difference curves at an intermediate internuclear distance (figure 2). This blue asymmetry is due to a blend of four quasi-molecular satellites in proximity to the atomic lines. Because of the blending, it can be accurately described *only* with a *unified* spectral line shape theory that is valid from the center to the far wing.

2. Density effects

The general behaviour of the profiles, particularly of the distribution of energy among multiple satellites, is simply related to the probability $W(m, h)$ of finding m perturbers in a volume τ near the perturbed atom (collision volume) at any instant.

When the perturber density is n, $W(m, h)$ is given by the Poisson distribution

$$W(m, h) = \frac{h^m}{m!} exp(-h) \tag{4}$$

[1] http://www.molpro.net

with $h = n\tau$.

The effects of two or more perturbers becomes quite significant at high densities where the peak of the line profile corresponds to the most probable distribution of perturbers.

To interpret what would be seen for an electrical discharge in a liquid, where the He density may be in excess of 10^{21} cm^{-3}, it is interesting to follow the evolution of the profiles as the perturber density increases from the regime of binary collisions to one where many more perturbers act simultaneously. The line core disappears completely when the probability of finding more than one perturber interacting at a time is large. The profile of the line then shifts by the mechanism of the redistribution of its oscillator strength in higher order satellites, rather than a shift of the "impact" line core.

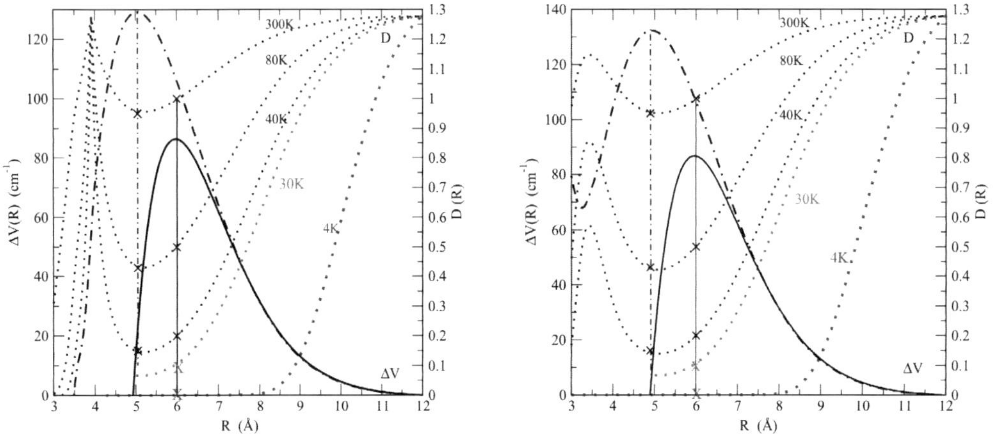

Figure 2. ΔV and the corresponding electric dipole moment $D(R)$ in atomic units for 300 to 4 K. Left: $3s\ ^3\Sigma_g \rightarrow 2p\ ^3\Pi_u$ (full line) and $3s\ ^3\Sigma_g \rightarrow 2p\ ^3\Sigma_u$ (slashed dotted line). Right: $3s\ ^3\Sigma_g \rightarrow 2p\ ^3\Pi_u$ (full line) and $3s\ ^3\Sigma_g \rightarrow 2p\ ^3\Sigma_u$ (slashed dotted line).

3. Temperature effects

While the position of the line satellites critically depends on the interaction potential, their strength depends on the radiative dipole moments, $D(R)$, in the internuclear region where the line satellite is formed [1]. In Allard *et al.* 1999 [1] we define $\tilde{d}_{ee'}(R(t))$ as a *modulated* dipole

$$D(R) \equiv \tilde{d}_{ee'}[R(t)] = d_{ee'}[R(t)]e^{-\frac{V_e[R(t)]}{2kT}} , \tag{5}$$

The values of $D(R)$ for the maxima of the difference potential energy ΔV are pointed out using $\times$ in figure 2. The decreasing strength of the satellites with temperature is illustrated in figure 4. They disappear for about $T \leq 10$ K when the transition moments $\tilde{d}_{ee'}[R(t)]$ get too small in the region of the maximum of the difference potential of $\Delta V(R)$ for all contributing transitions.

The simplest way of characterizing a line shape is in terms of the parameters: w, the full width at half maximum (HWHM), d, the shift and the asymmetry. We have studied the variations of these parameters versus the density for T=300, 80 and 40 K in Allard *et al.* 2011 [5]. As an example here, we will consider experimental spectra recorded at 4 K, (figure 3 of Li *et al.* 2009 [6]), to show that it is possible to determine the physical conditions of temperature and

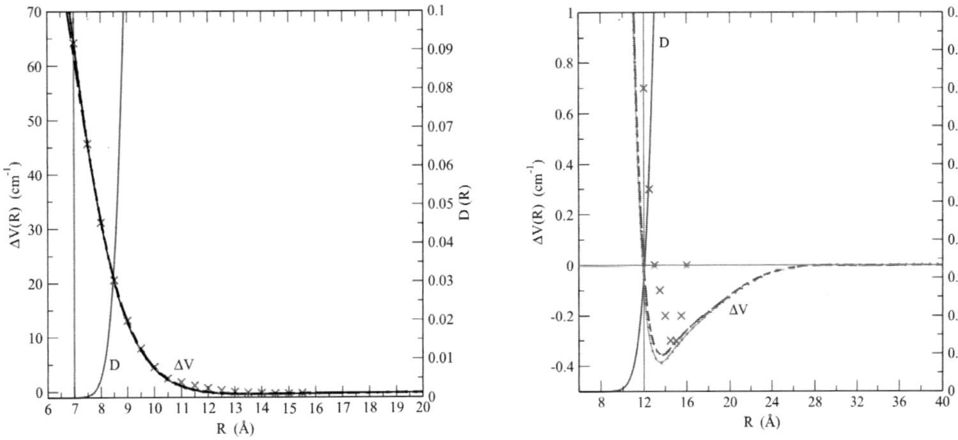

Figure 3. Same as figure 2 for T=4 K at long range distance. For comparison CI potential calculation (table 1 of [3]) for the $3s\ ^3\Sigma_g$ state is overplotted (red ×).

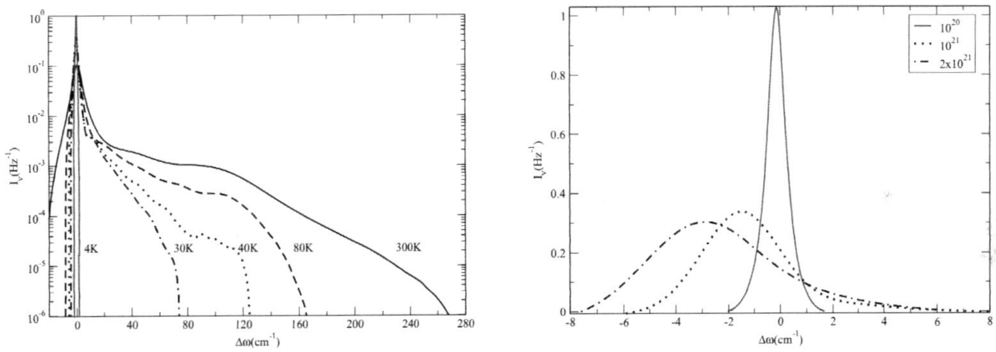

Figure 4. Evolution of the line profiles, with decreasing temperature for $n_{He} = 10^{20}$ cm^{-3} (left), with increasing helium density (right) at T=4 K.

pressure in the emitting region of an electric discharge in liquid helium. We determined the line parameters from these published spectra, and show them on the parameter curves in figure 5. The deduced values of T and n_{He} have to be consistent for the width (w) *and* the shift (d) to uniquely define the temperature and density close to the electrode tip. For the experiment at 1.4 MPa, the deduced helium density is not the same for the width *and* the shift, which means that the temperature must be higher than 40 K. A grid of the parameter curves shown in figure 5, should be done at different temperatures from about 60 K to 30 K, with a step size of 5 K. It would allow us to obtain the physical conditions in the medium close to the electrode tip for all of the experimental spectra. In the following we will consider the two extreme physical conditions of pressure reported in figure 3 of Li *et al.* 2009 [6]. At 0.1 MPa the profile is almost unshifted and the parent line at 706 nm is still apparent. This contrasts with

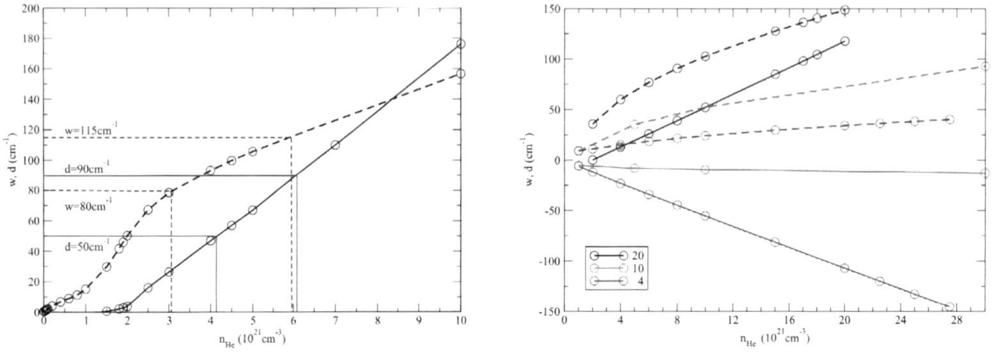

Figure 5. Variation of the width (dashed line) and shift (full line) of the unified profiles with He density (left: for 40K, right: from 20 K to 4 K).

2.3 MPa, where the maximum of the experimental spectrum has shifted to the position of the quasi-molecular satellites (figure 6). It is noteworthy that there is a close agreement between the experimental and theoretical profiles, and we conclude from this that the source temperature for these experiments varies from about 60 K to 30 K.

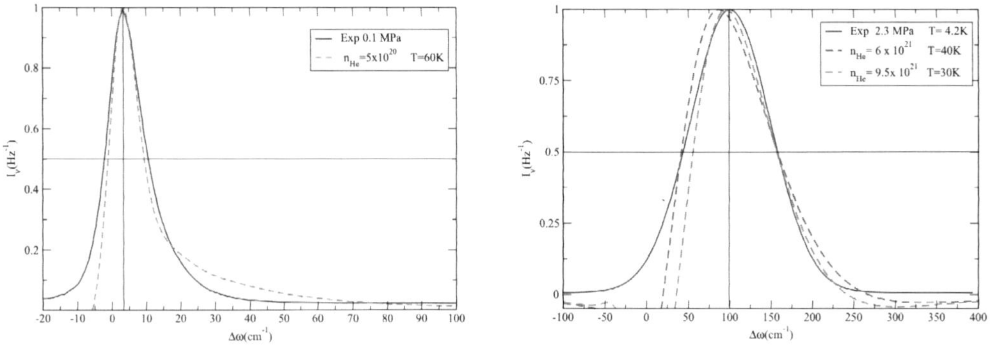

Figure 6. Experimental profile at T=4K compared to unified profiles.

4. Study of the theoretical profiles at 4 K in the quasistatic limit

Our unified line shape theory contains the impact as well the quasistatic limits [1], however in order to get a closer comparison with results presented in Bonifaci *et al.* 2012 [3] we will use specifically the quasistatic limit of Eq. 2.

To explain the negative shift obtained at very low temperature shown in figures 4 and 5, we have plotted in figure 3 the potential difference in the "active" region where the *modulated* dipole moment $\tilde{d}_{ee'}(R(t))$ at T=4 K is different from zero. This region is the same one identified

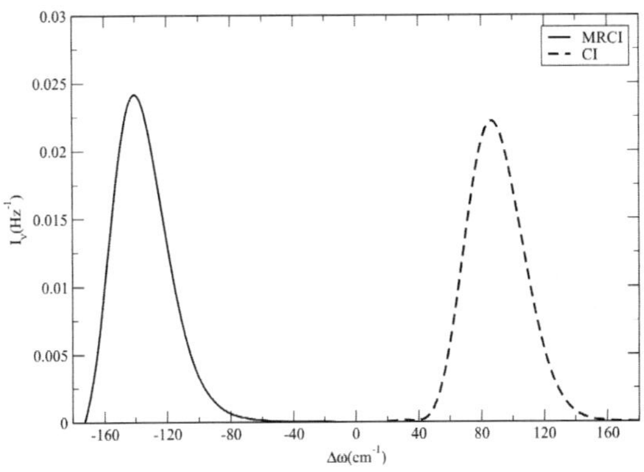

Figure 7. Quasistatic profiles at T=4 K, n_{He}=2.65× 10^{22} cm^{-3}. MRCI potentials [2] (full line) Truncated CI potentials [3](dashed line).

Table 1. Comparison of line parameter determinations from quasistatic profiles at T=4 K, for n_{He}=2.65× 10^{22} cm^{-3}.

	w (cm^{-1})	d (cm^{-1})
this work MRCI	39.5	-140
this work CI truncated	42.6	86.5
Bonifaci *et al.* 2012 [3]	140	200

"liquid helium" density profiles in figure 6 of [3]. Because they noticed that the $2p$ potentials for the triplet potentials appear to be essentially flat when $R \geq 5$ Å, we overplot in figure 3 their tabulated potential for the $3s$ $^3\Sigma_g$ state obtained with our basis set [2].

As can be seen in figure 3, ΔV is repulsive at intermediate range while very weakly attractive at long range with a well depth of 0.4 cm^{-1}. It should be noted that these values are close to the limit of *ab initio* calculations and that they can be expected to be sensitive to the methodology or basis sets employed in the calculations. Taking into account the fact that CI calculations of Bonifaci *et al.* [3] use the same reference functions as ours, this overall agreement is remarkable.

Unfortunately, table 1 of Bonifaci *et al.* [3] truncates R at R_{cut}=16 Å, which consequently neglects the critical long range part of the potential. It is well known that the shift of a spectral line profile is very sensitive to the long range part of the interaction potential, and for line shape calculations such truncation leads to unphysical results. We see that there is a dramatic effect on the resultant profiles in figure 7. The calculations have been done for n_{He}=2.65× 10^{22} cm^{-3} close to the asymptotic value of helium density for P=3.5 MPa, according to figure 6 of Bonifaci *et al.* 2012 [3]. The line parameter determinations from these quasistatic profiles are summarized in table 1. An examination of figure 7 and table 1 reveals that the width of the profiles are very similar because the intermediate range of the potentials are the same. This sensitivity to the long range potential leads to opposite shifts of the profile in the two calculations. Truncation of the long range part of the potentials produces an incorrect result.

We evaluated the line parameters of Bonifaci *et al.* 2012 [3] from the results shown in their figures 7 and 8. The width of a spectral line profile, has to be small at low temperature, the large theoretical values obtained for the width at 4 K reported in [3, 9] are consequently unphysical.

It is noteworthy that Path Integral Monte Carlo (PIMC) calculations by [7] are in close agreement with our theoretical approach for the study of absorption spectra of helium doped with sodium. This interesting result is also presented at this conference (Allard *et al.* 2012 [8]).

5. Conclusions

The positive shift observed in spectra of a corona discharge in liquid He must be due to quasi-molecular lines in the blue wing which arise from potential extrema. These satellites are sensitive to temperature because, at higher temperature, the perturbers are more likely to climb the excited-state potential barrier that generates the difference-potential maximum. Large values of the width and shift are characteristic of quasi-molecular features.

From the results described above compared with published experiments we infer that the temperature in the corona discharge is above 10 K. *The temperature and the density are not those in the He liquid.*

As it is well known, the shift of a spectral line profile is very sensitive to the long range part of the interaction potential. Even when there is an uncertainty in the long-range van der Waals well, truncating the long range part of the potentials leads to inaccurate values of the shift. On the other hand, the width of a spectral line profile, mainly due to collisions at intermediate distance, has to be small at low temperature.

As a final caution, we note from our experience that the Fourier Transform connecting temporal physics to observed spectra is a powerful tool for line profile analyses. However as we see here, it may give completely erroneous results when used with *ad hoc* computational assumptions without considering the underlying physical phenomena. Furthermore, unless the time series is finely sampled and very long, and the windowing function is carefully chosen for the problem at hand, spurious oscillations will appear on the FT which are unreal and easily misinterpreted.

Acknowledgments

One of us (NFA) would like to acknowledge PNPS for the financial support which allows the collaboration with the Laboratoire de Chimie et Physique Quantique in Toulouse

References

[1] Allard N F, Royer A, Kielkopf J F and Feautrier N 1999 *Phys. Rev. A* **60** 1021

[2] Allard N F, Deguilhem B, Gadéa F X, Bonifaci N and Denat A 2009 *Europhysics Letters* **88** 53002

[3] Bonifaci N, Aitken F, Atrazhev V M, Fiedler S L and Eloranta J 2012 *Phys. Rev. A* **85** 042706

[4] Allard N F, Monari A, Deguilhem B and Gadéa F X 2012 Comparative Study of Emission Spectra of He(3S)-He(2P) at 706 and 728 nm Due to the Triplet and Singlet Transitions 21th *ICSLS St Petersburg, Russia, June 3-10, 20012* (JPCS)

[5] Allard N F, Bonifaci N and Denat A 2011 *European Physical Journal D* **61** 365–372

[6] Li Z, Bonifaci N, Aitken F, Denat A, van Haeften K, Atrazhev V M and Shakhatov V A 2009 *Eur. Phys. J. Appl. Phys.* **47** 22821

[7] Nakayama A and Yamashita K 2001 *J. Chem. Phys.* **114** 780

[8] Allard N F, Nakayama A, Spiegelman F, Kielkopf J F and Stienkemeier F 2012 Absorption spectra of NaHe from white dwarfs to helium clusters 21th *ICSLS St Petersburg, Russia, June 3-10, 20012* (JPCS)

[9] Bonifaci N, Aitken F, Atrazhev V M, Von Haeften K and Eloranta J 2012 Shape of atomic lines emitted by liquid helium 21th *ICSLS, St Petersburg, Russia, June 3-10, 20012* (JPCS)

Journal of Physics: Conference Series **397** (2012) 012066 doi:10.1088/1742-6596/397/1/012066

Shape of atomic lines emitted by cryoplasma in Helium

N Bonifaci[1], F Aitken[1], Hai Van Nguyen[1], V M Atrazhev[2], K von Haeften[3], R Rincon[4].

[1]G2E.lab, CNRS et Universite Joseph Fourier, Grenoble, France
[2]Institute for High Energy Densities, AIHT RAS, Moscow, 25412, Russia
[3]Departement of physics and Astronomy, University of Leicester, UK
[4]Grupo de Espectroscopía de Plasmas (GEP) University of Cordoba, Cordoba, Spain

nelly.bonifaci@grnoble.cnrs.fr

Abstract. Experimental data on spectral shape of the line 706nm emitted by corona discharge in liquid and gaseous cryogenic helium are presented. The data have their explanation in a frame work of the bubble model. The long-range interaction between excited He* atoms and He atoms in the ground state was used for a description of the fluorescence measurements. The 'hump' like repulsive part of the inter-atomic potential at intermediate internuclear separations around 5 Å was calculated by using the full configuration interaction method as implemented in the MOLPRO code. The repulsion gives rise to the establishment of cavities around excited atoms in liquid helium and radius and profiles of the cavity boundaries were calculated for this potential and different pressures. The potential was used to simulate the atomic line fluorescence profile as a function of external pressure between 0.1 and 3.5 MPa. The fluorescence shows characteristic line shifts and widths as a function of pressure. By comparison with the theoretical predictions, we found evidence that for corona discharge-excited liquid helium the He* atoms reside in a cavity and hence emit from within the liquid phase. The measurements carried out at higher temperature 5.1K give spectra are characterised for the bubble states similar to the states were obtained in superfluid He II at 1.8 K

1. Introduction

Emission spectroscopy is a powerful tool to obtain information about parameters that characterize non-equilibrium discharge plasma (corona) in high pressures. Spectroscopic observations of the light emitted by ionized gases can be used to determine conditions surrounding the emitted atoms or molecules. Spectroscopic studies of internal excitations in liquid helium have a long history [1–6]. Experimental techniques, such as α-particle bombardment [1, 7, 8], high energy electron bombardment [2–4], corona discharge [5, 6, 9–11], strong field ionization by femtosecond laser pulses [12, 13], and synchrotron radiation [14–16], have been employed. The latter technique has been applied to study intrinsic excitations in liquid helium droplets. Common to all these techniques is the first step, which involves direct ionization of helium atoms creating the corresponding positive ions and electrons in the liquid. This can directly lead to the formation of positive He^{+}_{2} and subsequently through electron-ion recombination He_2^*, or through electron impact ionization/excitation in the liquid electronically excited helium atoms He* may also form. These species are formed both in their singlet and triplet electronic manifolds and they relax towards lower energy states by emitting photons in the visible and infrared regions as well as in the VUV when the system returns back to its electronic ground state.

Such radiators undergo strong perturbation by surrounding atoms if they are placed in very dense medium as a liquid He. Though, atomic lines were observed in a lot of experiments and their width depends on external pressure applied to a liquid. This is possible because the excited species are established to reside inside voids ranging from (7 – 15) Å ("bubble states"), which are slightly smaller than for free electron in the liquid (18.5 Å) [17]. Such structures are the result of the repulsive Pauli exchange repulsion between the Rydberg electron and the surrounding closed shell helium atoms. These experiments have been carried out in a superfluid liquid He II at temperatures less than 2 K. It is well-known that electrons injected in dense helium remain localized in bubbles both in normal liquid He I at higher temperatures and in cryogenic gaseous He at temperatures less 20 K [18, 19]. In this work, we present results of spectroscopic experiments carried out in liquid and gaseous helium at low temperatures 4.2 – 5.2 K. The spectral line 706 nm of the transition $3^3S \Rightarrow 2^3P$ is under the study. It will be shown that a shape (shift and width) of the line observed is strong different than in low density gas and it is related with bubble-like complex with excited atom in it.

2. Experimental facilities.

The experimental facilities used in our spectral investigations of corona discharges were described in details in [20]. The corona discharge (ionization of matter) occurs in vicinity of a tip cathode under action of high strength electric field. Tungsten needle electrode with point radii r_p = 2.5μm was used under dc high voltage. The point-plane inter-electrode distance was in the range of 0.65-0.8cm. The starting material was helium gas N60 (99.99990% pure, Air Liquide) which had an impurity concentration of about 0.1 ppm of oxygen. After the purification, the gas is liquefied in a cell housed in a cryostat. The temperature in the cell is measured by a germanium resistor and was fixed for each series of measurements. The measurements were carried out for different external pressures applied to the cell. The pressure increased until the spectral line was observed. Some measurements can be possible for the pressure 3.5 MPa. The voltage was supplied by a stabilized DC power supply (Spellman model RHSR/20PN60) giving either positive or negative tip polarity. The stabilized dc voltage (up to 20 kV) was connected to the tungsten point electrode. The Tektronix TDS540 oscilloscope or the Keithley 610C current meter was connected to the plane electrode.

An ionization zone near a tip electrode is a source of a light emitted by the corona. Gaseous and liquid Helium under pressures (0.1 -0.2) MPa at temperatures between of 4.2K and 5.2K was excited using the corona discharge both for negative and positive high voltages. The light emitted from the ionization zone of the discharge was analyzed. Both the helium cryostat and the copper beryllium cell were equipped with sapphire windows, which allowed the collection of light near the corona discharge zone by a short focal length lens. A second quartz lens was used to focus the light into a 300 mm focal length spectrograph (Acton Research Corporation; Spectra-Pro-300i with 150 gr/mm and 1200 gr/mm gratings). A liquid N_2 cooled charge coupled detector (CCD) was installed at the exit plane of the spectrograph (Princeton Instruments model 2D-CCDTKB-UV/AR). The spectral resolution with the 1200 gr/mm grating was 0.1 nm as determined from the spectral profiles of argon discharge lamp lines.

3. Results and discussion

The relevant atomic He* line observed in the experiments through fluorescence is the triplet HeI line 706 nm of the transition $3s^3S \rightarrow 2p^3P$. The line observed in liquid He is strongly different than the line observed in gaseous He, figure 1. Moreover, the shape of the line has strong pressure dependence in the LHe, figure 2. In figure 1 the experimental spectra obtained for 706 nm line in a negative corona under the same pressure and at different temperatures. Gas density at 570K is $4.4*10^{20}$ cm^{-3}. Liquid density at 4.2 K is $2*10^{22}$ cm^{-3}. The pressure dependence of the line width is a gas is a result of the linear relations between the gas density and the pressure. In the liquid variation of density with the pressure increasing is less than that in a gas. The pressure dependence of line width in a liquid has another reason.

Journal of Physics: Conference Series **397** (2012) 012066 doi:10.1088/1742-6596/397/1/012066

Pressure broadening of spectral lines depends on the surrounding matter density. The "impact" interaction of radiator with surrounding atoms results in the Lorentzian symmetric profile of spectral lines emitted by a discharge in low pressure gases [21, 22]. In this case the width $\Delta\lambda$ (Full Width Half Maximum) of a line and its shift S are proportional to gas density [23, 24]. The analysis of the line broadening and shift has been made in [25] for the Lennard-Jones potential of an interaction between excited atom and surrounded atoms in the ground state in the framework of the "impact" approximation. The shift sign ("red shift" for the shift toward longer wavelengths and "blue shift" for the shift toward shorter wavelengths) depends on character of radiator-perturbator interaction. The blue shift corresponds to significant repulsion. The "impact" interaction shifts the centre of line weaker than the broadening of the line. The ratio Shift/Width does not exceed 0.2 for "impact" distortion of the line shape. The low shift of line compare with its broadening is a feature of the "impact" broadening by pressure in a gas. Subsequent growth of the gas density is accompanied by distortion of the "impact" Lorentz profile of a line [26]. The asymmetry of a spectral line shape was described using "quasi-static" approach where perturbators are assumed immobile. The asymmetric shape of spectral lines has been observed in He gas under high pressures [27]. The lines had more intensive "blue wing" that was a result of repulsion between the excited atom and surrounding atoms. The number density of the liquid He is $2*10^{22}$ cm^{-3}, which is 45 times larger than gas density $4.4*10^{20}$ cm^{-3} at temperature 570 K and pressure 3.2 MPa. So a linear extrapolation of the "gaseous" $\Delta\lambda$ = 2.6 nm for the density $4.4*10^{20}$ cm^{-3} up to the liquid density leads to non-real magnitude of the line width in the liquid and cannot explain the experimental data, figure 2.

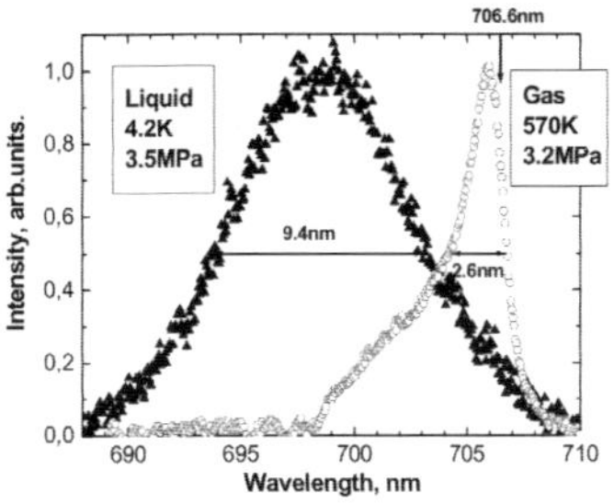

Figure 1. Line 706nm observed in liquid He at 4.2K and in gaseous He at 570K under the same pressure 3.5 MPa.

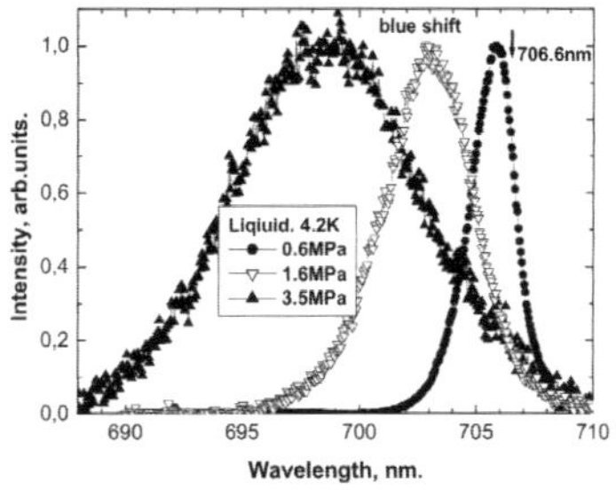

Figure 2. Line 706nm observed in liquid He at 4.2K under different pressures. Shape of the lines is symmetric one.

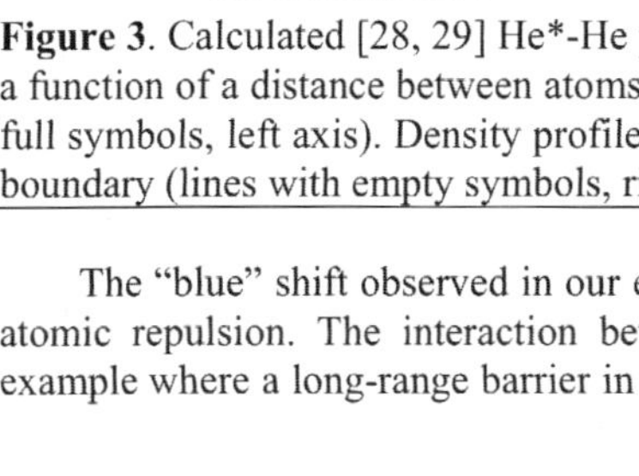

Figure 3. Calculated [28, 29] He*-He potential as a function of a distance between atoms (lines with full symbols, left axis). Density profile of a cavity boundary (lines with empty symbols, right axis).

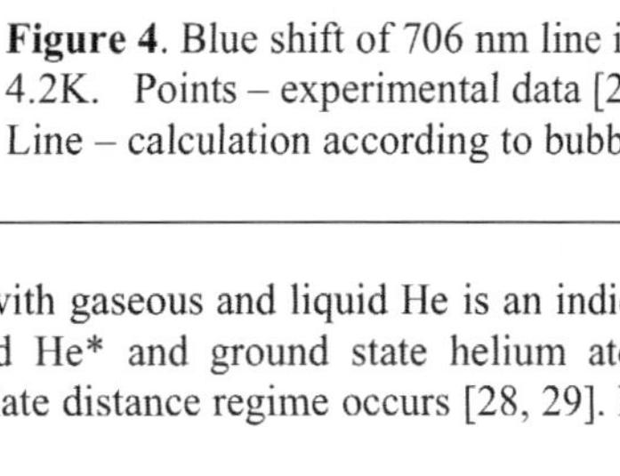

Figure 4. Blue shift of 706 nm line in liquid He at 4.2K. Points – experimental data [20].
Line – calculation according to bubble model [29]

The "blue" shift observed in our experiments with gaseous and liquid He is an indicator of inter-atomic repulsion. The interaction between excited He* and ground state helium atoms provides example where a long-range barrier in the intermediate distance regime occurs [28, 29]. Both He* and

He_2* species have been established to reside inside voids ranging from (7 – 15) Å ("bubble states"), which are slightly smaller than for free electron in the liquid (18.5 Å) [30-34]. Such structures are the result of the repulsive Pauli exchange interaction between the Rydberg electron and the surrounding closed shell helium atoms. Excited atom pushes away surrounding atoms. The bubble radius depends on external pressure the spectra are recorded systematically as a function of the pressure, figure 2. The long-range repulsion potential between $3s^3S$ excited atom and $1s^1S$ ground state atom calculated in [28, 29] is presented in figure 3 as a function of the distance between the atoms. The excited atom is located in the origin. Empty cavity surrounds the radiator. It is perturbed by weak part of a long-range tail of the potential, which occupies a non-zero density boundary of the cavity, figure 3. The boundary approaches to radiator with a pressure increasing and the perturbation leads to growth of shift and broadening of the emitted line. Such calculations give values of 706 nm line shift close to experimental, figure 4.

The phenomenon of localization of electron and excited atom in "bubble" was well known for liquid He at temperatures less then the temperature of the superfluid transition 2.17 K, i.e. in the superfluid Helium II. Our experiments at isotherm 4.2 K and high pressures showed that such phenomenon is possible in normal liquid Helium also. Localization of electrons injected in dense cryogenic He gas has been observed up to temperature 20 K [18, 19]. Our experiments have been made in the normal He I at 4.2 K showed that the localization of excited atoms is possible at that conditions.

It is natural that the localization of Rydberg-atoms should become weaker with growth of temperature and decreasing of gas density. The next step of our investigation is to discover the boundary of thermodynamic domain of the localization of excimers in cryogenic He gas. Moreover it is very interesting to study of atomic line shape in immediate vicinity of the critical point of Helium T_{cr}≈5.2K, P_{cr} ≈ 0.227MPa. Figure 5 represents the density-pressure diagram with low temperature isotherms and circumscribes the domain of the observations. The isotherm 5.1 K shows that the boiling of He occurs at the pressure 0.21 MPa at this temperature. A liquid is realized under higher pressures but the liquid has properties similar to a supercritical fluid. As example, note a problem of the surface tension for such "liquid". Figure 6 shows a shape of 706nm line emitted in liquid He at 5.1 K and different pressures. Shift of the lines increases with pressures and its magnitude is large enough to conclude that the radiator is in a cavity with rarefied gas inside. Such "pseudo-bubbles" were proposed to explain a nature of low mobility of electrons injected into dense cryogenic He gas [35].

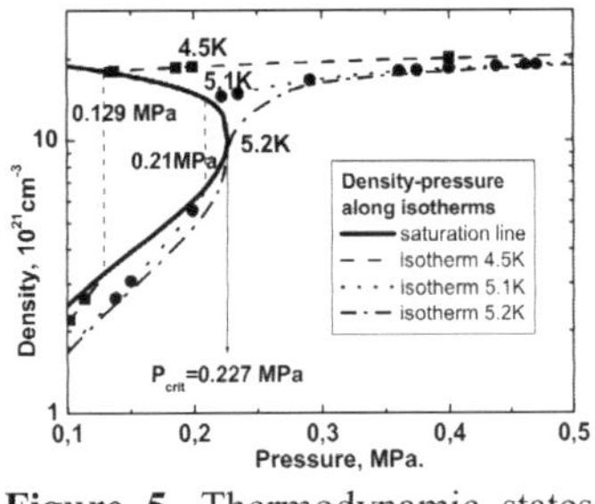

Figure 5. Thermodynamic states of Helium in cryogenic conditions. T_{cr} ≈ 5.2K, P_{cr} ≈ 0.227MPa. Symbols show states where spectroscopic observations were made.

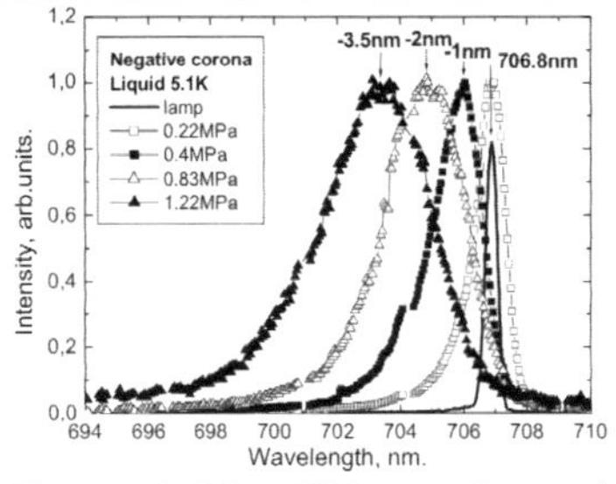

Figure 6. Line 706 nm observed in negative corona in liquid He at 5.1 K. Blue shift values are noted at intensity maxima. Isotherm 5.1 K is very close to the critical temperature 5.19 K.

4. Conclusions.

The experiments carried out in cryogenic Helium are described in the paper. The relevant $3s^3S$ excimer which is an origin of 706 nm HeI line is presented in the liquid He as "impurity" created a cavity surround it. The corona micro-discharge in liquid and cryogenic gaseous helium is realized in our experiments and it was served as a source of emitted atoms. Shift and shape of an atomic line 706 nm were measured for these conditions. The shape (broadening and shift) of the line emitted from the cryogenic He (liquid and dense gas) is different strongly of the line emitted from the gas at high temperature and pressure. The line has asymmetric shape with more intensive short-wavelength (blue) wing for a broadening characterized for high pressure gas. Blue shift (toward short-wavelengths) is small in a gas. Asymmetric shape of an atomic line 706 nm was recorded in our experiments in gaseous He at intermediate temperatures. Blue wing of the line recorded in negative corona was more intensive than its red wing. Spectral shift of the line was small for the range of applied pressure. The line shape observed in gaseous and liquid Helium for conditions closed to the boiling pressure at a fixed temperature was the same.

The spectra observed in cryogenic Helium have rather symmetric shape described by Gauss-function. Such spectra are characterized for light emitted by Rydberg-atoms localized inside of empty cavity (bubble). The bubble model has been developed for excimers in superfluid liquid He II at temperature 1.8 K. The model is modified for temperatures $\approx$ 4 K and gives good agreement with measured width and shift of the 706 nm line. Such structure is the result of the Pauli repulsive exchange interaction between the Rydberg electron of excited atom and surrounding helium atoms. The excited atom and atoms from its environment are far enough each other in the "bubble state". So, the long-length asymptotic of the repulsive potential is important in formation of this state. The He*-He pair interaction potential was calculated by using the full configuration interaction method as implemented in the MOLPRO code. The potential was used for calculation of the cavity shape and of perturbation of a radiator inside the cavity.

Thus, the spectra emitted corona discharge in liquid normal He I at 4.2 K are corresponded to bubble model which is developed for low temperature liquid He. Such limitation has physical basis because a temperature blurs of a static structure formed due to Pauli repulsion. In some respect the experimental evidence of the bubble-like spectra at 5.1 K is a significant step to develop of a theory of "pseudo-bubble" states of the Rydberg atoms in dense cryogenic supercritical fluid helium.

Acknowledgements
Russian participant of the work (V. M. A.) thanks the Russian Foundation for Basic Researches, grant 12-08-91052 NCNI_a for support.

References

[1] Surko C M and Reif F 1968 *Phys. Rev.* **175** 229
[2] Dennis W S, Durbin J E, Fitzsimmons W A, Heybey O, and Walters G K 1969 *Phys. Rev. Lett.* **23** 1083
[3] Soley F J and Fitzsimmons W A 1974 *Phys. Rev. Lett.* **32** 988
[4] Keto J W, Soley F J, Stockton M, and Fitzsimmons W A 1974 *Phys. Rev.* A **10** 872
[5] Zimmermann P H, Reichert J F, and Dahm A J 1977 *Phys. Rev.* B **15** 2630
[6] Goncharov V A and Levitov V I 1975 *Izvestiya Academii Nauk, Energetica i Transport* **12** 134
[7] Jortner J, Meyer L, Rice S A, and Wilson E G 1964 *Phys. Rev. Lett.* **12** 415
[8] Moss F E and Hereford H L 1963 *Phys. Rev. Lett.* **11** 63
[9] Kafanov S G, Parshin A Y, and Todoshenko I A 2000 *JETP* **91** 991
[10] Parshin A Y, Todoshenko I A, and Kafanov S G 2000 *Physica* B **91** 284
[11] Li Z-L, Bonifaci N, Aitken F, Denat A, von Haeften K, Atrazhev V M, and Shakhatov V A 2009 *Eur. Phys. J. Appl. Phys.* **47** 22821
[12] Benderskii A V, Zadoyan R, Schwentner N, and Apkarian V A 1999 *J. Chem. Phys.* **110** 1542
[13] Benderskii A V, Eloranta J, Zadoyan R, Schwentner N, and Apkarian V A 2002 *J. Chem. Phys.* **117** 1201
[14] von Haeften K, de Castro A R B, Joppien M, Moussavizdeh L, von Pietrowski R, and Moller T 1997 *Phys. Rev. Lett.* **78** 4371
[15] von Haeften K, Laarmann T, Wabnitz H, and Moller T 2002 *Phys. Rev. Lett.* **88** 233401
[16] von Haeften K, Laarmann T, Wabnitz H, and Moller T 2005 *J. Phys. B: At. Mol. Opt. Phys.* **38**

S373
[17] Eloranta J and Apkarian V A 2002 *J. Chem. Phys.* **117**, 10139
[18] Levine J and Sanders T M 1967 *Phys. Rev.* **154** 138
[19] Harrison H R, Sander L M, and Springett B E 1973 *J. Phys. B.* **6** 908
[20] Li Z-L, Bonifaci N, Aitken F, Denat A, von Haeften K, Atrazhev V M and Shakhatov V A 2009 *EEE Transactions on Dielectrics and Electrical Insulation* **16** 742
[21] Traving G 1968 *Interpretation of line broadening and line shift in Plasma Diagnostic (Chap. 2)*, edited by W. Lochte-Holtgreven, North-Holland Publishing Company, Amsterdam.
[22] Allard N and Kielkopf J 1982 *Rev. Mod. Phys.* **54** 1103
[23] Lindholm E 1945 *Ark. Fis.* A **32** 1
[24] Foley H M 1946 *Phys. Rev.* **69** 616
[25] Hindmarsh W R, Petford A D and Smith G 1967 *Proc. of the Royal Soc. Series* A 297
[26] Margenau H 1935 *Phys. Rev.* **48** 55
[27] Bonifaci N, Li Z-L, Denat A, von Haeften K, Atrazhev V M and Shakhatov V A 2011 *Eur. Phys. J. Appl. Phys.* **55** 13809
[28] Allard N F, Bonifaci N, and Denat A 2011 *Eur. Phys. J.* D **61** 365
[29] Bonifaci N, Aitken F, Atrazhev V M, S L, and Eloranta J 2012 *Phys. Rev* A **85** 042706
[30] Eloranta J and Apkarian V A 2002 *J. Chem. Phys.* **117** 10139
[31] Eloranta J, Schwentner N, and Apkarian V A 2002 *J. Chem. Phys.* **116** 4039
[32] Mateo D, Jin D, Barranco M, and Pi M 2011 *J. Chem. Phys.* **134** 044507
[33] Maris H J 2000 *J. Low Temp. Phys.* **120** 173
[34] Hickman A P, Steets W, and Lane N F 1975 *Phys. Rev.* B **12** 3705
[35] Khrapak A G and Iakubov I T 1979 *Sov. Phys. – Usp.* **22** 703

Absorption spectra of NaHe from white dwarfs to helium clusters

N F Allard[1,2], A Nakayama[3], F Spiegelman[4], J F Kielkopf[5] and F. Stienkemeier[6]

[1] Observatoire de Paris, GEPI, 61, Avenue de l'Observatoire, F-75014, Paris, France
[2] Institut d'Astrophysique, 98bis Boulevard Arago, F-75014, Paris, France
[3] Division of Chemistry, Graduate School of Science, Hokkaido University, Sapporo 060-0810, Japan
[4] Laboratoire de Chimie et Physique Quantique, Université Paul Sabatier, 118 route de Narbonne,Toulouse, France
[5] Department of Physics and Astronomy, University of Louisville, Louisville, Kentucky 40292 USA
[6] Physikalisches Institut, Universität Freiburg, Hermann-Herder-Str. 3, D-76104 Freiburg, Germany

E-mail: nicole.allard@obspm.fr

Abstract. The absorption spectrum of Na produced in dense He shows line features, broadening and shift due to extrema in the $3s$-$3p$ potential differences of the transitions which contribute to the two fine-structure components. We present here a comprehensive study of Na-He collisional profiles at high density, and temperatures from 5000 K, the temperature prevailing in the atmosphere of ultra-cool white dwarfs, down to 1 K. At extremely low temperatures, experimental measurements and theoretical calculations of helium doped with alkali metals have been the subject of active study. The alkali-metal atoms are weakly bound to helium nanodroplets, and it appears that the dopants are trapped in a shallow dimple on the cluster's surface, rather being solvated inside the cluster.

1. Theory

A unified theory of spectral line broadening (Allard *et al.* 1999 [1]) has been developed to calculate neutral atom spectra given the interaction and radiative transition moments for relevant states of the radiating atom with other atoms in its environment. This collisional theory goes beyond the Franck-Condon approximation but is quasi-classical for the ground state. The spectral line is computed from the Fourier transform (FT) of a dipole autocorrelation function. A pairwise additive assumption allows us to calculate the total profile $I(\Delta\omega)$ when all the perturbers interact as the FT of the Nth power of the autocorrelation function $\phi(s)$ of a unique atom-perturber pair. We obtain for a perturber density n_p

$$\Phi(s) = e^{-n_p g(s)}, \tag{1}$$

The dipole autocorrelation function $\Phi(s)$ is evaluated for a classical collision path with an average over all possible collisions. For a transition $\alpha = (i, f)$ from an initial state i to a final state f,

Published under licence by IOP Publishing Ltd

we have

$$g_\alpha(s) = \frac{1}{\sum_{e,e'}{}^{(\alpha)} |d_{ee'}|^2} \sum_{e,e'}^{(\alpha)} \int_0^{+\infty} 2\pi\rho d\rho \int_{-\infty}^{+\infty} dx\, \tilde{d}_{ee'}[R(0)]$$
$$[e^{\frac{i}{\hbar}\int_0^s dt\, V_{e'e}[R(t)]}\, \tilde{d}^*{}_{ee'}[R(s)] - \tilde{d}_{ee'}[R(0)]\,]. \quad (2)$$

Our calculations are based on the *ab initio* Na-He potentials of Nakayama and Yamashita 2001 [2] and radiative dipole moments of Pascale 1983 [3].

2. Quasi-molecular line satellites

The unified theory predicts that there will be line satellites centered periodically at frequencies corresponding to integer multiples of the extrema of the difference potential $\Delta V(R)$:

$$\Delta V(R) \equiv V_{e'e}[R(t)] = V_{e'}[R(t)] - V_e[R(t)]\,, \quad (3)$$

which represents the difference between the electronic energies of a quasi-molecular e-e' transition. In Allard *et al.* 1999 [1] we define $\tilde{d}_{ee'}(R(t))$ as a *modulated* dipole

$$D(R) \equiv \tilde{d}_{ee'}[R(t)] = d_{ee'}[R(t)]e^{-\frac{V_e[R(t)]}{2kT}}\,, \quad (4)$$

V_e is the lower state potential when we consider absorption profiles, or the upper state for the calculation of a profile in emission. Over regions where $V_e(R) < 0$ the factor $e^{-\frac{V_e(R)}{2kT}}$ accounts for bound states of the radiator-perturber pair, but in a classical approximation wherein the discrete bound states are replaced by a continuum; thus any band structure is smeared out. ΔV and variation of the D(R) with temperatures of Na-He modulated dipole are displayed in figures 1 and 3.

2.1. High temperature

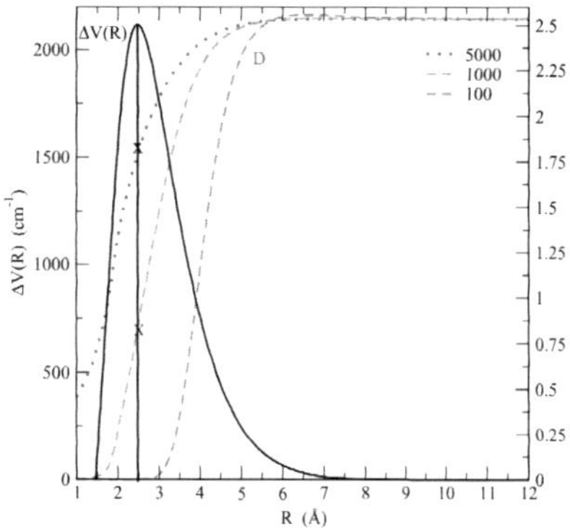

Figure 1. $\Delta V(R)$ and the temperature dependence of modulated dipole $D(R)$ corresponding to the $B - X$ transition.

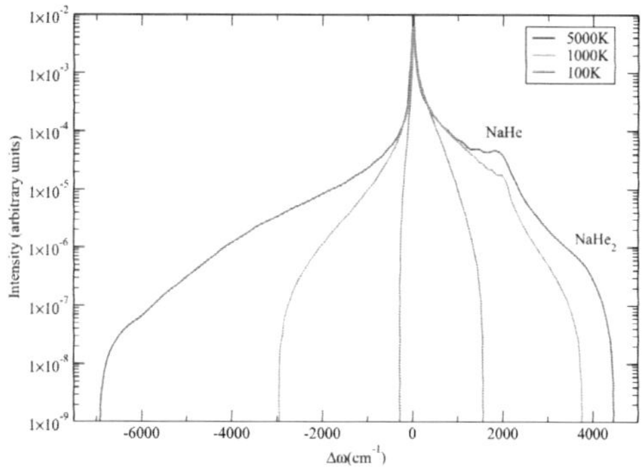

Figure 2. Temperature dependence of the D2 line profile (n_{He}= 10^{21}cm^{-3}).

The perturber density in the line-forming region of an ultra-cool white dwarf atmosphere can reach n_{He}=10^{21} cm^{-3}, and it then requires a unified line profile calculation [4]. Because of the large maximum in ΔV (figure 1), the line satellite at T=5000 K is very well-separated from the main line in the blue wing. It disappears for decreasing T when the transition moment $\tilde{d}_{ee'}[R(t)]$ gets very small. Quantum mechanical calculations of absorption profiles of the Na $3s$-$3p$ transition perturbed by He [5, 6] are in agreement with our previous results [7]. The disappearance of the Na-He line satellite is seen at T=158 K in calculations by Zhu *et al.* 2006 [5]. Figure 2 shows quasi-molecular line satellites, positioned at about ΔV for NaHe, $2\times$ ΔV for $NaHe_2$, ... The dependence on the average number of perturbers is expected on the basis of the Poisson distribution, which indicates the probability of finding a select number of uncorrelated perturbers in the collision volume. Multiple perturber effects such as these are now well known, and were first identified in experimental alkali spectra by Exton and Snow 1978 [8] and by Kielkopf and Allard 1979 [9].

2.2. Low temperature

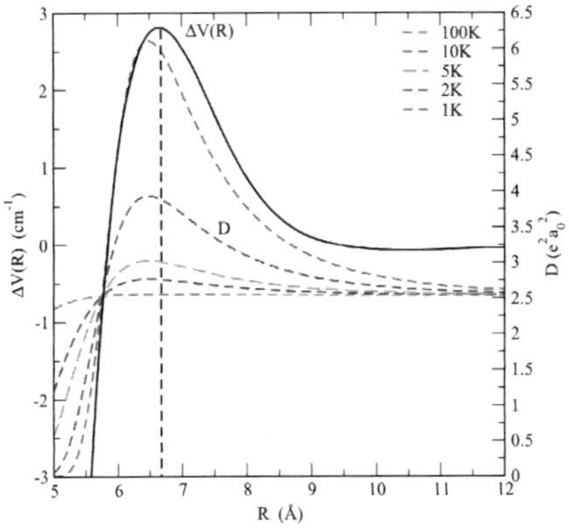

Figure 3. $\Delta V(R)$ and the temperature dependence of modulated dipole $D(R)$ for the D1 line.

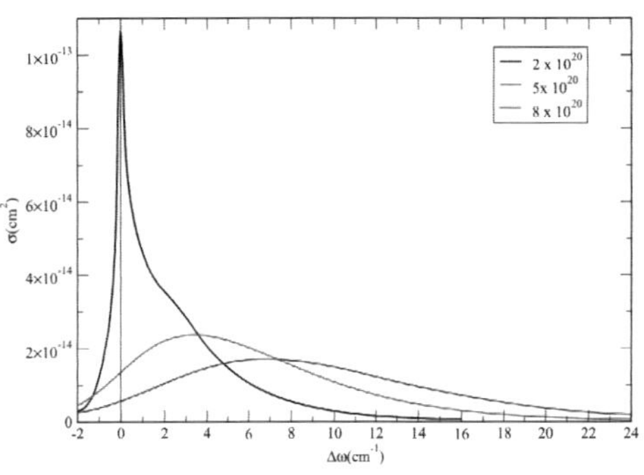

Figure 4. Variation with He density (cm^{-3}) of the D1 component at T=1 K.

While the $^2\Pi_{3/2}$ state is purely attractive, the $^2\Pi_{1/2}$ state has a small barrier at 6.72 Å due to the spin-orbit coupling, the height of which relative to the $^2P_{1/2}$ state is 1.62 cm^{-1} (figure 7 of Nakayama and Yamashita 2001 [2]). Because ΔV for the transition $3s\ X \rightarrow 3p\ A\ P_{1/2}$ is very small, there is no resolved quasi-molecular feature. It results an enhancement in the core of the line at $\Delta\omega = \Delta V = 3\text{cm}^{-1}$. At very low temperature and at the internuclear separation $R_e = 6.4$ Å close to the position where ΔV is an extremum (figure 3), the transition moment $\tilde{d}_{ee'}[R(t)]$ increases for decreasing temperature. This increasing of $D(R)$ can be explained by examining the potential energy curves given in figure 1 of Nakayama and Yamashita 2001 [2]. The ground state is repulsive at short range and very weakly attractive at intermediate range (6.6 Å) with a well depth of 1.22 cm^{-1}. Figure 4 shows clearly an asymmetrical shape of the D1 component due to the bound-free transition. Increasing the He density results in a shift of the D1 line towards the position of the extremum of ΔV. The intensity variation in figure 4 is evidence of the presence of high order effects, and as ΔV is very small, we observe a blend of these structures.

3. Pressure dependence of the sum of the D1 and D2 components

Absorption cross-sections for $3s$-$3p$ $P_{1/2}$ and $3s$-$3p$ $P_{3/2}$ transitions are shown in figure 5. At low density the ratio for the components is 2, as expected for resonance doublets of light alkalies. Whereas the amplitude of the D2 line decreases exponentially, the blue wing of the D1 line gains in strength relative to the parent line. For n_{He}=6 × 10^{20} cm^{-3} the peak amplitude of D1 becomes more important than the peak of D2 (figure 7) and for n_{He}= 9× 10^{20} cm^{-3} (figure 6) there is a blend of the two fine-structure components. The same behaviour is observed in the experimental spectra (figure 8). The two vertical lines indicate the positions of the unperturbed D1 and D2 lines. In order to check whether we have reached the quasistatic limit we compare in figure 6 the unified profile and the one obtained using the quasistatic limit of the autocorrelation function. As we can expect at this high density the profiles are comparable with a perfect agreement in the wing.

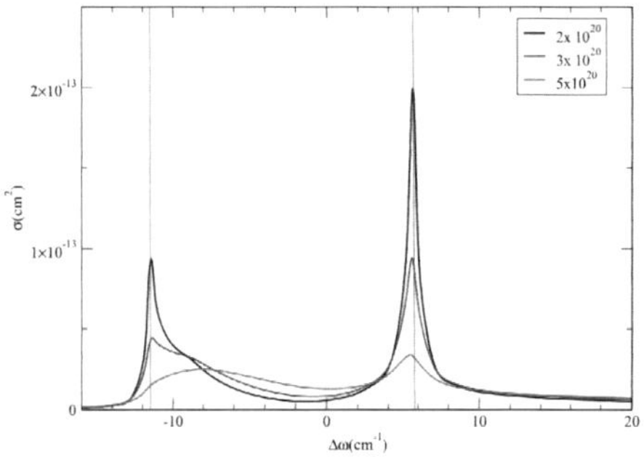

Figure 5. Sum of the absorption cross-sections for the D1 and D2 lines at T=1 K. Variation with helium atom density (cm^{-3}).

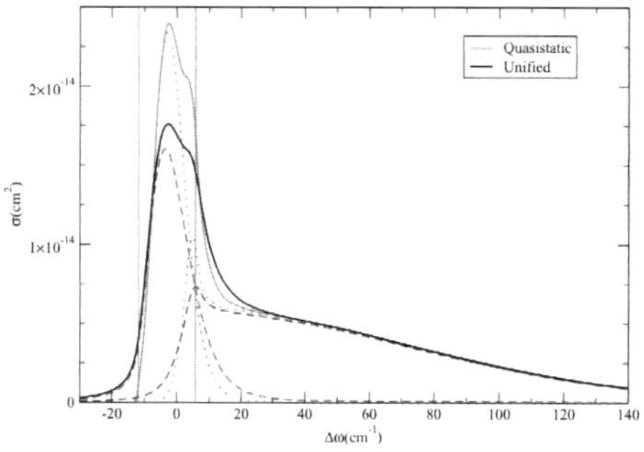

Figure 6. Comparison of profiles obtained in the unified theoretical approach and in the quasistatic limit at T=1 K, (n_{He}=9 × 10^{20} cm^{-3}).

($\Delta\omega$= ω -ω_0, where ω_0 is relative to the center of gravity of the two fine-structure components of the $3p$ state, neglecting the spin-orbit coupling.)

4. Comparison with Path Integral Monte Carlo calculations

In recent years special attention has been focused on experimental measurements and theoretical calculations of helium nanodroplets doped with alkali metals [10, 11, 12, 13, 14]. Several theoretical works have described absorption spectra of Na atoms attached to helium nanodroplets (see Hernando *et al.* 2010 [13] and references therein). The first physical interpretation of the spectra was done by Nakayama and Yamashita 2001 [2]. Using a diatomics-in-molecules (DIM) type of calculation for describing the interaction potential between Na and helium atoms, they performed quantum path integral Monte Carlo (PIMC) simulations to obtain the helium density profile around Na in the ground state and also the absorption spectra of Ak-He_n via the Franck-Condon (FC) approximation (Ak=Li, Na, K).

$$I_{eg}(\omega) \propto |M_{eg}|^2 \int dR \rho(R, R; \beta) \delta \left(V_e(R) - V_g(R) - \hbar\omega\right), \quad (5)$$

where V_g and V_e are the potential energies in the electronic ground and excited states, respectively. M_{eg} is the electronic transition dipole moment. The quantity $\rho(R, R; \beta)$ is the quantum mechanical thermal density matrix on the electronic ground state surface. One can, therefore, obtain the absorption spectrum by calculating the potential energy difference at each nuclear coordinate generated by path integral Monte Carlo calculations. Figure 9 shows the average helium density profiles in the same range of internuclear distance as in figure 3 where ΔV is significant. We can see that both theoretical approaches have a dramatic change in the region where the bound-free transition takes place. They successfully reproduced the overall shapes of the spectra and explained their features through a discussion of potential difference and number of He atoms interacting with the Na atom in the region which is, from our point of view, the collision volume.

In our unified approach we use averages of independent Na-He pair collisions, emphasizing the collisional process rather the collective quantum description of the interaction of a Na atom with the He cluster (which may be important for very small He atom numbers). However the present approach goes beyond the FC approximation and provides a better description close to the line center. It can be seen that despite those differences, the two methods provide similar results. The complex behavior observed experimentally (figure 8) reflects the fact that the radiating alkali atom experiences multiple encounters with helium atoms and our collisional analysis gives a good physical understanding of the phenomena. Using the average helium density profile for $NaHe_{300}$ in the quasistatic limit instead of the modulated dipole moment, we obtained the profile shown in figure 10 compared to a quasistatic profile for ($n_{He}=9 \times 10^{20}$ cm^{-3}). This interesting result demonstates that the active region is as well described by the average profile as it is by the modulated dipole moment. It explains why our two approaches are in such close agreement. They both describe a radiating alkali atom surrounded by an equilibrium distribution of He atoms, with a spectrum given by multiple perturber phenomena in the high He density and very low temperature of liquid He.

5. Experimental spectra

Large liquid helium clusters produced in a supersonic jet are doped with alkali atoms (Li, Na, K) and characterized by means of laser induced fluorescence (Stienkemeier *et al.* 1996 [11]). The main feature of the spectra is that two absorption maxima are observed. One is near the D2 line of the atomic transition and the other between the D1 and D2 lines. The spectrum broadens and the red maximum shifts slightly to the blue with increasing cluster size. The blue maximum is almost unshifted at all cluster sizes. The relative amplitude of the two absorption maxima changes slightly, that is, the red maximum becomes higher and dominant in larger clusters.

Acknowledgments

One of us (NFA) would like to acknowledge PNPS for the financial support which allows the collaboration with the Laboratoire de Chimie et Physique Quantique in Toulouse.

References

[1] Allard N F, Royer A, Kielkopf J F and Feautrier N 1999 *Phys. Rev. A* **60** 1021
[2] Nakayama A and Yamashita K 2001 *J. Chem. Phys.* **114** 780
[3] Pascale J 1983 *Phys. Rev. A* **28** 632
[4] Homeier D, Allard N, Johnas C M S, Hauschildt P H and Allard F 2007 *15th European Workshop on White Dwarfs* (*Astronomical Society of the Pacific Conference Series* vol 372) ed R Napiwotzki & M R Burleigh p 277 (*Preprint* 0707.4138)
[5] Zhu C, Babb J F and Dalgarno A 2006 *Phys. Rev. A* **73** 012506
[6] Alioua K, Bouledroua M, Allouche A R and Aubert-Frécon M 2008 *J. Phys. B: At. Mol. Opt. Phys.* **41** 175102
[7] Allard N F, Allard F, Hauschildt P H, Kielkopf J F and Machin L 2003 *A&A* **411** 473
[8] Exton R J and Snow W L 1978 *J. Quant. Spectros. Radiat. Transfer.* **20** 1
[9] Kielkopf J F and Allard N F 1979 *Phys. Rev. Lett.* **43** 196

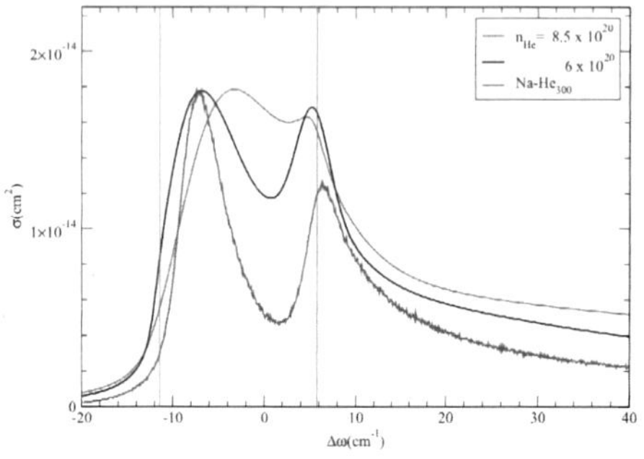

Figure 7. Unified profiles for (n_{He}=6 and 8.5× 10^{20}cm^{-3}) compared to PIMC calculations of [2]. ($\Delta\omega$= ω -ω_0, where ω_0 is relative to the center of gravity of the two fine-structure components).

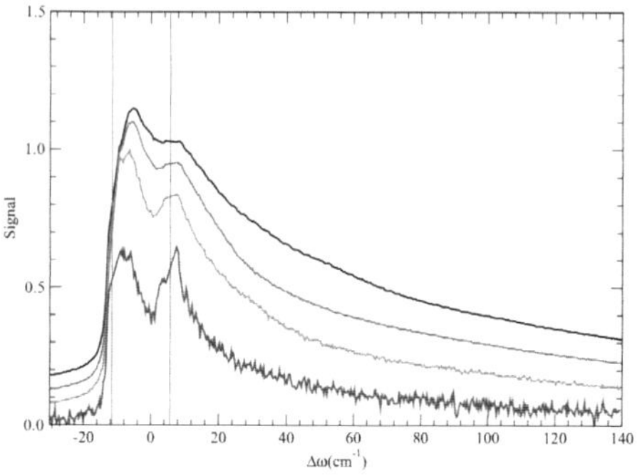

Figure 8. Absorption spectrum of Na attached to large helium clusters He$_N$ for different cluster sizes (N = 12500, 10000, 5000, 2000, 900; from top to bottom). For sake of clarity individual curves are displaced vertically. (Extracted from Stienkemeier *et al.* 1996 [11]).

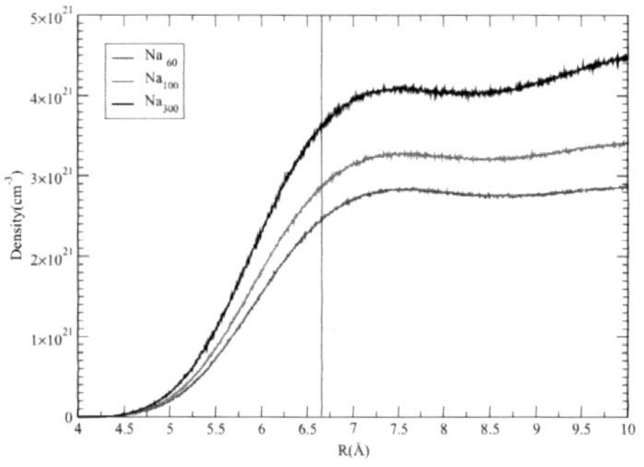

Figure 9. Average helium density profiles for the clusters at T= 0.5 K. Extracted partly from figure 2 of [2].

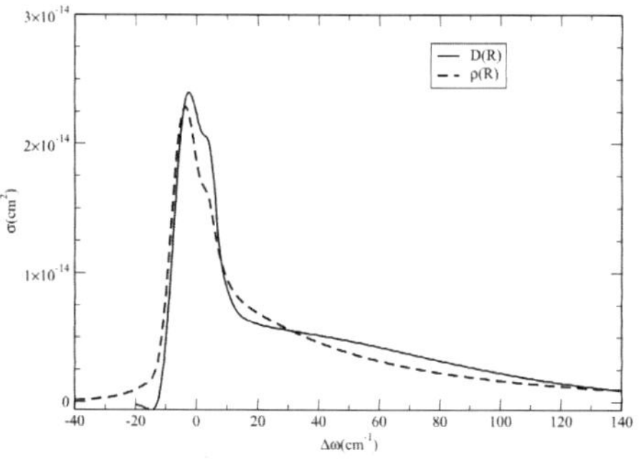

Figure 10. Quasistatic profiles obtained with average helium density profile and modulated dipole moment compared.

[10] Stienkemeier F, Higgins J, Ernst W E and Scoles G 1995 *Zeitschrift fur Physik B Condensed Matter* **98** 413–416
[11] Stienkemeier F, Higgins J, Callegari C, Kanorsky S I, Ernst G and Scoles W E 1996 *Zeitschrift fur Physik D Atoms Molecules Clusters* **38** 253
[12] Bünermann O, Droppelmann G, Hernando A, Mayol R and Stienkemeier F 2007 *J. Phys. Chem.A* **111** 12684–12694
[13] Hernando A, Barranco M, Mayol R, Pi M, Ancilotto F, Bünermann O and Stienkemeier F 2010 *Journal of Low Temperature Physics* **158** 105–111
[14] Mateo D, Hernando A, Barranco M, Mayol R and Pi M 2011 *Phys. Rev. B* **83** 174505

Journal of Physics: Conference Series **397** (2012) 012068 doi:10.1088/1742-6596/397/1/012068

Emission profiles of K-He exciplexes in cold helium gas

N F Allard[1,2]

[1] Observatoire de Paris, GEPI, UMR8111, CNRS, 61, Avenue de l'Observatoire, F-75014, Paris, France

[2] Institut d'Astrophysique, UMR7095, CNRS, Université Paris VI, 98bis Boulevard Arago, F-75014, Paris, France

E-mail: nicole.allard@obspm.fr

Abstract. Emission spectra of exciplexes composed of a light alkali atom in the first excited state and ^{4}He atoms have been observed in cryogenic gas in the spectral range from the atomic D lines to 6300 cm^{-1}. A unified semi-classical theory of line broadening has been used to determine the total profile from the center to the far wings of emission profiles of potassium perturbed by helium at low temperatures and high He density. The agreement of the theoretical peak positions of K*He$_n$ exciplexes compared to the experimental determinations is fairly good. Such comparisons provide a critical test of the calculated molecular potentials and the relevance of the theoretical approach which has been used.

1. Theory

A unified theory of spectral line broadening has been developed to calculate neutral atom spectra given the interaction and radiative transition moments for relevant states of the radiating atom with other atoms in its environment. Complete details and the derivation of the theory are given by [1]. In our theoretical work, the spectral line is computed from the Fourier transform (FT) of a dipole autocorrelation function. A pairwise additive assumption allows us to calculate the total profile $I(\Delta\omega)$ when all the perturbers interact as the FT of the Nth power of the autocorrelation function $\phi(s)$ of a unique atom-perturber pair. We obtain for a perturber density n_p

$$\Phi(s) = e^{-n_p g(s)}, \tag{1}$$

where decay of the autocorrelation function with time leads to atomic line broadening.

For a transition $\alpha = (i, f)$ from an initial state i to a final state f, we have

$$\begin{aligned} g_\alpha(s) &= \frac{1}{\sum_{e,e'}{}^{(\alpha)} |d_{ee'}|^2} \sum_{e,e'}{}^{(\alpha)} \int_0^{+\infty} 2\pi\rho d\rho \int_{-\infty}^{+\infty} dx\, \tilde{d}_{ee'}[\, R(0)\,] \\ & [e^{\frac{i}{\hbar}\int_0^s dt\, V_{e'e}[R(t)]}\, \tilde{d}^*{}_{ee'}[R(s)] - \tilde{d}_{ee'}[R(0)]\,]. \end{aligned} \tag{2}$$

The e and e' label the energy surfaces on which the interacting atoms approach the initial and final atomic states of the transition as $R \to \infty$ (R denotes the internuclear distance between the radiator and the perturber). $\Delta V(R)$, the difference potential, is given by

$$\Delta V(R) \equiv V_{e'e}[R(t)] = V_{e'}[R(t)] - V_e[R(t)]\ , \tag{3}$$

and represents the difference between the electronic energies of the quasi-molecular transition. The interatomic interactions are the main physical quantities needed for a good understanding of collisional processes.

In [1] we define $\tilde{d}_{ee'}(R(t))$ as a *modulated* dipole

$$D(R) \equiv \tilde{d}_{ee'}[R(t)] = d_{ee'}[R(t)]e^{-\frac{V_e[R(t)]}{2kT}} , \tag{4}$$

V_e is the lower state potential when we consider absorption profiles, or the upper state for the calculation of a profile in emission. Note that over regions where $V_e(R) < 0$, the factor $e^{-\frac{V_e(R)}{2kT}}$ accounts for bound states of the radiator-perturber pair, but in a classical approximation wherein the discrete bound states are replaced by a continuum; thus any band structure is smeared out.

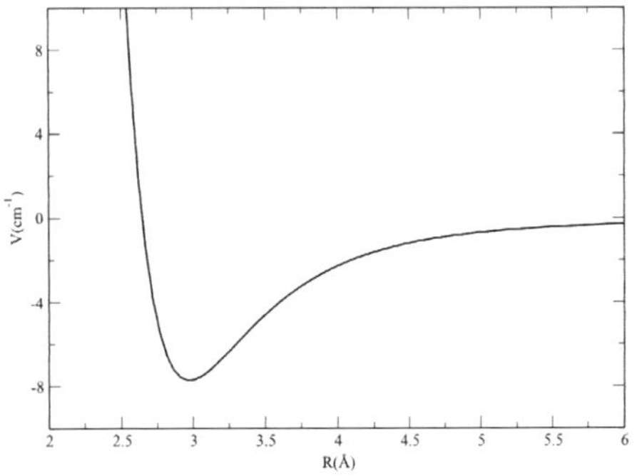

Figure 1. Potential curve for the $X^2\Sigma$ state of the He_2 molecule [2], (R_e= 2.96 Å).

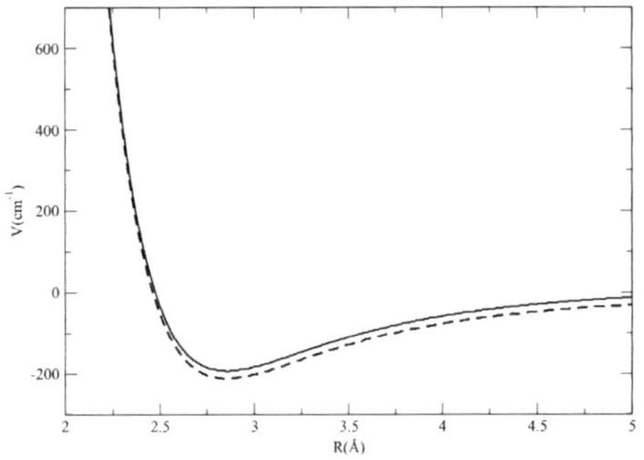

Figure 2. Potential curves for the $4p\,\Pi$ states of the K-He molecule [3], $A^2\Pi_{1/2}$ (full line) and $A^2\Pi_{3/2}$ (dashed line) (R_e=2.8 Å).

Laboratory spectra of K with H_2 and He [4, 5] agree exceptionally well with the semi-classical profiles based on molecular potentials of [3] and [6]. The comparison with theoretical profiles establishes the accuracy of the interaction potentials, which are difficult to compute *a priori*.

2. Emission profiles

While B states radiates in the blue wing, A states radiate in the red wing. Full quantum mechanical calculations in the binary approximation done by Zhu *et al.* 2006 [7] predict a K-He blue line satellite about 7080 nm in agreement with laboratory spectra whereas a broad plateau is obtained at around 830 nm on the red wing of emission profiles.

We present in figure 2 the excited-state potentials $A\ ^2\Pi$ as a function of internuclear distance. They have minimum values, $V_{\rm min}$=-212/193 cm^{-1}, situated at the equilibrium distance, R_e=2.8 Å respectively for $A^2\Pi_{3/2}$ and $A^2\Pi_{1/2}$ states. The well depth has six bound vibrational levels for zero angular momentum J=0. Table 1 of Zhu et al. *et al.* 2006 [7] gives the binding energies of the vibrational levels of KHe. The *ab initio* energy curve for the ground state of He_2 is shown in figure 1. The equilibrium distances for all these potentials are around 3 Å. The bound-free coefficients is a weighted sum of the emission rates of the individual rovibrational levels of the $A\ ^2\Pi$ state. In the high pressure limit the bound and continuum states are thermally populated and the emission coefficient is the sum of the bound-free and free-free coefficients. Figure 12 of [7] shows the variation with temperature between 158 and 3000 K of emission coefficients. At lower temperature it results an enhancement of the maximum at 830 nm, it arises, for bound upper states, through the Boltzmann factor in the relative population:

$$g(v,j) = (2J+1)e^{-\frac{V_e^{v,J}}{kT}} \tag{5}$$

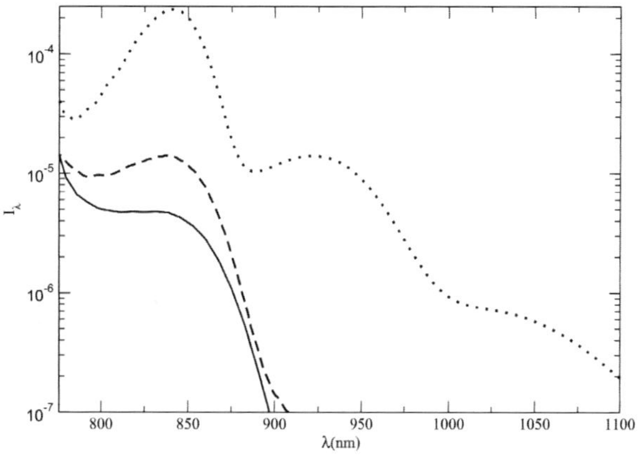

Figure 3. High pressure semi-classical profiles of the D_2 line at temperatures T=158 (full line), 100 (dashed line), and 50 K (dotted line). The helium density is $n_{He} = 9\times10^{19}$ cm^{-3}.

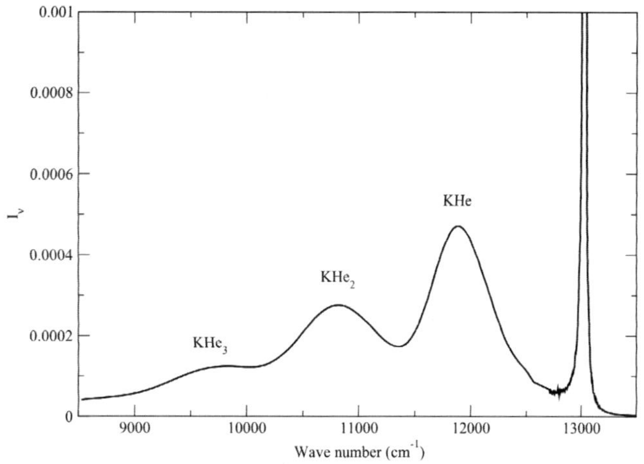

Figure 4. Emission profile at T=50 K and n_{He}=9×10^{19} cm^{-3}.

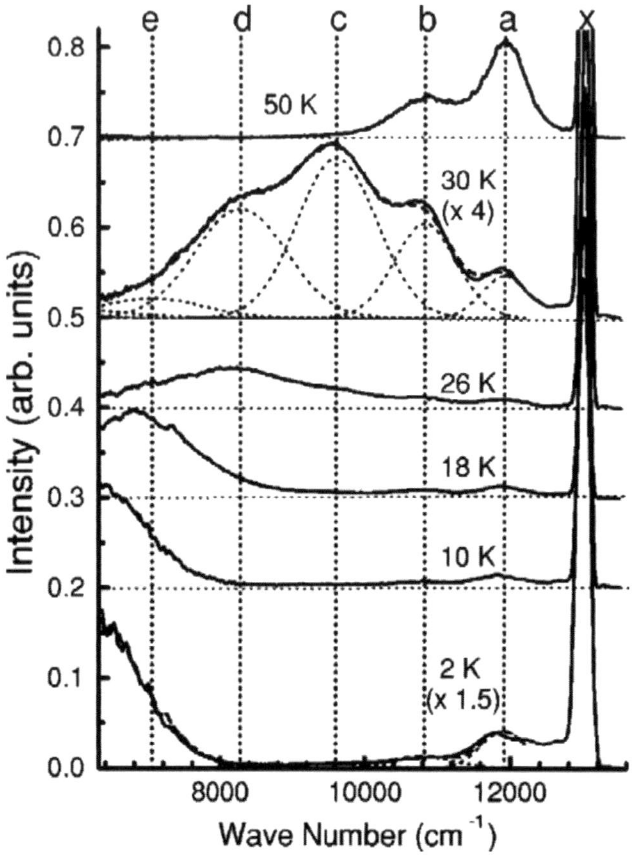

Figure 5. Emission spectra of KHe_n in the case of D_2 excitation at various temperatures in gaseous helium ($n_{He} = 9\times10^{19}$ cm^{-3}). Vertical lines labelled as *a-e* indicate the peak positions of five Gaussian curves for n=1-5 (extracted from Enomoto *et al.* 2004 [8]).

3. Semi-classical profiles

We present figure 3 the variation of the semi-classical emission profiles for T decreasing from T=158 K to 50 K. The intensity of the red wing increases as expected from Eqs. (4-5) corresponding to emission from the bottom of the $A\ ^2\Pi$-state well. At this small internuclear separation R_e the transition moment given by the *modulated dipole moment* gets very large at low temperature.

Broad peak was already observed in the NaHe wing spectrum of Havey *et al.* 1980 [9]. Such peaks are present in the red wing of the alkali-rare gas molecules when $D_e \gg kT$. For a temperature of 50 K , the thermal energy kT becomes 70 cm^{-1}. At high density, $n_{He} = 9\times10^{19}$ cm^{-3}, multiple-perturber effects appear (figures 3-4), not only we obtain the excimer K*He but also the exciplexes K^*He_2 and K^*He_3 (K* represents the potassium atom in the first excited p state).

4. Comparison with experimental spectra

Figure 5 shows emission spectra of Enomoto *et al.* 2004 [8] for KHe at various temperatures, the helium density is $n_{\mathrm{He}} = 9\times10^{19}$ cm^{-3}. The emission spectra are a superposition of spectral components of KHe$_n$ and each peak of the observed spectra corresponds to the spectrum for different n. In order to determine experimentally the peak positions and the widths of spectral components of K*He$_n$ they fitted superposition of five Gaussian curves.

The positions of the peaks of the emission bands (figure 4) and the experimental determinations of the observed peaks are summarized in table 1. The agreement of our theoretical peak position of KHe$_n$ exciplex compared to the experimental ones gives confidence in the determination of the potential of the $A^2\Pi$ state at short distance responsible of the red wing. The full width at half maximum (FWHM) is also very close, 700 cm^{-1} for the KHe satellite compare to 726 cm^{-1} determine from the Gaussian fits of the observed spectra.

Table 1. Comparison of theoretical peak positions at T=50 K with experimental determinations.

n	$\overline{\nu}_{theor}$ (cm^{-1})	$\overline{\nu}_{exp}$ (cm^{-1})
1	11884	11914
2	10824	10826
3	9823	9608

Acknowledgments

We are grateful to Dr J. Babb for kindly providing the KHe molecular potentials.

References

[1] Allard N F, Royer A, Kielkopf J F and Feautrier N 1999 *Phys. Rev. A* **60** 1021
[2] Allard N F, Gadéa F X, Monari A and Deguilhem B 2011 Theoretical analysis of the He$_2$ line at 585 Å *SF2A-2011: Proceedings of the Annual meeting of the French Society of Astronomy and Astrophysics* ed G Alecian, K Belkacem, S Collin, R Samadi and D Valls-Gabaud pp 435–437
[3] Santra R and Kirby K 2005 *J. Chem. Phys.* **123** 214309
[4] Kielkopf J F and Allard N F 2008 Atomic line shapes in stellar spectra 19th *International Conference on Spectral Line Valladolid, Spain, June 16-20, 2008* (New York: AIP Conference Proceeding)
[5] Kielkopf J F, Allard N F and Babb J F 2012 A comparison of laboratory spectra with a new theoretical study of potassium resonance lines perturbed by helium *ECLA-2011: Proceedings of the European Conference on Laboratory Astrophysics, European Astronomical Society* ed C Stehlé, C Joblin and L d'Hendecourt (Cambridge University Press) in press
[6] Allard N F, Spiegelman F and Kielkopf J F 2007 *A&A* **465** 1085
[7] Zhu C, Babb J F and Dalgarno A 2006 *Phys. Rev. A* **73** 012506
[8] Enomoto K, Hirano K, Kumakura M, Takahashi Y and Yabuzaki T 2004 *Phys. Rev. A* **69** 012501
[9] Havey M D, Frolking S E and Wright J J 1980 *Phys. Rev. Lett.* **45** 1783

Effect of X-ray Line Spectra Profile Fitting with Pearson VII, Pseudo-Voigt and Generalized Fermi Functions on Asphalt Binder Aromaticity and Crystallite Parameters

K Gebresellasie[1], J Shirokoff[2] and J C Lewis[1]
[1]Department of Physics and Physical Oceanography, Memorial University of Newfoundland, St. John's, Newfoundland, A1B 3X7, Canada
[2]Faculty of Engineering and Applied Science, Memorial University of Newfoundland, St. John's, Newfoundland, A1B 3X5, Canada

E-mail: john.lewis@mun.ca

Abstract. X-ray line spectra profile fitting using Pearson VII, pseudo-Voigt and generalized Fermi functions was performed on asphalt binders prior to the calculation of aromaticity and crystallite size parameters. The effects of these functions on the results are presented and discussed in terms of the peak profile fit parameters, the uncertainties in calculated values that can arise owing to peak shape, peak features in the pattern and crystallite size according to the asphalt models (Yen, modified Yen or Yen-Mullins) and theories. Interpretation of these results is important in terms of evaluating the performance of asphalt binders widely used in the application of transportation systems (roads, highways, airports).

Keywords: asphaltene; x-ray scattering; profile fitting
PACS: 81.05.-t; 81.40.-z; 81.70.-q.

1. Introduction

Asphalt binder is a black colored, sticky, viscous liquid (semi-solid) of varying chemical composition dependant on feedstock such as crude oil, bituminous coal, and natural sources. Asphalt binder production is mostly derived from petroleum refining (distillation) of crude oils. The three major components that are usually found in asphalt binders consist of asphaltenes, resins and oils (aromatic hydrocarbons) [1]. The molecular structure of asphalt binder is an important feature since it affects the physical and aging (oxidation) properties of asphalt as well as how the molecules may interact with each other and with additional components such as crushed stone often used in the construction of roads, highways, and airport runways, for example.

Asphaltenes are molecules that consist of carbon, hydrogen, nitrogen, oxygen, sulfur, and trace amounts of metals such as nickel and vanadium. The molecular weight (MW) of asphaltenes is on average about MW=750 Da with FWHM of 500 to 1000 Da [1,2]. They have been studied by techniques originally developed for colloids including x-ray diffraction (XRD) in order to quantify asphaltene molecular length scales [1-8]. Initial research by Yen et al created the Yen model in order to explain aromatic stacking and the mechanism from molecular to cluster states in asphaltenes [3-4]. Additional research led to further refinement of the Yen model and creation of the modified Yen model or Mullins-Yen model as a consequence of explaining the transformation of asphaltene molecules into asphaltene nanoaggregates and clusters of nanoaggregates [2,6-8]. The purpose of this paper is to focus on some of the recent spectral line shape data derived from XRD profile fits using three different mathematical functions on calculating aromaticity and crystallite parameters, and subsequently to link these data to asphalt science models.

2. Experimental

The asphalt binder samples (T2 to T7) were obtained from Northern Ontario, and Alberta, Canada. The asphalt grades and location (town, city) in Canada can be found elsewhere [9]. Thin films deposits of samplse (>1 mm) were prepared by coating onto glass slide holders followed by heating to 150°C for 10 minutes in a drying oven and air cooled at ambient (25°C).

Standard XRD methods previously developed by Warren, Franklin and Diamond and have been used to study petroleum asphaltenes with improvements in instrumentation and computer software [10]. The overall research program involved collecting XRD spectra that were obtained immediately after preparation and also after aging for 1 week and 6 weeks at low-temperature (-10°C) [9-10]. In this study various samples after 1 week of aging were measured by XRD and analyzed. The x-ray diffraction spectral scans were obtained by Rigaku D/Max-2200V-PC using monochromatic Cu K_α radiation at 40 kV and 40 mA. The scan range was 5° to 110° 2θ at a scan rate of 0.001° 2θ per second and detector count time of 5 seconds/step. The XRD instrument employed a divergence slit of 0.5° and receiving slit of 0.3mm.

The XRD spectra were then peak searched using a parabolic filter (99 raw data points , screened out K_α-2 peaks, peak location summit, threshold sigma 3.0, intensity cutoff 0.1%, range to find background 1.0, points to average background 7) over the angular range of 5° to 110°. The full width at half-maximum (FWHM) and profile fits were obtained by using either Pearson VII and pseudo-Voigt function (fixed background, exponent = 1.5 and Lorentzian = 0.5) over the ranges 5° to 35° 2θ and 60° to 110° 2θ on the XRD line spectra of interest. The XRD spectra were also modeled in Mathematica © software using a generalized Fermi function (GF) which was previously described in the literature [11].

3. Results and discussion

The spectral line shape of an XRD pattern from sample T3 is shown in figure 1 left, (full spectrum top, profile fit bottom), with several peaks: an overlapping peak in the low-angle range (0° to 30° 2θ) that contains both the γ-band (left side) and the (002)-graphene band (right side). The (002)-graphene band represents the spacing between the layers of a condensed aromatic system, while the γ-band defines the packing distance of saturated structures such as the aliphatic chains or condensed saturated rings. The two broad peaks which appear in the middle (30° to 60° 2θ) and high (60° to 110° 2θ) angle portions of the spectra are indexed as the (10) reflection and (11) reflection, analogous to the second (100) and third (110) most close-packed planes in face-centered-cubic (fcc) metal systems, that correspond to the first and second nearest neighbors in ring structured compounds. The method for obtaining the aromaticity f_a and crystallite parameters (aromatic sheet diameter L_a, cluster size L_c, interlayer distance between aromatic sheets d_M, interchain distance d_γ, and number of sheets per cluster M_e) were all obtained from the x-ray diffraction spectra according to standard methods [3-4,9-10].

Journal of Physics: Conference Series **397** (2012) 012069 doi:10.1088/1742-6596/397/1/012069

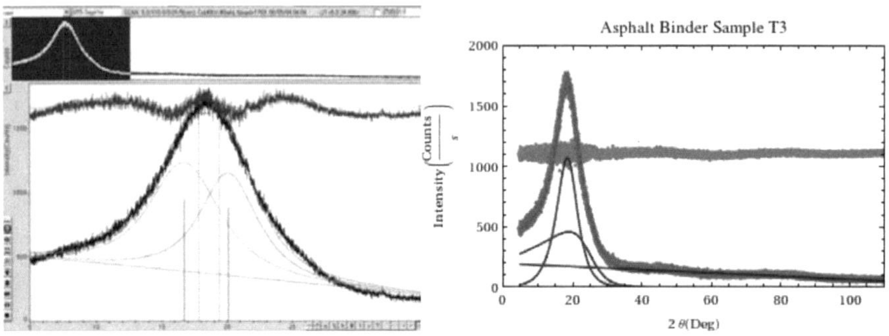

Figure 1. Left: X-ray diffraction spectra (top) and Pearson VII profile fit spectra and analysis curves (bottom) for sample T3; Right: GF-Mathematica © profile fit spectra for sample T3.

As previously described [9-10] and used in the GF results, the aromaticity f_a was determined by calculating the areas A of the resolved peaks for the γ and the (002)-graphene using the following formula:

$$f_a = C_A/C = C_A/(C_A + C_S) = A(\text{graphene})/A(\text{graphene}) + A(\gamma) \quad (1)$$

where C_S, C_A and C are the number of saturated, aromatic and total carbon atoms per structural unit, respectively [9-10]. Since the equation used for the calculation of the f_a is based only on the stack-cluster aromatic carbon to the (002)-graphene peak and is not due to all the aromatic carbons of asphaltenes, f_a does not represent a true aromaticity of the molecules but rather an estimate. A second issue arises from the use of the profile fitting program for Pearson VII versus pseudo-Voigt functions where the Pearson VII function favors larger γ-peak areas, $A(\gamma)$, relative to (002)-graphene peak areas, A(graphene). This second issue consequently calculates a higher aromaticity value when using the pseudo-Voigt function compared to the Pearson VII function calculations as seen in Table 1. A third issue arises when comparing calculations from the three functions as it concerns profile fitting the two major peaks in the low-angle spectrum (γ-peak and (002)-graphene peak) in the presence of additional peaks (i.e. paraffin peaks). Under these conditions equation (1) becomes questionable unless it is replaced with more variables to account for three or more peaks over the low-angle spectral range and validated by experimental results. Moreover, the fundamental diffraction method used in this case to calculate aromaticity should be questioned since it assumes that the γ-peak comes from all aliphatic carbons present (ie. paraffins and naphthenes) but in fact the γ-peak arises primarily from paraffins [10-12]. The generalized Fermi function results are poorly correlated to pseudo-Voigt and Pearson VII in the case of samples T6 and T7 presumably due to a lack of symmetry in the XRD data. The XRD data profiles in the low-angle spectral range of 5° to 10° 2θ varies from sample to sample and appears to be influence by the symmetry of the two major peaks (γ-peak and (002)-graphene) affecting the sensitivity of the profile fitting process. Also, x-ray scattering is influenced by variations in asphaltene size and number of molecules (~1.5 nm), nanoaggregates (~2 nm), and clusters (~5 nm) as reported in the modified Yen model [6]. Smaller asphaltene sizes and larger numbers of asphaltenes will lead to XRD line (peak) broadening affects. Since an increased volume fraction of small sized nanomaterial is known to alter stress-strain behavior, and strengthen solids (i.e. metals due to delocalized electrons) this becomes an important feature with respect to asphalt interfaces and interphase boundaries that will play a greater role in structure-property behavior of asphalt binder materials.

The interlayer distance between the aromatic sheets d_M, was calculated using Bragg's Law and the maximum of the (002) graphene band according to

$$d_M = \lambda/(2 \sin\theta) \quad (2)$$

where d_M is the interlayer distance, λ is the wavelength of the Cu K_α radiation and θ is Bragg's angle. The results in Table 1 indicate about the same d_M values using the pseudo-Voigt function however in the past asphaltene binder studies the values were higher for pseudo-Voigt due to consistently smaller measured 2θ values when compared to data generated using Pearson VII [10].

The distance between the saturated portions (aliphatic chains, condensed saturated rings) molecules or interchain layer distance is given by the relationship:

$$d_\gamma = 5\lambda/(8\ \sin\theta) \tag{3}$$

The interchain layer distance d_γ results in Table 1 trended towards higher values using the pseudo-Voigt function due to consistently smaller measured 2θ values when compared to the data generated by the Pearson VII and generalized Fermi. This trend in measuring crystallite parameters in equations (2) and (3) is related to the measuring functions (Pearson VII, pseudo-Voigt, generalized Fermi) in a low-angle spectral range that appears shape-related since the spectrum in this range is unsymmetrical.

Table 1. Aromaticity and crystallite parameters for asphalt binders calculated using Pearson VII (P), pseudo-Voigt (V), and generalized Fermi (GF) functions on line profile fits of the XRD spectra for various samples after 1 week of low-temperature aging.

	f_a			d_M (Å)			d_γ (Å)			L_a (Å)			L_c (Å)			M_e		
	P	V	GF	P	V	GF	P	V	GF	P	V	GF	P	V	GF	P	V	GF
T2	0.53	0.66	0.58	4.5	4.5	4.7	6.9	7.3	6.0	3.7	3.7	3.8	5.2	4.4	5.2	2.2	2.0	2.1
T3	0.53	0.60	0.52	4.5	4.5	4.2	7.0	7.2	6.0	3.3	3.3	3.3	5.3	4.6	4.7	2.2	2.0	2.1
T4	0.55	0.69	0.55	4.5	4.6	4.7	7.0	7.4	6.3	3.2	3.2	3.3	5.2	4.4	4.7	2.2	2.0	2.0
T5	0.49	0.57	0.59	4.5	4.5	4.8	6.9	7.1	5.7	3.4	3.4	3.3	6.2	5.4	5.7	2.4	2.2	2.2
T6	0.65	0.73	0.08	4.6	4.6	7.1	7.6	8.0	6.0	3.1	3.1	3.3	4.4	3.9	4.3	2.0	1.8	1.6
T7	0.52	0.53	0.28	4.5	4.5	5.2	6.9	6.9	5.9	3.3	3.3	3.1	5.9	6.3	5.2	2.3	2.4	1.9

The average diameter of the aromatic sheets is calculated from the following formula incorporating the Scherrer equation for measuring mean crystallite size:

$$L_a = 1.84\lambda/(\omega\ \cos\theta) = 0.92/B_{1/2} \tag{4}$$

where $B_{1/2}$ is the FWHM using the (11) band and ω is the bandwidth. The results calculated for average diameter of the aromatic sheets are about the same in value for all three profile fitting functions.

The average height of the stack of aromatic sheets perpendicular to the sheet plane was calculated using the following formula:

$$L_c = 0.9\omega\ \cos\theta = 0.45/B_{1/2} \tag{5}$$

where $B_{1/2}$ is the FWHM using the (002)-graphene band. The results calculated for L_c in Table 1 are mostly larger for Pearson VII values compared to pseudo-Voigt fitted profiles. These calculations are the result of smaller (002)-graphene band FWHM measured values using Pearson VII in all samples but one (sample T7) which has additional paraffin (wax) peaks in the low-angle spectrum and wax is detrimental to asphalt performance [9].

The number of aromatic sheets in a stacked cluster, M_e, was calculated from the values of L_c and d_M according to the following relationship:

$$M_e = (L_c/d_M + 1) \tag{6}$$

It is not surprising that the results for the number of aromatic sheets in a stacked cluster, M_e, are generally higher in all samples (except T7) analyzed using Pearson VII compared to pseudo-Voigt, since these results are were influenced more by the change in values of L_c than that found in d_M.

When comparing our data to those obtained by small angle neutron scattering (SANS) and small angle x-ray scattering (SAXS) we find our values to be slightly lower for L_c higher for d_M and lower for M_e [2]. The main reason for this discrepancy requires accounting for the solvated solution asphaltene samples used by the authors of [2] compared to the dry asphaltenes used in our studies and by Yen et al. in order to understand the more extended stacking of nanoaggregates in dry asphaltene samples, especially as reported for Yen's samples.

4. Conclusions

Aromaticity and average crystallite parameters in asphalt binders can be calculated using XRD data and profile fits with various functions (Pearson VII, pseudo-Voigt, and generalized Fermi). With these parameters it is possible to consider a hypothetical average molecular basis for evaluating the material on a macro-to-nano scale that will aid in the physical and chemical understanding of complex materials such as asphaltene as it relates to line broadening, stress-strain, interphase-interfaces, and applications in transportation systems (roads, highways, airports). However, the pseudo-Voigt profile has weak theoretical support in this context, and the Pearson VII, which has the benefit that it can interpolate smoothly between Gaussian and Lorentzian shapes (centrosymmetric), is nonetheless entirely empirical [10]. Unfortunately, the generalized Fermi function results are less well correlated to Pearson VII and pseudo-Voigt since they appear to be more influenced by asymmetry in the XRD spectra. The results build on the modified Yen model and asphaltene theories where spectral features are correlated with the micro-to-nano-crystallinity in asphaltenes.

References

[1] Mullins O C et al. Eds 2007 *Asphaltenes, Heavy Oils and Petroleomics*, Springer, New York
[2] Eyssautier J, Levitz P, Espinat P et al. 2011 *J. Phys. Chem B* **115** 6827-37
[3] Yen T F 1992 *Fuel Sci. Technol. Int.* **10** 723-33 (1992).
[4] Yen T F, Erdman J G and Pollack S S 1961 *Anal. Chem.* **33** 1587-94
[5] Tanaka R, Sato E, Hunt J E et al. 2004 *Energy Fuels* **18** 1118-25
[6] Mullins O C 2010 *Energy Fuels* **24** 2179-207
[7] Mullins O C et al. 2010 *Offshore Technology Conf*, **OTC 20464** 1
[8] Goual L, Sedghi M, Zeng H et al. 2011 *Fuel* **90** 2480-90
[9] Hesp S A M, Iliuta S and Shirokoff J 2007 *Energy Fuels* **21** 1112-21
[10] Shirokoff J, Lewis J C 2010 *20th Int. Conf. Spectral Line Shapes*, CP1290 AIP 274
[11] Aldea N, Tiusan C V, Barz B 2004 *J. Optoelectronics Adv. Materials* **6** 225-35
[12] Siddiqui M N, Ali M F and Shirokoff J 2002 *Fuel* **81** 51-8

Diagnostics of capillary light sources by means of line shape measurements and modeling

G Revalde, N Zorina, J Skudra and A Skudra

Institute of Atomic Physics and Spectroscopy, University of Latvia, Šķunu str. 4, Riga, Latvia, LV 1050

E-mail: gitar@latnet.lv

Abstract. In this work the Hg visible triplet spectral lines were measured by means of Fourier Transform spectrometer and deconvoluted by Tikhonov's regularisation method and line shape modelling. The lines were collected from Hg/Ar and Hg/Xe high frequency capillary lamps. In our previous work, the spatial plasma homogeneity of isotope Hg/Xe and Hg/Ar capillary lamps was investigated by means of tomography. The lamps were operated in different working positions – horizontally and vertically. In this work the spectral diagnostics of capillary light sources by means of spectral line shape measurements and modelling is presented to obtain plasma temperature. The results are compared with the results for spectral line shapes emitted from the spherical light sources. The splitting of spectral lines in magnetic field due to Zeeman effect was observed.

1. Introduction

The capillary light sources are obtained an increasing attention due to the need in compact high-intensity light sources for different applications where narrow line sources are needed [1]. Spatial homogeneity of the plasma and emission profile is important for the use of such light sources in precision measurements. In addition, capillary light sources are difficult to investigate due to the small dimensions of the source, and spectral line shapes are one of the possibilities to get useful information about plasma properties, for example, plasma temperature.

The Hg/Ar, Hg/Xe, Hg/Kr high frequency (HF) capillary discharge lamps (figure 1) were manufactured in our laboratory. In our previous work, the spatial distribution of atoms in Hg/Xe, Hg/Ar, Hg/Kr spherical and capillary lamps was investigated by means of tomography [2-4]. In [3] the radial profiles of excited atoms showed differences in dependence of the horizontal or vertical operation position. To understand the mechanisms of this phenomenon, more information about processes in capillary lamps is necessary. In this work the results of spectral line shape diagnostics of capillary light sources are presented. The diagnostics include measurements of the spectral line shapes by high resolution Fourier spectrometer and later deconvolution to get the real spectral line shapes. The real lines shapes are modelled to estimate the plasma

Figure 1. Design of the capillary electrodeless light source - lamp together with HF generator.

temperature and homogeneity. The results are compared with the results for spherical discharge lamps.

2. Experimental

The capillary light sources were filled with mercury isotope of 0.003 Torr pressure and argon, xenon or krypton of 2 Torr pressure. The length of the capillary was of 2 cm. The discharge size was of 500 μm in radius. The lamp contains spherical mercury reservoir to control the vapour pressure inside the lamp, and it was kept at the room temperature. The lamps were operated by HF electromagnetic field of 100 MHz frequency, capacitive coupled from outer electrodes located at the ends of the capillary [1].

In this work the emission spectral line shapes from the Hg visible triplet 7^3S_1-$6^3P_{0,1,2}$ transitions (with the wave lengths Hg 404,7 nm, 435.8 nm, 546.1 nm respectively) were investigated for the Hg/Ar and Hg/Kr capillary lamps, operated in three different positions, horizontally, vertically with the Hg reservoir up and vertically Hg reservoir down. The spectral line profiles were registered also from spherical lamps of 1 cm diameter for comparison.

The spectral line shapes were registered by means of the Fourier Transform spectrometer *Bruker IFS-125HR,* allowing to register lines in the wavelength region 330-2000 nm. Examples of the registered spectral line shapes are shown in figure 2.

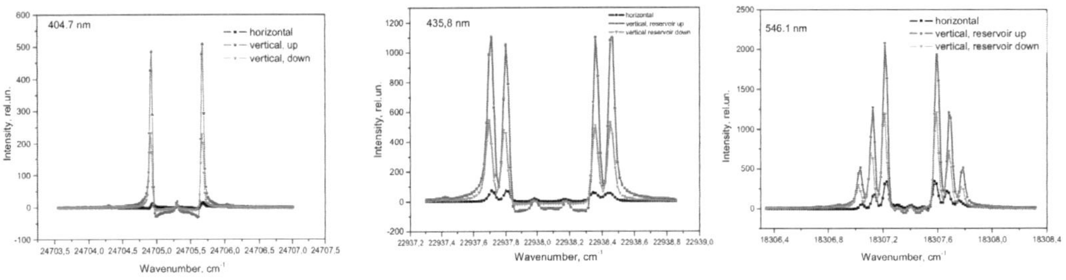

Figure 2. Experimental profiles of Hg visible triplet spectral lines 404.7 nm, 435.8 nm, 546.1 nm, emitted from the capillary light source. Light source was operated in three different operation positions: horizontally, vertically, Hg reservoir up, and vertically, Hg reservoir down. Line splitting is due to the magnetic field in the generator.

We can see that the FWHM of the experimental lines are about 0.06 cm^{-1}. In case of Fourier spectrometer the instrument broadening is on the same order as the width of real line shapes, emitted from low-pressure capillary discharge thus the deconvolution procedure has to be performed to get the real width without instrument distortion. The observed spectral lines are Zeeman split due to the magnetic field in the generator where the light sources are placed. The respective number of σ and π components can be observed. Clear difference of the intensities of the spectral lines in dependence of the operation position can be seen as already observed in our previous experiments [3].

3. Theoretical approach

3.1. Instrument function

The experimental instrument function of Fourier spectrometer is shown in figure 3. It was obtained registering 632.8 nm line from single mode He-Ne laser. For the detailed study of separate spectral line shapes, the influence of the approximation of instrument function of the Fourier spectrometer on the deconvolution results was estimated. The instrument function was approximated by Lorentz function, Gauss function, Voigt function and also the numerical data was taken for the calculations.

The best results were received for the approximation with the Lorentz function of 0.02 cm^{-1} FWHM which was used for the following calculations.

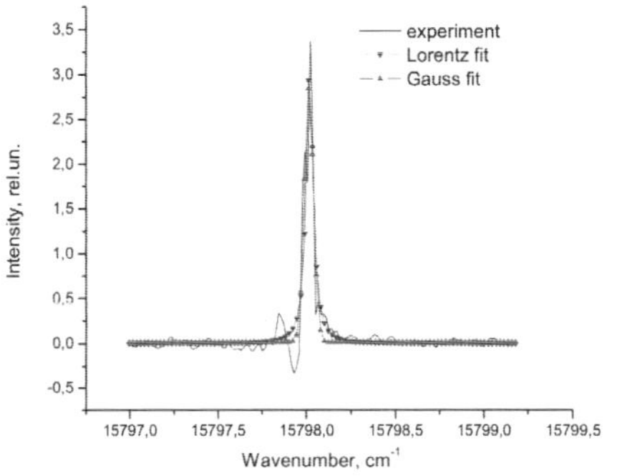

Figure 3. Instrument function of Fourier transform spectrometer and its approximation with Lorentz and Gauss functions.

3.2. Deconvolution method

Recovering the real spectral line profile from experimental line has always been an important problem in the high resolution spectroscopy. To study narrow emission lines like those emitted from capillary line sources, it is very important to deconvolute the instrument function to get the correct results for the plasma parameters. So-called problems of reduction to an ideal spectral device can be described by the first kind Fredholm integral equation:

$$\int_a^b A(v - v_0)y(v_0)dv_0 = f(v), \quad c \le v \le d, (1) \tag{1}$$

Equation (1) is a convolution of an unknown real profile $y\ (v_0)$ with instrument function $A(v,\ v_0)$, where a, b and c, d are the limits of the real and measured (experimental) profiles accordingly. $A(v,\ v_0)$ is the instrument function of the Fourier spectrometer in our case. As well known, equation (1) is an inverse ill-posed task, where direct solution is not possible. To obtain the real spectral line shape we use two methods: solving the inverse non-linear problem by the Tickhonov's regularization method [5,6] and non-linear multi-parameter modeling, described in detail in [7]. Using Tikhonov's algorithm, the initial ill-posed task (1) can be transformed to solution of following matrix equation:

$$\left(\tilde{A}^*\tilde{A} + \alpha E\right)Y = \tilde{A}^*\tilde{f}, \tag{2}$$

where, α – parameter of regularization; A_{ij} – elements of NxN size matrix A, which approximates kernel K(x,x); f_i –vectors– column with initial dates; Y_i –vector – column of solution. We use two different methods to determine the regularization parameter – method proposed by Kojdecki [6] and discrepancy method. After getting the real spectral line shape it was fitted to Gaussian profile. In the low-pressure discharge the main broadening mechanism is Doppler effect due to thermal motion of particles (the self-absorption is not present on the transitions under investigation). Thus the temperature of the radiating atoms was calculated from the FHWM of Gaussian profile [7].

4. Results and discussion

Figure 4 shows the example of the experimental and reconstructed line profiles using regularisation method. The wings of the real spectral line shape can be explained by the difficulties of the model to stabilise the solution by higher level of experimental uncertainties.

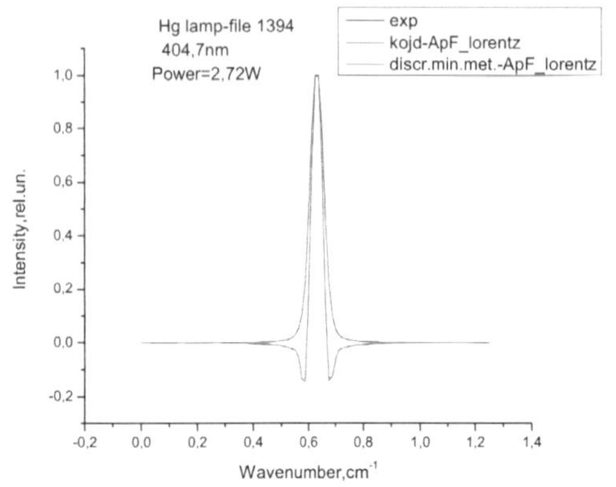

Figure 4. The experimental and reconstructed real Hg 404,7 nm spectral line profiles. Real line shape is reconstructed by two different methods for choosing the regularisation parameter – Kojdecki method and discrepancy method.

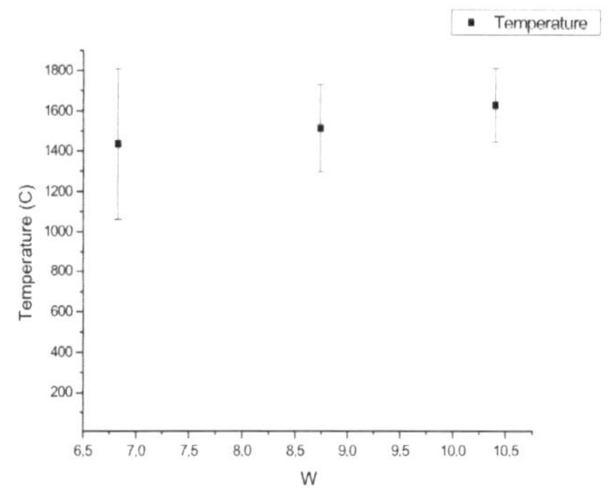

Figure 5. Calculated temperature of the emitting atoms in a spherical discharge in dependence of the discharge power.

For the capillary light sources, the temperature of emitting atoms was estimated to be about 600 K what is much lower than in spherical discharge lamps. The difference in line widths in dependence of observation positions could not be observed.

5. Conclusions

In this work we measured with Fourier spectrometer the profiles of Hg 404.7 nm, 435.8 nm, 546.1 nm lines, emitted from isotopic Hg/Ar and Hg/Xe capillary (500 μm diameter). The real spectral line shapes were obtained by means of the Tikhonov's regularisation method and mathematical modelling. The temperature of the emitting atoms was estimated. The calculations showed that the temperature of the atoms in the capillary light sources was about 600 K what is much lower than in a common spherical electrodeless discharge light source.

References

[1] Ganeev A et al 2003 *Spectrochem. Acta B* **58** 879
[2] Denisova N V 1998 *J.Phys.D: Appl.Phys* **31** 1888
[3] Denisova N, Revalde G, Skudra A and Skudra J 2011 *Jap. J. Appl. Phys.* **50** 08JB03
[4] Denisova N, Bogans E, Revalde G and Skudra J 2011 *Eur. Phys. J. Appl. Phys.* **56**, 24003
[5] Tikhonov A et al 1990 *Numerical solution's methods of the ill-posed problem* (Moscow: Nauka)
[6] Kojdecki M A 2000 New criterion of regularization parameter choice in Tikhonov's method, *Biuletyn WAT (Biul. Mil. Univ. Technol.),* V. 49. № 1 (569) 47–126
[7] Skudra A and Revalde G 1999 *J.Q.S.R.T.* **61** 717

Acknowledgments

The work was partially supported by ESF project "Spectrometric techniques for detection of heavy metal contaminants"(Nr.2009/0210/1DP/1.1.1.2.0/09/APIA/VIAA/100).